IMPULSIVE DIFFERENTIAL EQUATIONS

WORLD SCIENTIFIC SERIES ON NONLINEAR SCIENCE — SERIES A

Published Titles

Volume 1: From Order to Chaos
L. P. Kadanoff

Volume 6: Stability, Structures and Chaos in Nonlinear Synchronization Networks
V. S. Afraimovich, V. I. Nekorkin, G. V. Osipov, and V. D. Shalfeev
Edited by *A. V. Gaponov-Grekhov and M. I. Rabinovich*

Volume 7: Smooth Invariant Manifolds and Normal Forms
I. U. Bronstein and A. Ya. Kopanskii

Volume 12: Attractors of Quasiperiodically Forced Systems
T. Kapitaniak and J. Wojewoda

Forthcoming Titles

Volume 8: Dynamical Chaos: Models, Experiments, and Applications
V. S. Anishchenko

Volume 11: Nonlinear Dynamics of Interacting Populations
A. D. Bazykin

Volume 13: Chaos in Nonlinear Oscillations: Controlling and Synchronization
M. Lakshmanan and K. Murali

Volume 15: One-Dimensional Cellular Automata
B. Voorhees

Volume 16: Turbulence, Strange Attractors and Chaos
D. Ruelle

Volume 17: The Analysis of Complex Nonlinear Mechanical Systems: A Computer Algebra Assisted Approach
M. Lesser

Volume 18: Wave Propagation in Hydrodynamic Flows
A. L. Fabrikant and Yu. A. Stepanyants

WORLD SCIENTIFIC SERIES ON NONLINEAR SCIENCE
Series A Vol. 14
Series Editor: Leon O. Chua

IMPULSIVE DIFFERENTIAL EQUATIONS

A M Samoilenko & N A Perestyuk
Institute of Mathematics
National Academy of Science of Ukraine
3, Tereshchenkivska str
252601 Kiev, Ukraine

Translated from the Russian by
Yury Chapovsky

World Scientific
Singapore • New Jersey • London • Hong Kong

Published by

World Scientific Publishing Co. Pte. Ltd.

P O Box 128, Farrer Road, Singapore 9128

USA office: Suite 1B, 1060 Main Street, River Edge, NJ 07661

UK office: 57 Shelton Street, Covent Garden, London WC2H 9HE

IMPULSIVE DIFFERENTIAL EQUATIONS

ISBN 981-02-2416-8

Cover illustration: From, "On the Generation of Scroll Waves in a Three-Dimensional Discrete Active Medium," *International Journal of Bifurcation and Chaos*, Vol. 5, No. 1, February 1995, p. 318.

Printed in Singapore by Uto-Print

Preface

To describe mathematically an evolution of a real process with a short-term perturbation, it is sometimes convenient to neglect the duration of the perturbation and to consider these perturbations to be "instantaneous". For such an idealization, it becomes necessary to study dynamical systems with discontinuous trajectories or, as they might be called, differential equations with impulses.

By itself, it is not a new idea to make an exhaustive study of ordinary differential equations with impulses. Such problems were considered at the beginning of the development of nonlinear mechanics and attracted the attention of physicists because they gave the possibility to adequately describe processes in nonlinear oscillating systems. A well known example of such a problem is the model of a clock. Using this elegant example, N.M. Kruylov and N.N. Bogolyubov have shown in 1937 in their classical monograph *Introduction to Nonlinear Mechanics* that for a study of systems of differential equation with impulses, it is possible to apply approximation methods used in nonlinear mechanics.

The interest in systems with discontinuous trajectories have grown in recent years because of the needs of modern technology, where impulsive automatic control systems and impulsive computing systems became very important and are intensively developing broadening the scope of their applications in technical problems, heterogeneous by their physical nature and functional purpose. As a natural response to this, the number of mathematical works on impulsive differential equations has increased in different mathematical schools both in our country and abroad. However, the most systematic and in-depth studies were made in the Kiev school of Nonlinear Mechanics. It is the mathematicians of this school, who could broadly approach this problem, consider it in a general form, formulate and solve a number of problems which are important for applications but have not been studied before. With every reason one can say that, as a result of the efforts of this group of Kiev mathematicians, there arose a mathematical theory of impulsive ordinary differential equations, which has its methods, general and deep results, and specific problems.

This monograph is written by the representatives of the Kiev school of Nonlinear Mechanics, who fruitfully work in the area of impulsive differential equations, and give a systematic and sufficiently complete treatment of the subject.

It contains a sufficiently complete study of systems of impulsive linear differential equations. It was shown in the monograph that the classic theory of the first Lyapunov's method can be naturally carried over to the considered systems. This substantially enriches and develops the fundamental research in differential equations as such. Deep results are obtained for the stability of solutions of impulsive systems. Again (how many times!) one sees that the idea of the direct Lyapunov method is universal and can be applied not only to classic differential equations but to more general classes of mathematical objects.

In addition to this, the problem of the existence of integral sets of impulsive differential equations is solved and properties of the integral sets are investigated in the monograph. An important class of discontinuous almost periodic systems is studied and the problem of optimum control is solved for impulsive systems.

An indubitable merit of this book is that it contains a large number of worked out nontrivial examples which could serve as guidelines for solving other particular applied problems.

Undoubtedly, this monograph would be interesting not only to specialists in the theory of differential equations but to many high technology specialists who work in the areas of applied mathematics, computer technology, and automatic control. The book will also be useful to instructors of courses in differential equations. Already, it is impossible today not to offer elements of the theory of impulsive differential equations in special courses for students specializing in differential equations, theoretical and applied mechanics.

Academician *Yu.A. Mitropol'skii*

Contents

Chapter 1

General Description of Impulsive Differential Systems

1.1 Description of mathematical model

Let **M** be a phase space of a certain evolution process, i.e. the set of all possible states of the process. Let $x(t)$ denote the point that describes the state of the process at the time t. We assume that the process is finite dimensional, i.e. it is necessary only a finite number of parameters, say n, to describe its state at a fixed time. Having made such assumptions, a point $x(t)$ can be interpreted, for a fixed value of t, as n – dimensional vector of the Euclidean space $\mathbf{R}^n$ and **M** can be considered as a subset of $\mathbf{R}^n$. The topological product of the phase space **M** and the real axis **R**, $\mathbf{M} \times \mathbf{R}$, will be called the extended phase space of the considered evolution process. Let the evolution of the process be described by:

a) the system of differential equations

$$\frac{dx}{dt} = f(t,x), \qquad x \in \mathbf{M}, \quad t \in \mathbf{R}; \tag{1.1}$$

b) a certain set $\mathcal{T}_t$ given in the extended phase space;

c) an operator $\mathcal{A}_t$ defined on the set $\mathcal{T}_t$, which is mapped into the set $\mathcal{T'}_t = \mathcal{A}_t\mathcal{T}_t$ in the extended phase space.

The process itself goes as follows: the point $P_t = (t, x(t))$, starting at (t_0, x_0), moves along the curve $\{t, x(t)\}$ defined by the solution of system (1.1), $x(t) = x(t, t_0, x_0)$. It moves along the curve until the moment $t = t_1 > t_0$, at which the point $(t, x(t))$ meets the set $\mathcal{T}_t$. At the moment $t = t_1$, the point P_t is "instantaneously" transferred by the operator $\mathcal{A}_t$ from $P_{t_1} = (T_1, x(t_1))$ into $P_{t_1}^+ = \mathcal{A}_{t_1}P_{t_1} = (t_1, x^+(t_1)) \in \mathcal{T}'_{t_1}$ and then it continues to move but along the curve {t,x(t)} described by the solution of system (1.1), $x(t) = x(t, t_1, x^+(t_1))$. It moves along this curve until the moment $t_2 > t_1$, at which the point P_t again meets the set $\mathcal{T}_t$. At this moment, the operator $\mathcal{A}_t$ "instantaneously" transfers the point P_t from $P_{t_2} = (t_2, x(t_2))$ into $P_{t_2}^+ = \mathcal{A}_{t_2}P_{t_2} = (t_2, x^+(t_2)) \in \mathcal{T}'_{t_2}$. It then continues its motion along the curve $\{t, x(t)\}$ determined by the solution $x(t) = x(t, t_2, x^+(t_2))$ until it again meets the set $\mathcal{T}_t$ and so on.

The set of conditions a) – c) that characterize the evolution of the process will be called an impulsive differential system . We will call the curve described by the point P_t in the extended phase space, $\{t, x(t)\}$, an *integral curve* and the function $x = x(t)$ that gives this curve a *solution* of this system.

Write the impulsive differential system, i.e. the set of conditions a) – c) in a more concise form:

$$\frac{dx}{dt} = f(t,x), \qquad (t,x) \notin \mathcal{T}_t,$$
$$\Delta x\big|_{(t,x)\in\mathcal{T}_t} = \mathcal{A}_t x - x. \tag{1.2}$$

Hence, the solution of system (1.2) $x = \varphi(t)$ is a function that satisfies equation (1.1) outside the set $\mathcal{T}_t$ and has discontinuities of the first kind at the points of the set $\mathcal{T}_t$ with the jumps

$$\Delta x = \varphi(t+0) - \varphi(t-0) = \mathcal{A}_t\varphi(t-0) - \varphi(t-0). \tag{1.3}$$

A priori, system (1.2) could have three types of solutions:

1) the solutions that do not have an instant change – the integral curve of system (1.1) does not intersect the set $\mathcal{T}_t$ or the intersection points are fixed points of the operator $\mathcal{A}_t$;

2) the solutions that have a finite number of instant changes – the integral curve intersects the set $\mathcal{T}_t$ in a finite number of points that are not fixed points of the operator $\mathcal{A}_t$;

3) the solutions that change countably many times – the integral curve encounters the set $\mathcal{T}_t$ in a countable number of points that are not fixed by the operator $\mathcal{A}_t$.

Among the solutions, the integral curves of which intersect the set $\mathcal{T}_t$ in countably many points, let us choose solutions that are "swollowed" by $\mathcal{T}$ (remain in $\mathcal{T}$ starting from a certain time $t_1 > t_0$) or that have an accumulation point. The motion of the point P_t along a trajectory that lies in $\mathcal{T}_t$ starting from $t_1 > t_2$ is a sequence of jumps of the point P_t from (t_1, x_1) to $(t_1, \mathcal{A}_{t_1})$, then to $(t_1, \mathcal{A}^2_{t_1})$, then to $(t_1, \mathcal{A}^3_{t_1})$, and so on. If a trajectory has an accumulation point

in $\mathcal{T}_t$, then as t approaches certain $t_1 > t_0$, the point P_t meets and leaves the set $\mathcal{T}_t$ a countable number of times and, consequently, the motion can not be extended to the moment $t = t_1$. If such a motion describes a real process, then in the neighbourhood of the point $t = t_1$, there either appears a fundamentally new motion or the physical condition that implied such a motion becomes unrealistic as t approaches t_1 and should be replaced by another physical condition.

In the theory of impulsive systems, there are some problems similar to the ones considered in the theory of ordinary differential equations but there are also problems that are specific to the theory of impulsive systems. These problems depend greatly on the properties of the operator $\mathcal{A}_t$. For example, if the operator $\mathcal{A}_t$ is not supposed to be single-valued, then there are problems related to a study of the trajectories, for which the moving point can "instantaneously" split into several points when it meets the set $\mathcal{T}_t$. If the operator $\mathcal{A}_t$ is not one-to-one, we can study problems related to the motion, for which the points moving independently, on meeting the set $\mathcal{T}_t$, "instantaneously" become a single point. Very specific problems arise if we assume that for some $\mathfrak{M}_t \subset \mathcal{T}_t$, the set $\mathcal{A}_t\mathfrak{M}_t$ is empty.

With such an assumption, we can consider "mortal" systems: a representing point P_t, when it meets the set $\mathfrak{M}_t$, is mapped by the operator $\mathcal{T}_t$ into the empty set, i.e. according to Vogel, it "dies" and $\mathfrak{M}_t$ is a trajectory "death" set. For these systems, natural problems are to consider an average lifetime of a moving point, a probability of its "death" for $t_0 \le t \le T$, etc.

Unfortunately, it is not possible to have an indepth classification of impulsive differential systems with respect to their properties because of a wide variety of systems of differential equations that describe the evolution process between consecutive hits of the set $\mathcal{T}_t$ by the moving point, variety of the sets $\mathcal{T}_t$, and the mappings $\mathcal{A}_t : \mathcal{T} \to \mathcal{T}_t'$. Depending on the characteristics of the impulses, there are three essentially different classes (types) of the systems: 1) systems with impulses at fixed times; 2) systems with impulses at the moments when the point P_t meets surfaces $t = \tau_i(x)$ given in the extended phase space; 3) discontinuous dynamical systems.

Before going into a brief description of each of these classes, consider a simple example that would illustrate the diversity of the motions and trajectories in an impulsive system, and how essentially they depend on the operator $\mathcal{A}_t$ and the set $\mathcal{T}_t$.

Example. Let the phase space of the process be a line, the set $\mathcal{T}_t$ be given as

$$\mathcal{T}_t = \{(t,x) \in \mathbf{R}^2 \mid x = \tan^{-1}(\tan(t))\},$$

the operator $\mathcal{A}_t$ be given by the formula

$$\mathcal{A}_t(t,x) = (t, x^2 \operatorname{sign} x),$$

and system of differential equations (1.1) be

$$\frac{dx}{dt} = 0,$$

i.e. the impulsive differential system is

$$\begin{aligned} \frac{dx}{dt} &= 0, \qquad (t,x) \notin \mathcal{T}_t, \\ \Delta x|_{(t,x)\in\mathcal{T}_t} &= x^2 \operatorname{sign} x - x. \end{aligned} \tag{1.4}$$

Let us look at the integral curves and possible motions of this system. Any motion of this system that starts for $t = 0$ at a point x_0, $|x_0| \geq \frac{\pi}{2}$, is stationary because the integral curve of such a motion (which is the line $x = x_0$) does not meet the set $\mathcal{T}_t$ for any $t \geq 0$. The trajectory of this motion is the point x_0. For a motion that starts for $t = 0$ at the point x_0, $1 < |x_0| < \frac{\pi}{2}$, there is a finite number of impulses. The integral curve of this motion meets the set $\mathcal{T}_t$ a finite number of times. For every such motion, it is possible to find the time $t_1 = t_1(x_0)$, $0 < t_1 < \frac{\pi}{2}$, such that for $t > t_1(x_0)$, the integral curve belongs to the set $|x| \geq \frac{\pi}{2}$, and so there are no more impulses in the system. The trajectory of this motion is a finite number of points. For example, the trajectory that starts for $t = 0$ at the point $x = \sqrt{2}$ consists of two points, $x = \sqrt{2}$ and $x = 2$. If the trajectory starts at $x = \sqrt[8]{2}$, then the trajectory consists of four points, $\{\sqrt[8]{2}, \sqrt[4]{2}, \sqrt{2}, 2\}$. For a motion that starts at a point $x_0 \in]0,1[$ for $t = 0$, there are countably many impulses. The integral curve of this motion intersects the set $\mathcal{T}_t$ countably many times with $x(t, x_0) \to 0$ for $t \to \infty$. The trajectory is the set of a countable number of points of the interval $]0,1[$. For example, if the motion starts at the point $x = \frac{1}{2}$ for $t = 0$, the trajectory is the set of points $x = \frac{1}{2^n}$, $n = 0, 1, 2, \ldots$. The integral curves that go through the points $x = 0$ and $x = \pm 1$ also intersect the set $\mathcal{T}_t$ countably many times but there are no impulses and they are stationary. This is because the integral curves intersect the set $\mathcal{T}_t$ at points fixed by the operator $\mathcal{A}_t$.

For motions that start for $t = 0$ at a point of the interval $]-1,0[$, there are also countably many impulses, but unlike the previous case, the impulses occur during a finite period of time.

This example shows that in impulsive systems, there appears a phenomenon of *pulsation* at the set $\mathcal{T}_t$, i.e. during a small period of time the integral curve intersects the set $\mathcal{T}_t$ sufficiently (even infinitely, countably) many times.

Thus for the motions that start for $t = 0$ at points x_0 sufficiently close but greater than one, there are sufficiently many impulses occurring on the interval $\left[1, \frac{\pi}{2}\right]$ but their number is finite.

For a motion that start at a point of the interval $]-1,0[$ for $t = 0$, there are countably many impulses occurring on the interval $\left]\frac{\pi}{2}, \pi\right[$. The sequence of moments when the impulses occur has an accumulation point $t = \pi$ and so the solution that corresponds to this motion can not be extended to the interval $t \geq \pi$.

One can see from this example that besides a great number of types of motions and integral curves in impulsive systems, there can exist a confluence of two integral curves at a certain moment. Thus, for example, the curves that start at $x = \sqrt{2}$ and $x = 2$ for $t = 0$ merge into $x = 2$ at $t = \sqrt{2}$.

1.2 Systems with impulses at fixed times

If in a real process described by system (1.1), the impulses occur at fixed times, the mathematical model of this process will be given by the following impulsive system

$$\begin{gathered} \frac{dx}{dt} = f(x, y), \qquad t \neq \tau_i, \\ \Delta x|_{t=\tau_i} = I_i(x). \end{gathered} \tag{1.5}$$

In such a system, the set T_t is a sequence of hyperplanes of the extended phase space $t = \tau_i$, where $\{\tau_i\}$ is a given sequence of times (finite or infinite). In this case, it is sufficient to define the operator A_i only for $t = \tau_i$,i.e. to define its restrictions, $A_{t_i} : \mathbf{M} \to \mathbf{M}$, to the hyperplanes $t = \tau_i$. The most convenient is to define a sequence of the operators $A_i : \mathbf{M} \to \mathbf{M}$ by the following:

$$A_i : x \to A_i x = x + I_i(x). \tag{1.6}$$

A solution of system (1.5) is such a piecewise continuous function $x = \varphi(t)$ that has discontinuities of the first kind at $t = \tau_i$ such that $\varphi'(t) = f(t, \varphi(t))$ for all $t \neq \tau_i$ and, for $t = \tau_i$, satisfying the jump condition, i.e.

$$\Delta\varphi|_{t=\tau_i} = \varphi(\tau_i + 0) - \varphi(\tau_i - 0) = I_i(\varphi(\tau_i - 0)). \tag{1.7}$$

By the value of the function $\varphi(t)$ at a point t', we will understand the value of $\lim_{t \to t'-0} \varphi(t)$, i.e., if τ_i is a point of discontinuity of the first kind, then we will suppose that the function $\varphi(t)$ is continuous from the left and set

$$\varphi(\tau_i) = \varphi(\tau_i - 0) = \lim_{t \to \tau_i - 0} \varphi(t). \tag{1.8}$$

Following [93], we give some general theorems on properties of solutions of system (1.5). Suppose that the function $f(t, x)$ is defined in the whole space $(t; x) \in \mathbf{R}^{n+1}$. The case when it is given in a certain domain of this space is absolutely similar. Besides, we also assume that the collection of solutions of system (1.1) has the following properties.

1 (nonextensibility). Every solution $x(t)$ is a continuous function defined in an interval $]a, b[$ $(-\infty \le a < b \le \infty)$ that depends on the solution. If $a > -\infty$ ($b < -\infty$), then $\| x(a+0) \| = \infty$ (correspondingly, $\| x(b - 0) \| = \infty$).

2 (locality). If a function $x(t)$ defined on $a < t < b$ satisfies condition 1 and if for any $t_0 \in]a, b[$ there exists $\varepsilon > 0$ such that $x(t)$ coincides with

a solution on every interval $]t_0 - \varepsilon, t_0[$ and $]t_0, t_0 + \varepsilon[$, then $x(t)$ is also a solution.

3 (solvability of Cauchy problem). For any t_0, x_0, there exists at least one solution $x(t)$, $a < t < b$, such that $a < t_0 < b$ and $x(t_0) = x_0$.

These conditions, together with the subsequent Condition 4, are fulfilled, in particular, for system (1.1) if the function in the right hand side is continuous or satisfies Caratheodory criteria, and for contingency equations with usual assumptions. The operators $\mathcal{A}_i$ are not assumed to be single-valued, i.e., for $x \in \mathbf{R}^n$, $i \in K$, $\mathcal{A}_i x$ is not necessarily a nonempty subset of $\mathbf{R}^n$. The following theorem follows from the definition of an impulsive system and from the conditions imposed on solutions of system (1.1).

Theorem 1 *If solutions of system (1.1) satisfy conditions 1 – 3, then, for any $t_0 \in \mathbf{R}$ and $x_0 \in \mathbf{R}^n$, there exists at least one solution of impulsive system (1.6) , $x(t)$, $a < t < b$, such that $-\infty \leq a < t_0 \leq b \leq \infty$, $x(t_0) = x_0$ (for $t_0 < b$) or $x(t_0 - 0) = x_0$ (for $t_0 = b$) and:*

if $a < -\infty$, then either $\| x(a+0) \| = \infty$ or $a = \tau_i$, $x(a+0)$ exists (as a finite limit), and $x(a+0) \notin \mathcal{A}_i \mathbf{R}^n$;

if $b < \infty$, then either $\| x(b-0) \| = \infty$ or $b = \tau_i$, $x(b-0)$ exists and $\mathcal{A}_i x(b-0) = \emptyset$, where $\emptyset$ is the empty set.

Such a solution $x(t)$ is *inextensible.*

For any $\mathbf{M} \subset \mathbf{R}^n$, denote by $g(t, t_0)\mathbf{M}$ the set of values of all solutions of system (1.1), $x(t)$, such that $x(t_0) \in \mathbf{M}$. Such a set for solutions of system (1.5) is $G(t, t_0)\mathbf{M}$, where, for $t > t_0$, the mapping G is given by the formula

$$\begin{gathered} G(t,t_0)\mathbf{M} = g(t,\tau_i)\mathcal{A}_i(\tau_i,\tau_{i-1})\mathcal{A}_{i-1}\ldots\mathcal{A}_i g(\tau_j,t_0)\mathbf{M} \\ (\tau_i < t < \tau_{i+1}, \qquad i = j-1, j, \ldots, \qquad j = \min\{\, i \,|\, \tau_i \geq t_0 \,\}). \end{gathered} \tag{1.9}$$

If also for $t > t_0$, there are only a finite number of times τ_i and $\tau_m = \max\{\tau_i\}$, then formula (1.9) also holds for $\tau_m < t < \infty$ with $i = m$. In the sequel, we will not make this reservation. Solving system (1.5) for $t < t_0$, we obtain a similar formula, where $\mathcal{A}_i$ should be replaced by the naturally introduced mappings $\mathcal{A}_i^{-1}$. By introducing for a system with impulses the trajectory translation operator $G(t, t_0)$, it becomes possible to reformulate the conditions of boundness, stability, etc. for solutions of this system in terms of properties of this operator.

As a remark to the given theorem note that if, additionally, it is given that $\mathcal{A}_i x \neq \emptyset$, for $b < \infty$, we have $\| x(b-0) \| = \infty$. If instead, $\mathcal{A}_i \mathbf{R}^n = \mathbf{R}^n$ $(i \in K)$, then, for $a > -\infty$, we have $\| x(a+0) \| = \infty$. The case when $\mathcal{A}_i x^* = \emptyset$ corresponds, according to Vogel [172], to the "death" of the trajectory which came to the point x^* at τ_i, and hence the set $\{ x \mid \mathcal{A}_i x = \emptyset \}$ is a "death set" for the trajectories at time τ_i.

Thus, for example, considering the impulsive equation

$$\frac{dx}{dt} = 1, \qquad t \neq \tau_i, \qquad \Delta x|_{t=\tau_i} = \ln(2-x), \tag{1.10}$$

where $\tau_i = i$, $i = 1, 2, \ldots$, we see that the solution $x = \varphi(t)$, $\varphi(0) = 0$, can not be extended to the interval $]0, 2[$ and the time $t = 2$ is the time of death of this solution. Indeed, for $0 \leq t < 2$, this solution is given by $x = \varphi(t) = t$ (for $t = \tau_1 = 1$, $\varphi(\tau_1) = 1$), and so, there is no discontinuity at $t = \tau_1$ because $\ln(2 - \varphi(1)) = 0$. When $t = \tau_2 = 2$, $\varphi(2) = 2$ and the function $\ln(2-x)$ is undefined at the point $x = \varphi(2)$, hence this solution dies at the time $t = \tau_2$.

Theorem 2 *The Cauchy problem for impulsive system (1.5) has a unique solution for t increasing if and only if for any initial conditions, system (1.1) has this property for any $t_0 \neq \tau_i$ and, for any $t_0 = \tau_i$, $x_0 \in \mathcal{A}_i \mathbf{R}^n$, the set $\mathcal{A}_i x$ contains not more than one element. The Cauchy problem for system (1.5) has a unique solution for t decreasing if and only if system (1.1) has this property for any t_0 and every set $\mathcal{A}_i^{-1} x$ contains not more than one element.*

Thus even if the Cauchy problem for system (1.1) has a unique solution, solutions of impulsive system, while being extended, could split or merge as the operators $\mathcal{A}_i$ act on them.

In order to extend all the solution of system (1.5) infinitely forward (backward) with respect to time, it is necessary and sufficient that the solutions of system (1.1) have this property and, for all $i \in K$, $\mathcal{A}_i x \neq \emptyset$ (correspondingly, $\mathcal{A}_i \mathbf{R}^n = \mathbf{R}^n$). It is not true that if a solution of a Cauchy problem for system (1.1) can not be extended, for example to an interval $]t_0, t_0 + h[$, $h > 0$, the solution of the corresponding Cauchy problem for system (1.5) could not be extended to this interval.

For example, the solution $x = \varphi(t)$, $\varphi(0) = 0$, of the equation

$$\frac{dx}{dt} = 1 + x^2$$

can not be extended to the interval $\left[0, \frac{\pi}{2}\right]$ (this solution goes to infinity in finite time: $\varphi(t) = \tan(t) \to \infty$ for $t \to \frac{\pi}{2} - 0$). However, if we consider the solution $x = \varphi(t)$, $\varphi(0) = 0$, of the impulsive system

$$\frac{dx}{dt} = 1 + x^2, \qquad t \neq \tau_i, \qquad \Delta x|_{t=\tau_i} = -1, \qquad \tau_i = \frac{\pi i}{4},$$

it turns out that this solution can be extended for all $t \geq 0$. It is easy to see that this solution is a periodic function for $t \geq 0$, with the period $\frac{\pi}{4}$, defined for $t \in \left]0, \frac{\pi}{4}\right[$ by $\varphi(t) = \tan(t)$, i.e., for $t \in]\tau_i, \tau_{i+1}[$, $\varphi(t) = \tan\left(t - \frac{\pi i}{4}\right)$.

Suppose that solutions of system (1.1) have, in addition, the following property:

4 (local compactness). For any t_0, x_0, there exists such $\varepsilon > 0$ that if $|\bar{t}_0 - t_0| < \varepsilon$, $\|\bar{x}_0 - x_0\| \leq \varepsilon$, then any solution $x(t)$ such that $x(t_0) = \bar{x}_0$ exists on the interval $t_0 - \varepsilon \leq t \leq t_0 + \varepsilon$ and, for fixed t_0, x_0, ε, the set of these solutions is compact (in itself) with respect to the metric in $C[t_0 - \varepsilon, t_0 + \varepsilon]$.

We also assume that the mappings $\mathcal{A}_i$ are *upper semicontinuous*, i.e. for every i, the mapping $\mathcal{A}_i$ is locally bounded and, if $x_j \to \bar{x}$, $\mathcal{A}_i x_j \in y_j \to \bar{y}$ for $j \to \infty$, then $\bar{y} \in \mathcal{A}_i \bar{x}$.

Theorem 3 *With the stated conditions assumed to be satisfied, for given t_0, $x(t_0)$, and a nonempty compact set $K \subset \mathbf{R}^n$, let all the solutions of system (1.1), for which $x(t_0) \in K$, exist on some interval $t_0 \leq t \leq T$, $t_0 < T < \infty$. Then, for some $\varepsilon > 0$, any solution $\bar{x}(t)$, which satisfies the condition $\rho(\bar{x}(\bar{t}_0), K) \leq \varepsilon$, $|\bar{t}_0 - t_0| \leq \varepsilon$, exists on the whole interval $t_0 - \varepsilon \leq t \leq T$ and, for fixed t_0, K, T, ε, the set of these solutions is compact with respect to the uniform variance metric for discontinuous functions. If $t_0 = \tau_i$, then the same holds if it is additionally assumed that $\bar{t}_0 \leq t_0$.*

Note that if $t = \tau_i$, the interval $[t_0 - \varepsilon, T + \varepsilon]$ can be taken instead of $[t_0 - \varepsilon, T]$. Considering the interval $T_1 \leq t \leq t_0$, $T_1 < t_0$, for Theorem 3 to hold, we need to assume that the mappings $\mathcal{A}_t^{-1}$ are lower semicontinuous and replace the interval $[t_0 - \varepsilon, T]$ by $[T_1 - \varepsilon, t_0 + \varepsilon]$ for $t_0 + \tau_i$, and by $[T - 1 - \varepsilon, t_0]$ for $t_0 = \tau_i$.

Corollary 1 *If, additionally, the Cauchy problem for system (1.1) fas a unique solution and the mappings $\mathcal{A}_i$ are one-to-one, then a solution of impulsive system (1.5), $x(t, t_0, x_0)$, is continuous with respect to $t_0 \neq \tau_i$, x_0 on every closed interval of the t-axis, where the solution is defined. If $t_0 = \tau_i$, then the solution is continuous from the left.*

Corollary 2 *Suppose that the conditions of Theorem 3 are satisfied. Then, for every* t, $t_0 \leq t \leq T$, *the set* $G(t,t_0)K$ *is compact, continuously depends on* $t \neq \tau_i$, *and, for* $t = \tau_i$, *this dependence is continuous from the left and* $G(\tau_i + 0, t_0)K = A_i G(\tau_i, t_0)K$. *The dependence of* $G(t,t_0)K$ *on* K *is upper semicontinuous and uniform with respect to* t. *If system (1.1) possesses the Knezer connectedness property for the section of the integral funnel, all the sets* $A_i x$ *are connected, and* K *is connected, then the set* $G(t,t_0)K$ *is connected for every* $t \in [t_0, T]$.

Points of rest and absorbent sets. If it is not required that a solution of the Cauchy problem be unique, then there are different ways to define a point of rest. A point $x_0 \in \mathbf{R}^n$ is called a point of *possible rest* if the function $x(t) \equiv x_0$, $t \in \mathbf{R}^n$, is a solution of impulsive system (1.5). For this, it is necessary and sufficient that $x(t) = x_0$ be a solution of system (1.1) and $x_0 \in A_i x_0$ for all i. If it is given that, for any t_0, a solution of the Cauchy problem $x(t_0) = x_0$ for system (1.5) is unique for t increasing, then x_0 is called a point of compulsory rest. A point x_0 is a point of compulsory rest if and only if $x(t) \equiv x_0$ is a unique solution of system (1.1) and $A_i x_0 = x_0$ for all i.

A set $Q \subset \mathbf{R}^n$ is called *absorbent* if for any solution of system (1.5), $x(t)$, $a < t < b$, the condition $x(t_0) \in Q$ implies that $x(t) \in Q$ for all $t_0 < t < b$. A set Q will be absorbent if and only if $A_i Q \subseteq Q$ for all $i \in K$ and Q is locally absorbent for system (1.1) for any $t_0 \neq \tau_i$, $x(t_0) \in Q$ and $t_0 = \tau_i$, $x(t_0) \in A_i Q$.

If an absorbent set Q is closed, nonempty, and does not contain proper closed absorbent subsets, the set Q is called an *irreducible* (*minimal*) absorbent set. Thus, a set consisting of a point of compulsory rest is an irreducible absorbent set.

Theorem 4 *Any compact absorbent set has at least one irreducible absorbent subset.*

As opposed to a system of ordinary differential equations, for an impulsive system, an irreducible absorbent set must not be connected.

If an absorbent set Q is zero-dimensional and the right-hand side of system (1.1) is single valued and continuous, then

$$f(t,x) = 0, \qquad t \in \mathbf{R}, \tag{1.11}$$

for all $x \in Q$.

Suppose that the function $f(t,x)$ is single valued, continuous, and such that a solution of the Cauchy problem for system (1.1) is unique for t increasing. By S denote the set of all x such that identity (1.11) holds.

In order to obtain a maximal absorbent set $\tilde{Q} \subseteq S$, index all the times τ_i by an arbitrary sequence of numbers $i_1, t_2, \ldots$, which contains every index infinitely many times. Then set $Q_0 = S$, $Q_k = \{x \mid x \in Q_{i-1}, A_{i_k}x \subseteq Q_i\}$. Then the sequence of the sets Q_k is monotone and $\tilde{Q}$ coincides with the intersection of these sets.

Irreducible absorbent subsets Q are similar to points of rest of impulsive systems.

Continuity of solutions with respect to the initial conditions and the right hand side of the system of equations. We will give sufficient conditions that must be satisfied by system (1.5) for its solutions to depend continuously on initial conditions and the right hand sides of the equations.

In the sequel, we will need the following lemmas.

Lemma 1 *Let a nonnegative piecewise continuous function $u(t)$ satisfy for $t \geq t_0$ the inequality*

$$u(t) \leq C + \int_{t_0}^{t} v(\tau)u(\tau)dt + \sum_{t_0<\tau_i<t} \beta_i u(\tau_i), \tag{1.12}$$

where $C \geq 0$, $\beta_i \geq 0$, $v(\tau) > 0$, τ_i are the first kind discontinuity points of the function $u(t)$. Then the following estimate holds for the function $u(t)$,

$$u(t) \leq C \prod_{t_0<\tau_i<t} (1+\beta_i)e^{\int_{t_0}^{t} v(\tau)d\tau}. \tag{1.13}$$

Proof. We will use induction. On the interval $[t_0, t_1]$, inequality (1.12) has the form

$$u(t) \leq C + \int_{t_0}^{t} v(\tau)u(tau)d\tau,$$

and so by the Gronwell–Bellman inequality,

$$u(t) \leq Ce^{\int_{t_0}^{t} v(\tau)d\tau},$$

i.e. for $t \in [t_0, t_1]$, estimate (1.13) holds. Suppose it holds for $t \in]\tau_i, \tau_{i+1}]$, $i = 1, 2, \ldots, k-1$. Then, for $t \in]\tau_k, \tau_{k+1}]$, we have

$$u(t) \le C + \sum_{i=1}^{k} \beta_i C \prod_{j=1}^{i-1} (1+\beta_i) e^{\int_{t_0}^{\tau_i} v(\tau) d(\tau)} +$$

$$+ \sum_{i=1}^{k} \int_{\tau_{i-1}}^{\tau_i} v(\tau) C \prod_{j=1}^{i-1} (1+\beta_j) e^{\int_{t_0}^{\tau} v(\tau) d(\tau)} d(\tau) + \int_{\tau_k}^{t} v(\tau) u(\tau) d(\tau) =$$

$$= C[1 + \sum_{i=1}^{k} \prod_{j=1}^{i-1} (1+\beta_j) e^{\int_{t_0}^{\tau_i} v(\tau) d\tau} (1+\beta_i) -$$

$$- \sum_{i=1}^{k} \prod_{j=1}^{i-1} (1+\beta_j) e^{\int_{t_0}^{\tau_{i-1}} v(\tau) d(\tau)}] + \int_{\tau_k}^{t} v(\tau) u(\tau) d\tau =$$

$$= C \prod_{j=1}^{k} (1+\beta_j) e^{\int_{t_0}^{\tau_k} v(\tau) d\tau} + \int_{\tau_k}^{t} v(\tau) u(\tau) d\tau,$$

i.e., for $t \in]\tau_k, \tau_{k+1}]$, the function $u(t)$ satisfies the inequality

$$u(t) \le C_1 + \int_{\tau_k}^{t} v(\tau) u(\tau) d\tau,$$

where

$$C_1 = C \prod_{i=1}^{k} (1+\beta_i) e^{\int_{t_0}^{\tau_k} v(\tau) d\tau},$$

and so by the Gronwell–Bellman lemma for $t \in]\tau_k, \tau_{k+1}]$,

$$u(t) \le C_1 e^{\int_{\tau_k}^{t} v(\tau) d\tau},$$

and, finally,

$$u(t) \le C \prod_{t_0 < \tau_i < t} (1+\beta_i) e^{\int_{t_0}^{t} v(\tau) d\tau}.$$

□

Lemma 2 *Let a nonnegative piecewise continuous function $u(t)$, for $t \ge t_0$, satisfy the inequality*

$$u(t) \leq C + \int_{t_0}^{t} \gamma u(\tau) d\tau + \sum_{t_0 < \tau_i < t} \beta u(\tau_i), \tag{1.14}$$

where $C \geq 0$, $\beta \geq 0$, $\gamma > 0$, *and* τ_i *are the first kind discontinuity points of the function* $u(t)$. *Then, for the function* $u(t)$, *the following estimate holds,*

$$u(t) \leq C(1+\beta)^{i(t_0,t)} e^{\gamma(t-t_0)}, \tag{1.15}$$

where $i(t_0, t)$ *is the number of the points* τ *lying in the interval* $[t_0, t[$, *i.e.* $i(t_0, t) = i$ *if* $\tau_i < t \leq \tau_{i+1}$.

Proof. Inequality (1.15) immediately follows from estimate (1.13) for $v(t) = \gamma$, $\beta_i = \beta$.

□

Lemma 3 *If, for* $t \geq t_0$, *a nonnegative piecewise continuous function* $u(t)$ *satisfies the inequality*

$$u(t) \leq \alpha + \int_{t_0}^{t} [\beta + \gamma u(\tau)] d\tau + \prod_{t_0 < \tau_i < t} [\beta + \gamma u(\tau_i)], \tag{1.16}$$

where $\alpha \geq 0$, $\beta \geq 0$, $\gamma > 0$, *and* τ_i *are the first kind discontinuity points of the function* $u(t)$, *then the following estimate holds,*

$$u(t) \leq \left(\alpha + \frac{\beta}{\gamma}\right)(1+\gamma)^{i(t_0,t)} e^{\gamma(t-t_0)} - \frac{\beta}{\gamma}. \tag{1.17}$$

Proof. To prove that estimate (1.17) holds, one can use induction noting that if a function $u(t)$, continuous for $t \geq t_0$, satisfies the inequality

$$0 \leq u(t) \leq \alpha + \int_{t_0}^{t} (\beta + \gamma u(\tau)) d\tau, \tag{1.18}$$

where $\alpha \geq 0$, $\beta \geq 0$, $\gamma > 0$ are constants, then

$$u(t) \leq \left(\alpha + \frac{\beta}{\gamma}\right) e^{\gamma(t-t_0)} - \frac{\beta}{\gamma}. \tag{1.19}$$

However, it is easier to prove the lemma by using Lemma 1. Indeed, rewrite inequality (1.16) as

$$u(t)+\frac{\beta}{\gamma}\le\alpha+\frac{\beta}{\gamma}+\int_{t_0}^{t}\gamma\left(u(\tau)+\frac{\beta}{\gamma}\right)d\tau+\prod_{t_0<\tau_i<t}\gamma\left(u(\tau_i)+\frac{\beta}{\gamma}\right). \quad (1.20)$$

Then the function $v(t)=u(t)+\frac{\beta}{\gamma}$ satisfy inequality (1.14) if we set $C=\alpha+\frac{\beta}{\gamma}$ and replace β by γ. Consequently, it satisfies the inequality similar to (1.15), i.e.

$$u(t)+\frac{\beta}{\gamma}\le\left(\alpha+\frac{\beta}{\gamma}\right)(1+\gamma)^{i(t_0,t)}e^{\gamma(t-t_0)}.$$

Estimate (1.17) follows immediately from this inequality.

□

By using these lemmas, we find an estimate for the change of the solution of system (1.5), corresponding to a change of the initial conditions and the expressions in the right-hand sides of this system.

Suppose that the functions $f(t,x)$ and $I_i(x)$ are continuous with respect to their variables for $x\in\mathbf{M}$, $t\in T$, and satisfy the Lipschitz conditions with respect to x uniformly with respect to $t\in I$ and $i\in K$, i.e.

$$\begin{aligned}\|f(t,x')-f(t,x'')\|&\le L\|x'-x''\|,\\ \|I_i(x')-I_i(x'')\|&\le L\|x'-x''\|.\end{aligned} \quad (1.21)$$

Using equations (1.5), consider the system of equations

$$\begin{aligned}\frac{dy}{dt}&=f(t,y)+R(t,y), \qquad t\ne\tau_i,\\ \Delta y|_{t=\tau_i}&=I_i(y)+R(t,y),\end{aligned} \quad (1.22)$$

where the functions $R(t,y)$ and $R_i(y)$ are such that solutions of system (1.22) exist.

Suppose that for all $x\in\mathbf{M}$, $t\in I$,

$$\|R(t<y)\|<\eta, \qquad \|R_i(y)\|<\eta, \quad (1.23)$$

and consider solutions of systems (1.5) and (1.22), $x(t,x_0)$ and $y(t,y_0)$ respectively. Let these solutions be defined for $t_0\le t\le t_0+T$ and

$$\|x_0-y_0\|<\delta. \quad (1.24)$$

Theorem 5 *If the functions that define systems (1.5) and (1.22) satisfy inequalities (1.21) and (1.23), then, for all* $t_0 \le t \le t_0 + T$, *the following estimate holds for the solutions of equations (1.5) and (1.22),* $x(t, x_0)$ *and* $y(t, y_0)$, *with the initial values satisfying (1.24):*

$$\|x(t,x_0) - y(t,y_0)\| < \left(\delta + \frac{\eta}{L}\right)^{(1+L)^i(t_0,t)} e^{L(t-t_0)} - \frac{\eta}{L}. \tag{1.25}$$

Proof. For $t \in]\tau_i, \tau_{i+1}]$, the solution of system (1.5), $x(t, x_0)$, with $x(t_0, x_0) = x_0$, coincides with a solution of the system of ordinary differential equations

$$\frac{dx}{dt} = f(t,x).$$

Because every solution of this system $x = \varphi(t)$ can be written as

$$\varphi(t) = \varphi(\tau) + \int_\tau^t f(\sigma, \varphi(\sigma))d\sigma$$

for $t \in]\tau_i, \tau_{i+1}]$,

$$x(t,x_0) = x(\tau_i,x_0) + I_i(x(\tau_i,x_0)) + \int_{\tau_i}^t f(\tau, x(\tau,x_0))d\tau,$$

and so, for $t \in]t_0, t_0 + T]$, $x(t, x_0)$ can be expressed as

$$x(t,x_0) = x_0 + \sum_{t_0<\tau_i<t} I_i(x(\tau_i,x_0)) + \int_{t_0}^i f(\tau, x(\tau,x_0))d\tau. \tag{1.26}$$

A similar expression can be obtained for the solution of system (1.22), $y(t, y_0)$, i.e.

$$y(t,y_0) = y_0 + \sum_{t_0<\tau_i<t} \left[I_i(y(\tau_i,y_0)) + R_i(y(\tau_i,y_0))\right] +$$
$$+ \int_{t_0}^t \left[f(\tau, y(\tau,y_0)) + R(\tau, y(\tau,y)0))\right]d\tau. \tag{1.27}$$

Hence, for the norm of $x(t, x_0) - y(t, y_0)$, we have

$$\|x(t,x_0) - y(t,y_0)\| \le \|x_0 - y_0\| + \sum_{t_0<\tau_i<t} [\|I_i(x(\tau_i,x_0)) -$$
$$-I_i(y(\tau_i,y_0))\| + R_i(y(\tau_i,y_0))\|] + \int_{t_0}^t [\|f(\tau, x(\tau,x_0)) -$$
$$-f(\tau, y(\tau,y_0))\| + \|R(\tau,y_0))\|]d\tau.$$

From here, by using inequalities (1.21) and (1.23), we get

$$\|x(t,x_0) - y(t,y_0)\| \le \|x_0 - y_0\| + \sum_{t_0<\tau_i<t} [\eta + L\|x(\tau_i,x_0) -$$

$$-y(\tau_i,y_0)\|] + \int_{t_0}^{t} [\eta + L\|x(\tau,x_0) - y(\tau,y_0)\|]\,d\tau,$$

i.e. if we set

$$u(t) = \|x(t,x_0) - y(t,y_0)\|, \qquad a = \|x_0 - y_0\|, \qquad \beta = \eta, \qquad \gamma = L,$$

the norm of the difference $\| x(t,x_0) - y(t,y_0) \|$ satisfies the conditions of Lemma 3. By using this lemma, we have

$$\|x(t,x_0) - y(t,y_0)\| \le \left(\|x_0 - y_0\| + \frac{\eta}{L}\right)(1+L)^{i(t_0,t)}e^{L(t,t_0)} - \frac{\eta}{L}. \qquad (1.28)$$

Inequality (1.25) directly follows from (1.28) because the initial conditions x_0 and y_0 satisfy condition (1.24).

□

Let us consider particular cases of the proved theorem. Let $\eta = 0$. Then $x(t,x_0)$ and $y(t,y_0)$ are solutions of the same system of equations (1.5) but with different initial conditions. For such solutions, estimate (1.25) takes the form

$$\|x(t,x_0) - y(t,y_0)\| < \delta(1+L)^{i(t_0,t)}e^{L(t-t_0)},$$

i.e.

$$\|x(t,x_0) - y(t,y_0)\| < \delta(1+L)^p e^{LT} \qquad (1.29)$$

for all $t \in [t_0, t_0+T]$, where p is the number of points τ_i in the interval $[t_0, t_0+T]$.

It follows from inequality (1.29) that for an arbitrary $\varepsilon > 0$, there exists such a number $\delta = \delta(\varepsilon) = \varepsilon(1+L)^{-p}e^{-LT}$ that if $\| x_0 - y_0 \| < \delta$, then $\| x(t,x_0) - y(t,y_0) \| < \varepsilon$ for all $t \in [t_0, t_0+T]$. This means that if inequalities (1.21) are satisfied, the solutions of system (1.5) are continuous with respect to the initial conditions. Moreover, as follows from estimate (1.28), not only they are continuous but also satisfy the Lipschitz conditions with respect to x_0 and uniformly with respect to $t \in [t_0, t_0+T]$:

$$\|x(t,x_0) - y(t,y_0)\| \le (1+L)^p e^{LT}\|x_0 - y_0\|. \qquad (1.30)$$

If $\delta = 0$ and $\eta \neq 0$, then this becomes the case of constant perturbations, and so estimate (1.25) becomes

$$\|x(t, x_0) - y(t, y_0)\| < \frac{\eta}{L}((1+L)^{i(t_0,t)}e^{L(t,t_0)} - 1),$$

i.e.

$$\|x(t, x_0) - y(t, y_0)\| < \frac{\eta}{L}((1+L)^p e^{LT} - 1) \tag{1.31}$$

for all $t \in [t_0, t_0 + T]$.

It follows from (1.31) that for an arbitrary $\varepsilon > 0$, one can find such $\eta = \eta(\varepsilon) = \varepsilon L \left((1+L)^p e^{LT} - 1\right)^{-1}$ that if inequalities (1.23) hold, then

$$\|x(t, x_0) - y(t, y_0)\| < \varepsilon$$

for all $t \in [t_0, t_0 + T]$.

This statement claims that solutions of impulsive system (1.5) are continuous in a functional space of the right-hand sides. In particular, if the right-hand sides of equations (1.5) depend continuously on some parameter λ, then it follows from the obtained estimates that the solutions are continuous with respect to this parameter.

1.3 Systems with impulses at variable times

Systems that are more complicated than the systems with impulses at fixed times are the systems with impulses occurring when the integral curve intersects given surfaces in the extended phase space. If the surface, intersecting which the integral curve undergoes an impulsive effect, is given in the extended phase space by the equation $\Phi(t, x) = 0$, then this impulsive differential system can be written as

$$\begin{aligned} \frac{dx}{dt} &= f(t, x), \qquad \Phi(t, x) \neq 0, \\ \Delta x|_{\Phi(t,x)=0} &= I(t, x)|_{\Phi(t,x)=0}. \end{aligned} \tag{1.32}$$

The sets $\mathcal{T}_t$ and $\mathcal{T}_t'$, used to define an impulsive system, are given in this case as follows:

$$\mathcal{T}_t = \{(t; x) | \Phi(t, x) = 0\},$$
$$\mathcal{T}_t' = \{(t; x) | \Phi(t, \mathcal{A}_t^{-1}(t, x)) = 0\},$$

and the operator $\mathcal{A}_t; \mathcal{T}_t \to \mathcal{T}_t'$ acts by

$$\mathcal{A}_t : (t; x) \to (t; x + I(t, x)).$$

In the sequel, it will be convenient to assume that the equation $\Phi(t, x) = 0$ can be solved with respect to t and it has a countable number of solutions if the system is considered either on the whole real axis or for $t \geq t_0$, and it has a finite number of solutions if the system is considered on a finite time interval. Denote these solutions by $t = \tau_i(x)$ and index them by the set of integers (or by a subset of integers) in such a way that $\tau_i(x) \to \infty$ as $i \to \infty$ and $\tau_i(x) \to -\infty$ as $i \to -\infty$.

The restriction of the operator $\mathcal{A}_t'$ to the hypersurface $t = \tau_i(x)$ defines the following operator acting by

$$x \to \mathcal{A}_{\tau_i(x)} x \equiv x^+ = x + I(\tau_i(x), x) \equiv x + I_i(x). \tag{1.33}$$

An so equations (1.32) can also be written as

$$\begin{aligned} \frac{dx}{dt} &= f(t, x), \qquad t \neq \tau_i(x), \\ \Delta x|_{t=\tau_i(x)} &= I_i(x). \end{aligned} \tag{1.34}$$

As opposed to systems (1.5), solutions of equations (1.34) are although piecewise continuous but the points of discontinuity here depend on the solution, i.e. every solution has its own points of discontinuity. This makes the study of such systems considerably more difficult.

One of the difficulties is that there are pulsations of the solution at the surfaces $t = \tau_i(x)$. It is because of that, the solutions very often can not be extended over a sufficiently large interval. This phenomenon has already been mentioned in one of the considered examples. Often the pulsation is the main cause for the solution to leave the domain of definition of the impulsive system or the region where it is being considered. The following impulsive system can serve as an illustration of this,

$$\frac{dx}{dt} = -x, \qquad t \neq \tau_i(x), \qquad \Delta x|_{t=\tau_i(x)} = \alpha x, \tag{1.35}$$

where $\alpha \in \mathbf{R}$, $\tau_i(x) = \arctan x + i\pi$, $i = 0, 1, 2, \ldots$.

If $\alpha > 0$, the integral curve of an arbitrary solution of this impulsive system, $x = \varphi(t)$, $\varphi(0) = x_0 \leq 0$, intersects every curve $t = \tau_i(x)$ only once, and the integral curve of the solution $x = \varphi(t)$, $\varphi(0) = x_0 > 0$ meets the curve $t = \tan x$ countably many times, i.e. the solution $x = \varphi(t)$ has pulsations. In this case, the integral curve of every solution goes to infinity during the time not greater than $\frac{\pi}{2}$.

If in this example $-1 < \alpha < 0$, the integral curve of the solution $x = \varphi(t)$, such that $\varphi(0) \geq 0$, intersects every curve $t = \tau_i(x)$ at one point. The solutions $x = \varphi(t)$ with $\varphi(0) < 0$ have pulsations – the integral curve of each such solution meets the curve $t = \pi + \arctan x$ approaching the line $x = 0$ as $t \to \pi - 0$ countably many times. Also in this case, every solution $x = \varphi(t)$, $\varphi(0) < 0$, can not be extended to the interval $[0, a]$, where $a \geq \pi$. The integral curve of every such solution approaches the point $(\pi, 0)$ as $t \to \pi - 0$.

In what follows, we will be considering impulsive differential systems with no pulsations, in other words, the system, the solutions of which intersect each hypersurface $t - \tau_i(x)$ only once. A sufficient condition for the system to have no pulsations is given by the following lemma.

Lemma 4 *Let the functions $f(t, x)$, $I_i(x)$, and $\tau_i(x)$ that define the system of equations (1.34) be continuous for $(t, x) \in I \times D$, $\tau_i(x)$ be continuously differentiable with respect to $x \in D$, and*

$$\max_{(t;x)\in D} \|f(t, x)\| \leq C, \qquad \max_{x\in D} \left\| \frac{\partial \tau_i(x)}{\partial x} \right\| \leq N, \tag{1.36}$$

where D is a compact subspace of the phase space, C and N are positive numbers. Moreover, suppose that the inequality

$$\max_{0\leq\sigma\leq 1}\left\langle\frac{\partial\tau_i(x+\sigma I_i(x))}{\partial x}, I_i(x)\right\rangle \leq 0 \tag{1.37}$$

holds for all $x \in D$.

Then there is such a positive number N_0 that, for all $N \leq N_0$, the integral curve of any solution of system (1.34), $x(t)$, which belongs to the domain D for $t_0 \leq t \leq t_T$ ($[t_0, t_0+T] \subseteq I$), intersects each hypersurface $t = x\tau_i(x)$ for t in $[t_0, t_0+T]$ only once.

Proof. To prove the lemma it is sufficient to show that, for sufficiently small values of the constant N, any solution $x(t)$ of system (1.34), which starts at the point $x_0 + I_i(x_0)$ for $t = \tau_i(x_0) + 0$ and lying in the domain D at least for $\tau_i(x_0) < t < \bar{t}_i$, where $\bar{t}_i = \max_{x\in D}\tau_i(x)$, does not meet the surface $t = \tau_i(x)$ for $t > \tau_i(x_0)$.

Suppose the contrary. Let a solution $x(t)$, which starts at the point $x_0 + I_i(x_0)$ for $t = \tau_i(x_0) + 0$, intersect the surface $t = \tau_i(x)$ in a point (t_i^*, x^*), $t_i^* = \tau_i(x^*)$. It is clear that $t_i^* > \tau_i(x_0)$ and also the solution $x(t)$ is continuous in this interval. Moreover,

$$x^* = x_0 + I_i(x_0) + \int_{\tau_i(x_0)}^{t_i^*} f(\tau, \varphi(\tau))d\tau. \tag{1.38}$$

Considering the difference $t_i^* - \tau_i(x_0)$, we have

$$\begin{aligned} t_i^* - \tau_i(x_0) &= \tau_i(x^*) - \tau_i(x_0) = \tau_i(x_0) - \tau_i(x_0 + I_i(x_0)) + \\ &+ \tau_i(x_0 + I_i(x_0)) - \tau_i(x_0) = \int_0^1 \left\langle \frac{\partial\tau_i}{\partial x}(x_0 + I_i(x_0) + \sigma h), h\right\rangle d\sigma + \\ &+ \int_0^1 \left\langle \frac{\partial\tau_i}{\partial x}(x_0 + \sigma I_i(x_0)), I_i(x_0)\right\rangle d\sigma, \end{aligned} \tag{1.39}$$

where

$$h = \int_{\tau_i(x_0)}^{t_i^*} f(\tau, x(\tau))d\tau. \tag{1.40}$$

It follows from inequalities (1.36) that the first summand in the right-hand side of (1.39), by Cauchy–Schwarz inequality, admits the estimate

$$\int_0^1 \left\langle \frac{\partial\tau_i}{\partial x}(x_0 + I_i(x_0) + \sigma h), h\right\rangle d\sigma \leq NC(t_i^* - \tau_i(x_0)),$$

and thus,

$$(1 - NC)(t_i^* - \tau_i(x_0)) \leq \int_0^1 \left\langle \frac{\partial \tau_i}{\partial x}(x_0 + \sigma I_i(x_0)), I_i(x_0) \right\rangle d\sigma. \qquad (1.41)$$

To finish the proof it remains to choose N_0 such that $CN_0 < 1$ because, in this case, inequality (1.41) can not hold because of condition (1.37).

□

The following lemma allows to exclude the phenomenon of pulsation at the surfaces $t = \tau_i(x)$ for solutions of system (1.34) in the case when the functions $\tau_i(x)$ are not continuously differentiable.

Lemma 5 *Let, in system (1.34), the functions $f(t, x)$ and $I_i(x)$ satisfy the conditions of the previous lemma and the functions $\tau_i(x)$ satisfy the Lipschitz condition*

$$|\tau_i(x') - \tau_i(x'')| \leq N\|x' - x''\|, \qquad x', x'' \in D. \qquad (1.42)$$

Moreover, let

$$\tau_i(x) \geq \tau_i(x + I_i(x)). \qquad (1.43)$$

for all $x \in D$. Then there exists such a positive number N_0 that, for all $n \leq N_0$, the integral curve of any solution of system (1.34), $x(t)$, which belongs to the domain D for $t_0 < t \leq t_0 + T$, intersects each hypersurface $t = \tau_i(x)$ only once for t in $[t_0, t_0 + T]$.

Proof. The proof of this lemma is similar to the proof of Lemma 4 but, instead of (1.40), we have

$$\begin{aligned} |\tau_i(x^*) - \tau_i(x + I_i(x))| &\leq N \int_{\tau_i(x_0)}^{\tau_i(x^*)} \|f(\tau, x(\tau))\| d\tau \leq \\ &\leq NC(\tau_i(x^*) - \tau_i(x_0)). \end{aligned}$$

Hence,

$$\begin{aligned} t_i^* - \tau_i(x_0) = \tau_i(x^*) - \tau_i(x_0) = \tau_i(x^*) - \tau_i(x_0 + I_i(x_0)) + \\ + \tau_i(x_0 + I_i(x_0)) - \tau_i(x_0) \leq NC(\tau_i(x^*) - \tau_i(x_0)) + \\ + \tau_i(x_0 + I_i(x_0)) - \tau_i(x_0), \end{aligned}$$

i.e.

$$(1 - NC)(t_i^* - \tau_i(x_0)) \leq \tau_i(x_0 + I_i(x_0)) - \tau_i(x_0). \qquad (1.44)$$

Now, if N is so small that $1 - NC > 0$, then (1.44) can not hold because of (1.43), and this finishes the proof.

□

In the previous section we have shown that the solutions of system (1.5), which experience an impulse at fixed times, continuously depend on the initial conditions if the functions $f(t,x)$ and $I_i(x)$ satisfy a Lipschitz condition. Moreover, if $x(t,x_0)$ and $x(t,y_0)$ are two solutions of equations (1.5) defined for $t \in [t_0, t_0+T]$, then

$$\|x(t,x_0) - x(t,y_0)\| \le (1+N)^p e^{NT} \|x_0 - y_0\|$$

for all t from this interval. Here N is the Lipschitz constant, p is the number of the points τ_i in the interval $[t_0, t_0+T]$. This means that the dependence of the solutions is uniform with respect to $t \in [t_0, t_0+T]$.

Solutions of impulsive system (1.34) do not possess this property. That can be easily seen from the following simple example. Let

$$\frac{dx}{dt} = 0, \qquad t \ne \tau_i(x), \qquad \Delta x|_{t=\tau_i(x)} = b, \qquad \tau_i(x) = 2i - x, \qquad b > 0$$

and consider two solutions of this system that start for $t = 0$ at the points $x = 1$ and $x = 1 + \alpha$, where α is an arbitrarily small number. Denote these solutions by $\varphi(t)$ and $\psi(t)$ correspondingly, i.e. $\varphi(0) = 1$ and $\psi(o) = 1 + \alpha$. No matter how small α may be, the solutions in $[0,2]$ do not depend continuously on x_0 because this interval contains $[1 - |\alpha|, 1]$ if $\alpha > 0$ (or $[1, 1+|\alpha|]$ if $\alpha < 0$), where $|\varphi(t) - \psi(t)| = b$. But, outside of this interval, the difference $|\varphi(t) - \psi(t)|$ can be made arbitrarily small by making α sufficiently small.

However, if we exclude from the interval $[t_0, t_0+T]$ sufficiently small neighborhoods of the points where the integral curve intersects the surfaces $t = \tau_i(x)$, then the solutions will depend uniformly with respect to the remaining values of the independent variable.

Suppose that the functions $f(t,x)$, $I_i(x)$, $\tau_i(x)$ in (1.34) are continuous for $(t,x) \in I \times D$ and the inequalities

$$\|f(t,x)\| \le C, \qquad \|f(t,x') - f(t,x'')\| \le L\|x' - x''\|,$$
$$\|I_i(x') - L_i(x'')\| \le L\|x' - x''\|, \qquad |\tau_i(x') - \tau_i(x'')| \le N\|x' - x''\| \tag{1.45}$$

hold for all $t \in I$, x, x', $x" \in D$.

Let $x(t, x_0)$ and $x(t, y_0)$ be two solutions of (1.34), which belong to the domain D for all $t \in [t_0, t_0 + T]$. Suppose that each of these solutions intersects every hypersurface $t = \tau_i(x)$ only once and denote by $\tau_i^{x_0}$, $\tau_i^{y_0}$ the corresponding times when these solutions intersect the surfaces $t = \tau_i(x)$.

Lemma 6 *If the stated above conditions are satisfied and $NC < 1$, then*

$$\|x(t, x_0) - x(t, y_0)\| \leq \left(1 + \frac{L}{1 - NC}\right)^p e^{LT} \|x_0 - y_0\| \qquad (1.46)$$

for all $t \in]\bar{\tau}_i, \tau_{i+1}]$, *where* $\tau_i = \min(\tau_i^{x_0}, \tau_i^{y_0})$, $\bar{\tau}_i = \max(\tau_i^{x_0}, \tau_i^{y_0})$.

Proof. Without loss of generality, we can suppose that the hyperplanes $t = t_0$ and $t = t_0 + T$ do not intersect the hypersurfaces $t = \tau_i(x)$. Write the solutions $x(t, x_0)$ and $x(t, y_0)$ in the integral form

$$x(t, x_0) = x_0 + \int_{t_0}^{t} f(\sigma, x(\sigma, x_0)) d\sigma + \sum_{t_0 < \tau_i^{x_0} < t} I_i(x(\tau_i^{x_0}, x_0)),$$

$$x(t, y_0) = y_0 + \int_{t_0}^{t} f(\sigma, x(\sigma, y_0)) d\sigma + \sum_{t_0 < \tau_i^{y_0} < t} I_i(x(\tau_i^{y_0}, y_0)).$$

Make an estimate for the difference of these solutions:

$$\|x(t, x_0) - x(t, y_0)\| \leq \|x_0 - y_0\| + L \int_{t_0}^{t} \|x(\sigma, x_0) - x(\sigma, y_0)\| d\sigma +$$

$$+ \left\| \sum_{t_0 < \tau_i^{x_0} < t} I_i(x(\tau_i^{x_0}, x_0)) - \sum_{t_0 < \tau_i^{y_0} < t} I(x(\tau_i^{y_0}, y_0)) \right\|. \qquad (1.47)$$

If $t \in [t_0, t_0 + T] \setminus \cup_i [\underline{\tau}_i, \bar{\tau}_i]$, where the union is taken over those i, for which $\tau_i^{x_0}$, $\tau_i^{y_0} \in [t_0, t_0 + T[$, then

$$\left\| \sum_{t_0 < \tau_i^{x_0} < t} I_i(x(\tau_i^{x_0}, x_0)) - \sum_{t_0 < \tau_i^{y_0} < t} i_i(x(\tau_i^{y_0}, y_0)) \right\| \leq$$

$$\leq L \sum_{t_0 < \underline{\tau}_i < t} (\|(\underline{\tau}_i, x_0) - x(\underline{\tau}_i, y_0)\| + \|x(\tau_i^{x_0}, z_0) - x(\tau_i^{y_0}, z_0)\|),$$

where $z_0 = x_0$ if $\underline{\tau}_i = \tau_i^{y_0}$, and $z_0 = y_0$ if $\underline{\tau}_i = \tau_i^{x_0}$. The second term in the last inequality can be estimated as

$$\|x(\tau_i^{x_0}, z_0) - x(\tau_i^{y_0}, z_0)\| \leq \int_{\underline{\tau}_i}^{\overline{\tau}_i} \|f(\sigma, x(\sigma, z_0))\| d\sigma \leq$$

$$\leq C|\tau_i^{x_0} - \tau_i^{y_0}| \leq \frac{NC}{1 - NC} \|x(\underline{\tau}_i, x_0) - x(\underline{\tau}_i, y_0)\|$$

because

$$\begin{aligned} |\tau_i^{x_0} - \tau_i^{y_0}| &\leq N\|x(\tau_i^{x_0}, x_0) - x(\tau_i^{y_0}, y_0)\| \leq N\|x(\underline{\tau}_i, x_0) - \\ &-x(\underline{\tau}_i, y_0)\| + N\|x(\tau_i^{x_0}, z_0) - x(\tau_i^{y_0}, z_0)\| \leq \\ &\leq N\|x(\tau_i, x_0) - x(\underline{\tau}_i, y_0)\| + NC|\tau_i^{x_0} - \tau_i^{y_0}|. \end{aligned}$$

By the assumptions of the lemma, it follows from (1.47) that

$$\|x(t, x_0) - x(t, y_0)\| \leq \|x_0 - y_0\| + L \int_{t_0}^{t} \|x(\sigma, x_0) -$$

$$-x(\sigma, y_0)\| d\sigma + \frac{L}{1 - NC} \sum_{t_0 < \underline{\tau}_i < t} \|x(\underline{\tau}_i, x_0) - x(\underline{\tau}_i, y_0)\|.$$

So, from Lemma 1, we have

$$\|x(t, x_0) - x(t, y_0)\| \leq \left(1 + \frac{L}{NC}\right)^p e^{LT} \|x_0 - y_0\|$$

for all $t \in [t_0, t_0 + T] \setminus \cup_i [\underline{\tau}_i, \overline{\tau}_i]$, p is the number of the points τ_i (or $\overline{\tau}_i$, which is the same thing) in the interval $[t_0, t_0 + T]$.

□

Theorem 6 *Let the functions $f(t, x)$, $I_i(x)$, and $\tau_i(x)$ in system (1.34) satisfy inequalities (1.45), inequality (1.45) hold, and $NC < 1$. If a solution of (1.34), $x(t, x_0)$, is defined for $t \in [t_0, t_0 + T]$, then this solution depends continuously on the initial condition x_0 in the following sense: for an arbitrary $\epsilon > 0$, there exists $\delta = \delta(\epsilon) > 0$ such that for any other solution $x(t, y_0)$ of (1.34), the inequality $\|x_0 - y_0\| < \delta$ implies*

$$\|x(t, x_0) - x(t, y_0)\| < \varepsilon \tag{1.48}$$

for all $t \in [t_0, t_0 + T]$, satisfying $|t - \tau_i^{x_0}| > 0$, where $\tau_i^{x_0}$ are the times, at which the integral curve of the solution $x(t, x_0)$ intersects the hypersurfaces $t = \tau_i(x)$.

Proof. It follows from the conditions of the theorem that there are no pulsations of the solutions of system (1.34) at the surface $t = \tau_i(x)$ and that the conditions of Lemma 6 hold. By Lemma 6, two solutions of system (1.34) satisfy estimate (1.46). Fix an arbitrary $\epsilon > 0$ and choose $\delta_1 > 0$ so small that

$$\max_{\|x_0 - y_0\| \le \delta_1} |\tau_i^{x_0} - \tau_i^{y_0}| < \varepsilon$$

for all i such that $\tau_i^{x_0} \in [t_0, t_0 + T]$. For δ, we take the number $\delta = \min\left(\delta_1, \epsilon\left(1 + \frac{L}{1-NC}\right)^{-p} e^{-LT}\right)$, where p is the number of the points $\tau_i^{x_0}$ that belong to the interval $[t_0, t_0 + T]$. If y_0 is such that $\|x_0 - y_0\| < \delta$, then, according to (1.46), the solutions $x(t, x_0)$, $x(t, y_0)$ satisfy inequality (1.48) for all $t \in [t_0, t_0 + T]$ with $|t - \tau_i^{x_0}| > \epsilon$.

□

We point out one more general property of solutions of system (1.34) that is not possessed by solutions of equations with impulses at fixed times.

For equations (1.5), it is possible that two different solution merge into one at some moment $t = \tau_i$ when the impulse effect occurs because the mapping $g_i : x \to x + I_i(x)$ is not one-to-one. If a point y is the image under this mapping of at least two points x_1 and x_2, then the integral curves that pass through the points x_1 and x_2 at $t = \tau_i$ merge into a single integral curve for $t > \tau_i$. It is possible tò have this phenomenon for system (1.34) even if the mapping g_i is one-to-one.

Let $\sigma_i(x) = \tau_i(g_i^{-1}(x))$, where g_i^{-1} is the mapping inverse to g_i, and suppose that the inequality $\tau_i(x) < \sigma_i(x)$ holds for all x belonging to the domain of definition of system (1.34). Each solution of system (1.34) that starts at a point (t_0, x_0) from the domain $\tau_i(x) < t < \sigma_i(x)$ and intersects the surface $t = \sigma_i(x)$ for $t > t_0$, say in the point $(\sigma_i(\bar{x}), \bar{x})$, for $t > \sigma_i(\bar{x})$, merge with the solution that passes through the point $g_i^{-1}(\bar{x})$ at $t = \sigma_i(\bar{x})$.

By studying how the surfaces $T = \tau_i(x)$ and $\tau = \sigma_i(x)$ are located, it is possible to see whether there are phenomena of pulsation in system (1.34) or not. A fortiori, if the inequality $\sigma_i(x) < \tau_i(x)$ has solutions in the domain of definition of impulsive system (1.34), then there are pulsations at the surface $t = \tau_i(x)$.

At the end of this section, we introduce the notion of a variational system, corresponding to the impulsive equations (1.34).

Suppose that the functions $f(t, x)$, $\tau_i(x)$, and $I_i(x)$ from (1.34) are continuously differentiable with respect to x. Let $x = \varphi(t)$ be a solution of this system, τ_i^0 – times at which this solution meets the surfaces $t = \tau_i(x)$. The corresponding to this solution variational equations will be

$$\frac{dz}{dt} = \frac{\partial f(t, \varphi(t))}{\partial x} z, \qquad t \neq \tau_i^0,$$
$$\Delta z|_{t=\tau_i^0} = \left(P_i - Q_i + \frac{\partial I_i(\varphi(\tau_i^0))}{\partial x}(E + P_i)\right) z, \tag{1.49}$$

where the matrices P_i and Q_i are defined as follows,

$$P_i\, z = \frac{1}{1 - \left\langle \frac{\partial \tau_i}{\partial x}(\varphi(\tau_i^0)), f(\tau_i^0, \varphi(\tau_i^0)) \right\rangle} \left\{ f(\tau_i^0, \varphi(\tau_i^0)) \left\langle \frac{\partial \tau_i(\varphi(\tau_i^0))}{\partial x}, z \right\rangle \right\},$$
$$Q_i\, z = \frac{1}{1 - \left\langle \frac{\partial \tau_i}{\partial x}(\varphi(\tau_i^0)), f(\tau_i^0, \varphi(\tau_i^0)) \right\rangle} \left\{ f(\tau_i^0, \varphi(\tau_i^0 +)) \left\langle \frac{\partial \tau_i(\varphi(\tau_i^0))}{\partial x}, z \right\rangle \right\}.$$

Note that, if the impulses occur at fixed times or the solution $x = \varphi(t)$ is such that $f(\tau_i^0, \varphi(\tau_i^0)) = 0$, then the matrices $P_i = 0$, $Q_i = 0$, and the variational equations (1.49) take a simpler form.

It is known [116] how important variational equations are for studying differential equations. They describe the behavior of the derivatives of solutions with respect to initial conditions.

Denote by $\varphi(t, x_0)$, $\|x_0 - x_0^*\| < \delta$ the family of solutions of system (1.34) with the initial condition x_0, which belongs to a δ-neighborhood of a fixed point x_0^*. Let τ_i^* be the times at which the integral curve of the solution $x = \varphi(t, x_0^*)$ meets the surface $t = \tau_i(x)$. If there are no pulsations of the solutions at the surface $t = \tau_i(x)$ and the functions defining system (1.34) are smooth, then the equations $t = \tau_i(\varphi(t, x_0))$ have solutions $t = \tau_i^*(x_0)$, $\tau_i^*(x_0^0) = \tau_i^0$, that, for sufficiently small δ and $\tau_i^*(x_0)$, are smooth as functions of the variable x_0. It turns out that the matrix function $z = \dfrac{\partial \varphi(t, x_0)}{\partial x_0}$ is described by system (1.49).

1.4 Discontinuous dynamical systems

Consider a class of impulsive differential systems that are also called discontinuous dynamical systems. A *discontinuous dynamical system* is given by the relations a) – c), where differential equation (1.1) does not explicitly depend on t, the sets $\mathcal{T}_t$ and $\mathcal{T}_t'$ are subsets of the phase space, and $\mathcal{A}_t = \mathcal{A}$ for all $t \in \mathbf{R}$. Set $\mathcal{T}_t = \Gamma$, $\mathcal{T}_t' = \Gamma_0$, and $\mathcal{A} : \Gamma \to \Gamma_0$, then a discontinuous dynamical system can be written as

$$\begin{aligned} \frac{dx}{dt} &= f(x), \qquad x \notin \Gamma, \\ \Delta x|_{x\in\Gamma} &= \mathcal{A}x - x = I(x). \end{aligned} \tag{1.50}$$

A phase point of such a system moves between two consecutive intersections with the set Γ along one of the trajectories of the differential equations

$$\frac{dx}{dt} = f(x)$$

and, at the moment of intersection with Γ, the point $x(t)$ is "instantaneously" mapped by the operator $\mathcal{A}$ into the point $y = \mathcal{A}x$ of the set Γ_0.

To get interesting effects in the systems under consideration, the moving point needs to intersect the set Γ relatively often. If, for example, the set Γ is not dense everywhere in the phase space, then there is a good chance that a large part of the motions will not be subject to an impulsive effect. That is why we will assume that the set is a compact (locally compact) manifold in the phase space and its dimension is one less than the dimension of the phase space.

It is interesting to study discontinuous dynamical systems (1.50), for which the differential equation is defined inside the *ball* (or for all $x \in \mathbf{R}$)

$$\mathcal{I}_r = \{x \in \mathbf{R}^n | \, \|x\| \leq r\}, \tag{1.51}$$

the set Γ is the *boundary* of this ball, i.e. the sphere

$$\Gamma = S_r = \{x \in \mathbf{R}^n | \, \|x\| = r\}, \tag{1.52}$$

and the set Γ_0 is a subset of $\mathcal{I}_r$, for example, a sphere with radius $r_1 < r$ and center at the origin.

For such a system, the set Γ is a reflector of the orbits.

It is clear that, in such a discontinuous dynamical system, if a motion that started at a point $x_0 \in \mathcal{I}_r \backslash S_r$ does not experience an impulsive effect, then the closure of the trajectory of such a motion is a compact set and, consequently, such a motion is Lagrange stable and the limit sets of this motion belong to the ball $\mathcal{I}_r$. Evidently, the converse statement is false, i.e. if the limit sets of a certain motion belong to the ball $\mathcal{I}_r$, then this does not mean that this motion did not experience an impulsive effect during its evolution.

Consider some examples of discontinuous dynamical systems.

Example 1. Let the phase point (x, y) move in the plane $\mathbf{R}^2$ and its motion, between two consecutive intersections with the set Γ, be described by the differential system

$$\frac{dx}{dt} = 0, \qquad \frac{dy}{dt} = -1.$$

Let Γ be the x-coordinate axis and the set Γ_0 be the semicircle $x^2 + (y - 2)^2 = 1$, $y < 2$. The map $\mathcal{A} : \Gamma \to \Gamma_0$ maps the point $(x, 0)$ into the point of the semicircle, which lies in on the line passing through the points $(x, 0)$ and $(0, 2)$.

All the solutions of this impulsive system, which start at the points $(x_0, y_0) \in \mathbf{R}^2$, $y < 0$, do not experience an impulsive effect. The solutions that start at the points (o, y_0), $y_0 \geq 0$, are periodic $(0 \leq y_0 \leq 1)$ or become periodic after the time $y_0 - 1$ $(y_0 > 1)$. Other solutions experience an impulsive effect countably many times and, for $t \to \infty$, approach the periodic solutions, the trajectory of which is the line segment OB that lies in the y-axis. Thus the segment OB attracts the trajectories of all the solutions that start at points with nonnegative y-coordinate.

Example 2. Consider a discontinuous dynamical system in the plane $\mathbf{R}^2$ defined by the differential system

$$\frac{dx}{dt} = -x, \qquad \frac{dy}{dt} = -\alpha y, \qquad \alpha > 0, \tag{1.53}$$

the sets

$$\Gamma_0 = \{(x; y) \in \mathbf{R}^2 | x^2 + y^2 = 4\}, \qquad \Gamma = \{(x; y) \in \mathbf{R}^2 | 5x^2 + y^2 = 8\}, \tag{1.54}$$

and the mappings $\mathcal{A} : \Gamma \to \Gamma_0$ that takes each point of the ellipse Γ into the point of the circle Γ_0, which lies on the same line passing through the point on the ellipse and the origin. Clearly, $\mathcal{A}$ is a one-to-one continuous mapping.

In this system, the points move along the trajectories of equations (1.53), i.e. along the branches of the curves

$$y = C|x|^{\alpha}. \tag{1.55}$$

All the motions that start in the region $5x^2+y^2=8$ approach the stationary point $(0,0)$ without an impulsive effect because the trajectories of these motions do not intersect the set Γ. All the motions that start at points of the region $5x^2+y^2>8$ are subject to an impulsive effect.

Let us study these motions and their trajectories. Let $0<\alpha<1$. Because of the symmetry, it is sufficient to study the motions that start in the region $5x^2+y^2>8$, $x \geq 0$, $y \geq 0$. The motions with the trajectories intersecting the ellipse Γ at the points with the x-coordinate $0 \leq x \leq 1$, after having experienced once an impulsive effect, get into the region $5x^2+y^2<8$ and approach the origin for $t \to \infty$. The trajectories of these motions have discontinuities at one point.

The trajectories that start at points of the curve $y=\sqrt{3}|x|^{\alpha}$, $x>1$, go along this curve to the point $(0,0)$. These motions have a continuous trajectory because it does not intersect Γ at the point $(1,3)$, which is a stationary point of the mapping $\mathcal{A}$.

The motions, the trajectories of which intersect Γ at a point with the x-coordinate $1<x<2\sqrt{\frac{2}{5}}$, have an impulsive effect countably many times as they approach to the point $(1,3)$ arbitrarily close. These motions have trajectories with countably many points of discontinuity. The point $(1,\sqrt{3})$ has the property of being an attractor for all the motions starting in the region

$$5x^2+y^2>8, \qquad x>1, \qquad 0<y<\sqrt{3}x^{\alpha}.$$

Finally, the motions, which start at points with $y=0$, $x>2\sqrt{\frac{2}{5}}$, become periodic by the time the trajectory gets to the point $\left(2\sqrt{\frac{2}{5}},0\right)$ for the first time and after that they follow along the line segment $y=0$, $x \in \left(2\sqrt{\frac{2}{5}},2\right)$ periodically jumping from the point $\left(2\sqrt{\frac{2}{5}},0\right)$ to the point $(2,0)$.

In the same way one can study the motions and their trajectories for $\alpha>1$.

As it was for $0<\alpha<1$, the motions, which start in the region $5x^2+y^2>8$, eventually approach the point $(0,0)$ without an impulsive effect.

All the motions, which start for $t=0$ in the region $5x^2+y^2>8$, can be divided into three types.

a) The motions that do not experience an impulsive effect. These are the motions that start in the set $|y| = \sqrt{3}|x|^\alpha$. The trajectories of these motions intersect the set Γ in the stationary points of the operator $\mathcal{A}$.

b) The motions that undergo an impulsive effect once are those, which start at points of the region $|y| > \sqrt{3}|x|^\alpha$.

c) The motions that are subject to a countable number of impulsive effects. They start at points of the region $|y| < \sqrt{3}|x|^\alpha$.

Among the latter motions, the ones that start at the points $y = 0$, $2\sqrt{\frac{2}{5}} < |x| \leq 2$ should be singled out. These motions are periodic. But unlike the case when $0 < \alpha < 1$, for $\alpha > 1$, these motions have a strong attractor. The segment of the x-axis $2\sqrt{\frac{2}{5}} < |x| \leq 2$, i.e. the trajectories of these motions, are attractors for many other trajectories. Thus, the segment $2\sqrt{\frac{2}{5}} < |x| \leq 2$, $y = 0$ has the set $|y| < \sqrt{3}x^\alpha$, $5x^2 + y^2 > 8$ as an attracted region, and the region of attraction of the segment $-2 \leq x < -2\sqrt{\frac{2}{5}}$, $y = 0$ is the set $|y| < \sqrt{3}|x|^\alpha$, $x < 0$, $5x^2 + y^2 > 8$. For $\alpha = 1$, the nature of the motion changes only in the region $|y| < \sqrt{3}|x|$, $5x^2 + y^2 > 8$. Here all the motion become eventually periodic and the trajectories move along the segment of the line $y = Cx$, which lies between the ellipse $5x^2 + y^2 = 8$ and the circle $x^2 + y^2 = 4$.

As it was shown in works of T. Pavlidis and V. Rozhko, discontinuous dynamical systems possess many of the properties of classical dynamical systems but they also possess properties that classical dynamical systems do not have. For example, it is known that for a dynamical system given in a plane, there are no motions, the trajectory of which is dense in some region of the plane, i.e. there are no motions, which are Poisson stable and different from a stationary point or a periodic motion. For discontinuous dynamical systems, such motions are possible.

Example 3. Consider a dynamical system with the motions described by the equations

$$\frac{dx}{dt} = \alpha x - \beta y, \qquad \frac{dy}{dt} = \beta x + \alpha y, \qquad \alpha < 0, \qquad \beta > 0, \tag{1.56}$$

the sets

$$\Gamma = \{(x;y) \in \mathbf{R}^2 | x^2 + y^2 = r_1^2\}, \qquad \Gamma_0 = \{(x;y) \in \mathbf{R}^2 | x^2 + y^2 = r_2^2\}, \quad r_2 > r_1 \tag{1.57}$$

and the mapping $\mathcal{A} : \Gamma \to \Gamma_0$, which maps every point $(x, y) \in \Gamma$ to the point $(x_1, y_1) \in \Gamma_0$ lying on the line that passes through (x, y) and the origin. Passing to polar coordinates

$$x = \rho \cos \varphi, \qquad y = \rho \sin \varphi, \tag{1.58}$$

we get the system

$$\frac{d\rho}{dt} = \alpha\rho, \qquad \frac{d\varphi}{dt} = \beta, \tag{1.59}$$

$\Gamma = \{(\rho, \varphi) | \rho = r_1\}$, $\Gamma_0 = \{(\rho, \varphi) | \rho = r_2\}$, and $\mathcal{A} : (r_1, \varphi) \to (r, \varphi)$.

Since the motions in this system follow the logarithmic spiral

$$\rho = \rho_0^{\frac{\alpha}{\beta}\varphi}, \qquad \alpha < 0, \qquad \beta > 0,$$

all the motions started in the region $\rho > r_2$ eventually go to the stationary point $\rho = 0$.

The phase point of the motion that started in the region $\rho > r_2$ eventually gets on the circle $\rho = r_2$, and so to study this system it is sufficient to consider the behavior of the solutions that start in the circle $\rho = r_2$.

Let a motion start at a point (r_2, φ_0). The phase point of this motion follows the spiral $\rho = r_2 e^{\frac{\alpha}{\beta}(\varphi - \varphi_0)}$ until it gets on the circle $\rho = r_1$. The time, at which it gets on the circle $\rho = r_1$, can be calculated from the equation $r_1 = r_2 e^{\alpha t}$, i.e. $t_1 = \frac{1}{\alpha} \ln \frac{r_1}{r_2}$, and the point of intersection is (r_1, φ_1), where

$$\varphi_1 = \varphi_0 + \beta t_1 = \varphi_0 + \frac{\beta}{\alpha} \ln \frac{r_1}{r_2}.$$

After that the phase point is mapped by the operator $\mathcal{A}$ onto the circle Γ_0 at the point $(r_2, \varphi_1) = (r_2, \varphi_0 + \Delta)$, where $\Delta = \frac{\beta}{\alpha} \ln \frac{r_1}{r_2}$. After that the motion continues along the spiral $\rho = r_2 e^{\frac{\alpha}{\beta}(\varphi - \varphi_1)}$ until the phase point will meet Γ again. It is easy to see that, after the operator $\mathcal{A}$ maps the phase point the second time, it gets onto the circle Γ_0 into the point (r_2, φ_2), where $\varphi_2 = \varphi_0 + 2\Delta$, and, after the nth mapping, it gets into the point $(r_2, \varphi_0 + n\Delta)$. Consequently, in order to know the complete motion, it is necessary to study the arrangement of the points

$$\varphi_n = \varphi_0 + n\Delta. \tag{1.60}$$

on the circle Γ_0. Clearly, if Δ is rationally commensurable with 2π, i.e. $\Delta = 2\pi\frac{p}{q}$, where p and q are mutually prime natural numbers, then the point $\varphi_q = \varphi_0 + q\Delta$ coincides with $\varphi_0 \pmod{2\pi}$, and so the trajectory "closes" and the motion will be periodic. Because φ_0 is arbitrary, we can assert that, if the number $\frac{\beta}{\alpha}\ln\frac{r_1}{r_2}$ is commensurable with 2π, then all the motions that start in the circle Γ_0 are periodic. Let Δ not be commensurable with 2π.

Lemma 7 *[13]. The images of any point of the circle under rotations at an angle Δ, which is not rationally commensurable with 2π, form a set, which is everywhere dense in the circle.*

Proof. Let us subdivide the circle into k intervals of equal length $\frac{2\pi}{k}$. Note that, because Δ is not commensurable with 2π, φ_n will not coincide with φ_0 for any $n \in \mathbf{N}$. There are two points in the first $k+1$ points

$$\varphi_0, \varphi_0 + \Delta, \varphi_0 + 2\Delta, \ldots \pmod{2\pi} \tag{1.61}$$

that lie in the same interval. Let these be the points $\varphi_0 + m\Delta$ and $\varphi_0 + r\Delta$, $m > r$. Set $s = m - r$. The difference between the rotation angle $s\Delta$ and a multiple of 2π is less than $\frac{2\pi}{k}$. In the sequence of the points $\varphi_0, \varphi_0 + s\Delta, \varphi_0 + 2s\Delta, \ldots \pmod{2\pi}$, the distance between two adjacent points is the same and is less than $\frac{2\pi}{k}$. Take an arbitrary $\epsilon > 0$. By choosing k sufficiently large, we can make $\frac{2\pi}{k} < \epsilon$. So, in any ϵ-neighborhood of an arbitrary point of the circle, there are points of the sequence $\{\varphi_0 + \mathbf{N}s\Delta\}$.

□

It follows from this lemma that if, in the dynamical system under consideration, the number $\frac{\beta}{\alpha}\ln\frac{r_1}{r_2}$ is not rationally commensurable with 2π, then the trajectory of any solution of this system, which starts in the region $x^2 + y^2 > r_1$, fills the annulus $r_1^2 \leq x^2 + y^2 < r_2^2$ densely everywhere.

1.5 Motion of an impulsive oscillator under the effect of an impulsive force

In many problems of theory of oscillation, we have to study the behavior of an oscillator subject to an impulsive force. Such problems arise, for example, when studying a model for a clock [12, 15].

A model for this type of problems is a discontinuous dynamical system, the motions of which are described by the linear second order differential equation

$$\ddot{x} + 2\lambda\dot{x} + \omega^2 x = 0, \qquad \lambda^2 < \omega^2,$$

with an impulse effect when the point $(x, \dot{x})$ in the phase plane passes through $x = x_0$.

A number of results for such an oscillator with different kind of impulsive forces are contained in the known monographs [12, 24].

There, the following two main assumptions about the impulsive force are used:

1) the impulsive force acts at the moment when the point $(x, \dot{x})$ passes through $x = x_0$ with nonnegative velocity increasing the impulse by a constant

$$m\dot{x}(t+0) - m\dot{x}(t-0) = mI_0 = \text{const};$$

2) the impulsive force acts when the point $(x, \dot{x})$ intersects $x = x_0$ increasing the kinetic energy by a constant

$$\frac{m\dot{x}^2(t+0)}{2} - \frac{m\dot{x}^2(t-0)}{2} = mI = \text{const}.$$

Consider some examples of these oscillators.

1. At the beginning, let us study motions of the following impulsive system:

$$\begin{aligned} \ddot{x} + \omega^2 x = 0, &\qquad x \neq x_0 > 0, \\ \Delta\dot{x}|_{x=x_0} = I, &\qquad I > 0. \end{aligned} \tag{1.62}$$

Such a system has been extensively studied in [147].

Because the phase plane $(x, \dot{x})$ is a union of the ellipses

$$\omega^2 x^2 + \dot{x}^2 = c^2,$$

it is easy to describe the trajectories of system (1.62). The trajectory that passes through the point (x_0, I) consists of the ellipse $\dot{x}^2 + \omega^2 x^2 = \omega^2 x_0^2$, which lies in the half-plane $x \leq x_0$, and the arc of the ellipse $\dot{x}^2 + \omega^2 x^2 = I_0^2 + \omega^2 x_0^2$, lying in the half-plane $x > x_0$. The trajectories that pass through the region $D_1 : \dot{x}^2 + \omega^2 x^2 < \omega^2 x_0^2$ are closed curves, and the motions, which follow these curves, are not subject to an impulsive effect. The trajectory that passes through the point $(x_0, \frac{I}{2})$ consists of the arc of the ellipse $\dot{x}^2 + \omega^2 x^2 = \frac{I^2}{4} + \omega^2 x_0^2$, which lies in the half-plane $x > x_0$. The trajectories that start in the region D_2, except for the considered trajectory, consist of arcs of two ellipses lying in the half-plane $x > x_0$, namely the arc of the ellipse $x^2 + \omega^2 x^2 = (y_0 + I)^2 + \omega^2 x_0^2$ and the arc of the ellipse $x^2 + \omega^2 x^2 = y_0^2 + \omega^2 x_0^2$. The trajectories that start at a point (x_0, y_0), which lies outside of the region $D_1 \cup D_2$, consists of the arc of the ellipse $x^2 + \omega^2 x^2 = y_0^2 + \omega^2 x_0^2$, which lies in the half-plane $x > x_0$, and the arc of the ellipse $x^2 + \omega^2 x^2 = (I - y_0)^2 + \omega^2 x_0^2$ in the half-plane $x < x_0$.

All the motions in this system are periodic but not with the same period.

2. By using the phase plane method, it is possible to study motions of a damped oscillator with an impulsive force acting at the moment when the phase point $(x, \dot{x}$ crosses the line $x = 0$ with a nonnegative velocity. If we assume that, as the result of the impulsive effect the velocity is increased by a quantity $I_0(\dot{x})$, which depends on the velocity of the point at the time of the impulsive effect, then the equations for the motion of an oscillator will become

$$\begin{gathered} \ddot{x} + 2\lambda\dot{x} + \omega^2 x = 0, \qquad x \neq 0, \\ \Delta\dot{x}|_{x=0} = \begin{cases} I_0(\dot{x}), & \text{for} \quad \dot{x} \geq 0, \\ 0, & \text{for} \quad \dot{x} < 0. \end{cases} \end{gathered} \tag{1.63}$$

By introducing the function $y = \dot{x}$, (1.62) can be rewritten as

$$\begin{gathered} \frac{dx}{dt} = y, \qquad \frac{dy}{dt} = -\omega^2 x - 2\lambda y, \qquad x \neq 0, \\ \Delta y|_{x=0} = I(y), \end{gathered} \tag{1.64}$$

where

$$I(y) = \begin{cases} I_0(y), & \text{for} \quad \dot{y} \geq 0, \\ 0, & \text{for} \quad \dot{y} < 0. \end{cases}$$

Assume that, for $y \geq 0$, $I_0(y) \geq 0$. This means that the impulsive effect does not decrease the velocity of the point.

Since the point moves along spirals of the system

$$\frac{dx}{dt} = y, \qquad \frac{dy}{dt} = -\omega^2 x - 2\lambda y, \tag{1.65}$$

the trajectories of these motions go either to infinity or the origin, or they are attracted by disconnected limit cycles if there are any. Let us find out whether these cycles exist.

Find the trajectories of system (1.65), i.e. the integral curves of the equation

$$ydy + (2\lambda y + \omega^2 x)dx = 0.$$

By solving it, we get

$$y^2 + 2\lambda xy + \omega^2 x^2 = C^2 e^{\frac{2\lambda}{\nu}\left(\arctan\frac{y-\lambda x}{\nu x} + k\pi\right)}, \tag{1.66}$$

where $\nu = \sqrt{\varrho^2 - \lambda^2}$, $k \in \mathbf{Z}$. The y-coordinates of the points, at which the spirals (1.66) intersect the half-line $x = 0$, $y \geq 0$, can be found from the equation

$$y^2 = C^2 e^{\frac{2\lambda}{\nu}\left(-\frac{\pi}{2} + 2k\pi\right)}. \tag{1.67}$$

Let us look at the motion that starts at the point $0, y_0)$, where

$$y_0 = C_0 e^{-\frac{\lambda}{\nu}\frac{\pi}{2}}. \tag{1.68}$$

After making one turn on the spiral, the phase point will get onto the line $x = 0$, $y \geq 0$ at the point with y-coordinate

$$y_1 = y_0 e^{-\frac{\lambda}{\nu}2\pi}, \tag{1.69}$$

and after that it will jump to a point of this line with y-coordinate $y_1^+ = y_1 + I_0(y_1)$. After having made one more turn on the spiral, the phase point will get onto the semi-line $x = 0$, $y \geq 0$ with y-coordinate

$$y_2 = y_1^+ e^{\frac{\lambda}{\nu}2\pi},$$

and after that, it will jump to the point of this semi-line with y-coordinate equal to $y_2^+ = y_2 + I_0(y_2)$ and so on.

Denote by h the mapping of the semi-line $y \geq 0$ into itself given by

$$h : R^+ \to R^+, \qquad h(y) = ye^{-\bar{\Delta}} + I_0(ye^{-\bar{\Delta}}), \tag{1.70}$$

where $\bar{\Delta} = \frac{\lambda}{\nu}2\pi$.

The fixed points of this mapping, i.e. the points with $y_0 \geq 0$ such that $h(y_0) = y_0$, give rise to disconnected cycles of system (1.64), periodic motions along which are subject to one impulsive effect in the period.

The stability of the disconnected limit cycles of system (1.64) is determined by the stability of the fixed points of the mapping $h^k(y)$.

For example, if the equation

$$ye^{-\bar{\Delta}} + I_0(ye^{-\bar{\Delta}}) = y \tag{1.71}$$

has a solution $y = y^*$, then the cycle of system (1.64) will be asymptotically stable if

$$\left|1 + \frac{dI_0}{dy}(y^* e^{-\bar{\Delta}})\right| e^{-\bar{\Delta}} < 1, \tag{1.72}$$

and will be unstable if

$$\left|1 + \frac{dI_0}{dy}(y^* e^{-\bar{\Delta}})\right| e^{-\bar{\Delta}} > 1. \tag{1.73}$$

In particular, if in system (1.64), $I_0(y) = I_0 > 0$, then it has a disconnected asymptotically stable limit cycle, the motions along which experience one impulsive effect in the period. Indeed, in this case, equation (1.71) has the form

$$ye^{-\bar{\Delta}} + I_0 = y.$$

It has a solution $y = I_0(1 - e^{-\bar{\Delta}} + I_0 = y$, which satisfy inequality (1.72). This disconnected cycle consists of a part of the spiral

$$y^2 + 2\lambda xy + \omega^2 x^2 = \frac{I_0^2}{\left(1 - e^{-\frac{\lambda}{\nu}2\pi}\right)^2} e^{\frac{2\lambda}{\nu}\left(\arctan\frac{y+\lambda x}{\nu x} - \frac{\pi}{2}\right)}.$$

In the case under consideration, system (1.64) does not have other limit cycles. The equation $h^k y = y$, in this case, becomes

$$ye^{-k\bar{\Delta}} + I_0 \sum_{m=0}^{k-1} e^{-m\bar{\Delta}} = y$$

and it has a unique solution

$$y = \frac{I_0}{1 - e^{-\bar{\Delta}}}$$

for any $k = 1, 2, \ldots$.

3. Let us find out whether, for a damped oscillator, there exist disconnected cycles in the case when the impulsive effect increases the kinetic energy by a constant value. Such an oscillator is described by the system

$$\ddot{x} + 2\lambda\dot{x} + \omega^2 x = 0, \qquad x \neq 0,$$
$$\Delta\dot{x}|_{x=0} = \begin{cases} I_0(\dot{x}), & \text{for } \dot{x} > 0, \\ 0, & \text{for } \dot{x} \leq 0. \end{cases} \tag{1.74}$$

By writing the condition imposed on the impulse as

$$\Delta\dot{x}(\Delta\dot{x} + 2\dot{x}) = I_0^2$$

and solving this equation with respect to $\Delta\dot{x}$, write (1.74) as

$$\frac{dx}{dt} = y, \qquad \frac{dy}{dt} = -\omega^2 x - 2\lambda y, \qquad x \neq 0,$$
$$\Delta y|_{x=0} = \begin{cases} -y + \sqrt{y^2 + I_0^2}, & \text{if } y > 0, \\ 0, & \text{if } y \leq 0. \end{cases}$$

We get a system of the form (1.64) with

$$I_0(y) = -y + \sqrt{y^2 + I_0^2}.$$

Equation (1.71) becomes

$$\sqrt{y^2 e^{-2\bar{\Delta}} + I_0^2} = y. \tag{1.75}$$

It has a unique solution

$$y^* = \frac{I_0}{\sqrt{1 - e^{-2\bar{\Delta}}}}.$$

Hence, system (1.74) has a limit cycle, which consists of one piece of the spiral.

The derivative of the function $I_0(y)$,

$$\frac{d}{dy} I_0(y) = -1 + \frac{y}{\sqrt{y^2 + I_0^2}},$$

at the point y^*, satisfies

$$-1 < \frac{dI_0(y^*)}{dy} < 0$$

and so inequality (1.72) holds, hence the limit cycle is asymptotically stable.

System (1.74) does not have other limit cycles because the equation $h^k(y) = y$, in this case, becomes

$$\sqrt{y^2 e^{-2k\bar{\Delta}} + I_0^2 \sum_{m=0}^{k-1} e^{-2m\bar{\Delta}}} = y$$

and, for any $k = 1, 2, \ldots$, this equation has only one solution $y = y^*$.

4. Suppose that, in system (1.74), an impulse occurs each time when the phase point intersects the line $x = 0$ and, as a result of the impulse, the kinetic energy of the system is increased by the same value I_0^2. A linear oscillator with such an impulsive effect can be described by the equation

$$\begin{gathered} \ddot{x} + 2\lambda\dot{x} + \omega^2 x = 0, \qquad x \neq 0, \\ \Delta\dot{x}|_{x=0} = I_0^2 \end{gathered} \tag{1.76}$$

or, which is the same thing, by the system

$$\begin{gathered} \frac{dx}{dt} = y, \qquad \frac{dy}{dt} = -\omega^2 x - 2\lambda y \qquad x \neq 0, \\ \Delta y|_{x=0} = \begin{cases} -y + \sqrt{y^2 + I_0^2}, & \text{if } \ y \geq 0, \\ -y - \sqrt{y^2 + I_0^2}, & \text{if } \ y < 0. \end{cases} \end{gathered} \tag{1.77}$$

In order not to repeat the reasoning used in the previous example, we shall study system (1.77) by using another method, which might be of interest by itself.

In system (1.77), change the variables (x, y) to (a, φ) by

$$x = a\sin\varphi, \qquad y = a(-\lambda\sin\varphi + \nu\cos\varphi), \qquad \nu^2 = \omega^2 - \lambda^2. \tag{1.78}$$

If $\sin\varphi \neq 0$, we have

$$\begin{gathered} \sin\varphi\frac{da}{dt} + a\cos\varphi\frac{d\varphi}{dt} = a(-\lambda\sin\varphi + \nu\cos\varphi), \\ (-\lambda\sin\varphi + \nu\cos\varphi)\frac{da}{dt} - a(\lambda\cos\varphi + \nu\sin\varphi)\frac{d\varphi}{dt} = \\ = -2\lambda a(-\lambda\sin\varphi + \nu\cos\varphi) - \omega^2 a\sin\varphi. \end{gathered}$$

By solving this system with respect to $\dfrac{da}{dt}$ and $\dfrac{d\varphi}{dt}$, we obtain

$$\frac{da}{dt} = -\lambda a, \qquad \frac{d\varphi}{dt} = \nu \qquad \varphi \neq k\pi.$$

For $x = a\sin\varphi = 0$, i.e. for $\varphi = k\pi$, the following equalities must hold

$$\begin{gathered}(a+\Delta a)\sin(k\pi+\Delta\varphi)=0\\ (a+\Delta a)\nu\cos(k\pi+\Delta\varphi)-\nu a(-1)^m=\\ =\begin{cases}-a\nu+\sqrt{a^2\nu^2-I_0^2}, & \text{if} \quad \varphi=2m\pi,\\ -a\nu-\sqrt{a^2\nu^2-I_0^2}, & \text{if} \quad \varphi=\pi+2m\pi.\end{cases}\end{gathered}$$

From here we find that

$$\Delta a|_{\varphi=k\pi}=-a+\sqrt{a^2+\frac{I_0^2}{\nu^2}}, \qquad \Delta\varphi|_{\varphi=k\pi}=0.$$

So after the change of variables (1.78), system (1.77) becomes

$$\begin{gathered}\frac{da}{dt}=-\lambda a, \qquad \frac{d\varphi}{dt}=\nu, \qquad \varphi\neq k\pi,\\ \Delta a|_{\varphi=k\pi}=-a+\sqrt{a^2+\frac{I_0^2}{\nu^2}}.\end{gathered} \tag{1.79}$$

Periodic solutions of the equation

$$\begin{gathered}\frac{da}{dt}=-\frac{\lambda}{\nu}a \qquad \varphi\neq k\pi,\\ \Delta a|_{\varphi=k\pi}=-a+\sqrt{a^2+\frac{I_0^2}{\nu^2}}\end{gathered} \tag{1.80}$$

determine disconnected cycles of system (1.77).

Let us study the existence of 2π-periodic solutions of equation (1.70). These solutions are

$$a(\varphi)=a_0e^{-\frac{\lambda}{\nu}(\varphi-\varphi_0)}$$

and they are 2π-periodic if

$$h(a_0)=\sqrt{\left(a_0^2+\frac{I_0^2}{\nu^2}\right)e^{-2\bar{\Delta}}+\frac{I_0^2}{\nu^2}e^{-\bar{\Delta}}}=a_0. \tag{1.81}$$

This equation has a unique solution

$$a_0^*=\frac{I_0e^{-\bar{\Delta}}}{\nu\sqrt{1-e^{-2\bar{\Delta}}}},$$

and so equation (1.80) has a unique 2π-periodic solution

$$a(\varphi) = \begin{cases} \sqrt{a_0^{*2} + \frac{I_0^2}{\nu^2} e^{-\frac{\lambda}{\nu}\varphi}}, & 0 < \varphi < \pi, \\ \sqrt{\left(a_0^{*2} + \frac{I_0^2}{\nu^2}\right) e^{-2\bar{\Delta}} + \frac{I_0^2}{\nu^2} e^{-\frac{\lambda}{\nu}(\varphi-\pi)}}, & \pi < \varphi < 2\pi. \end{cases}$$

Because, for $a_0 = a_0^*$, the derivative of the function $h(a_0)$,

$$\frac{dh(a_0^*)}{da} = \frac{a_0^* e^{-4\bar{\Delta}}}{\sqrt{\left(a_0^{*2} + \frac{I_0^2}{\nu^2}\right) e^{-2\bar{\Delta}} + \frac{I_0^2}{\nu^2} e^{-\bar{\Delta}}}} = \frac{a_0^* e^{-4\bar{\Delta}}}{a_0^*} = e^{-4\bar{\Delta}},$$

is less than one, the periodic solution $a = a(\varphi)$ is asymptotically stable.

It is not difficult to see that equation (1.80) does not have periodic solutions with the period $2p\pi$ for $p > 1$. Indeed, the points a_0, which generate such a solution, are solutions of the equation

$$h^k(a_0) = a_0, \qquad h^k(a_0) = h(h^{k-1}(a_0)),$$

where $h(a)$ is determined according to (1.81). This equation becomes

$$\sqrt{a_0^2 e^{-2p\bar{\Delta}} + \frac{I_0^2}{\nu^2} \sum_{j=0}^{p-1} e^{-2j\Delta} e^{-\Delta}} = a_0.$$

It has a unique solution $a_0 = a_0^*$.

Thus, equation (1.77) has a unique asymptotically stable disconnected limit cycle.

Chapter 2

Linear Systems

2.1 General properties of solutions of linear systems

In this section we will establish some simple properties of linear homogeneous impulsive differential systems

$$\frac{dx}{dt} = A(t)x, \qquad t \neq \tau_i, \qquad \Delta x|_{t=\tau_i} = B_i x, \tag{2.1}$$

as well as linear nonhomogeneous systems.

Here $A(t)$ is $n \times n$-matrix constant on the interval $]a, b[$, B_i are constant matrices, $\tau_i \in I$ are fixed times indexed by a set or a subset of integers such that $\tau_i < \tau_{i+1}$.

One of the main results of the theory of linear systems of type (2.1) is the following theorem.

Theorem 7 *Let an interval $[t_0, t_0+h] \subset I$ contain a finite number of points τ_i. Then, for any $x_0 \in \mathbf{R}^n$, a solution of system (2.1) $x(t, x_0)$, $x(t_0, x_0) = x_0$, exists for all $t \in [t_0, t_0+h]$. Moreover if, for all i such that $\tau_i \in [t_0, t_0+h]$, the matrices $E + B_i$ are nonsingular, then $x(t, x_0) \neq x(t, y_0)$ for all $t \in [t_0, t_0+h]$ if $x_0 \neq y_0$.*

Proof. Let $t_0 < \tau_j < \tau_{j+1} < \cdots < \tau_{j+k} \leq t_0 + h$. By Picard theorem, a solution $x = \varphi_j(t)$, $\varphi_j(t_0) = x_0$, of system

$$\frac{dx}{dt} = A(t)x \tag{2.2}$$

exists on $[t_0, \tau_j]$ for any $x_0 \in \mathbf{R}^n$ and is unique. Set $x(t, x_0) = \varphi_j(t)$ for $t \in [t_0, \tau_j]$. From system (2.1), we get for $t = \tau_j$,

$$x(\tau_j + 0, x_0) = (E + B_j)x(\tau_j, x_0) = x_j^+.$$

Again by using the Picard theorem, we see that there exists on $[\tau_j, \tau_{j+1}]$ a unique solution of system (2.2) $x = \varphi_{j+1}(t)$, $\varphi_{j+1}(\tau_j) = x_j^+$. Hence the solution of impulsive system (2.1) $x(t, x_0)$ can be extended to $t = \tau_{j+1}$ by setting

$$x(t, x_0) = \varphi_{j+1}(t) \qquad \text{for} \qquad \tau_j < t \leq \tau_{j+1},$$

and $x(\tau_{j+2}(t) + 0, x_0) = (E + B_{j+1})x(\tau_j, x_0)$ for $t = \tau_{j+1} + 0$.

Denoting by $\varphi_{j+2}(t)$ the solution of system (2.2) such that $\varphi_{j+2}(\tau_{j+1}) = x_{j+1}^+ = (E + B_{j+1})x(\tau_{j+1}, x_0)$, we can extend the solution $x(t, x_0)$ until the time τ_{j+2} by setting $x(t, x_0) = \varphi_{j+2}(t)$ for $\tau_{j+1} < t \leq \tau_{j+2}$ and so on.

Because, by the condition of the theorem, the interval $[t_0, t_0+h]$ contains only a finite number of points τ_j, by using this procedure, we can construct a solution $x(t, x_0)$ on the whole interval.

We have shown a way to construct a solution assuming that $t_0 < \tau_j$. If $t_0 = \tau_j$, then we use the same construction procedure as we did for $t_0 < \tau_j$ but, setting $x(t_0+0, x_0) = (E+B_j)x_0$, we construct the function $\varphi_{j+1}(t)$ to be a solution of system (2.2), which satisfies the initial condition $\varphi_{j+1}(\tau_j) = x_j^+ = (E+B_j)x_0$.

To prove the second statement of the theorem note that if $x(\tau_i+0, x_0) \neq x(\tau_i+0, y_0)$, then by Picard theorem, we also have that $x(t, x_0) \neq x(t, y_0)$ for all $\tau_i < t \leq \tau_{i+1}$, $i = j, j+1, \ldots, j+k$. Because $x(\tau_i+0, x_0) - x(\tau_i + 0, y_0) = (E+B_i)(x(\tau_i, x_0) - x(\tau_i, y_0))$, it follows from $x(\tau_i, x_0) \neq x(\tau_i, y_0)$ that $x(\tau_i+0, x_0) \neq x(\tau_i+0, y_0)$ if the matrix $(E+B_i)$ is nonsingular. This implies that $x(t, x_0) \neq x(t, y_0)$ for all $t \in [t_0, t_0+h]$ as long as $x_0 \neq y_0$ and $\det(E+B_i) \neq 0$ for $i = j, j+1, \ldots, j+k$.

□

If $[t_0 - h, t_0] \subset I$ and, for all i such that the $\tau_i \in [t_0-h, t_0]$, the matrices $E+B_i$ are nonsingular, then the solution $x(t, x_0)$ can be uniquely extended over the interval $[t_0-h, t_0]$. If, for some of these $\tau_i = \tau_{i_1}$, the matrix $E+B_i$ is singular, then the solution $x(t, x_0)$ can be extended in a unique way to the left only until the time τ_{i_1}. To extend the solution over the interval $[t_0-h, \tau_{i_1}]$ is either impossible or, if it can be done, there is no uniqueness.

Indeed, let τ_{i_1} be the time nearest to t_0 such that the matrix $E+B_{i_1}$ is singular. Let r be the rank of the matrix $E+B_{i_1}$. The solution $x(t, x_0)$ can be uniquely extended over the interval $]\tau_{i_1}, t_0]$. Denote $x^+ = x(\tau_{i_1}+0, x_0)$. The linear operator defined by the matrix $(E+B_{i_1})$ gives rise to a projection from the space $\mathbf{R}^n$ into a linear subspace $\mathbf{R}^r$. If x^+ is contained in the image of this projection, then the solution $x(t, x_0)$ can be extended to the left of the point τ_{i_1}. This extension is not unique because the equation $(E+B_{i_1})x = x^+$ has infinitely many solutions. If x^+ does not belong to the image of this projection, then the equation $(E+B_{i_1})x = x^+$ does not have solutions, and so the solution $x(t, x_0)$ can not be extended to the left of the point τ_{i_1}.

Hence, if some matrix $(E+B_i)$ is singular for an index i such that $\tau_i \in [t_0-h, t_0]$, then a number of solutions $x(t, x_0)$, $x(t_0, x_0) = x_0$, can not be extended to the left of the point τ_i and other split into a set of solutions at the time τ_i.

Let τ_i be the time nearest to t_0, $\tau_i < t_0$, such that the matrix $E+B_i$

is singular. Every solution $x(t, x_0)$, $x(t_0, x_0) = x_0$, of the original system can be extended over the whole interval $[t_0, t_0 + h]$ but at $t = \tau_i$ many such solutions merge into a single solution. The set of points $x(\tau_i, x_0)$, $x_0 \in \mathbf{R}^n$ form the space $\mathbf{R}^n$ and the set of the points $(E+B_i)x(\tau_i, x_0)$ form a subspace of the dimension $n - r$, where r is the rank of the matrix $E + B_i$. Besides at the time τ_i, new solutions are being "born". They are defined for $t > \tau_i$, i.e. they are the solutions $x(t, y)$, $x(\tau_i + 0, y) = y$, for which the initial point has the property that the algebraic system of equations $(E + B_i)x = y$ does not have solutions.

In the sequel, we will be studying only the systems, for which the following conditions hold:

1) any compact interval $[a, b]$ contains only a finite number of the points τ_i;

2) for all i such that $\tau_i \in I$, the matrices $E + B_i$ are nonsingular.

Under these assumptions, the following statement is true.

Theorem 8 *The set $\mathcal{X}$ of all the solutions of linear homogeneous impulsive differential system (2.1), which are defined on the interval $[a, b]$, form n-dimensional vector space.*

Proof. Let $\varphi_1(t)$ and $\varphi_2(t)$ be solutions of system (2.1), defined on the interval $[a, b]$ and c_1, c_2 be real (complex) numbers. It is easy to see that the linear combination $c_1\varphi_1(t) + c_2\varphi_2(t)$ is also a solution of system (2.1). This means that solutions of (2.1) form a linear space.

To show that its dimension is n, we show that it is isomorphic to the phase space of this equation, i.e. to the space $\mathbf{R}^n$.

Indeed, let $t \in [a, b]$. Consider a mapping $g_t : \mathcal{X} \to \mathbf{R}^n$, $g_t\varphi = \varphi(t)$, which maps a solution $\varphi(t) \in \mathcal{X}$ into its value at a time t.

The mapping $g_t : \mathcal{X} \to \mathbf{R}^n$ is linear (the value of the sum is the sum of the values). Its image is the whole space $\mathbf{R}^n$ because, by Theorem 7, for any $x_0 \in \mathbf{R}^n$, there exists a solution $\varphi(t) = x(t, x_0)$, $\varphi(t_0) = x_0$. The kernel of the mapping g_t equals to zero because, by Theorem 7, $\varphi(t_0) = 0$ implies that $\varphi(t) \equiv 0$.

Hence g_t is an isomorphism of $\mathcal{X}$ onto $\mathbf{R}^n$.

□

A basis in the space $\mathcal{X}$ is called a *fundamental system* of solutions of (2.1).

This theorem has a number of important corollaries:

1) system (2.1) has a fundamental system of n solutions $\varphi_1(t)$, $\varphi_2(t)$,..., $\varphi_n(t)$;

2) every solution of system (2.1) is a linear combination of solutions of the fundamental system;

3) any $n+1$ solutions of (2.1) are linearly dependent.

Denote by $X(t)$ the matrix, the columns of which are the solutions of system (2.1) that form a fundamental system. The matrix $X(t)$ will be called a *fundamental matrix* of system (2.1). It is clear that the function

$$x(t) = X(t)c \tag{2.3}$$

is a solution of system (2.1) for any constant vector c and, if c ranges over the whole space $\mathbf{R}^n$, then the family of functions (2.3) forms a space. It follows from the definition of the matrix $X(t)$ that it satisfies the matrix impulsive equation

$$\frac{dX}{dt} = A(t)X, \qquad t \neq \tau_i, \qquad \Delta X|_{t=\tau_i} = B_i X. \tag{2.4}$$

It is also clear that any nondegenerate solution of matrix system (2.4) is a fundamental matrix of system (2.1). All nondegenerate solutions of system (2.4) are given by the formula $X(t) = X_0(t)C$, where $X_0(t)$ is some nondegenerate solution of (2.4) and C is an arbitrary nonsingular matrix. A nondegenerate solution of system (2.4) $X(t)$, which satisfies the condition $X(t_0) = E$ will be called a *matriciant* of system (2.1) and denoted by $X(t, t_0)$.

Let $U(t,\tau)$ be a solution of the matrix Cauchy problem

$$\frac{dU}{dt} = A(t)U, \qquad U(\tau,\tau) = E, \tag{2.5}$$

i.e. a matriciant of system (2.2). Then any solution $X(t)$ of matrix system (2.2) can be represented as

$$\begin{aligned} X(t) &= U(t,\tau_{j+k})(E+B_{j+k})U(\tau_{j+k},\tau_{j+k-1})(E+B_{j+k-1})\dots \\ &\quad \dots(E+B_j)U(\tau_j,t_0)X(t_0), \\ &\quad \tau_{j-1} < t_0 \le \tau_j < \tau_{j+k} < t \le \tau_{j+k+1}. \end{aligned} \tag{2.6}$$

In particular, for the matriciant $X(t, x_0)$, we have

$$\begin{aligned} X(t,t_0) &= U(t,\tau_{j+k})(E+B_{j+k})U(\tau_{j+k},\tau_{j+k-1})(E+B_{j+k-1})\dots \\ &\dots(E+B_j)U(\tau_j,t_0), \\ &\tau_{j-1} < t_0 \le \tau_j < \tau_{j+k} < t \le \tau_{j+k-1} \end{aligned}$$

or

$$\begin{aligned} X(t,t_0) &= U(t,\tau_{j+k})(E+B_{j+k})\prod_{\nu=k}^{1} U(\tau_{j+\nu},\tau_{j+\nu-1})\times \\ &\times(E+B_{j+\nu-1})U(\tau_j,t_0). \end{aligned} \tag{2.7}$$

By using the Osrogradskiĭ-Liouville formula, we get from (2.6)

$$\begin{aligned} \det X(t) &= \det U(t,\tau_{j+k})\det(E+B_{j+k})\prod_{\nu=k}^{1}\det U(\tau_{j+\nu},\tau_{j+\nu-1})\times \\ &\times\det(E+B_{j+\nu-1})\det U(\tau_j,t_0)\det X(t_0) = \\ &= e^{\int_{\tau_{j+1}}^{t}\mathrm{Sp}\,A(\sigma)d\sigma}\det(E+B_{j+k})\prod_{\nu=k}^{1} e^{\int_{\tau_{j+\nu-1}}^{\tau_{j+\nu}}\mathrm{Sp}\,A(\sigma)d\sigma}\times \\ &\times\det(E+B_{j+\nu-1})e^{\int_{t_0}^{\tau_j}\mathrm{Sp}\,A(\sigma)d\sigma}\det X(t_0), \end{aligned}$$

i.e.

$$\begin{aligned} \det X(t) &= \det X(t_0)e^{\int_{t_0}^{t}\mathrm{Sp}\,A(\sigma)d\sigma}\prod_{\nu=1}^{k+1}\det(E+B_{j+\nu-1}), \\ &\tau_{j-1} < t_0 \le \tau_j \le \tau_{j+k} < t \le \tau_{j+k+1}. \end{aligned} \tag{2.8}$$

Because the matrices $E + B_i$ are nonsingular, it follows from (2.8) that the matrix $X(t)$ is nonsingular if $X(t_0)$ is such.

For a nonsingular matrix $X(t)$, the inverse matrix $X^{-1}(t)$ is given by the following

$$\begin{aligned} X^{-1}(t) &= X^{-1}(t_0)U^{-1}(\tau_j,t_0)(E+B_j)^{-1}\dots \\ &\dots U^{-1}(\tau_{j+k},\tau_{j+k-1})(E+B_{j+k})^{-1}U^{-1}(t,\tau_{j+k}) = \\ &= X^{-1}(t_0)U^{-1}(\tau_j,t_0)\prod_{\nu=1}^{k}(E+B_{j+\nu-1})^{-1}U^{-1}(\tau_{j+\nu},\tau_{j+\nu-1})\times \\ &\times(E+B_{j+k})^{-1}U^{-1}(t,\tau_{j+k}), \end{aligned}$$

$$\tau_{j-1} < t_0 \le \tau_j < \tau_{j+k} < t \le \tau_{j+k+1},$$

and the product

$$\begin{aligned} X(t)X^{-1}(\sigma) &= U(t,\tau_{j+k})\prod_{\nu=k}^{s+1}(E+B_{j+\nu})U(\tau_{j+\nu},\tau_{j+\nu-1})\times \\ &\quad \times(E+B_{j+k})U(\tau_{j+s},\sigma) \end{aligned}$$

$$\tau_{j+s-1} < \sigma < \tau_{j+s} < \tau_{j+k} < t \le \tau_{j+k+1}.$$

In particular, for the matriciant $X(t,t_0)$, we have

$$\begin{aligned} X^{-1}(t,t_0) &= U^{-1}(\tau_j,t_0)\prod_{\nu=1}^{k}(E+B_{j+\nu+1})^{-1}U^{-1}(\tau_{j+\nu},\tau_{j+\nu-1})\times \\ &\quad \times(E+B_{j+k})^{-1}U^{-1}(t,\tau_{j+k}) \end{aligned} \tag{2.9}$$

$$\begin{aligned} & X(t,t_0)X^{-1}(\sigma,t_0) = \\ & = U(t,\tau_{j+k})\prod_{\nu=K}^{s+1}(E+B_{j+\nu})U(\tau_{j+\nu},\tau_{j+\nu-1})(E+B_{j+s})U(\tau_{j+s},\tau) = \\ & = X(t,\sigma), \end{aligned} \tag{2.10}$$

$$\tau_{j-1} < t_0 \le \tau_j \le \tau_{j+s-1} < \sigma \le \tau_{j+s} < \tau_{j+k} < t \le \tau_{j+k+1}.$$

If $\tau_i < \sigma \le t \le \tau_{i+1}$, then $X(t,t_0)X^{-1}(\sigma,t_0) = U(t,\sigma)$. Note that, by using the matriciant, any solution $x(t,x_0)$, $x(t_0,x_0) = x_0$, can be written as

$$x(t,x_0) = X(t,t_0)x_0. \tag{2.11}$$

System of equations

$$\frac{dx}{dt} = A(t)x + f(t), \qquad t \ne \tau_i, \qquad \Delta x|_{t=\tau_i} = B_i x + a_i, \tag{2.12}$$

where the matrices $A(t)$, B_i, and the times τ_i are the same as in system (2.1), and the function $f(t)$ is a continuous (piecewise continuous) function on the interval I, a_i are constant vectors, will be called linear *nonhomogeneous* impulsive differential system.

The relationship between nonhomogeneous system (2.12) and the corresponding homogeneous system (2.1) is given by the following theorem.

Theorem 9 *If $x = \varphi(t)$ is a solution of system (2.1) and $x = \psi(t)$ is a solution of system (2.12), then the function $x = \varphi(t) + \psi(t)$ is a solution of system (2.12). Conversely, if $x = \varphi_1(t)$ and $x = \varphi_2(t)$ are solutions of nonhomogeneous system (2.12), then the function $x = \varphi_1(t) - \varphi_2(t)$ is a solution of system (2.1).*

This theorem can be verified directly.

In the sequel we will use a linear change of variables for systems (2.1) and (2.12).

Theorem 10 *Let S be a nonsingular matrix, continuously differentiable with respect to $t \in [a,b]/\{\tau_i\}$. Then, under the linear change of variables,*

$$x = S(t)y, \tag{2.13}$$

equations (2.12) become

$$\begin{aligned} \frac{dy}{dt} &= S^{-1}(t)\left[A(t)S(t) - \frac{dS}{dt}\right]y + S^{-1}(t)f(t), \qquad t \neq \tau_i, \\ \Delta y|_{t=\tau_i} &= S^{-1}(\tau_i + 0)(-\Delta S + B_i S)y|_{t=\tau_i} + S^{-1}(\tau_i + 0)a_i. \end{aligned} \tag{2.14}$$

In particular, if $S(t)$ is a fundamental matrix of the system

$$\frac{dx}{dt} = P(t)x, \qquad t \neq \tau_i, \qquad \Delta x|_{t=\tau_i} = I_i x, \tag{2.15}$$

where the matrices $E + I_i$ are nonsingular, then system (2.14) becomes

$$\begin{aligned} \frac{dy}{dt} &= S^{-1}(t)(A(t) - P(t))S(t)y + S^{-1}(t)f(t), \qquad t \neq \tau_i, \\ \Delta y|_{t=\tau_i} &= S^{-1}(\tau_i + 0)(B_i - I_i)Sy|_{t=\tau_i} + S^{-1}(\tau_i + 0)a_i. \end{aligned} \tag{2.16}$$

If $A(t) = P(t)$, $B_i = I_i$, $S(t) = X(t)$, where $X(t)$ is a fundamental matrix of system (2.1), change of variables (2.13) is called "variation of parameters" because the constant vector c in (2.3) is replaced by a variable vector y. In this case system (2.12) becomes

$$\frac{dy}{dt} = X^{-1}(t)f(t), \qquad t \neq \tau_i, \qquad \Delta y|_{t=\tau_i} = X^{-1}(\tau_i + 0)a_i, \tag{2.17}$$

which can be easily integrated. Taking into account that $X(\tau_i + 0) = (E + B_i)X(\tau_i)$, the jump condition (2.17) can also be written in the form

$$\Delta y|_{t=\tau_i} = X^{-1}(\tau_i)(E + B_i)^{-1}a_i. \tag{2.18}$$

From (2.17) we can find that, for $t \geq t_0$,

$$y(t) = c + \int_{t_0}^{t} X^{-1}(\tau)f(\tau)d\tau + \sum_{t_0<\tau_i<t} X^{-1}(\tau_i)(E+B_i)^{-1}a_i, \qquad (2.19)$$

where $c = y(t_0)$ is a constant vector.

Corollary 3 *Let $X(t)$ be a fundamental matrix of system (2.1) with nonsingular matrices $E + B_i$. Then every solution of system (2.12), for $t \geq t_0$, is given by the formula*

$$x(t) = X(t)\left[c + \int_{t_0}^{t} X^{-1}(t)f(t)d\tau + \sum_{t_0<\tau_i<t} X^{-1}(\tau_i)(E+B_i)^{-1}a_i\right]. \qquad (2.20)$$

In particular, if $X(t) = X(t, t_0)$ is a matriciant of system (2.1), then any solution of (2.12) $x(t, x_0)$, $x(t_0, x_0) = x_0$, for $t \geq t_0$, can be written as

$$x(t, x_0) = X(t, t_0)x_0 + \int_{t_0}^{t} X(t,\tau)f(\tau)d\tau + \sum_{t_0<\tau_i<t} X(t, \tau_i + 0)a_i. \qquad (2.21)$$

These two formulas show that if solutions of the corresponding homogeneous system are known, then solutions of system (2.12) can be found in quadratures.

2.2 Linear systems with constant coefficients

Let in system (2.1) the matrices $A(t)$, B_i be constant. Then we have a *linear system with constant coefficients*

$$\frac{dx}{dt} = Ax, \qquad t \neq \tau_i, \qquad \Delta x|_{t=\tau_i} = Bx. \tag{2.22}$$

Suppose that the times τ_i are indexed by the set of natural numbers such that $\tau_i \to +\infty$ for $i \to \infty$. Without loss of generality, we can assume that $\tau_1 > t_0$.

Write any solution $x(t, x_0)$, $x(t_0, x_0) = x_0$, of system (2.22) as

$$x(t, x_0) = X(t, t_0)x_0, \tag{2.23}$$

where

$$X(t, t_0) = e^{A(t-\tau_i)} \prod_{t_0 < \tau_\nu < t} (E + B)e^{A(\tau_\nu - \tau_{\nu-1})}, \qquad \tau_0 = t_0. \tag{2.24}$$

It is difficult to make any conclusions from (2.24) about the structure and behavior of the matrix $X(t, t_0)$ for $t > t_0$ and, hence, about the structure and behavior of solutions of (2.22) for arbitrary matrices A and B, in other words, there is no such an elegant description of properties of solutions of an impulsive system in terms of the eigenvalues of the matrix of the system as we have for a system of ordinary differential equations. The reason is that solutions of system (2.22) are not invariant with respect to shifts because, due to the times of an impulsive effect $t = \tau_i$, system (2.22) is not autonomous.

However, in some cases expression (2.24) can be simplified and it becomes possible to get information about the behavior of solutions of system (2.22). For example, if the matrices A and B commute, then the matrix exponent e^{At} commutes with the matrix B and equality (2.23) can be written as

$$x(t, x_0) = e^{A(t-t_0)}(E + B)^{i(t,t_0)}x_0, \tag{2.25}$$

where $i(t, t_0)$ is the number of points τ_i, which belong to the segment $[t_0, t[$, i.e. $i(t, t_0) = k$ if $\tau_k < t < \tau_{k+1}$. From this, it can be seen that as $t \to \infty$, the behavior of solutions of system (2.22) depends on the eigenvalues of the matrices A and B, and on the properties of the sequence $\{\tau_i\}$. In particular, if the times τ_i are equidistant, $\tau_i = \tau_1 + (i-1)\theta$, and the matrix $E + B$ is nonsingular, then from (2.25) we get that

$$x(t, x_0) = e^{A(t-t_0)}(E + B)e^{\mathrm{Ln}(E+B)\left[\frac{t-\tau_1}{\theta}\right]}x_0, \qquad t > \tau_1$$

or

$$x(t, x_0) = e^{A(\tau_1 - t_0)}(E + B)e^{-\operatorname{Ln}(E+B)\{\frac{t-\tau_1}{\theta}\}}e^{(A+\frac{1}{\theta}\operatorname{Ln}(E+B))(t_1-\tau_1)}x_0. \quad (2.26)$$

Consequently, if the real parts of all eigenvalues of the matrix $A + \frac{1}{\theta}\ln(E+B)$ are negative, then all solutions of system (2.22) go to zero for $t \to \infty$. If among the eigenvalues, there is at least one with the positive real part, then there are solutions, which are unbounded as $t \to \infty$.

If in system (2.22) the matrices A and B commute, then we can at once look for particular solutions expressed as

$$x(t, x_0) = e^{\lambda t}(E + B)^{i(0,t)}x_0, \quad (2.27)$$

where λ is an eigenvalue of the matrix A and x_0 is the eigenvector corresponding to this eigenvalue.

Indeed substituting (2.27) into (2.22) for $t \neq \tau_i$, we get

$$\lambda e^{\lambda t}(E + B)^{i(0,t)}x_0 = Ae^{\lambda t}(E + B)^{i(0,t)}x_0,$$

$$(\lambda E - A)(E + B)^{i(0,t)}x_0 = 0.$$

Because the matrices A and B commute, this equation can be written as

$$(E + B)^{i(0,t)}(\lambda E - A)x_0 = 0, \quad (2.28)$$

which is equivalent to the equation $(\lambda E - A)x_0 = 0$ if the matrix $E + B$ is nonsingular. Thus the construction of a fundamental system of solutions of equations (2.22) is the same as for a system of ordinary differential equations.

Example. Let us solve the impulsive differential system

$$\begin{aligned} \frac{dx_1}{dt} &= 3x_1 - x_2, & \Delta x_1|_{t=\tau_i} &= -\frac{x_1}{2} - \frac{x_2}{3}, \\ & & t &\neq \tau_i, \\ \frac{dx_2}{dt} &= 2x_1, & \Delta x_2|_{t=\tau_i} &= \frac{2}{3}x_1 - \frac{3}{2}x_2, \end{aligned} \quad (2.29)$$

where the sequence $\{\tau_i\}$ is given.

In this case

$$x = \begin{pmatrix} x_1 \\ x_2 \end{pmatrix}, \quad A = \begin{pmatrix} 3 & -1 \\ 2 & 0 \end{pmatrix}, \quad B = \begin{pmatrix} -\frac{1}{2} & -\frac{1}{3} \\ \frac{2}{3} & -\frac{3}{2} \end{pmatrix}.$$

By direct computation, we see that $AB = BA$. The eigenvalues of the matrix A are 1 and 2, and it is easy to find two linearly independent solutions of the system $\dot{x} = Ax$:

$$\bar{x}^{(1)}(t) = \begin{pmatrix} e^t \\ 2e^t \end{pmatrix}, \bar{x}^{(2)}(t) = \begin{pmatrix} e^{2t} \\ e^{2t} \end{pmatrix}.$$

Hence the vector-functions

$$x^{(1)}(t) = e^t(E+B)^{i(0,t)} \begin{pmatrix} 1 \\ 2 \end{pmatrix}, \qquad x^{(2)}(t) = e^{2t}(E+B)^{i(0,t)} \begin{pmatrix} 1 \\ 1 \end{pmatrix}, \tag{2.30}$$

where

$$E + B = \begin{pmatrix} \frac{1}{2} & -\frac{1}{3} \\ \frac{2}{3} & -\frac{1}{2} \end{pmatrix},$$

are linearly independent solutions of impulsive system (2.29).

Thus all the solutions of system (2.29) for $t \geq 0$ are given by the formula

$$x(t) = \begin{pmatrix} x_1(t) \\ x_2(t) \end{pmatrix} = (E+B)^{i(0,t)} \left(c_1 e^t \begin{pmatrix} 1 \\ 2 \end{pmatrix} + c_2 e^{2t} \begin{pmatrix} 1 \\ 1 \end{pmatrix} \right).$$

One can check that, if $i(0,t) = 2k$, then

$$(E+B)^{i(0,t)} = \begin{pmatrix} \frac{1}{6\,2k} & 0 \\ 0 & \frac{1}{6\,2k} \end{pmatrix},$$

and, if $i(0,t) = 2k+1$, then

$$(E+B)^{i(0,t)} = \begin{pmatrix} \frac{1}{2 \cdot 6^{2k}} & -\frac{1}{3 \cdot 6^{2k}} \\ \frac{2}{3 \cdot 6^{2k}} & -\frac{1}{2 \cdot 6^{2k}} \end{pmatrix}.$$

Finally, the solutions of system (2.29) will be

$$x_1(t) = \frac{1}{6^{2k}}(c_1 e^t + c_2 e^{2t}), \qquad x_2(t) = \frac{1}{6^{2k}}(2c_1 e^t + c_2 e^{2t})$$

if $\tau_{2k} < t \leq \tau_{2k+1}$, and

$$x_1(t) = \frac{1}{2 \cdot 6^{2k}}(c_1 e^t + c_2 e^{2t}) - \frac{1}{3 \cdot 6^{2k}}(2c_1 e^t + c_2 e^{2t}),$$

$$x_2(t) = \frac{2}{3 \cdot 6^{2k}}(c_1 e^t + c_2 e^{2t}) - \frac{1}{2 \cdot 6^{2k}}(2c_1 e^t + c_2 e^{2t})$$

if $\tau_{2k+1} < t \le \tau_{2k+2}$.

Let the times τ_i be such that $\tau_1 > 0$ and

$$0 < \theta_1 \le \tau_{i+1} - \tau_i \le \theta_2 \tag{2.31}$$

for some positive θ_1 and θ_2, $i = 1, 2, \ldots$.

If θ_2 is sufficiently small, i.e. the impulse effects occur relatively often, all the solutions of equations (2.29) approach zero as $t \to \infty$. This will happen if $6^{-i(0,t)}e^{2t} \to 0$ for $t \to \infty$. But $i(0,t) \ge \frac{t}{\theta_2}$, and so

$$6^{-i(0,t)}e^{2t} \le e^{\frac{t}{-\theta_2}\ln 6}e^{2t} = e^{-\frac{t}{\theta_2}(\ln 6 - 2\theta_2)}.$$

Thus, if $0 < \theta_1 \le \theta_2 < \ln\sqrt{6}$, then all the solutions of (2.29) go to zero as $t \to \infty$.

This example shows that, although the eigenvalues of the matrix of the differential system are positive, it is possible, due to impulsive effects, that all solutions of the original system be bounded and even go to zero as $t \to \infty$.

2.3 Stability of solutions of linear impulsive systems

Let us study the question of stability of solutions of the linear homogeneous impulsive system

$$\frac{dx}{dt} = A(t)x, \qquad t \neq \tau_i, \qquad \Delta x|_{t=\tau_i} = B_i x, \tag{2.32}$$

where the matrix $A(t)$ is continuous and bounded for $t \geq t_0$, the matrices B_i, $i = 1, 2, \ldots$, are bounded uniformly with respect to $i \in \mathrm{N}$, the times τ_i are indexed by the set of natural numbers, $t_0 < \tau_1 < \ldots < \tau_i < \tau_{i+1} < \ldots$, and $\tau_i \to \infty$ as $i \to \infty$.

Let $X(t, t_0)$ be a matriciant of system (2.32),

$$X(t, t_0) = U(t, \tau_j)(E + B_j) \prod_{\nu=j-1}^{1} U(\tau_{\nu+1}, \tau_\nu)(E + B_\nu)U(\tau, t_0), \tag{2.33}$$

where $\tau_i < t \leq \tau_{i+1}$, $U(t, \sigma)$, $U(\sigma, \sigma) = E$, is a matriciant of the differential system

$$\frac{dx}{dt} = A(t)x. \tag{2.34}$$

Because the difference $x(t, x_0) - x(t, y_0)$ of any two solutions of system (2.32) can be written as

$$x(t, x_0) - x(t, y_0) = X(t, t_0)(x_0 - y_0), \tag{2.35}$$

one can see that the solutions of system (2.32) will be stable or not depending on the behavior of the matriciant $X(t, t_0)$ as $t \to \infty$.

If the matriciant $X(t, t_0) = \{q_{ij}(t)\}$ is bounded for $t \geq t_0$, i.e.

$$\|X(t, t_0)\| = \sum_{\alpha,\beta=1}^{n} |q_{\alpha\beta}(t)| \leq M_0 < \infty$$

for all $t \geq t_0$, then, for all $t \geq t_0$ and any solution $x(t, x_0)$ of system (2.32), we have the inequality

$$\|x(t, x_0) - x(t, y_0)\| \leq \|X(t, t_0)\| \, \|x_0 - y_0\| \leq M_0 \|x_0 - y_0\|,$$

from where it follows that $\|x(t, x_0) - x(t, y_0)\| < \varepsilon$ for $t \geq t_0$ as soon as $\|x_0 - y_0\| < \delta = \dfrac{\varepsilon}{M_0}$. This means that the solution $x(t, x_0)$ is stable.

Suppose that $\lim_{t\to\infty} \|X(t,t_0)\| = 0$. In this case the matrix $X(t,t_0)$ is bounded for $t \geq t_0$ and so the solution $x(t,x_0)$ is stable. Moreover, it follows from (2.35) that

$$\lim_{t\to\infty} \|x(t,x_0) - x(t,y_0)\| = 0$$

for any solution $x(t,y_0)$, i.e. the solution $x(t,x_0)$ is asymptotically stable.

Let the matrix $X(t,t_0)$ be unbounded for $t \geq t_0$, i.e. there exists an infinitely increasing sequence of numbers $t_0 \leq t_1 < t_2 < \ldots$ such that $\lim_{k\to\infty} \|X(t_k,t_0)\| = \infty$. In this case, among the elements of the matrix $X(t,t_0)$ there exists at least one element $q_{\alpha\beta}(t)$ such that $\lim_{k\to\infty} |q_{\alpha\beta}(t_k)| = \infty$. Consider a solution of (2.32) $x(t,x_0^\star)$, which passes through the point $x_0^\star$ at $t = t_0$,

$$x_{10}^\star = x_{10},\ x_{20}^\star = x_{20}, \ldots, x_{\beta 0}^\star \neq x_{\beta 0},\ x_{\beta+10}^\star = x_{\beta+10}, \ldots, x_{n0}^\star = x_{n0}.$$

For this solution we have

$$x_\alpha(t,x_0^\star) - x_\alpha(t,x_0) = q_{\alpha\beta}(t)(x_{\beta 0}^\star - x_{\beta 0}),$$

and so

$$\lim_{k\to\infty} |x_\alpha(t_k,x_0^\star) - x_\alpha(t_k,x_0)| = \infty.$$

No matter how small the difference $x_{\beta 0}^\star - x_{\beta 0}$ is, the function $x_\alpha(t,x_0^\star) - x_\alpha(t,x_0)$ will be unbounded as $t \to \infty$ and so the difference $x(t,x_0^\star) - x(t,x_0)$ will also be unbounded. This means that the solution of system (2.32) $x(t,x_0)$ will be unstable.

We have shown that boundedness of the matriciant $X(t,t_0)$ for all $t \geq t_0$ is a sufficient condition for stability, the equality $\lim_{t\to\infty} X(t,t_0) = 0$ is a sufficient condition for asymptotical stability, and, if the matrix $X(t,t_0)$ is unbounded for $t \geq t_0$, then any solution of system (2.32) is unstable.

One can also show that the conditions given above are not only sufficient but also necessary for any solution of system (2.32) to be stable, asymptotically stable, unstable correspondingly.

Hence we have the following

Theorem 11 *For a solution of impulsive differential system (2.32) $x(t,x_0)$ to be stable, asymptotically stable, unstable, it is necessary and sufficient that the matriciant $X(t,t_0)$ (and, hence, any fundamental matrix) of this system be bounded for $t \geq t_0$, satisfy the condition*

$$\lim_{t\to\infty} X(t,t_0) = 0,$$

be unbounded correspondingly.

Because the matrix $X(t, t_0)$ does not depend on the initial value of the solution $x(t, x_0)$ of system (2.32), the solutions of impulsive system (2.32) are either simultaneously stable or unstable. Thus, depending on whether solutions of (2.32) are stable, asymptotically stable, or unstable, the system is called *stable, asymptotically stable*, or *unstable.*

For example, the first order system

$$\frac{dx}{dt} = a(t)x, \qquad t \neq \tau_i, \qquad \Delta x|_{t=\tau_i} = b_i x$$

is stable, asymptotically stable, or unstable depending on whether the expression

$$\int_{t_0}^{t} a(\sigma)d\sigma + \sum_{t_0<\tau_i<t} \ln|1 + b_i|$$

is bounded, goes to $-\infty$ as $t \to \infty$, or unbounded for $t \geq t_0$.

Suppose that the matrices $A(t)$ and B_i in (2.32) can be represented as $A(t) = A + P(t)$, $B_i = B + I_i$, where A and B are constant matrices. Then system (2.32) can be written in the form

$$\frac{dx}{dt} = Ax + P(t)x, \qquad t \neq \tau_i, \qquad \Delta x|_{t=\tau_i} = Bx + I_i x. \tag{2.36}$$

Together with system (2.36), let us consider the system

$$\frac{dx}{dt} = Ax, \qquad t \neq \tau_i, \qquad \Delta x|_{t=\tau_i} = Bx. \tag{2.37}$$

We have the following theorem.

Theorem 12 *If solutions of system (2.37) are stable, then solutions of system (2.36) will also be stable if*

$$\int_{t_0}^{\infty} \|P(t)\|dt < \infty \qquad \textit{and} \qquad \prod_{\tau_i>t_0} (1 + \|I_i\|) < \infty. \tag{2.38}$$

Proof. The matriciant $X(t, t_0)$ of system with constant coefficients (2.37) is

$$X(t, t_0) = e^{A(t-\tau_i)} \prod_{t_0<\tau_\nu<\tau_i} (E + B)e^{A(\tau_\nu - \tau_{\nu-1})}, \qquad \tau_0 = t_0. \tag{2.39}$$

Because the matrix $E+B$ is nonsingular, the matrix $X(t,t_0)$ is also nonsingular and

$$X(t,t_0)X^{-1}(\sigma,t_0) = e^{A(t-\tau_i)} \prod_{\sigma<\tau_j<\tau_i} (E+B)e^{A(\tau_j-\tau_{j-1})}(E+B)e^{A(\tau_{k+1}-\sigma)}, \tag{2.40}$$

$$\tau_i < t \leq \tau_{i+1}, \qquad \tau_k < \sigma < \tau_{k+1}, \qquad k < i.$$

Because solutions of system (2.37) are stable, it follows from (2.40) that there exists such a positive number K that

$$\|X(t,t_0)\| \leq K, \qquad \|X(t,t_0)X^{-1}(\tau,t_0)\| \leq K, \qquad t_0 \leq \tau \leq t. \tag{2.41}$$

Because any solution $x(t,x_0)$, $x(t_0,x_0) = x_0$, of system (2.36) can be written as

$$\begin{aligned} x(t,x_0) &= X(t,t_0)x_0 + \int_{t_0}^{t} X(t,\sigma)P(\sigma)x(\sigma,x_0)d\sigma + \\ &+ \sum_{t_0<\tau_i<t} X(t,\tau_i)I_i x(\tau_i,x_0), \end{aligned} \tag{2.42}$$

for any two solution of (2.36), $x(t,x_0)$ and $x(t,y_0)$, by using (2.41), we have

$$\begin{aligned} \|x(t,x_0)-x(t,y_0)\| \leq\ & K\|x_0-y_0\| + \\ &+ \int_{t_0}^{t} K\|P(\sigma)\|\,\|x(\sigma,x_0)-x(\sigma,y_0)\|d\sigma + \\ &+ \sum_{t_0<\tau_i<t} K\|I_i\|\,\|x(\tau_i,x_0)-x(\tau_i,y_0)\|. \end{aligned}$$

From Lemma 1 we get the estimate

$$\begin{aligned} &\|x(t,x_0)-x(t,y_0)\| \leq \\ &\leq K \prod_{t_0<\tau_i<t} (1+K\|I_i\|)e^{\int_{t_0}^{t} K\|P(\tau)\|d\tau}\|x_0-y_0\| \end{aligned} \tag{2.43}$$

for all $t \geq t_0$. Since the product $\prod_{\tau_i>t_0}(1+\|I_i\|)$ is convergent and $1+K\|I_i\| \leq (1+\|I_i\|)^K$, it follows that the product $\prod_{\tau_i>t_0}(1+K\|I_i\|)$ converges. Hence, finally, we get

$$\|x(t,x_0)-x(t,y_0)\| \leq K_1\|x_0-y_0\|, \qquad t \geq t_0, \tag{2.44}$$

where

$$K_1 = \prod_{\tau_i > t_0} (1 + K\|I_i\|) e^{\int_{t_0}^{\infty} K\|P(t)\|dt}.$$

And so, from inequality (2.44), we see that the solutions of system (2.36) are stable.

□

2.4 Characteristic exponents of functions and matrices of functions

To study stability properties of solutions of linear impulsive systems, we need to find the behavior of the function $f(t) = \|X(t, t_0)\|$ as $t \to \infty$. This can be done by comparing $f(t)$ with a function, the behavior of which for $t \to \infty$ is known. As a standard function one can take a function depending on a parameter in a way that, as $t \to \infty$, it goes to zero, infinity, or a certain value for different values of the parameter. As an example, one could take $e^{\lambda(t-t_0)}$, $(t - t_0)^l$, etc.

Following A.M. Lyapunov, we compare the functions $f(t)$ defined for $t \geq t_0$ with the function $e^{\lambda}t$.

The number

$$\lambda \equiv \chi[f] = \overline{\lim_{t\to\infty}} \frac{\ln |f(t)|}{t} \tag{2.45}$$

is called a *characteristic exponent* of the function $f(t)$.

It follows from this definition that if λ is a characteristic exponent of $f(t)$, then for an arbitrary $\varepsilon > 0$, we have

$$\overline{\lim_{t\to\infty}} |f(t)|e^{-(\lambda+\epsilon)t} = 0, \qquad \overline{\lim_{t\to\infty}} |f(t)|e^{-(\lambda-\epsilon)t} = \infty. \tag{2.46}$$

Conversely, if there is λ such that, for any $\varepsilon > 0$, the first of relations (2.46) holds, then $\chi[f] \leq \lambda$; if the second relation holds, then $\chi[f] \geq \lambda$; and, finally, if both of (2.46) hold, then $\chi[f] = \lambda$.

We give some properties of characteristic exponents of functions without a proof which can be found, for example, in [38].

1. The characteristic exponent of the sum of functions $f_\nu(t)$ ($\nu = 1, 2, \ldots, m$) is not greater than the maximum of the characteristic exponents (if they are finite) and equals to the greatest characteristic exponent of the term if such a term is unique.

2. The characteristic exponent of the product of a finite number of functions $f_\nu(t)$ ($\nu = 1, 2, \ldots, m$) is not grater than the sum of the characteristic exponents.

   We call a characteristic exponent *strict* if there exists a finite limit

$$\chi[f] = \lim_{t\to\infty} \frac{1}{t} \ln |f(t)|. \tag{2.47}$$

3. If a function $f(t)$ has a strict characteristic exponent, then

$$\chi[f] + \chi\left[\frac{1}{f}\right] = 0. \tag{2.48}$$

4. If a function $f(t)$ has a strict characteristic exponent, then the characteristic exponent of the product of the functions $f(t)$ and $g(t)$ equals to the sum of characteristic exponents of these functions.

   Following A.M. Lyapunov, by the integral of the function $f(t), t \geq t_0$, we will understand the function

$$F(t) = \int_{t_0}^{t} f(\sigma)d\sigma, \qquad \chi[f] \geq 0,$$

   and

$$F(t) = \int_{t}^{\infty} f(\sigma)d\sigma, \qquad \chi[f] < 0.$$

5. The characteristic exponent of the integral of a function $f(t)$ is nor greater than the characteristic exponent of $f(t)$.

   A *characteristic exponent* of a matrix $F(t) = \{f_{jk}(t)\}$ defined for $t \geq t_0$ is called the greatest characteristic exponent of an element of this matrix, $\chi[F] = \max_{j,k} \chi[f_{jk}]$.

6. The characteristic exponent of the sum of a finite number of matrices is not greater than the greatest of the characteristic exponents of these matrices.

7. The characteristic exponent of the product of a finite number of matrices is not greater than the sum of the characteristic exponents of these matrices.

Consider the linear impulsive differential system

$$\frac{dx}{dt} = A(t)x, \qquad t \neq \tau_i, \qquad \Delta x|_{t=\tau_i} = B_i x, \tag{2.49}$$

where $A(t)$ is a continuous matrix bounded for $t \geq t_0$, B_i are matrices uniformly bounded with respect to $i \in \mathbf{N}$, the times τ_i, $t_0 < \tau_1 < \tau_2 < \ldots$ are such that $\lim_{t\to\infty} \tau_i = \infty$.

In the sequel we will need the following statement (analogue of the Wazjevsky inequality).

Theorem 13 *For any solution $x(t, t_0)$, $x(t_0, x_0) = x_0$, of linear impulsive differential system (2.49) and $t \geq t_0$, the following inequality holds*

$$\prod_{t_0<\tau_i<t} \lambda_i e^{\int_{t_0}^{t} \lambda(\sigma)d\sigma} \|x_0\| \leq \|x(t, x_0)\| \leq \prod_{t_0<\tau_i<t} \Lambda_i e^{\int_{t_0}^{t} \Lambda(\sigma)d\sigma} \|x_0\|, \qquad (2.50)$$

where $\lambda(t)$ and $\Lambda(t)$ are the greatest and the least eigenvalue of the matrix $\hat{A}(t) = \frac{1}{2}(A(t) + A^T(t))$ respectively, $A^T(t)$ is the transpose of the matrix $A(T)$, λ_i^2 and Λ_i^2 are the greatest and the least eigenvalue of the matrices $(E + B_i^T)(E + B_i)$, $i = 1, 2, \ldots$, respectively, $\|x\|^2 = \langle x, x\rangle$.

Proof. If $x(t, t_0)= 0$, then (2.50) holds. Let $x(t, t_0)= x(t)$ be a nontrivial solution of system (2.49). Then for $t \neq \tau_i$, we have

$$\begin{aligned} \frac{d}{dt}\|x(t)\|^2 &= \frac{d}{dt}\langle x(t), x(t)\rangle = 2\left\langle \frac{dx}{dt}, x(t)\right\rangle = \\ &= 2\langle A(t)x(t), x(t)\rangle = 2\langle \hat{A}(t)x(t), x(t)\rangle. \end{aligned} \qquad (2.51)$$

Because the matrix $\hat{A}(t)$ is symmetric,

$$\lambda(t)\langle x(t), x(t)\rangle \leq \langle \hat{A}(t)x(t), x(t)\rangle \leq \Lambda(t)\langle x(t), x(t)\rangle, \qquad (2.52)$$

where $\lambda(t)$ and $\Lambda(t)$ are the greatest and the least eigenvalue of the matrix $\hat{A}(t)$. Hence it follows from (2.51) that for $t \neq \tau_i$,

$$2\lambda(t)\|x(t)\|^2 \leq \frac{d}{dt}\|x(t)\|^2 \leq 2\Lambda(t)\|x(t)\|^2. \qquad (2.53)$$

Thus if $\tau_i < t \leq \tau_{i+1}$, we have from (2.53) that

$$e^{2\int_{\tau_i}^{t} \lambda(\sigma)d\sigma} \|x(\tau_i + 0)\|^2 \leq \|x(t)\| \leq e^{2\int_{\tau_i}^{t} \Lambda(\sigma)d\sigma} \|x(\tau_i + 0)\|^2. \qquad (2.54)$$

Because $x(\tau_i + 0) = (E + B_i)x(\tau_i)$, we see that

$$\begin{aligned} \lambda_i^2\|x(\tau_i)\|^2 &\leq \|x(\tau_i + 0)\|^2 = \langle (E + B_i)x(\tau_i), (E + B_i)x(\tau_i)\rangle = \\ &= \langle (E + B_i^T)(E + B_i)x(\tau_i), x(\tau_i)\rangle \leq \Lambda_i^2\|x(\tau_i)\|^2. \end{aligned}$$

Using this inequality it will follow from (2.54) that

$$\lambda_i^2 e^{2\int_{\tau_i}^{t} \lambda(\sigma)d\sigma} \|x(\tau_i)\|^2 \leq \|x(t)\|^2 \leq \Lambda_i^2 e^{2\int_{\tau_i}^{t} \Lambda(\sigma)d\sigma} \|x(\tau_i)\|^2, \qquad (2.55)$$

for $\tau_i < t \leq \tau_{i+1}$.

By using induction, it is easy to see from (2.55) that for all $i = 1, 2, \ldots$,

$$\prod_{t_0<\tau_i<t} \lambda_i^2 e^{2\int_{t_0}^t \lambda(\sigma)d\sigma} \|x_0\|^2 \leq \|x(t, x_0)\|^2 \leq \prod_{t_0<\tau_i<t} \Lambda_i^2 e^{2\int_{t_0}^t \Lambda(\sigma)d\sigma} \|x_0\|^2. \tag{2.56}$$

The needed estimate (2.50) follows from this inequality.

□

We use this estimate to prove

Theorem 14 *Let, for the matrices B_i in system (2.49), the following relation hold,*

$$\inf_i |\det(E + B_i)| \geq \delta > 0, \tag{2.57}$$

and the times τ_i be such that the limit

$$\lim_{t\to\infty} \frac{i(t_0, t)}{t} = p \tag{2.58}$$

exists and is finite. Here $i(t_0, t)$ is the number of the points τ_i, which belong to the interval $[t_0, t[$.

Then any nontrivial solution $x(t, t_0)$ of system (2.49) has a finite characteristic exponent.

Proof. By using the previous theorem, we see that every solution $x(t, t_0)$ of system (2.49) admits estimate (2.50) for $t \geq t_0$. For a nontrivial solution $x(t, t_0)$, $x_0 \neq 0$, we have

$$\overline{\lim_{t\to\infty}} \frac{1}{t}\left(\int_{t_0}^t \lambda(\sigma)d\sigma + \sum_{t_0<\tau_i<t} \ln \lambda_i\right) \leq \chi[x(t, x_0)] \leq$$

$$\leq \overline{\lim_{t\to\infty}} \frac{1}{t}\left(\int_{t_0}^t \Lambda(\sigma)d\sigma + \sum_{t_0<\tau_i<t} \ln \Lambda_i\right).$$

Since inequalities (2.57) and (2.58) hold, by using the former inequality, we can get that

$$\overline{\lim_{t\to\infty}} \frac{1}{t}\int_{t_0}^t \lambda(\sigma)d\sigma + p\lambda_0 \leq \chi[x(t, x_0)] \leq \overline{\lim_{t\to\infty}} \frac{1}{t}\int_{t_0}^t \Lambda(\sigma)d\sigma + p\Lambda_0, \tag{2.59}$$

where

$$\lambda_0 = \inf_i \ln \lambda_i, \qquad \Lambda_0 = \sup_i \ln \Lambda_i.$$

Because the matrices B_i are uniformly bounded and (2.57) holds, the values of λ_0 and Λ_0 are finite and this finishes the proof.

□

The set of all characteristic exponents of nontrivial solutions of a linear differential system is called a *spectrum* of this system.

Theorem 15 *If system (2.49) satisfies the conditions of the previous theorem, then its spectrum consists of a finite number of elements* $\lambda_1 < \lambda_2 < \ldots < \lambda_m$, $m \leq n$.

The proof follows from the observation that the vector-functions that have different characteristic exponents are linearly independent and system (2.49) has at most n linearly independent solutions.

Note that, if the greatest of the characteristic exponents of linear system (2.49) is negative, then this system is asymptotically stable.

Indeed, let $x(t, x_0)$ be a nontrivial solution of system (2.49) and $\lambda = \max_j \lambda_j < 0$. Choose $\varepsilon > 0$ so small that $\lambda + \varepsilon < 0$. Because $\chi[x(t, x_0)] < \lambda + \varepsilon$, we have $\|x(t, t_0)\| e^{-(\lambda+\varepsilon)t} \to 0$ for $t \to \infty$, and so $\|x(t, t_0)\| \to 0$ for $t \to \infty$. This implies that system (2.49) is asymptotically stable.

We give one more sufficient condition for system (2.49) to be asymptotically stable.

Theorem 16 *Let system (2.49) satisfy the conditions of Theorem 14, the greatest eigenvalue of the matrix* $\hat{A}(t)$ *satisfy*

$$\Lambda(t) \leq \gamma \tag{2.60}$$

for all $t \geq t_0$, *for all* $i = 1, 2, \ldots$, *the greatest and the least eigenvalues of the matrix* $(E + B_i^T)(E + B_i)$ *be such that*

$$\Lambda_i^2 \leq \alpha^2. \tag{2.61}$$

If

$$\gamma + p \ln \alpha < 0, \tag{2.62}$$

then all solutions of system (2.49) are asymptotically stable.

Proof. The conditions of the theorem imply that we can get the following estimate for a nontrivial solution of system (2.49) by using the analogue of Wazjevsky inequality,

$$\begin{aligned} \chi[x(t,x_0)] &\leq \overline{\lim_{t\to\infty}} \frac{1}{t}\left[\int_{t_0}^{t} \Lambda(\sigma)d\sigma + \sum_{t_0<\tau_i<t} \ln \Lambda_i\right] \leq \\ &\leq \gamma + p\ln\alpha. \end{aligned} \tag{2.63}$$

Because by the conditions of the theorem $\gamma + p\ln\alpha < 0$, the theorem is proved.

□

Suppose that system of equation (2.49) is such that relations (2.57) and (2.58) hold and

$$-\infty < \lambda_1 < \lambda_2 \ldots < \lambda_m < \infty \tag{2.64}$$

is the spectrum of this system.

Let $X(t)$ be a fundamental matrix of system (2.49) and

$$\sigma_x = \sum_{\nu=1}^{n} \chi[x^{(\nu)}(t)] = \sum_{\alpha=1}^{m} n_\alpha \lambda_\alpha \tag{2.65}$$

be the sum of the characteristic exponents of all the solutions from $X(t)$, where n_α ($n_\alpha \geq 1$) is the number of the solutions with the characteristic exponent λ_α, contained in $X(t)$.

Consider the Wronskian $W(t) = \det X(t)$. Calculating the determinant we get

$$W(t) = \sum_{(p_1,\ldots,p_n)} (-1)^{\varkappa} x_{p_1^1}(t) \ldots x_{p_n^n}(t), \tag{2.66}$$

where the summation is taken over all the permutations $(p_1, \ldots, p_n)$ of n elements and $(-1)^{\varkappa}$ is the signature of the permutation.

From (2.66) we get

$$\chi[W(t)] \leq \max_{(p_1,\cdots,p_n)} (\chi[x_{p_1^1}(t)] + \ldots + \chi[x_{p_n^n}(t)]) \leq \sigma_x. \tag{2.67}$$

Moreover, by applying the analogue of Liouville-Osrogradskiï formula

$$W(t) = W(t_0) e^{\int_{t_0}^{t} \mathrm{Sp}\, A(\sigma)d\sigma} \prod_{t_0<\tau_i<t} \det(E + B_i),$$

we get

$$\chi[W(t)] = \overline{\lim_{t\to\infty}} \frac{1}{t} \left(\int_{t_0}^{t} \text{Sp } A(\sigma)d\sigma + \sum_{t_0<\tau_i<t} \ln|\det(E+B_i)| \right). \tag{2.68}$$

From this, the Lyapunov inequality follows,

$$\sigma_X \geq \overline{\lim_{t\to\infty}} \frac{1}{t} \left(\int_{t_0}^{t} \text{Sp } A(\sigma)d\sigma + \sum_{t_0<\tau_i<t} \ln|\det(E+B_i)| \right). \tag{2.69}$$

If in system (2.45), all the matrices B_i are equal, i.e. $B_i = B$ for all i, then the Lyapunov inequality becomes

$$\sigma_X \geq \overline{\lim_{t\to\infty}} \frac{1}{t} \int_{t_0}^{t} \text{Sp } A(\sigma)d\sigma + p\sum_{\nu=1}^{n} \ln|1+\lambda_\nu(B)|, \tag{2.70}$$

where $\lambda_\nu(B)$ are the eigenvalues of the matrix B.

Also note that, if for a fundamental matrix of system (2.49), the Lyapunov equality

$$\sigma_X = \overline{\lim_{t\to\infty}} \frac{1}{t} \left(\int_{t_0}^{t} \text{Sp } A(\sigma)d\sigma + \sum_{t_0<\tau_i<t} \ln|\det(E+B_i)| \right), \tag{2.71}$$

holds, then this matrix is normal in the sense that the sum of its characteristic exponents is the least comparing with other fundamental matrices of this system.

Let

$$\sigma = \sum_{\nu=1}^{m} n_\nu \alpha_\nu \tag{2.72}$$

be the sum (counting the multiplicities) of the solutions of system(2.49), which are contained in certain normal fundamental system.

System (2.49) is called *Lyapunov regular* [76] if

$$\sigma = \lim_{t\to\infty} \frac{1}{t} \left(\int_{t_0}^{t} \text{Sp } A(\sigma)d\sigma + \sum_{t_0<\tau_i<t} \ln|\det(E+B_i)| \right). \tag{2.73}$$

Necessary and sufficient conditions for system (2.49) to be regular are given by the following statement.

Theorem 17 *Linear impulsive system (2.49) is regular if and only if the limit*

$$\lim_{t\to\infty} \frac{1}{t} \left(\int_{t_0}^{t} \mathrm{Sp}\ A(\sigma) d\sigma + \sum_{t_0<\tau_i<t} \ln|\det(E+B_i)| \right) = s \tag{2.74}$$

exists and the Lyapunov equality

$$\sigma = s. \tag{2.75}$$

holds.

Proof. It is evident that these conditions are sufficient for regularity of system (2.49). Let us prove that they are also necessary. Suppose that system (2.49) is regular and let

$$\underline{s} = \underline{\lim}_{t\to\infty} \frac{1}{t} \left(\int_{t_0}^{t} \mathrm{Sp}\ A(\sigma) d\sigma + \sum_{t_0<\tau_i<t} \ln|\det(E+B_i)| \right),$$

$$\overline{s} = \overline{\lim}_{t\to\infty} \frac{1}{t} \left(\int_{t_0}^{t} \mathrm{Sp}\ A(\sigma) d\sigma + \sum_{t_0<\tau_i<t} \ln|\det(E+B_i)| \right).$$

From the definition of a regular system and by using Lyapunov inequality, we have that $\overline{s} \le \sigma \le \underline{s}$. But $\underline{s} \le \overline{s}$, and so $\underline{s} = \overline{s} = s = \sigma$.

□

2.5 Adjoint systems. Perron theorem

Consider an impulsive differential system

$$\frac{dx}{dt} = A(t)x, \qquad t \neq \tau_i, \qquad \Delta x|_{t=\tau_i} = B_i x. \tag{2.76}$$

We assume that the matrix $A(t)$ is continuous and bounded for $t \geq t_0$, the matrices B_i are bounded uniformly with respect to $i \in \mathbf{N}$ and

$$\inf_i |\det(E + B_i)| \geq \delta > 0, \tag{2.77}$$

the times τ_i, $t_0 < \tau_1 < \tau_2 < \ldots$ are indexed by the set of natural numbers with $\tau_i \to \infty$ as $i \to \infty$, and there exists a finite upper limit

$$\overline{\lim_{t\to\infty}} \frac{i(t, t_0)}{t} = p < \infty. \tag{2.78}$$

The system of impulsive differential equations

$$\frac{dy}{dt} = -A^T(t)y, \qquad t \neq \tau_i, \qquad \Delta y|_{t=\tau_i} = -(E + B_i^T)^{-1} B_i^T y, \tag{2.79}$$

where $A^T(t)$ and B_i^T are transposed matrices of $A(t)$ and B_i, is called the system *adjoint* to (2.76).

By using the equality

$$(E - B_i(E + B_i)^{-1})^{-1} B_i (E + B_i)^{-1} = B_i,$$

it is easy to see that the system adjoint to (2.79) is system (2.76). These systems are called *mutually adjoint.*

Lemma 8 *For any two solutions $x(t)$ and $y(t)$ of mutually adjoint systems (2.76) and (2.79), we have*

$$\langle x(t), y(t) \rangle = c, \tag{2.80}$$

where c is some constant. For fundamental matrices of solutions of these systems $X(t)$ and $Y(t)$, the following holds:

$$Y^T(t)X(t) = C, \tag{2.81}$$

where C is a constant matrix.

If (2.81) holds, the matrix C is nonsingular, and $X(t)$ is a fundamental matrix of system (2.76), then $Y(t)$ is a fundamental matrix of adjoint system (2.79).

Proof. The fact that the scalar product of solutions of adjoint systems is constant on every interval $]\tau_i, \tau_{i+1}]$ follows because, on this interval, we have

$$\frac{dx}{dt} = A(t)x, \qquad \frac{dy}{dt} = -A^T(t)y \tag{2.82}$$

and hence, for $\tau_i < t \leq \tau_{i+1}$, we get

$$\langle \frac{dx}{dt}, y\rangle = \langle A(t)x, y\rangle, \qquad \langle x, \frac{dy}{dt}\rangle = -\langle x, A^T(t)y\rangle$$

and adding them obtain

$$\frac{d}{dt}\langle x(t), (t)\rangle = 0, \qquad \langle x(t), y(t)\rangle = c_i. \tag{2.83}$$

It remains to show that the constants c_i are all equal, i.e. $c_i = c$ for all $i = 1$, $2, \ldots$. This is equivalent to proving that

$$\Delta\langle x(t), y(t)\rangle|_{t=\tau_i} = 0. \tag{2.84}$$

We have

$$\begin{aligned}\Delta\langle x(t), y(t)\rangle|_{t=\tau_i} &= \langle x(\tau_i + 0), y(\tau_i + 0)\rangle - \langle x(\tau_i), y(\tau_i)\rangle = \\ &= \langle (E + B_i)x(\tau_i), (E - (E + B_i^T)^{-1}B_i^T)y(\tau_i)\rangle - \langle x(\tau_i), y(\tau_i)\rangle \\ &= \langle x(\tau_i), ((E + B_i^T) - B_i^T)y(\tau_i)\rangle - \langle x(\tau_i), y(\tau_i)\rangle = 0.\end{aligned}$$

It follows from here that $c_{i+1} = c_i$ for all $i = 1, 2, \ldots$, and so $\langle x(t), y(t)\rangle = c$ for all $t \geq t_0$.

Because the fundamental matrices $X(t)$ and $Y(t)$ satisfy the corresponding impulsive matrix equations

$$\frac{dX}{dt} = A(t)X, \qquad t \neq \tau_i, \qquad \Delta X|_{t=\tau_i} = B_i X;$$

$$\frac{dY}{dt} = -A^T(t)Y, \qquad t \neq \tau_i, \qquad \Delta Y|_{t=\tau_i} = -(E + B_i^T)^{-1}B_i^T Y$$

and, for $t = \tau_i$, the matrix $Y^T(t)$ satisfies

$$\frac{dY^T}{dt} = -Y^T A(t),$$

the proof of equality (2.80) is similar to that of (2.81).

We will prove the last statement of the lemma. Suppose that identity (2.81) holds, where C is a nonsingular matrix and $X(t)$ is a fundamental matrix of system (2.76). Let us show that $Y(t)$ is a fundamental matrix of system (2.79).

It follows from (2.81) that

$$Y(t) = (X^T(t))^{-1}C^T. \tag{2.85}$$

The matrix $X^T(t)$ satisfies the system of equations

$$\frac{dX^T}{dt} = X^T A^T(t), \qquad t \neq \tau_i, \qquad \Delta X^T|_{t=\tau_i} = X^T B_i^T. \tag{2.86}$$

Hence, by using (2.85) and the expression for the derivative of the inverse of a matrix, we get

$$\begin{aligned}\frac{dY}{dt} &= -(X^T)^{-1}\frac{dX}{dt}(X^T)^{-1}C^T = -(X^T)^{-1}X^T A^T(t)(X^T)^{-1}C^T = \\ &= -A^T(t)Y.\end{aligned} \tag{2.87}$$

Besides

$$\det Y(t) = \det(X^T(t))^{-1}\det C^T = \det X^{-1}(t)\det C \neq 0. \tag{2.88}$$

For $t = \tau_i$,

$$\begin{aligned}Y(\tau_i+0) &= (X^T(\tau_i+0))^{-1}C^T = (E+B_i^T)^{-1}X^{-1}(\tau_i)C^T = \\ &= (E+B_i^T)^{-1}Y(\tau_i)\end{aligned}$$

or

$$\Delta Y|_{t=\tau_i} = -Y(\tau_i) + (E+B_i^T)^{-1}Y(\tau_i) = -(E+B_i^T)^{-1}B_i^T Y. \tag{2.89}$$

Taking into account relations (2.88), we see from (2.87) and (2.89) that $Y(t)$ is a fundamental matrix of system (2.79). Note that if $X(t)$ and $Y(t)$ are matriciants of systems (2.76) and (2.79) respectively, then it follows from (2.85) that

$$Y(t) = (X^T(t))^{-1}. \tag{2.90}$$

□

Theorem 18 *(Perron) Linear impulsive system (2.76) is regular if and only if its complete spectrum (i.e. counting the multiplicities of the characteristic exponents)*

$$\alpha_1 \leq \alpha_2 \leq \ldots \leq \alpha_n \tag{2.91}$$

and the complete spectrum of the adjoint system

$$\beta_1 \geq \beta_2 \geq \ldots \geq \beta_n \tag{2.92}$$

are symmetric with respect to zero, i.e. if

$$\alpha_k + \beta_k = 0 \qquad (k = 1, 2, \ldots, n). \tag{2.93}$$

Proof. Let system (2.76) be regular and $X(t) = \{x_{jk}(t)\}$ be its normal fundamental matrix which consists of the solutions $x^{(k)}(t)$ such that $\chi[x^k(t)] = \alpha_k$, where the numbers α_k satisfy (2.91). The matrix

$$Y(t) = |X^{-1}(t)|^T = \{y_{jk}(t)\} \tag{2.94}$$

is a fundamental matrix of adjoint system (2.79) and $Y^T(t)X(t) = E$. Noting that $\langle y^{(k)}(t), x^{(k)}(t)\rangle = 1$ we find that $\chi[1] = 0 = \chi[y^{(k)}(t)] + \chi[x^{(k)}(t)]$ i.e.

$$\beta_k + \alpha_k \geq 0. \tag{2.95}$$

Besides, if $X_{jk}(t)$ is the algebraic complement of x_{jk} in $\det X(t)$, then

$$y_{jk}(t) = \frac{1}{\det X(t)}\{X_{jk}(t)\},$$

where

$$\det X(t) = \det X(t_0)e^{\int_{t_0}^{t} \mathrm{Sp}\, A(\sigma)d\sigma} \prod_{t_0<\tau_i<t} \det(E + B_i) \neq 0.$$

From this it follows that

$$\chi[y_{jk}(t)] \leq \chi[e^{-\int_{t_0}^{t} \mathrm{Sp}\, A(\sigma)d\sigma - \sum_{t_0<\tau_i<t} \ln|\det(E+B_i)|}] + \chi[X_{jk}(t)].$$

Because system (2.76) is regular, the Lyapunov equality holds:

$$\sigma = \sum_{k=1}^{n} \alpha_k = \lim_{t\to\infty} \frac{1}{t}\left[\int_{t_0}^{t} \mathrm{Sp}\, A(\sigma)d\sigma + \sum_{t_0<\tau_i<t} \ln|\det(E + B_i)|\right],$$

and so

$$\chi[e^{-\int_{t_0}^t \mathrm{Sp}\, A(\sigma)d\sigma - \sum_{t_0<\tau_i<t} \ln|\det(E+B_i)|}] = -\sigma.$$

Clearly $\chi[X_{jk}(t)] \le \sigma - \alpha_k$ because, when we calculate X_{jk}, the kth column, which contains the coordinates of the solution $x^{(k)}(t)$, is crossed out in the determinant $X(t)$. Thus

$$\chi[y_{jk}(t)] \le -\sigma + \sigma - \alpha_k = -\alpha_k$$

and so $\beta_k = \max_\nu \chi[y_{\nu k}(t)] \le -\alpha_k$, i.e.

$$\beta_k + \alpha_k \le 0. \tag{2.96}$$

By comparing this inequality with (2.95), we get

$$\alpha_k + \beta_k = 0 \qquad (k = 1, 2, \dots, n). \tag{2.97}$$

It remains to show that the fundamental matrix $Y(t)$ is normal and, consequently, the numbers $\beta_1, \beta_2, \dots \beta_n$ realize the whole spectrum of the adjoint system. Indeed by (2.97),

$$\begin{aligned}
\sigma_Y &= \sum_{k=1}^n \beta_k = -\sum_{k=1}^n \alpha_k = \\
&= -\lim_{t\to\infty} \frac{1}{t}\left[\int_{t_0}^t \mathrm{Sp}\, A(\sigma)d\sigma + \sum_{t_0<\tau<t} \ln|\det(E+B_i)|\right] = \\
&= \lim_{t\to\infty} \frac{1}{t}\left(\int_{t_0}^t \mathrm{Sp}\,(-A^T(\sigma))d\sigma + \sum_{t_0<\tau_i<t} \ln|\det(E-(E+B_i^T)^{-1}B_i^T)|\right).
\end{aligned}$$

Hence, the Lyapunov equality holds for the fundamental matrix $Y(t)$ of adjoint system (2.79) and this matrix is normal.

Now we show that if (2.93) holds, then system (2.76) is regular.

By Lyapunov inequality

$$\begin{aligned}
\sigma_X &= \sum_{k=1}^n \alpha_k \ge \overline{\lim_{t\to\infty}} \frac{1}{t}\left(\int_{t_0}^t \mathrm{Sp}\, A(\sigma)d\sigma + \sum_{t_0<\tau_i<t} \ln|\det(E+B_i)|\right) = \overline{s}, \\
\sigma_Y &= \sum_{k=1}^n \beta_k \ge
\end{aligned}$$

$$\geq \overline{\lim_{t\to\infty}} \frac{1}{t}\left(\int_{t_0}^{t} \mathrm{Sp}\,(-A^T(\sigma))d\sigma + \sum_{t_0<\tau_i<t} \ln|\det(E+B_i^T)^{-1}|\right) =$$

$$= \underline{\lim}_{t\to+\infty} \frac{1}{t}\left(\int_{t_0}^{t} \mathrm{Sp}\ A(\sigma)d\sigma + \sum_{t_0<\tau_i<t} \ln|\det(E+B_i)|\right) = -\underline{s}.$$

Adding these inequalities and using (2.93) we get that $\underline{s} = \overline{s}$. So there exists the limit

$$s = \lim_{t\to\infty} \frac{1}{t}\left(\int_{t_0}^{t} \mathrm{Sp}\ A(\sigma)d\sigma + \sum_{t_0<\tau_i<t} \ln|\det(E+B_i)|\right).$$

Moreover, we have $\sum_{k=1}^{n} \alpha_k = s$. Indeed, if $\sum_{k=1}^{n} \alpha_k > s$, by using $\sum_{k=1}^{n} \beta_k \geq -s$, we would have $\sum_{k=1}^{n}(\alpha_k + \beta_k) > 0$ which is impossible.

Applying Theorem 17 we see that system (2.76) is regular.

□

In particular it follows from Theorem 2.76 that:

1) a system adjoint to a regular system is also regular;

2) if system (2.76) is regular and $X(t)$ is its normal fundamental matrix, then $Y(t) = (X^{-1}(t))^T$ is a normal fundamental matrix of adjoint system (2.79).

Suppose that in system (2.76) the matrices $A(t) = \{a_{\alpha\beta}(t)\}$ and $B_i = \{b_i^{\alpha\beta}\}$ are triangular. We will be considering only lower triangular matrices: $a_{\alpha\beta}(t) = 0$, $b_i^{\alpha\beta} = 0$ for $\beta > \alpha$ and for all $t \geq t_0$, $i = 1, 2, \ldots$. With these assumptions system (2.76) can be written as

$$\frac{dx_\alpha}{dt} = \sum_{\beta\leq\alpha} a_{\alpha\beta}(t)x_\beta, \qquad t \neq \tau_i, \qquad \Delta x_\alpha|_{t=\tau_i} = \sum_{\beta\leq\alpha} b_i^{\alpha\beta} x_\beta(\tau_i), \tag{2.98}$$

and condition (2.77) becomes

$$\inf_i \min_{1\leq\alpha\leq n} |1 + b_i^{\alpha\alpha}| \geq \delta > 0. \tag{2.99}$$

Theorem 19 *Linear triangular impulsive differential system (2.76) is regular if and only if there exist finite limits*

$$\lim_{t\to\infty} \frac{1}{t}\left(\int_{t_0}^{t} a_{kk}(\tau)d\tau + \sum_{t_0<\tau_i<t} \ln|1+b_i^{kk}|\right) = \mu_k, \tag{2.100}$$

$$k = 1, 2, \ldots, n.$$

Proof. Let system (2.98) be regular. Denote

$$\overline{\mu}_k = \overline{\lim_{t\to\infty}} \frac{1}{t}\left(\int_{t_0}^{t} a_{kk}(\tau)d\tau + \sum_{t_0<\tau_i<t} \ln|1+b_i^{kk}|\right),$$

$$\underline{\mu}_k = \underline{\lim_{t\to\infty}} \frac{1}{t}\left(\int_{t_0}^{t} a_{kk}(\tau)d\tau + \sum_{t_0<\tau_i<t} \ln|1+b_i^{kk}|\right)$$

and let

$$s = \lim_{t\to\infty} \frac{1}{t}\left(\int_{t_0}^{t} \mathrm{Sp}\ A(\tau)d\tau + \sum_{t_0<\tau_i<t} \ln|\det(E+B_i)|\right).$$

Set

$$A_k(t) = e^{\int_{t_0}^{t} a_{kk}(\tau)d\tau} \prod_{t_0<\tau_i<t} (1+b_i^{kk}),\ k = 1, 2, \ldots, n. \tag{2.101}$$

Because system (2.98) can be successively integrated, it is easy to see that its fundamental matrix $X(t,t_0) = \{x_{\alpha\beta}(t)\}$, $X(t_0,t_0) = E$, where

$$x_{\alpha\beta}(t) = 0 \quad (\alpha < \beta), \qquad x_{\alpha\alpha}(t) = A_\alpha(t),$$

$$\begin{aligned} x_{\alpha\beta}(t) &= A_\alpha(t)\left(\int_{t_0}^{t} A_\alpha^{-1}(\tau)\sum_{j=\beta}^{\alpha-1} a_{\alpha j}(\tau)x_{j\beta}(\tau)d\tau + \right. \\ &\left. + \sum_{t_0<\tau_i<t} A_\alpha^{-1}(\tau_i)\sum_{j=\beta}^{\alpha-1} b_i^{\alpha j}x_{j\beta}(\tau_i)\right), \end{aligned}$$

$$\alpha > \beta, \qquad \alpha, \beta = 1, 2, \ldots, n.$$

Without loss of generality we can assume that $X(t,t_0)$ is a normal fundamental matrix because, if not, a normal fundamental matrix can be obtained from $X(t,t_0)$ by multiplying it by a lower triangular matrix with identities on the main diagonal.

The matrix $Y(t,t_0) = (X^{-1}(t,t_0))^T = \{y_{\alpha\beta}(t)\}$ is a normal fundamental matrix adjoint with respect to system (2.98). It is clear that

$$y_{\alpha\alpha}(t) = A_\alpha^{-1}(t). \tag{2.102}$$

Let $\chi[x^k(t)] = \max_j \chi[x_{jk}(t)] = \alpha_k$ and $\chi[y^{(k)}] = \max_j \chi[y_{jk}(t)] = \beta_k$ ($k = 1$, 2, ..., n), where, by Perron theorem, $\alpha_k + \beta_k = 0$ ($k = 1, 2, \ldots, n$).

From the structure of the matrix $X(t, t_0)$, we see that

$$\alpha_k \geq \chi[A_k(t)] = \overline{\mu}_k, \qquad \beta_k \geq \chi[A_k^{-1}(t)] = -\underline{\mu}_k, \qquad k = 1, 2, \ldots, n.$$

By adding these two inequalities, we get

$$\begin{aligned} \underline{\mu}_k = \overline{\mu}_k = \mu_k &= \lim_{t\to\infty} \frac{1}{t}\left(\int_{t_0}^t a_{kk}(\tau)d\tau + \sum_{t_0<\tau_i<t} \ln|1 + b_i^{kk}|\right), \\ & k = 1, 2, \ldots, n. \end{aligned} \tag{2.103}$$

Hence, we have shown that existence of finite limits (2.100) is a necessary condition for a linear triangular system to be regular. Now we show that these conditions are sufficient.

Suppose that for any $k = 1, 2, \ldots, n$, limits (2.100) exist and finite.

Let $Z(t) = \{z_{jk}(t)\}$ be a matrix, the elements of which are the functions

$$z_{jk}(t) = 0, \qquad j < k; \qquad z_{kk} = A_k(t),$$

$$\begin{aligned} z_{jk}(t) &= A_j(t)\left(\int_{t_{jk}}^t A_j^{-1}(\tau)\sum_{\nu=k}^{j-1} a_{j\nu}(\tau)z_{\nu k}(\tau)d\tau + \right. \\ &+ \left. \operatorname{sign}(t - t_{jk}) \sum_{t_{jk}\lesseqgtr\tau_i\lesseqgtr t} A_i^{-1}(\tau_i)\sum_{\nu=k}^{j-1} b_i^{j\nu} z_{\nu k}(\tau_i)\right) \end{aligned} \tag{2.104}$$

where $j, k = 1, 2, \ldots, n$, $t_{jk} = t_0$ if $\mu_j \leq \mu_k$ and $t_{jk} \to \infty$ if $\mu_j > \mu_k$ and the summation in the second term is taken over $t_0 < \tau_i < t$ if $\mu_j \leq \mu_k$ and with respect to $\tau_i \geq t$ if $\mu_j > \mu_k$.

It is easy to see that $Z(t)$ is a fundamental matrix of system (2.98) and so $Z(t) = X(t)C$, where C is some constant lower triangular matrix with the elements on the main diagonal equal to one.

From the construction of the matrix $Z(t)$, it follows that

$$\chi[z_{kk}(t)] = \chi[A_k(t)] = \mu_k, \qquad \chi[z_{jk}(t)] \leq \mu_k, \qquad k = 1, 2, \ldots, n.$$

and so

$$\alpha_k = \chi[z^k(t)] = \max_i \chi[z_{jk}(t)] \leq \mu_k, \qquad k = 1, 2, \ldots, n.$$

By setting

$$s = \lim_{t\to\infty} \frac{1}{t} \left(\int_{t_0}^{t} \sum_{k=1}^{n} a_{kk}(\tau)d\tau + \sum_{t_0<\tau_i<t} \sum_{k=1}^{n} \ln|1+b_i^{kk}| \right) = \sum_{k=1}^{n} \mu_k$$

and applying the Lyapunov inequality, we get

$$s = \sum_{k=1}^{n} \mu_k \geq \sum_{k=1}^{n} \alpha_k \geq s,$$

i.e.

$$\alpha_k = \mu_k \qquad (k = 1, 2, \dots, n) \qquad \text{and} \qquad \sum_{k=1}^{n} \alpha_k = s.$$

So the constructed fundamental matrix is normal and hence system (2.98) is regular.

□

In particular it follows from this theorem that if impulsive triangular system (2.98) is regular, then the values

$$\alpha_k = \lim_{t\to\infty} \frac{1}{t} \left(\int_{t_0}^{t} a_{kk}(\tau)d\tau + \sum_{t_0<\tau_i<t} \ln|1+b_i^{kk}| \right) \tag{2.105}$$

give the spectrum of this system.

At the end of this section, we prove the Perron theorem on existence of a linear transformation that reduces impulsive linear system (2.76) to a triangular form.

Theorem 20 *Any linear homogeneous differential impulsive system (2.76) can be reduced by an orthogonal transformation $x = U(t)y$ to the form*

$$\frac{dy}{dt} = Q(t)y, \qquad t \neq \tau_i, \qquad \Delta y|_{t=\tau_i} = \Lambda_i y, \tag{2.106}$$

where the matrices $Q(t)$ and Λ_i are triangular (both are either upper triangular or lower triangular).

Proof. Let $x_1(t)$, $x_2(t)$, ..., $x_n(t)$ are linearly independent solutions of system (2.76). Set

$$v_1(t) = x_1(t), \qquad u_1(t) = \frac{v_1(t)}{\|v_1(t)\|}$$

and define recurrently

$$v_k(t) = x_k(t) - \sum_{j=1}^{k-1} \langle x_k(t), u_j(t) \rangle u_j(t), \qquad u_k(t) = \frac{v_k(t)}{\|v_k(t)\|}$$

for $k = 2, 3, \ldots, n$. Because the solutions $x_1(t)$, $x_2(t)$, $\ldots$, $x_k(t)$, $k = 1, 2, \ldots, n$ are linearli independent, we have that $\|v_k(t)\| \neq 0$ for $t \geq t_0$ and so the vectors $u_k(t)$, $k = 1, 2, \ldots, n$ are defined for $t \geq t_0$. Clearly these vectors are continuously differentiable for $t \geq t_0$, $t \neq \tau_i$, and it is easily seen that

$$\langle u_j(t), u_k(t) \rangle = \delta_{jk}, \qquad j, k = 1, 2, \ldots, n,$$

where

$$\delta_{jk} = \begin{cases} 1, & \text{if} \quad j = k, \\ 0, & \text{if} \quad j \neq k. \end{cases}$$

Also note that $u_1(t), u_2(t), \ldots, u_k(t), k = 1, 2, \ldots, n$, depend only on the vectors $x_1(t), x_2(t), \ldots, x_k(t)$. Consequently, if $U(t) = (u_1(t), \ldots u_n(t))$ and $X(t) = (x_1(t), \ldots, x_n(t))$, then $U(t) = X(t)S(t)$, where $S(t)$ is a triangular matrix. It is clear that the matrix $S(t)$ is nonsingular and continuously differentiable for $t \geq t_0$ and $t \neq \tau_i$.

Let us make the change of variables

$$x = U(t)y. \tag{2.107}$$

in system (2.76). For $t \neq \tau_i$ we get

$$\frac{dy}{dt} = U^{-1}(t)\left(A(t)U(t) - \frac{dU}{dt}\right) y \equiv Q(t)y.$$

By differentiating (2.107) at $t \neq \tau_i$, we find that

$$\frac{dU}{dt} = \frac{dX}{dt}S(t) + X(t)\frac{dS}{dt} = A(t)U(t) + U(t)S^{-1}(t)\frac{dS}{dt}$$

and consequently

$$Q(t) = U^{-1}(t)\left(A(t)U(t) - \frac{dU}{dt}\right) = -S^{-1}(t)\frac{dS}{dt}.$$

Because the matrix $S(t)$ is triangular, the matrix $S^{-1}(t)$ will also be triangular and so is the matrix $Q(t)$.

For $t = \tau_i$, according to (2.107), we have

$$U(\tau_i + 0)y(\tau_i + 0) = (E + B_i)U(\tau_i)y(\tau_i)$$

or

$$U(\tau_i + 0)\Delta y = (-U(\tau_i + 0) + U(\tau_i) + B_iU(\tau_i))y(\tau_i).$$

From this we find that

$$\Delta y = S^1(\tau_i + 0)X^{-1}(\tau_i + 0)(-X(\tau_i + 0)S(\tau_i + 0) + \\ +X(\tau_i + 0)S(\tau_i))y(\tau_i) = (-E + S^{-1}(\tau_i + 0)S(\tau_i))y(\tau_i)$$

or

$$\Delta y|_{t=\tau_i} = \Lambda_i y,$$

where $\Lambda_i = S^{-1}(\tau_i + 0)S(\tau_i) - E$ is a triangular matrix.

□

It is easy to see that if triangular system (2.106) is regular, then system (2.76) is regular.

Proposition 1 *If linear system (2.76) is regular, then for every its nontrivial solution $x = x(t)$ there exists a strict characteristic exponent*

$$\alpha = \lim_{t\to\infty} \frac{1}{t} \ln \|x(t)\|. \tag{2.108}$$

Proof. Indeed, let $x(t)$ be contained in some fundamental system of solutions of equation (2.76), for example, let $x(t) = x_1(t)$ be the first column of the fundamental matrix $X(t)$. Then, by Theorem 19,

$$\lim_{t\to\infty} \frac{1}{t} \left(\int_{t_0}^{t} q_{11}(\tau)d\tau + \sum_{t_0<\tau_i<t} \ln|1 + \lambda_i^{11}| \right) = \lim_{t\to\infty} \frac{1}{t} \ln \|v_1(t)\|,$$

where $q_{11}(t)$ and λ_i^{11} are the corresponding elements of the matrices $Q(t)$ and Λ_i. But one can see from the construction that $v_k(t)v_1(t) = x_1(t) = x(t)$, and so limit (2.108) exists.

2.6 Reducible systems

Let $L(t)$ be a matrix bounded, piecewise continuous for $t \geq t_0$ with the first kind discontinuities at $t = \tau_i$, $\tau_i \to \infty$ for $i \to \infty$. We call $L(t)$ a *Lyapunov matrix* if $\frac{dL}{dt}$ is continuous for $t \in [t_0, \infty[\setminus \{\tau_i\}$, bounded and $|\det L(t)| \geq m > 0$.

It can be shown that if $L(t)$ is a Lyapunov matrix, then $L^{-1}(t)$ is also a Lyapunov matrix.

Consider an impulsive differential system

$$\frac{dx}{dt} = A(t)x, \qquad t \neq \tau_i, \qquad \Delta x|_{t=\tau_i} = B_i x, \tag{2.109}$$

where $A(t)$ is a matrix continuous for $t \geq t_0$, all $i \in \mathbf{N}$, $\det(E + B_i) \neq 0$.

Following Lyapunov we call system (2.109) *reducible* if it can be reduced by the linear transformation

$$x = L(t)y \tag{2.110}$$

to the system

$$\frac{dy}{dt} = Py \tag{2.111}$$

with a constant matrix P.

We give necessary and sufficient conditions for a linear impulsive system to be reducible.

Theorem 21 *(Erugin). Linear impulsive differential system (2.109) is reducible if and only if a fundamental matrix of this system can be represented as*

$$X(t) = L(t)e^{Pt}, \tag{2.112}$$

where $L(t)$ is a Lyapunov matrix and P is a constant matrix.

Proof. Let system (2.109) be reducible to system (2.111) by Lyapunov transformation (2.110). Then a fundamental matrix of this system $X(t)$ and a fundamental matrix of system (2.111) satisfy

$$X(t) = L(t)e^{Pt}C$$

with a nonsingular matrix C. Setting $C = E$ we obtain (2.112).

Conversely, suppose that relation (2.79) holds for system (2.109) and show that in this case system (2.109) is reducible. To do that we apply Lyapunov transformation (2.110) to equations (2.109) with the matrix $L(t) = X(t)e^{-Pt}$. We have

$$\begin{aligned} X(t)e^{-Pt}\frac{dy}{dt} + \frac{dX}{dt}e^{-Pt}y &- X(t)e^{-Pt}Py = A(t)X(t)E^{-Pt}y, \\ t &\neq \tau_i, \\ \Delta X|_{t=\tau_i}e^{-P\tau_i} &= B_iX(\tau_i)e^{-P\tau_i}. \end{aligned} \tag{2.113}$$

Because the matrix $X(t)$ satisfies the relations

$$\frac{dX}{dt} = A(t), \qquad t \neq \tau_i, \qquad \Delta X|_{t=\tau_i} = B_iX,$$

it follows from (2.113) that

$$\frac{dy}{dt} = Py,$$

i.e. system (2.109) is reducible.

□

It is interesting when equation (2.109) can be reduced to system (2.111) with the zero matrix, i.e. with the matrix $P = 0$. Sufficient conditions are given by the following theorem.

Theorem 22 *If all solutions of system (2.109) are bounded for $t \geq t_0$ and*

$$\int_{t_0}^{t} \text{Sp } A(\sigma)d\sigma + \sum_{t_0<\tau_i<t} \ln|\det(E + B_i)| \geq a > -\infty \tag{2.114}$$

for all $t \geq t_0$, a being a constant, the system (2.109) can be reduced by a Lyapunov transformation to a system with the zero matrix.

Proof. To prove the theorem it is sufficient to show that if the conditions of the theorem hold, then the fundamental matrix $X(t)$ is a Lyapunov matrix. Indeed, $X(t)$ is a matrix continuously differentiable for $t \geq t_0$, $t \neq \tau_i$. Because all solutions of (2.109) are bounded for $t \geq t_0$, $X(t)$ and $\frac{dX}{dt}$ are bounded for $t \neq \tau_i$ since

$$\left\|\frac{dX}{dt}\right\| \leq \|A(t)\|\,\|X(t)\|.$$

Moreover, by using the analogue of Osrogradskiĭ-Liouville formula

$$\det X(t) = \det X(t_0) e^{\int_{t_0}^{t} \mathrm{Sp}\, A(\sigma)d\sigma} \prod_{t_0<\tau_i<t} \det(E+B_i),$$

and the conditions of the theorem for $t \geq t_0$, we get

$$\begin{aligned} |\det X(t)| &= |\det X(t_0)| e^{\int_{t_0}^{t} \mathrm{Sp}\, A(\sigma)d\sigma + \sum_{t_0<\tau_i<t} \ln|\det(E+B_i)|} \geq \\ &\geq |\det X(t_0)| e^{a} = m > 0. \end{aligned} \tag{2.115}$$

So $X(t)$ is a Lyapunov matrix. The change of variables $x = X(t)y$ transforms (2.109) into the system $\dfrac{dy}{dt} = 0$.

□

From this theorem, one can easily deduce

Theorem 23 *If the matrices $A(t)$, B_i and the times τ_i in system (2.109) are such that*

$$\int_{t_0}^{t} \|A(t)\| dt = c_1 < \infty, \qquad \sum_{\tau_i > t_0} \|B_i\| = c_2 < \infty, \tag{2.116}$$

then this system can be reduces to a system with a zero matrix.

Proof. Indeed any solution of (2.109) $x(t, x_0)$, $x(t_0, x_0) = x_0$, can be written as

$$x(t, x_0) = x_0 + \int_{t_0}^{t} A(\sigma) x(\sigma, x_0) d\sigma + \sum_{t_0<\tau_i<t} B_i x(\tau_i, x_0). \tag{2.117}$$

By applying Lemma 1 for $t \geq t_0$, we get the estimate

$$\begin{aligned} \|x(t, x_0)\| &\leq \prod_{t_0<\tau_i<t} (1 + \|B_i\|) e^{\int_{t_0}^{t} \|A(\sigma)\| d\tau} \|x_0\| \leq \\ &\leq \prod_{\tau_i > t_0} (1 + \|B_i\|) e^{c_1} \|x_0\|. \end{aligned} \tag{2.118}$$

Because convergence of the series $\sum_{\tau_i > \tau_0} \|B_i\|$ implies convergence of the product $\prod_{\tau_i > \tau_0} (1 + \|B_i\|)$, it follows from (2.118) that all solutions of system

(2.109) are bounded. Besides, conditions (2.116) imply that the quantity

$$\left| \int_{t_0}^{t} \text{Sp } A(\sigma) d\sigma + \sum_{t_0 < \tau_i < t} \ln |\det(E + B_i)| \right| \leq \int_{t_0}^{\infty} |\text{Sp } A(\sigma)| d\sigma +$$

$$+ \sum_{\tau_i < t_0} |\ln |\det(E + B_i)|| \leq c < \infty.$$

Hence all the conditions of the previous theorem hold and so if conditions (2.116) hold, then system (2.109) can be reduced to the system with a zero matrix.

The following theorem gives a relation between regular and reducible systems.

Theorem 24 *Any reducible linear impulsive system (2.109) is regular.*

Proof. Let system (2.109) be reducible. Then there exists a Lyapunov matrix $L(t)$ such that a normal fundamental matrix of system (2.109) $X(t, t_0)$ can be written in the form

$$X(t, t_0) = L(t) e^{B(t - t_0)}$$

with some constant matrix B. By using the analogue of Osrogradskiï-Liouville formula for impulsive systems, we have

$$\det X(t_0, t_0) e^{\int_{t_0}^{t} \text{Sp } A(\sigma) d\sigma} \prod_{t_0 < \tau_i < t} \det(E + B_i) = \det L(t) e^{\text{Sp } B(t - t_0)}$$

or

$$e^{\int_{t_0}^{t} \text{Sp } A(\sigma) d\sigma + \sum_{t_0 < \tau_i < t} \ln |\det(E + B_i)|} = |\det X(t_0, t_0)|^{-1} |\det L(t)| e^{\text{Sp } B(t - t_0)}.$$

From this we see that the limit

$$s = \lim_{t \to \infty} \frac{1}{t} \left(\int_{t_0}^{t} \text{Sp } A(\sigma) d\sigma + \sum_{t_0 < \tau_i < t} \ln |\det(E + B_i)| \right) = \text{Sp } B$$

exists. Besides, let σ_X and σ_Y be the sums of characteristic exponents of the solutions contained in the fundamental matrices $X(t, t_0)$ and $e^{B(t - t_0)}$.

Because a Lyapunov transformation does not change the characteristic exponents and the characteristic exponents of the matrix $e^{B(t-t_0)}$ are the real parts of the eigenvalues of the matrix B, the Lyapunov equality

$$\sigma_X = \sigma_Y = \sum_{\nu=1}^{n} Re\, \lambda_\nu(B) = \mathrm{Sp}\ B = s$$

holds. Hence system (2.109) is regular if it is reducible.

□

2.7 Linear periodic impulsive systems

A linear impulsive differential system

$$\frac{dx}{dt} = A(t)x, \qquad t \neq \tau_i, \qquad \Delta x|_{t=\tau_i} = B_i x \tag{2.119}$$

is called *periodic* with period T or T-*periodic* if the matrix $A(t)$ is T-periodic and there is a natural number p such that

$$B_{i+p} = B_i, \qquad \tau_{i+p} = \tau_i + T \tag{2.120}$$

for all $i \in \mathbf{Z}$.

In this section we assume that the matrix $A(t)$ is continuous (piecewise continuous with discontinuities of the first kind at $t = \tau_i$), the matrices $E + B_i$ are nonsingular, and the times τ_i are indexed by integers such that $0 < \tau_1 < \ldots < \tau_p < T$.

Let $X(t)$ be a fundamental matrix of periodic system (2.119) such that $X(0) = E$, i.e. a matriciant of the system. Because system (2.119) is periodic, it is easy to see that $X(t+T)$ is also a fundamental matrix of system (2.119) and

$$X(t+T) = X(t)X(T), \tag{2.121}$$

where

$$X(T) = U(T, \tau_p) \prod_{\nu=p-1}^{1} (E + B_{\nu+1})U(\tau_{\nu+1}, \tau_\nu)(E + B_1)U(\tau_1, 0) \tag{2.122}$$

is a monodromy matrix and $U(t, \sigma)$, $U(\sigma, \sigma) = E$ is a matriciant of differential system (2.119).

The eigenvalues of the matrix $X^T(T)$ are called *multipliers* of system (2.119).

Theorem 25 *For any multiplier ρ there exists a nontrivial solution of periodic system (2.119) $x = \varphi(t)$, which satisfies the condition*

$$\varphi(t+T) = \rho\varphi(t). \tag{2.123}$$

Conversely, if for a nontrivial solution $x = \varphi(t)$ and some number ρ relation (2.123) holds, then ρ is a multiplier of this system.

Proof. Take as a solution of system (2.119) such $\varphi(t)$ that the vector $\varphi(0)$ is an eigenvector of the monodromy matrix, corresponding to the eigenvalue ρ. We have

$$X(T)\varphi(0) = \rho\varphi(0), \qquad \varphi(t) = X(T)\varphi(0).$$

From this it follows that

$$\begin{aligned}\varphi(t+T) &= X(t+T)\varphi(0) = X(t)X(T)\varphi(0) = \\ &= X(t)\rho\varphi(0) = \rho\varphi(t),\end{aligned}$$

i.e. condition (2.123) holds.

Now suppose that, for a certain nontrivial solution $\varphi(t) = X(t)\varphi(0)$, equality (2.123) holds, i.e.

$$X(t+T)\varphi(0) = \rho X(t)\varphi(0), \qquad X(t)X(T)\varphi(0) = \rho X(t)\varphi(0),$$

and so

$$(X(T) - \rho E)\varphi(0) = 0.$$

This implies that ρ is a root of the equation $\det(X(T) - \rho E)) = 0$, i.e. ρ is a multiplier of system (2.119).

□

An important corollary follows from this theorem.

Corollary 4 *A linear T-periodic system (2.119) has a nontrivial kT- periodic solution if and only if the kth power of at least one of its multipliers equals to one.*

Proof. Indeed it follows from (2.119) that, if ρ is a multiplier, then there exists a solution of (2.119) such that $\varphi(t+kT) = \rho^k\varphi(t)$. If $\rho^k = 1$, then the solution $\varphi(t)$ is kT-periodic.

Conversely, if system (2.119) has a kT-periodic solution $x = \varphi(t)$, $\varphi(t + kT) = \varphi(t)$, then

$$\varphi(t+kT) = X(t+kT)\varphi(0) = X(t)\varphi(0) = \varphi(t)$$

and so,

$$(X(kT) - E)\varphi(0) = 0. \tag{2.124}$$

Because $X(kT) = (X(T))^k$, it follows from (2.124) that $\varphi(0)$ is an eigenvector of the matrix $X(T)^k$. The eigenvalues of the matrix $X^k(T)$ are ρ_j^k, where ρ_j are multipliers, so

$$X(kT)\varphi(0) = \rho_j^k \varphi(0)$$

for some $j = 1, \ldots, n$, and consequently

$$\varphi(t + kT) = X(t)X^k(T)\varphi(0) = \rho^k X(t)\varphi(0) = \rho^k \varphi(t).$$

But $\varphi(t + kT) = \varphi(t)$ by the assumption, and thus $\rho^k = 1$.

□

We will prove that linear periodic system (2.119) is reducible to a system with constant coefficients.

Theorem 26 *Linear periodic impulsive differential system (2.119) can be reduced to a system with constant coefficients by a linear nonsingular piecewise continuous periodic Lyapunov transformation of variables.*

Proof. Let us show that the matriciant of system (2.119) can be written in the Floquet form

$$X(t) = \Phi(t)e^{Pt}, \tag{2.125}$$

where $\Phi(t)$ is a T-periodic Lyapunov matrix continuously differentiable for $\tau_i < t < \tau_{i+1}$, P is a constant matrix. To do that, set

$$\frac{1}{T} \operatorname{Ln} X(T) = P \tag{2.126}$$

and write

$$X(t) = X(t)e^{-Pt}e^{Pt} = \Phi(t)e^{Pt},$$

where

$$\Phi(t) = X(T)e^{-Pt}. \tag{2.127}$$

Verify that the matrix $\Phi(t)$ is T-periodic. We have

$$\begin{aligned} \Phi(t+T) &= X(t+T)e^{-P(t+T)} = X(t)X(T)e^{-PT}e^{-Pt} = \\ &= X(t)e^{-Pt} = \Phi(t). \end{aligned}$$

It follows from the definition of $\Phi(t)$ that, for $\tau_i < t < \tau_{i+1}$, this matrix is continuously differentiable, bounded for all $t \in \mathbf{R}$ because of its periodicity, and nonsingular with $\min_{0 \le t \le T} |\det \Phi(t)| = m > 0$. So $\Phi(t)$ is a Lyapunov matrix.

By Erugin theorem, it follows from (2.125) that the periodic Lyapunov transformation

$$x = \Phi(t)y \tag{2.128}$$

transforms system (2.119) to the system with constant coefficients

$$\frac{dx}{dt} = Py. \tag{2.129}$$

□

Because of (2.126), multipliers of system (2.119) ρ_j and the eigenvalues of the matrix P, λ_j, satisfy

$$\frac{1}{T}\ln|\rho_j| = \Re\,\lambda_j(P), \tag{2.130}$$

and so, if $|\rho_i| \leq 1$, we have $\Re\lambda_j \leq 0$. From this the following criterion for stability of solutions of periodic system (2.119) follows.

Theorem 27 *All solutions of linear periodic impulsive system (2.119) are stable if and only if all the multipliers of this system satisfy $|\rho_j| \leq 1$, $j = 1, 2, \ldots, n$, and the multipliers, for which $|\rho_j| = 1$, being considered as eigenvalues of a monodromy matrix, have prime elementary divisors. All solutions are asymptotically stable if and only if all the multipliers satisfy $|\rho_j| < 1$.*

As an example let us study stability of motion of a physical pendulum [58], which is a solid body that can freely rotate in a certain vertical plane around a suspension point. Suppose that the suspension point executes a vertical periodic motion and its displacement is given during the period T by the function

$$f(t) = \begin{cases} at, & \text{if} \quad 0 \leq t \leq \frac{T}{4}, \\ \frac{aT}{2} - at, & \text{if} \quad \frac{T}{4} \leq t \leq \frac{3}{4}T, \\ at - aT, & \text{if} \quad \frac{3}{4}T \leq t \leq T, \end{cases}$$

the positive direction is taken to be downwards. If the oscillations of the pendulum are small, the equation of the motion can be written as

$$\begin{aligned} \ddot{x} \pm \frac{lg}{k_0^2}x &= 0, \qquad t \neq \tau_i, \\ \dot{x}|_{t=\tau_i} &= (-1)^{i+1}\left(\pm\frac{2al}{k_0^2}\right)x, \end{aligned} \tag{2.131}$$

where x is the angular displacement, t – time, l – the distance between the center of mass and the suspension point, g – acceleration of gravity, $\tau_i = \frac{T}{4} + (i-1)\frac{T}{2}$ $(i = 0, \pm 1, \pm 2, \dots)$. The plus sign in (2.131) corresponds to an oscillation near the lower equilibrium point and the minus sign – to an oscillation near the upper equilibrium point.

Let ω be the ratio of the period of excitation and the period of the intrinsic oscillations of the pendulum, and b be the dimensionless impulse intensity of the motion of the suspension point, $\omega = \frac{T\sqrt{lg}}{2\pi k_0}$, $b = \frac{alT}{\pi k_0^2}$. By replacing T with $\frac{T}{2\pi}t$, equations (2.131) can be written as

$$\ddot{x} \pm \omega^2 x = 0, \qquad t \neq \tau_i, \qquad \Delta\dot{x}|_{t=\tau_i} = (-1)^{i+1}(\pm b)x,$$
$$\tau_i = \frac{\pi}{2} + (i-1)\pi, \qquad i \in \mathbf{Z} \tag{2.132}$$

or, finally, as the system

$$\dot{x} = \omega y, \qquad \dot{y} = \pm\omega x \qquad \text{for} \qquad t \neq \tau_i,$$
$$\Delta y|_{t=\tau_i} = (-1)^{i+1}\left(\pm\frac{b}{\omega}\right)x. \tag{2.133}$$

Because system (2.133) is periodic with the period 2π $(\tau_{i+2} - \tau_i = 2\pi,\ I_{i+2} = I_i)$, stability of solutions is determined by the properties of the monodromy matrix. Let us find this matrix.

A normal fundamental matrix of differential system (2.133) has the form

$$U(t,\tau) = \begin{pmatrix} \cos\omega(t-\tau) & \sin\omega(t-\tau) \\ -\sin\omega(t-\tau) & \cos\omega(t-\tau) \end{pmatrix} \tag{2.134}$$

for the motion near the lower equilibrium point and

$$U(t,\tau) = \begin{pmatrix} \cosh\omega(t-\tau) & \sinh\omega(t-\tau) \\ \sinh\omega(t-\tau) & \cosh\omega(t-\tau) \end{pmatrix} \tag{2.135}$$

for the motion near the upper equilibrium point. So a monodromy matrix can be written as

$$\begin{pmatrix} 1 & 0 \\ -\frac{b}{\omega} & 1 \end{pmatrix} \begin{pmatrix} \cos\omega\pi & \sin\omega\pi \\ -\sin\omega\pi & \cos\omega\pi \end{pmatrix} \begin{pmatrix} 1 & 0 \\ \frac{b}{\omega} & 1 \end{pmatrix} \begin{pmatrix} \cos\omega\pi & \sin\omega\pi \\ -\sin\omega\pi & \cos\omega\pi \end{pmatrix} =$$

$$= \begin{pmatrix} \cos 2\omega\pi + \dfrac{b}{2\omega}\sin 2\omega\pi & \sin 2\omega\pi + \dfrac{b}{\omega}\sin^2\omega\pi \\ \dfrac{b}{\omega}\sin^2\omega\pi - \sin 2\omega\pi - \dfrac{b^2}{2\omega^2}\sin 2\omega\pi & \cos 2\omega\pi - \dfrac{b^2}{\omega^2}\sin^2\omega\pi - \dfrac{b}{2\omega}\sin 2\omega\pi \end{pmatrix} \tag{2.136}$$

if the motion considered at the lower equilibrium and

$$\begin{pmatrix} 1 & 0 \\ \dfrac{b}{\omega} & 1 \end{pmatrix} \begin{pmatrix} \cosh\omega\pi & \sinh\omega\pi \\ \sinh\omega\pi & \cosh\omega\pi \end{pmatrix} \begin{pmatrix} 1 & 0 \\ -\dfrac{b}{\omega} & 1 \end{pmatrix} \begin{pmatrix} \cosh\omega\pi & \sinh\omega\pi \\ \sinh\omega\pi & \cosh\omega\pi \end{pmatrix} =$$

$$= \begin{pmatrix} \cosh 2\omega\pi - \dfrac{b}{2\omega}\sinh 2\omega\pi & \sinh 2\omega\pi - \dfrac{b}{\omega}\sinh^2\omega\pi \\ \dfrac{b}{\omega}\sinh^2\omega\pi + \sinh 2\omega\pi - \dfrac{b^2}{2\omega^2}\sinh^2 2\omega\pi & \cosh 2\omega\pi + \dfrac{b}{2\omega}\sinh 2\omega\pi - \dfrac{b^2}{\omega^2}\sin^2\omega\pi \end{pmatrix} \tag{2.137}$$

for the motion near the upper equilibrium.

The eigenvalues of monodromy matrices (2.136) and (2.137) are calculated as roots of the equations

$$\lambda^2 - 2\left(\cos 2\omega\pi - \frac{b^2}{2\omega^2}\sin^2\omega\pi\right)\lambda + 1 = 0, \tag{2.138}$$

$$\lambda^2 - 2\left(\cosh 2\omega\pi - \frac{b^2}{2\omega^2}\sinh^2\omega\pi\right)\lambda + 1 = 0. \tag{2.139}$$

Because these equations are reciprocal, both roots can not lie inside the unit circle, i.e. the trivial solution of system (2.133) can not be asymptotically stable. But the roots of these equations could lie on the unit circle. Two roots of (2.138) lie on the unit circle if and only if

$$\left|\cos 2\omega\pi - \frac{b^2}{2\omega^2}\sin^2\omega\pi\right| \leq 1, \tag{2.140}$$

and the roots of (2.139) satisfy this condition if

$$\left|\cosh 2\omega\pi - \frac{b^2}{2\omega^2}\sinh^2\omega\pi\right| \leq 1. \tag{2.141}$$

If we have strict inequality in (2.140), which will be for

$$0 < \frac{b^2}{4\omega^2} \tan^2 \pi\omega < 1, \tag{2.142}$$

then the roots of equation (2.138) will be complex conjugate and hence simple. If

$$\cos 2\pi\omega - \frac{b^2}{2\omega^2} \sin^2 \omega\pi = 1,$$

which will be for $\sin^2 \pi\omega = 0$, i.e. when ω is an integer, then equation (2.138) will have the root $\lambda_{1,2} = 1$ and, considered as an eigenvalue of the monodromy matrix, it corresponds to prime elementary divisors. If $\cos 2\pi\omega - \frac{b^2}{2\omega^2} \sin^2 \pi\omega = -1$, i.e. $4\omega^2 = b^2 \tan^2 \pi\omega$, equation (2.138) has the double root $\lambda = -1$ and the corresponding divisors will not be prime if considered as an eigenvalue of matrix (2.136).

To summarize we see that the lower equilibrium of the pendulum, the motion of which is given by system (2.132) (with the plus sign), will be stable if and only if

$$\frac{b^2}{4\omega^2} \tan^2 \pi\omega < 1. \tag{2.143}$$

Similarly, it follows from inequality (2.141) that the upper equilibrium point of the pendulum described by equations (2.132) (the minus sign is taken in (2.132)) will be stable if the following inequality holds

$$1 < \frac{b^2}{4\omega^2} < \cot^2 \omega\pi. \tag{2.144}$$

2.8 Linear Hamiltonian systems of impulsive differential equations

A system of linear impulsive differential equations

$$\frac{dx}{dt} = \mathcal{J}A(t)x, \qquad t \neq \tau_i \qquad \Delta x|_{t=\tau_i} = \mathcal{J}B_i x, \tag{2.145}$$

where $x = (x_1, x_2, \ldots, x_{2n})$, $A(t)$ and B_i are symmetric $2n \times 2n$-matrices, $\mathcal{J}$ is a so-called symplectic identity

$$\mathcal{J} = \mathcal{J}_{2n} = \begin{pmatrix} 0 & E_n \\ -E_n & 0 \end{pmatrix},$$

E_n is a unit $n \times n$-matrix, is called *Hamiltonian* or *canonical* if the matrices B_i are such that $(\mathcal{J}B_i)^2 = 0$ for all $i \in \mathbf{Z}$.

In particular, if $n = 1$, then system (2.145) will be Hamiltonian if the elements of the matrix

$$B_i = \begin{pmatrix} \alpha_i & \gamma_i \\ \gamma_i & \beta_i \end{pmatrix}$$

satisfy the equality $\alpha_i\beta_i = \gamma_i^2$.

Note that the trace of the matrices that define system (2.145) equals to zero, i.e. $\operatorname{Sp}\mathcal{J}A(t) = \operatorname{Sp}\mathcal{J}B_i = 0$. This is because the trace of any symmetric $2n \times 2n$-matrix $\operatorname{Sp}(\mathcal{J}A) = 0$. Indeed, if

$$A = \begin{pmatrix} A_1 & A_2 \\ A_2 & A_3 \end{pmatrix},$$

where A_1, A_2, A_3 are $n \times n$-matrices, then

$$\mathcal{J}A = \begin{pmatrix} 0 & E_n \\ -E_n & 0 \end{pmatrix} \begin{pmatrix} A_1 & A_2 \\ A_2 & A_3 \end{pmatrix} = \begin{pmatrix} A_2 & A_3 \\ -A_1 & -A_2 \end{pmatrix},$$

and so $\operatorname{Sp}(\mathcal{J}A) = 0$.

Solutions of linear Hamiltonian system (2.145) have an important property given in the following lemma.

Lemma 9 *For any two solutions of linear Hamiltonian impulsive differential system (2.145) $x(t)$ and $y(t)$, their symplectic scalar product, i.e.*

$$\langle x(t), \mathcal{J}y(t)\rangle \tag{2.146}$$

is constant. Similarly, if $X(t)$ and $Y(t)$ are matrix solutions of linear Hamiltonian system (2.145), then

$$X^T(t)\mathcal{J}Y(t) = C, \tag{2.147}$$

where C is a constant $(2n \times 2n)$-matrix.

Proof. If $x(t)$ and $y(t)$ are solutions of equation (2.145), then for $t \neq \tau_i$,

$$\begin{aligned}\frac{d}{dt}\langle x(t), \mathcal{J}y(t)\rangle &= \left\langle \frac{dx}{dt}, \mathcal{J}y(t)\right\rangle + \left\langle x(t), \mathcal{J}\frac{dy}{dt}\right\rangle = \\ &= \langle \mathcal{J}A(t)x, \mathcal{J}y(t)\rangle + \langle x(t), \mathcal{J}^2A(t)y\rangle = \\ &= \langle \mathcal{J}^T\mathcal{J}A(t)x, y\rangle + \langle x(t), \mathcal{J}^2A(t)y\rangle.\end{aligned}$$

Noting that $\mathcal{J}^T = -\mathcal{J}$ and $\mathcal{J}^2 = -E_{2n}$, we get from this equality that

$$\frac{d}{dt}\langle x(t), \mathcal{J}y(t)\rangle = \langle A(t)x, y\rangle - \langle x, A(t)y\rangle.$$

Since the matrix $A(t)$ is symmetric, we get

$$\frac{d}{dt}\langle x(t), \mathcal{J}y(t)\rangle = 0, \qquad \langle x(t), \mathcal{J}y(t)\rangle = C_i \tag{2.148}$$

for all $t \in]\tau_i, \tau_{i+1}[$.

It remains to show that all the constants C_i are equal, i.e. independent of i. For $t = \tau_i$ we have

$$\begin{aligned}\langle x(\tau_i+0), \mathcal{J}y(\tau_i+0)\rangle &= \langle (E+\mathcal{J}B_i)x(\tau_i), \mathcal{J}(E+\mathcal{J}B_i)y(\tau_i)\rangle = \\ &= \langle x(\tau_i), \mathcal{J}y(\tau_i)\rangle + \langle x(\tau_i), \mathcal{J}^2B_iy(\tau_i)\rangle + \\ &\quad +\langle \mathcal{J}B_ix, \mathcal{J}y\rangle + \langle \mathcal{J}B_ix, \mathcal{J}^2B_iy\rangle = \\ &= \langle x(\tau_i, \mathcal{J}y(\tau_i)\rangle - \langle x(\tau_i), B_iy(\tau_i)\rangle + \\ &\quad +\langle B_ix, y\rangle - \langle B_i\mathcal{J}B_ix(\tau_i), y(\tau_i)\rangle.\end{aligned}$$

From $(\mathcal{J}B_i)^2 = 0$ it follows that $B_i\mathcal{J}B_i = 0$ and so by using the symmetry of the matrices B_i, we finally get

$$\langle x(\tau_i+0), \mathcal{J}y(\tau_i+0)\rangle = \langle x(\tau_i), \mathcal{J}y(\tau_i)\rangle.$$

Thus we have shown that, for all $t \in \mathbf{R}$,

$$\langle x(t), \mathcal{J}y(t)\rangle = \text{const.}$$

In the same way we can prove that equality (2.147) holds.

□

Now suppose that system (2.145) is T-periodic, i.e. the matrix $A(t)$ of this system satisfies $A(t+T) = A(t)$ and there exists a natural number p such that

$$B_{i+p} = B_i, \qquad \tau_{i+p} = \tau_i + T \tag{2.149}$$

for all $i \in \mathbf{Z}$.

Before studying properties of a periodic Hamiltonian system, recall some properties of *reciprocal* algebraic equations, i.e. equations of the type

$$F(\lambda) \equiv \lambda^n + a_1\lambda^{n-1} + \ldots + a_{n-1}\lambda + 1 = 0, \tag{2.150}$$

where the coefficients a_j satisfy

$$a_j = a_{n-j} \qquad (j = 1, 2, \ldots, n-1). \tag{2.151}$$

From this it follows that, for a reciprocal polynomial $F(\lambda)$, the following identity holds:

$$F\left(\frac{1}{\lambda}\right) = \frac{1}{\lambda^n}F(\lambda). \tag{2.152}$$

Conversely, if for a polynomial $F(\lambda)$, (2.152) holds, then this polynomial is reciprocal.

Lemma 10 *If $\lambda = \lambda_0$ is a root of a reciprocal equation (2.150), then $\lambda = \dfrac{1}{\lambda_0}$ is also a root of this equation of the same multiplicity. Besides, if this equation has the root $\lambda = 1$, then the multiplicity of this root is even; if $\lambda = -1$ is a root of this equation, then its multiplicity is even if n is even, and odd if n is odd.*

A proof of this lemma can be found, for example, in [38].

□

For a periodic Hamiltonian impulsive differential system (2.145) we have

Theorem 28 *If linear Hamiltonian system (2.145) is T-periodic, then the characteristic equation*

$$F(\lambda) = \det(\lambda E - X(T)) = 0,$$

where $X(T)$ is a monodromy matrix, is reciprocal.

Proof. Let $X(t)$, $X(0) = E$, be a normal fundamental matrix of solutions of periodic system (2.145). By Lemma 9, $X^T(t)\mathcal{J}X(t) = C$. By setting $t = 0$, we find that $C = \mathcal{J}$ and so $X^T(t)\mathcal{J}X(t) = \mathcal{J}$. From this we get $X^*(T)\mathcal{J}X(T) = \mathcal{J}$. Because $\det \mathcal{J} = \det \mathcal{J}^{-1} = 1$, we have

$$\begin{aligned} F\left(\frac{1}{\lambda}\right) &= \det\left(\frac{1}{\lambda}E - X(\omega)\right) = \frac{1}{\lambda^{2n}}\det(E - \lambda X(T)) = \\ &= \frac{1}{\lambda^{2n}}\det(E - \lambda X^T(T)) = \\ &= \frac{1}{\lambda^{2n}}\det(\mathcal{J}E\mathcal{J}^{-1} - \lambda\mathcal{J}X^{-1}(T)\mathcal{J}^{-1}) = \\ &= \frac{1}{\lambda^{2n}}\det\mathcal{J}\det(E - \lambda X^{-1}(T))\det\mathcal{J}^{-1} = \\ &= \frac{1}{\lambda^{2n}}\det X^{-1}(T)\det(\lambda E - X(T)) = \\ &= \frac{1}{\lambda^{2n}}F(\lambda)\det X^{-1}(T). \end{aligned} \tag{2.153}$$

To end the proof it remains to show that $\det X^{-1}(T) = 1$. The matrix $X(T)$ has the form

$$X(T) = U(T, \tau_p)\prod_{i=p}^{1}(E + \mathcal{J}B_i)U(\tau_i, \tau_{i-1}), \qquad \tau_0 = 0,$$

where $U(t, \tau)$ is a matriciant of the system

$$\frac{dx}{dt} = \mathcal{J}A(t)x.$$

Because $\mathrm{Sp}\,\mathcal{J}A(t) = 0$, by Liouville-Osrogradskiĭ formula,

$$\det U(t, \tau) = \det U(\tau, \tau)e^{\int_\tau^t \mathrm{Sp}\,A(\sigma)d\sigma} = \det U(\tau, \tau) = 1,$$

and consequently

$$\det X(T) = \prod_{i=1}^{p}\det(E + \mathcal{J}B_i).$$

The matrices B_i satisfy the condition $(\mathcal{J}B_i)^2 = 0$ and so we can show that $\det(E + \mathcal{J}B_i) = 1$. Indeed the matrix $E + \mathcal{J}B_i$ can be written as $E + \mathcal{J}B_i = e^{\mathcal{J}B_i}$, and so $E + \mathcal{J}B_it = e^{\mathcal{J}B_it}$. But $e^{\mathcal{J}B_it}$ is a matriciant of the system

$$\frac{dx}{dt} = \mathcal{J}B_ix,$$

hence

$$\det(E+\mathcal{J}B_it)=\det e^{\mathcal{J}B_it}=\det Ee^{\int_0^t \mathrm{Sp}\,\mathcal{J}B_i d\tau}=\det e^{\mathrm{Sp}\,\mathcal{J}B_it}.$$

From this, because $\mathrm{Sp}\,\mathcal{J}B_i=0$, we have $\det(E+\mathcal{J}B_it)=1$. In particular, if $t=1$, we get $\det(E+\mathcal{J}B_i)=1$.

Thus $\det X(T)=1$ and so $\det X^{-1}(T)=1$. Hence from (2.153) we have that

$$F\left(\frac{1}{\lambda}\right)=\frac{1}{\lambda^{2n}}F(\lambda),$$

which proves that the equation $F(\lambda)=0$ is reciprocal.

□

From this theorem it follows that linear Hamiltonian periodic system (2.145) can not be asymptotically stable.

Theorem 29 *A linear Hamiltonian periodic system is stable if and only if all its multipliers ρ_j lie on the unit circle and have prime elementary divisors.*

Consider the impulsive system

$$\begin{aligned}\ddot{x}+A_1(t)x &= 0, \quad t\neq\tau_i,\\ \Delta\dot{x}|_{t=\tau_i} &= B_i^0x,\end{aligned} \tag{2.154}$$

where $A_1(t)$ is a symmetric T-periodic matrix, B_i^0 are symmetric matrices such that $B_{i+p}^0=B_i^0$ for all $i\in\mathbf{Z}$ and some natural p, the times τ_i are such that $\tau_{i+p}=\tau_i+T$.

Write equation (2.154) as the system

$$\begin{aligned}\frac{dx}{dt} &= y, \qquad t\neq\tau_i,\\ \frac{dy}{dt} &= -A_1(t)x, \qquad \Delta y|_{t=\tau_i}=B_i^0x,\end{aligned} \tag{2.155}$$

or

$$\begin{aligned}\frac{d}{dt}\begin{pmatrix}x\\y\end{pmatrix} &= \mathcal{J}A(t)\begin{pmatrix}x\\y\end{pmatrix}, \qquad t\neq\tau_i,\\ \Delta\begin{pmatrix}x\\y\end{pmatrix}\bigg|_{t=\tau_i} &= \mathcal{J}B_i\begin{pmatrix}x\\y\end{pmatrix},\end{aligned} \tag{2.156}$$

where

$$A(t) = \begin{pmatrix} A_1(t) & 0 \\ 0 & E_n \end{pmatrix}, \qquad B_i = \begin{pmatrix} -B_i^0 & 0 \\ 0 & 0 \end{pmatrix}.$$

Because the matrices $A(t)$ and B_i are symmetric and

$$(\mathcal{J}B_i)^2 = \begin{pmatrix} 0 & 0 \\ B_i^0 & 0 \end{pmatrix} \begin{pmatrix} 0 & 0 \\ B_i^0 & 0 \end{pmatrix} = \begin{pmatrix} 0 & 0 \\ 0 & 0 \end{pmatrix},$$

the following theorem is true.

Theorem 30 *The characteristic equation of a monodromy matrix of system (2.156) is reciprocal.*

□

As an example consider the linear Hamiltonian system of two impulsive equations:

$$\begin{aligned} \frac{dx}{dt} &= ax + by, & \Delta x|_{t=\tau_i} &= \alpha x + \beta y, \\ & & t &\neq \tau_i, \\ \frac{dy}{dt} &= cx - ay, & \Delta y|_{t=\tau_i} &= \gamma x - \alpha y, \end{aligned} \tag{2.157}$$

where a, b, c, α, β, γ are real numbers and $\alpha^2 + \beta\gamma = 0$, and the sequence τ_i is such that

$$\tau_{i+1} = \tau_i + T.$$

System (2.157) can be written in the form

$$\frac{dz}{dt} = \mathcal{J}Az, \qquad t \neq \tau_i, \qquad \Delta z|_{t=\tau_i} = \mathcal{J}Bz,$$

where

$$z = \begin{pmatrix} x \\ y \end{pmatrix}, \qquad A = \begin{pmatrix} -c & a \\ a & b \end{pmatrix}, \qquad B = \begin{pmatrix} -\gamma & \alpha \\ \alpha & \beta \end{pmatrix},$$

$$\mathcal{J} = \begin{pmatrix} 0 & 1 \\ -1 & 0 \end{pmatrix}.$$

Now let us study stability of solutions of system (2.157). Because this system is T-periodic, we need only to find eigenvalues of its monodromy matrix.

Let $a^2 + bc > 0$. In this case the matriciant of the system

$$\begin{aligned} \frac{dx}{dt} &= ax + bx, \\ \frac{dy}{dt} &= cx - ay \end{aligned} \tag{2.158}$$

will be

$$U(t,\tau) =$$

$$= \begin{pmatrix} \cosh\omega(t-\tau) + \frac{a}{\omega}\sinh(t-\tau) & \frac{b}{\omega}\sinh(t-\tau) \\ \frac{c}{\omega}\sinh\omega(t-\tau) & \cosh\omega(t-\tau) - \frac{a}{\omega}\sinh\omega(t-\tau) \end{pmatrix},$$

where $\omega = \sqrt{a^2+bc}$. So the monodromy matrix is

$$X(T) = \begin{pmatrix} \cosh\omega T + \frac{a}{\omega}\sinh\omega T & \frac{b}{\omega}\sinh\omega T \\ \frac{c}{\omega}\sinh\omega T & \cosh\omega T - \frac{a}{\omega}\sinh\omega T \end{pmatrix} \times$$

$$\times \begin{pmatrix} 1+\alpha & \beta \\ \gamma & 1-\alpha \end{pmatrix}.$$

The eigenvalues of this matrix are the roots of

$$\lambda^2 - \mathrm{Sp}\ X(T) + \det X(T) = 0. \tag{2.159}$$

It follows from the direct calculations that

$$\mathrm{Sp}\ X(T) = 2(\cosh\omega T + \frac{2a\alpha + b\gamma + c\beta}{2\omega}\sinh\omega T),$$

and $\det X(T) = 1$ because the determinant of each of the factors equal to one.

Hence equation (2.159) becomes

$$\lambda^2 - 2\left(\cosh\omega T + \frac{2a\alpha + b\gamma + c\beta}{2\omega}\sinh\omega T\right)\lambda + 1 = 0. \tag{2.160}$$

Solutions of the system will be stable if the roots of (2.160) are distinct with the absolute value equal to one. This will be if

$$\left|\cosh\omega T + \frac{2a\alpha + b\gamma + c\beta}{2\omega}\sinh\omega T\right| < 1. \tag{2.161}$$

If

$$\left|\cosh\omega T + \frac{2a\alpha + b\gamma + c\beta}{2\omega}\sinh\omega T\right| \geq 1, \tag{2.162}$$

then the solutions of system (2.157) will be unstable if $a^2 + bc > 0$.

Let $a^2 + bc < 0$. Then the matriciant of system (2.158) will be

$$U(t,\tau) = \begin{pmatrix} \cos\omega(t-\tau) + \frac{a}{\omega}\sin(t-\tau) & \frac{b}{\omega}\sin(t-\tau) \\ \frac{c}{\omega}\sin\omega(t-\tau) & \cos\omega(t-\tau) - \frac{a}{\omega}\sin\omega(t-\tau) \end{pmatrix},$$

where $\omega = \sqrt{-(a^2+bc)}$.

Consequently the monodromy matrix is

$$X(T) = \begin{pmatrix} \cos\omega T + \frac{a}{\omega}\sin\omega T & \frac{b}{\omega}\sin\omega T \\ \frac{c}{\omega}\sin\omega T & \cos\omega T - \frac{a}{\omega}\sin\omega T \end{pmatrix} \times$$
$$\times \begin{pmatrix} 1+\alpha & \beta \\ \gamma & 1-\alpha \end{pmatrix}.$$

and its eigenvalues are roots of the equation

$$\lambda^2 - 2\left(\cos\omega T + \frac{2a\alpha + b\gamma + c\beta}{2\omega}\sin\omega T\right)\lambda + 1 = 0. \tag{2.163}$$

Hence, if in system (2.157) $a^2 + bc < 0$, $\alpha^2 + \beta\gamma = 0$, its solutions are stable if

$$\left|\cos\omega T + \frac{2a\alpha + b\gamma + c\beta}{2\omega}\sin\omega T\right| < 1. \tag{2.164}$$

and unstable if

$$\left|\cos\omega T + \frac{2a\alpha + b\gamma + c\beta}{2\omega}\sin\omega T\right| \geq 1, \tag{2.165}$$

Let $a^2 + bc = 0$. The matriciant of system (2.158) is

$$U(t,\tau) = \begin{pmatrix} 1 + a(t-\tau) & b(t-\tau) \\ c(t-\tau) & 1 - a(t-\tau) \end{pmatrix},$$

and the monodromy matrix is

$$X(T) = \begin{pmatrix} 1+aT & bT \\ cT & 1-aT \end{pmatrix}\begin{pmatrix} 1+\alpha & \beta \\ \gamma & 1-\alpha \end{pmatrix}.$$

The characteristic equation of this matrix will be

$$\lambda^2 - 2\left(1 + a\alpha + \frac{b\gamma + c\beta}{2}\right)T\lambda + 1 = 0.$$

So, if the parameters of this system satisfy the conditions

$$a^2 + bc = 0, \quad \alpha^2 + \beta\gamma = 0, \quad \left|1 + \alpha a + \frac{b\gamma + c\beta}{2}\right| < 1,$$

the solutions of this system will be stable and they will be unstable if

$$\left|1 + a\alpha + \frac{b\gamma + c\beta}{2}\right| \geq 1.$$

2.9 Periodic solutions of a certain second order equation

Consider the impulsive differential system

$$\begin{aligned} \ddot{x} + \omega^2 x &= 0, \qquad t \neq \tau_i, \\ \Delta \dot{x}|_{t=\tau_i} &= ax + b\dot{x}. \end{aligned} \tag{2.166}$$

We will assume that the times τ_i are equally spaced. Let us find out whether this system could have periodic solutions and if so, in what way the frequency of the impulses depends on parameters of the system.

Write (2.166) in the form

$$\begin{aligned} \frac{dx}{dt} = \omega y, \qquad \frac{dy}{dt} &= -\omega x, \qquad t \neq \tau_i, \\ \Delta y|_{t=\tau_i} &= \frac{a}{\omega} x + by. \end{aligned} \tag{2.167}$$

Because the times τ_i are equally spaced, system (2.167) is periodic with the period $T = \tau_{i+1} - \tau_i$ and it is possible to determine whether there exist periodic solutions by calculating its multipliers.

The fundamental matrix of the system

$$\frac{dx}{dt} = \omega y, \qquad \frac{dy}{dt} = -\omega x \tag{2.168}$$

is the matrix

$$X(t, \tau) = \begin{pmatrix} \cos\omega(t-\tau) & \sin\omega(t-\tau) \\ -\sin\omega(t-\tau) & \cos\omega(t-\tau) \end{pmatrix},$$

and so the monodromy matrix for equations (2.167) will be

$$\begin{aligned} X(T) &= \begin{pmatrix} \cos\omega T & \sin\omega T \\ -\sin\omega T & \cos\omega T \end{pmatrix} \begin{pmatrix} 1 & 0 \\ \frac{a}{\omega} & b+1 \end{pmatrix} = \\ &= \begin{pmatrix} \cos\omega T + \frac{a}{\omega}\sin\omega T & (b+1)\sin\omega T \\ -\sin\omega T + \frac{a}{\omega}\cos\omega T & (b+1)\cos\omega T \end{pmatrix}. \end{aligned} \tag{2.169}$$

It should be noted that so far we are considering periodic solutions of system (2.167) that experience an impulsive effect once in a period, i.e. T-periodic solutions.

Let us write the characteristic equation for the monodromy matrix

$$\rho^2 - \left(\frac{a}{\omega}\sin\omega T + (b+2)\cos\omega T\right)\rho + b + 1 = 0. \tag{2.170}$$

System (2.167) has T-periodic solutions if 1 is a root of equation (2.170), i.e. when

$$b + 2 = \frac{a}{\omega}\sin\omega T + (b+2)\cos\omega T$$

or

$$b + 2 = \frac{a}{\omega}\cot\frac{\omega T}{2} \qquad \text{either} \qquad \sin\frac{\omega T}{2} = 0. \tag{2.171}$$

If $\sin\frac{\omega T}{2} = 0$, then

$$X(T) = \begin{pmatrix} 1 & 0 \\ \frac{a}{\omega} & b+1 \end{pmatrix}$$

and the initial points of periodic equations are calculated from the system of equations

$$\begin{pmatrix} 0 & 0 \\ \frac{a}{\omega} & b \end{pmatrix}\begin{pmatrix} x_0 \\ y_0 \end{pmatrix} = 0$$

or

$$\frac{a}{\omega}x_0 + by_0 = 0. \tag{2.172}$$

But every solution of (2.167) that starts at a point on the line $\frac{a}{\omega}x + by = 0$ at time $t = \tau_i - 0$ does not experience an impulsive effect and is described by the functions

$$\begin{aligned} x(t) &= x_0\cos\omega(t-\tau_i) &+& \quad y_0\sin(t-\tau_i), \\ y(t) &= -x_0\sin\omega(t-\tau_i) &+& \quad y\cos\omega(t-\tau_i), \\ \frac{a}{\omega}x_0 + by_0 &&=& \quad 0. \end{aligned}$$

Hence, if $\omega(\tau_{i+1} - \tau_i) = 2\pi k$, $k = 1, 2, \ldots$, then equation (2.166) has a one-parameter family of continuous periodic solutions

$$\begin{aligned} x(t) = x_0\cos\omega(t-\tau_i) &+ \quad \frac{x_0}{\omega}\sin\omega(t-\tau_i), \\ ax_0 + bx_0 &= \quad 0. \end{aligned} \tag{2.173}$$

Let $\sin\frac{\omega T}{2} \neq 0$. Then system (2.167) has periodic solutions if

$$b + 2 = \frac{a}{\omega}\cot\frac{\omega T}{2}. \tag{2.174}$$

Initial values of periodic solutions are determined from the system

$$(X(T)-E)\begin{pmatrix} x_0 \\ y_0 \end{pmatrix} = 0,$$

which, in this case, is equivalent to the equation

$$\left(\frac{a}{\omega}\cot\frac{\omega T}{2} - 1\right)x_0 + (b+1)\cot\frac{\omega T}{2}y_0 = 0. \tag{2.175}$$

Taking into account conditions (2.174), equation (2.175) can be written for $a \neq 0$ as

$$(b+1)\left(x_0 + \frac{(b+2)\omega}{a}y_0\right) = 0. \tag{2.176}$$

Now it is easy to see that, if $a \neq 0$ and $b \neq -1$, system (2.167) has a family of periodic solutions that pass through the points (x_0, y_0) at $t = \tau_i - 0$ such that

$$ax_0 + (b+2)\omega y_0 = 0. \tag{2.177}$$

This will hold if impulses occur at τ_i such that $\tau_{i+1} - \tau_i = T$, where T satisfies (2.174).

If $a = 0$ and $b = -2$, then system (2.167) has a family of periodic solutions that pass through the points of the line

$$x_0 + \cot\frac{\omega T}{2}y_0 = 0.$$

at $t = \tau_i$ and, if $a = 0$ and $b \neq -2$, then, if $\sin\frac{\omega T}{2} \neq 0$, system (14.2) does not have periodic solutions that would undergo an impulsive effect once in a period.

If $b = -1$ and the period of the impulses is T such that $\cot\frac{\omega T}{2} = \frac{\omega}{a}$, then all solutions of system ((2.167) are T-periodic.

Stability of these periodic equations of system (2.167) is determined by the roots of equation (2.170). In all the cases when periodic solutions exist, the roots of (2.170) are given by

$$\rho_1 = 1, \qquad \rho_2 = b+1,$$

and hence periodic solutions of equations (2.167) will be stable if

$$-2 \leq b \leq 0. \tag{2.178}$$

Let us consider existence of periodic solutions of equation (2.166) that experience an impulsive effect twice in a period. In this case, the monodromy matrix is

$$X(2T) = \begin{pmatrix} \cos\omega T & \sin\omega T \\ -\sin\omega T & \cos\omega T \end{pmatrix} \begin{pmatrix} 1 & 0 \\ \frac{a}{\omega} & b+1 \end{pmatrix} \times$$
$$\times \begin{pmatrix} \cos\omega t & \sin\omega t \\ -\sin\omega t & \cos\omega t \end{pmatrix} \begin{pmatrix} 1 & 0 \\ \frac{a}{\omega} & b+1 \end{pmatrix} =$$
$$= \begin{pmatrix} \left(\cos\omega T + \frac{a}{\omega}\sin\omega T\right)^2 + (b+1)\sin\omega T\left(-\sin\omega T + \frac{a}{\omega}\cos\omega T\right) & \left(\cos\omega T + \frac{a}{\omega}\sin\omega T\right)(b+1) + (b+1)^2\sin\omega T\cos\omega T \\ \left(\cos\omega T + \frac{a}{\omega}\sin\omega T + (b+1)\cos\omega T\right)\left(-\sin\omega T + \frac{a}{\omega}\cos\omega T\right) & (b+1)\sin\omega T\left(-\sin\omega T + \frac{a}{\omega}\cos\omega T\right) + (b+1)^2\cos^2\omega T \end{pmatrix} \tag{2.179}$$

and its characteristic equation is

$$\rho^2 - \left[(b+1)^2\cos^2\omega T + 2(b+1)\left(-\sin\omega T + \frac{a}{\omega}\cos\omega T\right)\times\right.$$
$$\left.\times\sin\omega T + \left(\cos\omega T + \frac{a}{\omega}\sin\omega T\right)^2\right]\rho + (1+b)^2 = 0. \tag{2.180}$$

Equation (2.180) has 1 as a root if

$$\left((b+2)^2 - \frac{a^2}{\omega^2}\right)\sin^2\omega T = \frac{2a}{\omega}(b+2)\sin\omega T\cos\omega T, \tag{2.181}$$

i.e. when either one of the equations hold:

$$\begin{aligned} \sin\omega T &= 0 \\ a &= \omega(b+2)\tan\frac{\omega T}{2} \\ a &= -\omega(b+2)\cot\frac{\omega T}{2}. \end{aligned} \tag{2.182}$$

Let us consider each of these three cases.

1. Let $\sin\omega T = 0$, i.e. $\omega T = k\pi$, $k = 1, 2, \ldots$. The system used to determine initial conditions for periodic solutions

$$(X(2T) - E)\begin{pmatrix} x_0 \\ y_0 \end{pmatrix} = 0 \tag{2.183}$$

is equivalent to the equation

$$(b+2)\left(\frac{a}{\omega}x_0 + by_0\right) = 0. \tag{2.184}$$

In the case when $b = -2$, all solutions of system (2.167) are periodic with the period $T = \frac{2\pi k}{\omega}$ if the period of the impulses is $T = \frac{k\pi}{\omega}$, $k = 1, 2, \ldots$.

If $b \neq -2$, then periodic solutions pass through the points (x_0, y_0) which lie on the line $ax + b\omega y = 0$. Such periodic solutions, as we have seen, are continuous.

2. Suppose that $a = \omega(b+2)\tan\frac{\omega T}{2}$. Then system (2.173) is equivalent to

$$\left[\frac{a}{\omega}(b+3) + \left(\frac{a^2}{\omega^2} - b - 1\right)\tan\omega T\right] x_0 +$$
$$+\left[\frac{a}{\omega}(b+1)\tan\omega T + (b+1)(b+2)\right] y_0 = 0$$

or

$$(b+1)(ax_0 + \omega(b+2)y_0) = 0. \tag{2.185}$$

From this equation it follows that, if $b = -1$ in system (2.167) and the period of the impulses equals to $T = \frac{\omega}{2}\left(k\pi + \arctan\frac{a}{\omega}\right)$, $k = 1, 2, \ldots$, then all solutions of system (2.167) are periodic with the period $2T$. If $b \neq -1$, then periodic solutions are induced by points on the line $ax + \omega(b+2)y = 0$. Their period $2T$ can be calculated from the equation $a = \omega(b+2)\tan\frac{\omega T}{2}$. These $2T$ periodic solutions, as it follows from the previous considerations, are T-periodic.

3. It can be determined from the third equation of (2.182) whether there exist $2T$-periodic solutions, which are not T-periodic. In this case, system (2.183) is equivalent to the equation

$$(b+1)(ax_0 + \omega(b+2)y_0) = 0,$$

i.e. to the same equation (2.185).

From this we can make the following conclusion: if the period of the impulses T is such that $a = -\omega(b+2)\cot\frac{\omega T}{2}$, then system (2.167) does not have T-periodic solutions but it has $2T$-periodic solutions and, if $b = -1$, any point (x_0, y_0) of the plane can be taken as an initial point to get such a periodic solution. If $b \neq -1$, these solutions start at points of the line $ax + \omega(b+2)y = 0$.

Now consider the case when $a^2 = \omega^2(b+2)^2$, i.e. when we could have periodic solutions for the period of the impulses T such that $\cos\omega T = 0$, $\tan\frac{\omega T}{2} = \pm 1$. In this case, initial values of the periodic solutions are determined from the equation

$$(b+1)\left((b+2)x_0 + \frac{a}{\omega}y_0\right) = 0.$$

If we assume that $a \geq 0$, then this equation can be written as

$$(b+1)((b+2)x_0 + |b+2|y_0) = 0.$$

From this it follows that all solutions will be periodic if $b = -1$, $a = \omega$ or $b = -2$, $a = 0$. In other cases, $2T$-periodic solutions will be those that start in a point of the set

$$x_0 + \text{sign}(b+2)y_0 = 0$$

at the time when an impulse occurs.

In the case when there exist $2T$-periodic solutions, equation (2.180) will have solutions $\rho_1 = 1$, $\rho_2 = (1+b)^2$ and it is easy to see that the considered solutions of system (2.167) will be stable if $-2 \leq b < 0$.

Chapter 3

Stability of solutions

3.1 Linear systems with constant and almost constant matrices

In section 2.3 we gave two theorems on stability of solutions of impulsive systems. In this section we will prove theorems on stability of solutions of linear systems with constant matrices.

Consider the system

$$\frac{dx}{dt} = Ax, \qquad t \neq \tau_i, \qquad \Delta x|_{t=\tau_i} = Bx, \tag{3.1}$$

where the matrices A and B are constant matrices, the times of the impulsive effects τ_i are indexed by natural numbers in the increasing order such that $\tau_i \to +\infty$ as $i \to \infty$. Without loss of generality we can assume that $\tau_i > t_0$, where t_0 is the initial time.

As it was shown in Section 2.1, any solution of equations (3.1), $x(t, x_0)$, $x(t_0, x_0) = x_0$, are given by the formula

$$x(t, x_0) = X(t, t_0)x_0, \tag{3.2}$$

where

$$X(t, t_0) = e^{A(t-\tau_i)} \prod_{t_0<\tau_j<\tau_i} (E + B)e^{A(\tau_j-\tau_{j-1})},$$

$$\tau_0 = t_0, \qquad \tau_i < t \leq \tau_{i+1}. \tag{3.3}$$

As can be seen from this formula, it is not possible in the general case to give necessary and sufficient conditions for stability of solutions of system (3.1) in terms of eigenvalues of the matrix of this system, which was possible for system of ordinary differential equations with constant coefficients. However, there are particular cases when this can be done. For example, if the times τ_i are equally spaced, i.e.

$$\tau_{i+1} - \tau_i = \theta > 0, \qquad i \in \mathbf{N}, \tag{3.4}$$

then the stability of solutions of (3.1) is completely determined by the eigenvalues of the matrix $(E+B)e^{A\theta}$ because, in this case, system (3.1) is periodic ($p = 1, T = \theta$) (at least for $t \geq \tau_1$) and this matrix is its monodromy matrix. Hence we get the following

Theorem 31 *Let in system (3.1) the times τ_i of the impulsive effect satisfy (3.4). Then the solutions of this system are:*

a) *stable if and only if the eigenvalues of the matrix* $(E+B)e^{A\theta}$ *lie in the unit disk of the complex plane,* $|\lambda| \leq 1$, *and the elementary divisors which correspond to the eigenvalues that lie on the border if this circle are prime;*

b) *asymptotically stable if and only if all eigenvalues of the matrix* $(E+B)e^{A\theta}$ *satisfy the inequality* $|\lambda| < 1$.

□

An interesting case is when the matrices A and B commute, i.e. $AB = BA$. In this case the matrix exponent commutes with the matrix B and the matriciant $X(t,t_0)$ can be represented as

$$X(t,t_0) = e^{A(t-t_0)}(E+B)^{i(t_0,t)}. \tag{3.5}$$

Suppose that the times of the impulsive effect satisfy the condition

$$0 < \theta_1 \leq \tau_{i+1} - \tau_i \leq \theta_2. \tag{3.6}$$

Let

$$\alpha = \max_j \lambda_j(A), \qquad \beta = \max_j(1 + \Re\lambda_1(B)).$$

Then, for $t \geq t_0$, the matrix $e^{A(t-t_0)}$ satisfies the estimate

$$\|e^{A(t-t_0)}\| \leq K_1 e^{(\alpha+\epsilon)(t-t_0)}, \qquad \epsilon > 0,\ K_1 \geq 1, \tag{3.7}$$

and for the matrix $(E+B)e^{i(t_0,t)}$, we have

$$\|(E+B)^{i(t_0,t)}\| \leq K_2(\beta+\epsilon)^{i(t_0,t)}, \qquad K_2 = K_2(\epsilon) \geq 1. \tag{3.8}$$

By using these inequalities, we obtain

$$\begin{aligned} \|X(t,t_0)\| &\leq K_1K_2e^{(\alpha+\epsilon)(t-t_0)}(\beta+\epsilon)^{i(t_0,t)} \leq \\ &\leq K_1K_2e^{(\alpha+\epsilon)\theta_0}(e^{\alpha+\epsilon)\theta_0}(\beta+\epsilon))^{i(t_0,t)}, \end{aligned} \tag{3.9}$$

where

$$\theta_0 = \begin{cases} \theta_1, & \text{if} \quad \alpha \geq 0, \\ \theta_2, & \text{if} \quad \alpha < 0. \end{cases} \tag{3.10}$$

It follows from (3.9) that if $e^{\alpha\theta_0\beta} < 1$, then $\|X(t,t_0)\| \to 0$ for $t \to \infty$ because the number $\epsilon > 0$ can be chosen to be arbitrarily small, namely to be such that

$$e^{(\alpha+\epsilon)\theta_0}(\beta+\epsilon) < 1.$$

Thus we have proved the following theorem.

Theorem 32 *Let in system (3.1) the matrices A and B commute and the times τ_i satisfy (3.6). If*

$$\max_j \lambda_j(A) + \frac{1}{\theta_0} \ln \max_j |1 + \lambda_j(B)| < 0, \tag{3.11}$$

where θ_0 is given by relation (3.10), then the solutions of system (3.1) are asymptotically stable.

□

It should be noted that if the matrices A and B commute, the matrix $E + B$ is nonsingular, and the times (3.4), then without loss of generality we can assume that $\tau_i = t_0 + i\theta$, $i = 1, 2, \dots$. In this case the matriciant $X(t, t_0)$ can be written as

$$X(t, t_0) = e^{\left(A + \frac{1}{\theta}\ln(E+B)\right)\left[\frac{t-t_0}{\theta}\right]\theta} e^{A\theta\left\{\frac{t-t_0}{\theta}\right\}}, \tag{3.12}$$

where $\left[\dfrac{t-t_0}{\theta}\right]$ and $\left\{\dfrac{t-t_0}{\theta}\right\}$ are the integer and fractional parts of the number $\dfrac{t-t_0}{\theta}$, and so necessary and sufficient conditions for asymptotical stability of the solutions of system (3.1) can be easily expressed in terms of the eigenvalues of the matrix

$$A + \frac{1}{\theta}\ln(E + B). \tag{3.13}$$

Such a transition from a study of the eigenvalues of the matrix $e^{A\theta}(E + B)$ to a study of the eigenvalues of matrix (3.13) can be used if it is easier to calculate the matrix $\ln(E + B)$ than the matrix $e^{A\theta}$.

Example. Let the linear oscillator

$$\ddot{x} + 2\lambda x + \omega^2 x = 0, \qquad \omega^2 > \lambda^2,$$

be subject to an impulsive effect at the moment when the phase point in the phase plane $(x, \dot{x})$ intersects the line $x = 0$ with nonnegative velocity. Suppose that as the result of an impulsive effect, the impulse of the system is increased by a quantity proportional to the velocity at the moment when the impulse occurs.

Thus the motion of the oscillator is given by the equations

$$\ddot{x} + 2\lambda\dot{x} + \omega^2 x = 0, \qquad x \neq 0,$$
$$\Delta\dot{x}|_{x=0} = \begin{cases} bx, & \text{if} \quad \dot{x} \geq 0, \\ 0, & \text{if} \quad \dot{x} < 0 \end{cases}$$

or

$$\begin{aligned} \dot{x} &= \omega y, \qquad x \neq 0, \\ \dot{y} &= -\omega x - 2\lambda y, \\ \Delta y|_{x=0} &= \begin{cases} by, & \text{if} \quad y \geq 0, \\ 0, & \text{if} \quad y < 0. \end{cases} \end{aligned} \tag{3.14}$$

Let $(x(t), y(t))$ be a solution of system (3.1) that passes through a point (x_0, y_0) at $t = 0$. Denote by τ_i the solutions of the equation $x(t) = 0$ such that $y(\tau_i) > 0$, i.e. τ_i are the points of discontinuity of the function $y(t)$.

Clearly, the solution $(x(t), y(t))$ of system (3.1) is also a solution of the system

$$\begin{aligned} \dot{x} &= \omega y, \qquad x \neq 0 \\ \dot{y} &= -\omega x - 2\lambda y, \\ \Delta y|_{t=\tau_i} &= by, \end{aligned}$$

which we will write in the following form

$$\begin{aligned} \dot{x} &= \omega y, \qquad t \neq \tau_i, \\ \Delta x|_{t=\tau_i} &= bx, \\ \dot{y} &= \omega x - 2\lambda y, \\ \Delta y|_{t=\tau_i} &= by. \end{aligned}$$

It can be written in this form because $x(\tau_i) = 0$. If we write system (3.15) in matrix notations, then the matrices A and B of this system,

$$A = \begin{pmatrix} 0 & \omega \\ -\omega & -2\lambda \end{pmatrix}, \qquad B = \begin{pmatrix} b & 0 \\ 0 & b \end{pmatrix},$$

will commute.

It is easy to see that, for $b > -1$, τ_i are equally spaced:

$$\tau_{i+1} - \tau_i = \frac{2\pi}{\sqrt{\omega^2 - \lambda^2}}. \tag{3.15}$$

So the stability of system (3.15) is completely determined by matrix (3.13), which in this case is

$$\begin{pmatrix} \frac{\sqrt{\omega^2-\lambda^2}}{2\pi}\ln(1+b) & \omega \\ -\omega & \frac{-2\lambda+\sqrt{\omega^2-\lambda^2}}{2\pi}\ln(1+b) \end{pmatrix}.$$

Its eigenvalues are

$$\rho_{1,2} = -\lambda + \frac{\sqrt{\omega^2-\lambda^2}}{2\pi}\ln(1+b) \pm i\sqrt{\omega^2-\lambda^2}. \tag{3.16}$$

Consequently, for $b > -1$ the solutions of system (3.15) will be stable if and only if

$$-\lambda + \frac{\sqrt{\omega^2-\lambda^2}}{2\pi}\ln(1+b) = 0,$$

they will be asymptotically stable if and only if

$$-\lambda + \frac{\sqrt{\omega^2-\lambda^2}}{2\pi}\ln(1+b) < 0,$$

and unstable if

$$-\lambda + \frac{\sqrt{\omega^2-\lambda^2}}{2\pi}\ln(1+b) > 0.$$

In the case when $b = -1$, any solution of system (3.14) is taken by the first impulse into the origin. If $b < -1$, then

$$\tau_{i+1} - \tau_i = \frac{\pi}{\sqrt{\omega^2-\lambda^2}}. \tag{3.17}$$

and matrix (3.13) has the form

$$\begin{pmatrix} \frac{\sqrt{\omega^2-\lambda^2}}{\pi}(\ln|1+b| + i(2k+1)\pi) & \omega \\ -\omega & -2\lambda + \frac{\sqrt{\omega^2-\lambda^2}}{\pi}(\ln|1+b| + (2k+1)i\pi \end{pmatrix}.$$

Its eigenvalues are

$$\rho_{1,2} = -\lambda + \frac{\sqrt{\omega^2-\lambda^2}}{\pi}\ln|1+b| + 2ik\sqrt{\omega^2-\lambda^2},$$
$$k = 0, \qquad \pm1, \pm2, \ldots,$$

and so the stability of the solutions of system (3.15) is determined by the sign of the expression

$$-\lambda + \frac{\sqrt{\omega^2 - \lambda^2}}{\pi} \ln|1+b|.$$

If it equals to zero, then solutions of (3.15) are stable, if this expression is less than zero, then they are asymptotically stable, and if it is greater than zero, then the solutions of (3.15) are unstable.

It should be noted that when the solution $(x(t), y(t))$ is stable, then it is periodic.

Now we will prove two more theorems on stability of solutions of (3.1). Without loss of generality, we can assume that the matrix A is written in the real canonical form, i.e. $A = \mathrm{diag}\{D_1, \ldots, D_m\}$, where

$$D_j = \lambda_j E + \epsilon_1 Z, \qquad Z = \begin{Bmatrix} 0 & 1 & 0 & \ldots & 0 & 0 \\ 0 & 0 & 1 & \ldots & 0 & 0 \\ \ldots & \ldots & \ldots & \ldots & \ldots & \ldots \\ 0 & 0 & 0 & \ldots & 0 & 1 \\ 0 & 0 & 0 & \ldots & 0 & 0 \end{Bmatrix}$$

if λ_j is a real eigenvalue of A, and

$$D_j = J_{2p_j}(\lambda_j, \bar{\lambda}_j) = \mathrm{diag}\{S_2, \ldots, S_2\} + \epsilon_1 Z_{2p_j},$$

where

$$Z = \begin{Bmatrix} 0 & E_2 & 0 & \ldots & 0 & 0 \\ 0 & 0 & E_2 & \ldots & 0 & 0 \\ 0 & 0 & 0 & \ldots & 0 & E_2 \\ 0 & 0 & 0 & \ldots & 0 & 0 \end{Bmatrix}, \quad S_2 = \begin{pmatrix} \alpha & \beta \\ -\beta & \alpha \end{pmatrix}, \quad E_2 = \begin{pmatrix} 1 & 0 \\ 0 & 1 \end{pmatrix},$$

$\lambda_j, \bar{\lambda}_j = \alpha \pm i\beta$ if $\lambda_j, \bar{\lambda}_j$ is a pair of complex conjugate eigenvalues of the matrix A.

Taking this real canonical form of the matrix A, we assume that the parameter ϵ_1 that characterizes this form is sufficiently small.

We will give a sufficiently general condition which would imply asymptotic stability of solutions of system (3.1). We will need the following statement [25].

Lemma 11 *Let A be a matrix in the real canonical form, $\lambda_1, \ldots, \lambda_n$ be its eigenvalues, $\epsilon_1 > 0$, $\gamma = \max_j \Re\lambda_j$. Then, for all $t \geq 0$,*

$$\|e^{At}x\| \leq e^{(\epsilon_1+\gamma)t}\|x\|, \tag{3.18}$$

where

$$\|x\|^2 = \langle x, x\rangle = \sum_{j=1}^{n} x_j^2.$$

□

Theorem 33 *Let in system (3.1) the matrix be in the real canonical form and the times τ_i are such that the limit*

$$\lim_{T\to\infty} \frac{i(t, t+T)}{T} = p, \tag{3.19}$$

where $i(t, t+T)$ is the number of the points of the sequence $\{\tau_i\}$ that belong to the interval $[t, t+T]$, exists and is uniform with respect to $t \geq t_0$. Set

$$\gamma = \max_j \Re\, \lambda_j(A), \qquad \alpha^2 = \max_j \lambda_j((E+B)^T(E+B)).$$

If the inequality

$$\gamma + p \ln \alpha < 0 \tag{3.20}$$

holds, then the solutions of system (3.1) are asymptotically stable.

Proof. For any two solutions of equations (3.1), $x(t)$, $x(t_0) = x_0$, and $y(t)$, $y(t_0) = y_0$, we have

$$x(t) - y(t) = e^{A(t-\tau_i)} \prod_{t_0<\tau_i<t} (E+B)e^{A(\tau_j-\tau_{j-1})}(x_0 - y_0).$$

From the conditions of the theorem and Lemma 11, we see that

$$\|x(t) - y(t)\| \leq e^{(\epsilon_1+\gamma)(t-t_0)}\alpha^{i(t_0,t)}\|x_0 - y_0\|, \qquad t \geq t_0.$$

Since we assumed that limit (3.19) exists, it follows that, for any $\epsilon_2 > 0$ there exists such $K = K(\epsilon_2)$ that

$$\alpha^{i(t_0,t)} \leq Ke^{(\epsilon_2+p\ln\alpha)(t-t_0)}$$

and, consequently,

$$\|x(t) - y(t)\| \leq Ke^{(\epsilon_1+\epsilon_2+\gamma+p\ln\alpha)(t-t_0)}\|x_0 - y_0\|, \qquad t \geq t_0.$$

This finishes the proof since, by using (3.20), we have that $\|x(t) - y(t)\| \to 0$ for $t \to \infty$ because the numbers ϵ_1 and ϵ_2 can be chosen to be sufficiently small.

□

Theorem 34 *Let the matrices A and B commute, the matrix $E + B$ be nonsingular, and the sequence $\{\tau_i\}$ be such that limit (3.19) exists and is uniform with respect to $t \geq t_0$. Then*

1) if the real parts of all the eigenvalues of the matrix $\Lambda = A + p\ln(E+B)$ are negative, then solutions of equations (3.1) are asymptotically stable;

2) if the real part of at least one eigenvalue of the matrix Λ is positive, then solutions of system (3.1) are unstable.

Proof. Because the matrices A and B commute, for any two solutions of equations (3.1), $x(t)$ and $y(t)$, we have

$$x(t) - y(t) = e^{A(t-t_0)}(E+B)^{i(t_0,t)}(x_0 - y_0),$$

or

$$x(t) - y(t) = e^{(A+p\ \ln(E+B)(t-t_0)}(E+B)^{i(t_0,t)-p(t-t_0)}\|x_0 - y_0\|. \tag{3.21}$$

Since limit (3.19) exists, there is a constant $K_1 > 0$ such that

$$\|(E+B)^{i(t_0,t)-p(t-t_0)}\| \leq K_1 e^{\epsilon(t-t_0)} \tag{3.22}$$

for all $t \geq t_0$, $\epsilon > 0$.

If the real parts of the eigenvalues of the matrix Λ are negative, there exist such numbers $K > 0$ and $\gamma > 0$ that

$$\|e^{\Lambda t}\| \leq K e^{-\gamma t}, \qquad t \geq 0.$$

By using this inequality, we get from (3.21) that

$$\|x(t) - y(t)\| \leq K K_1 e^{-(\gamma-\epsilon)(t-t_0)}\|x_0 - y_0\| \qquad t \geq t_0.$$

This estimate for the difference of the two solutions of (3.1) proves the first part of the theorem. To prove the second statement, we will need the following lemma.

Lemma 12 *If there is an eigenvalue of the matrix Λ, λ_0, the real part of which is positive, then for any $x_0 \in \mathbf{R}^n$ and any its neighborhood, there exists a vector y_0 from this neighborhood such that*

$$\|e^{\Lambda t}(x_0 - y_0)\| \geq K e^{\Re \lambda_0 t}, \qquad t \geq 0. \tag{3.23}$$

Proof of Lemma. Indeed, let T be such a matrix that $T\Lambda T^{-1} = J$, where J is the Jordan form of the matrix Λ. Then

$$e^{\Lambda t}(x_0 - y_0) = T^{-1} e^{Jt} T(x_0 - y_0). \tag{3.24}$$

By using the block structure of the matrix e^{Jt}, we can find such its diagonal element $l_{rr} = e^{\lambda_0 t}$ that all the elements l_{rj} of the rth column of the matrix e^{Jt} equal to zero for $r \neq j$.

Choose y_0 in such a way that the only coordinate of the vector $T(x_0 - y_0)$ different from zero be the rth coordinate. Take y_0 to be

$$y_0 = x_0 - T^{-1} z_0,$$

where z_0 is a fixed vector with the only r^{th} coordinate different from zero.

Having chosen y_0 in such a way, we have, according to (3.24),

$$e^{\Lambda t}(x_0 - y_0) = T^{-1} e^{Jt} z_0$$

or

$$e^{Jt} z_0 = T e^{\Lambda t}(x_0 - y_0).$$

Because the only r^{th} coordinate of the vector $e^{Jt} z_0$ is nonzero and equals to $e^{\lambda_0 t} z^*$,

$$\|e^{Jt} z_0\| = |e^{\lambda_0 t} z^*| = \|T e^{\Lambda T}(x_0 - y_0)\| \leq \|T\| \, \|e^{\Lambda t}(x_0 - y_0)\|.$$

Finally we have the estimate

$$\|e^{\Lambda t}(x_0 - y_0)\| \geq \frac{1}{\|T\|} e^{\Re \lambda_0 t} |z^*|,$$

which leads to inequality (3.23) if

$$\frac{|z^*|}{\|T\|} = K.$$

□

Now we will finish proving the theorem.

Because the matrices A and B commute, equality (3.21) can be written in the form

$$(E+B)^{p(t-t_0)-i(t_0,t)}(x(t)-y(t)) = e^{(A+p\ln(E+B))(t-t_0)}(x_0-y_0).$$

For a given vector x_0, we choose a vector y_0 as in Lemma 12 to get

$$\|(E+B)^{p(t-t_0)-i(t_0,t)}(x(t)-y(t))\| \geq Ke^{\Re\lambda_0(t-t_0)}, \qquad t \geq t_0, \qquad (3.25)$$

where $K > 0$, λ_0 is an eigenvalue of the matrix Λ, the real part of which is positive.

So from (3.22) and (3.25) we obtain

$$\begin{aligned} Ke^{\Re\lambda_0(t-t_0)} &\leq \|(E+B)^{p(t-t_0)-i(t_0,t)}(x(t)-y(t))\| \leq \\ &\leq K_1 e^{\epsilon(t-t_0)}\|x(t)-y(t)\| \end{aligned}$$

or

$$\|x(t)-y(t)\| \geq \frac{K}{K_1}e^{(\Re\lambda_0-\epsilon)(t-t_0)}, \qquad t \geq t_0,$$

and this ends the proof of the theorem.

□

Now consider equations (3.1) together with the system

$$\begin{aligned} \frac{dx}{dt} &= Ax + P(t)x, \qquad t \neq \tau_i, \\ \Delta x|_{t=\tau_i} &= Bx + I_i x, \end{aligned} \qquad (3.26)$$

where $P(t)$ is a continuous (piecewise continuous) matrix for $t \geq t_0$, I_i are constant matrices.

In section 2.3 we have obtained the conditions that must be satisfied by the matrices $P(t)$ and I_i so that the stability of solutions of (3.1) would imply stability of solutions of system (3.26).

Theorem 35 *If solutions of system (3.1) are exponentially stable, the matrices $P(t)$ and I_i satisfy the inequalities*

$$\|P(t)\| \leq c, \qquad \|I_i\| \leq c \qquad (3.27)$$

for sufficiently large t and i, c is a sufficiently small positive number, and the times τ_i satisfy condition (3.6), then solutions of equations (3.26) are exponentially stable.

Proof. Let $X(t,\tau)$ be a matriciant of system (3.1). By the conditions of the theorem, there are positive K and γ such that

$$\|X(t,\tau)\| \leq Ke^{-\gamma(t-\tau)}, \qquad t \geq \tau. \tag{3.28}$$

For any two solutions $x(t)$, $x(t_0) = x_0$, and $y(t)$, $y(t_0) = y_0$, of system (3.26) the following identity holds

$$\begin{aligned} x(t) - y(t) &= X(t,t_0)(x_0 - y_0) + \int_{t_0}^{t} X(t,\tau)P(\tau)(x(\tau) - \\ &\quad -y(\tau))d\tau + \sum_{t_0<\tau_i<t} X(t,\tau_i)I_i(x(\tau_i) - y(\tau_i)), \end{aligned}$$

by which, using inequality (3.28), we have

$$\begin{aligned} e^{\gamma(t-t_0)}\|x(t) - y(t)\| &\leq K\|x_0 - y_0\| + \\ &\quad + \int_{t_0}^{t} Ke^{\gamma(\tau-t_0)}\|P(\tau)\|\,\|x(\tau) - y(\tau)\|d\tau + \\ &\quad + \sum_{t_0<\tau_i<t} Ke^{\gamma(\tau_i-t_0)}\|I_i\|\,\|x(\tau_i) - y(\tau_i)\|. \end{aligned}$$

Let $T > t_0$ be such that for $t > T$ and $\tau_i > T$, inequalities (3.27) hold simultaneously. Then by using (3.27) it follows from this inequality that for $t > T$ we have

$$\begin{aligned} e^{\gamma(t-t_0)}\|x(t) - y(t)\| &\leq C + \int_{T}^{t} Kce^{\gamma(\tau-t_0)}\|x(\tau) - y(\tau)\|d\tau + \\ &\quad + \sum_{T\leq\tau_i<t} Kce^{\gamma(\tau_i-t_0)}\|x(\tau_i) - y(\tau_i)\|, \end{aligned}$$

where

$$\begin{aligned} C &= K\|x_0 - y_0\| + \int_{t_0}^{T} Ke^{\gamma(\tau-t_0)}\|P(\tau)\|\,\|x(\tau) - y(\tau)\|d\tau + \\ &\quad + \sum_{t_0<\tau_i<T} Ke^{\gamma(\tau_i-t_0)}\|I_i\|\,\|x(\tau_i) - y(\tau_i)\|. \end{aligned}$$

So,

$$\begin{aligned} e^{\gamma(t-t_0)}\|x(t) - y(t)\| &\leq C + \int_{t_0}^{t} Kce^{\gamma(\tau-t_0)}\|x(\tau) - y(\tau)\|d\tau + \\ &\quad + \sum_{t_0<\tau_i<t} Kce^{\gamma(\tau_i-t_0)}\|x(\tau_i) - y(\tau_i)\|. \end{aligned}$$

By Lemma 1 we can write

$$e^{\gamma(t-t_0)}\|x(t)-y(t)\| \leq C(1+Kc)^{i(t_0,t)}e^{Kc(t-t_0)},$$

and applying condition (3.6) we get

$$\|x(t)-y(t)\| \leq Ce^{-\left(\gamma-Kc-\frac{1}{\theta_1}\ln(1+Kc)\right)(t-t_0)}. \tag{3.29}$$

If the constant c is so small that

$$\gamma - Kc - \frac{1}{\theta_1}\ln(1+Kc) > 0,$$

it will follow from (3.29) that $\|x(t)-y(t)\| \to 0$ for $t \to \infty$.

□

By using this theorem and the approach used to prove the previous theorem, it is not difficult to see that the following statement hold.

Theorem 36 *Let in equations (3.26) the matrix A be taken in its real canonical form, the times τ_i satisfy relation (3.19), and*

$$\gamma = \max_j Re\, \lambda_i(A), \qquad \alpha^2 = \max_j \lambda_j((E+B)^T(E+B)).$$

If $\gamma + p\ln\alpha < 0$ and the matrices $P(t)$ and I_i satisfy inequalities (3.27), then solutions of system (3.26) are asymptotically stable.

□

Finally we remark that the statement of the theorem is still true if, instead of condition (3.19), we require that the times τ_i satisfy inequalities (3.6). In this case, instead of the inequality $\gamma + p\ln\alpha < 0$, the following inequality must hold:

$$\gamma + \frac{1}{\theta}\ln\alpha < 0,$$

where

$$\theta = \begin{cases} \theta_1, & \text{if } \quad \alpha \geq 1, \\ \theta_2, & \text{if } \ 0 < \alpha < 1. \end{cases}$$

Also note that conditions (3.27) always hold if $p(t) \to 0$ for $t \to \infty$ and $I_i \to 0$ for $i \to \infty$.

3.2 Stability criterion based on first order approximation

Let us study stability of the nonlinear system of impulsive differential equations

$$\frac{dx}{dt} = f(t,x), \qquad t \neq \tau_i,$$
$$\Delta x|_{t=\tau_i} = I_i^0(x). \tag{3.30}$$

The problem of stability of a solution $x = \varphi(t)$ of system (3.30) can be reduced to a study of stability of the zero solution of some other system. To get this system, let us make a change of variables in (3.30) by setting $x = y + \varphi(t)$. As the result, system (3.30) will be

$$\frac{dx}{dt} = F(t,y), \qquad t \neq \tau_i,$$
$$\Delta y|_{t=\tau_i} = I_i^1(y), \tag{3.31}$$

where

$$F(t,y) = f(t, y+\varphi(t)) - f(t,\varphi(t)), \qquad F(t,0) \equiv 0,$$
$$I_i^{(1)}(y) = I_i^0(y+\varphi(\tau_i)) - I_i^0(\varphi(\tau_i)), \qquad I_i^{(1)}(0) = 0,$$

and the solution $x = \varphi(t)$ will become $y = 0$.

Thus without loss of generality we will suppose that system (3.30) has the zero solution and we will study stability of this solution, i.e. we assume that the functions $f(t,x)$ and $I_i(x)$ satisfy

$$f(t,0) = 0, \qquad I_i(0) = 0 \tag{3.32}$$

for all $t \geq t_0$, $i = 1, 2, \ldots$.

Write the functions $f(t,x)$ and $I_i^0(x)$ as

$$f(t,x) = A(t)x + g(t,x), \qquad I_i^0(x) = B_i x + I_i(x),$$

where $A(t)$ and B_i are matrices, and $g(t,x)$ and $I_i(x)$ satisfy the conditions $g(t,0) = 0$ and $I_i(0) = 0$, and rewrite system (3.30) in the form

$$\frac{dx}{dt} = A(t)x + g(t,x), \qquad t \neq \tau_i,$$
$$\Delta x|_{t=\tau_i} = B_i x + I_i(x). \tag{3.33}$$

Together with equations (3.33), we will be considering the linear system

$$\frac{dx}{dt} = A(t)x, \qquad t \neq \tau_i, \Delta x|_{t=\tau_i} = B_i x \tag{3.34}$$

and call it the system of the first order approximation of system (3.33).

We also assume that the times of impulsive effects τ_i are indexed by the set of natural numbers taken with the natural order and

$$\tau_{i+1} - \tau_i \geq \theta \tag{3.35}$$

for some $\theta > 0$.

Theorem 37 *Let the matriciant $X(t, \tau)$ of system (3.34) admit the estimate*

$$\|X(t,\tau)\| \leq Ke^{-\gamma(t-\tau)}, \qquad K \geq 1, \qquad \gamma > 0 \tag{3.36}$$

for all t and τ, $t_0 \leq \tau \leq t$, and the functions $g(t,x)$ and $I_i(x)$ satisfy the inequalities

$$\|g(t,x)\| \leq a\|x\|, \qquad \|I_i(x)\| \leq a\|x\| \tag{3.37}$$

for all $t \geq t_0$, $i = 1, 2, \ldots$, $\|x\| \leq h$, $h > 0$. Then, for sufficiently small values of a, the zero solution of equations (3.33) is asymptotically stable.

Proof. Every solution of (3.33) can be written as

$$\begin{aligned} x(t,x_0) &= X(t,t_0)x_0 + \int_{t_0}^{t} X(t,\tau)g(\tau, x(\tau,x_0))d\tau + \\ &+ \sum_{t_0<\tau_i<t} X(t,\tau_i)I_i(x(\tau_i,x_0)). \end{aligned} \tag{3.38}$$

By using inequalities (3.36) and (3.37), we get

$$\begin{aligned} \|x(t,x_0)\| &\leq Ke^{-\gamma(t-t_0)}\|x_0\| + \int_{t_0}^{t} Ke^{-\gamma(t-\tau)}a\|x(\tau,x_0)\|d\tau + \\ &+ \sum_{t_0<\tau_i<t} Ke^{-\gamma(t-\tau_i)}a\|x(\tau_i,x_0)\| \end{aligned}$$

or

$$\begin{aligned} e^{\gamma(t-t_0)}\|x(t,x_0)\| &\leq K\|x_0\| + \int_{t_0}^{t} Kae^{\gamma(\tau-t_0)}\|x(\tau,x_0)\|d\tau + \\ &+ \sum_{t_0<\tau_i<t} Kae^{\gamma(\tau_i-t_0)}\|x(\tau_i,x_0)\|. \end{aligned}$$

By applying Lemma 1, we obtain

$$e^{\gamma(t-t_0)}\|x(t,x_0)\| \leq K\|x_0\|(1+Ka)^{i(t_0,t)}e^{Ka(t-t_0)}.$$

Because the times τ_i satisfy inequality (3.35), it follows from (3.38) that

$$\|x(t,x_0)\| \leq Ke^{-(\gamma-Ka-\frac{1}{\theta}\ln(1+Ka))(t-t_0)}\|x_0\|. \tag{3.39}$$

So, if a is so small that

$$\gamma - Ka - \frac{1}{\theta}\ln(1+Ka) > 0,$$

any solution $x(t,x_0)$, $\|x_0\| < \frac{h}{K}$, of equations (3.33) is defined for all $t \geq t_0$ and $\lim_{t\to\infty}\|x(t,x_0)\| = 0$, i.e. the zero solution of equations (3.33) is asymptotically stable.

□

Theorem 38 *Let the largest eigenvalue of the matrix* $\hat{A}(t) = \frac{1}{2}(A(t) + A^T(t))$ *satisfy* $\Lambda(t) \leq \gamma$ *for all* $t \geq t_0$*, the largest eigenvalue of the matrix* $(E+B_i^T)(E+B_i)$ *be such that* $\Lambda_i \leq \alpha^2$ *for all* $i = 1, 2, \ldots$*, the limit*

$$\lim_{T\to\infty} = \frac{i(t,t+T)}{T} = p \tag{3.40}$$

exist and be uniform with respect to $t \geq t_0$*. Then, if*

$$\gamma + p\ln\alpha < 0, \tag{3.41}$$

the trivial solution of system (3.33) is asymptotically stable if the functions $g(t,x)$ *and* $I_i(x)$ *satisfy inequalities (3.37) with a sufficiently small constant* a.

Proof. By using the analog of Wazhevsky inequality, it is easy to get the estimate

$$\|x(t)\| \leq \alpha^{i(t_0,t)}e^{\gamma(t-t_0)}\|x_0\| \tag{3.42}$$

for any solution of equations (3.34). Because limit (3.40) exists, using inequalities (3.41) and (3.42) it is possible to find such $K \geq 1$ and $\mu > 0$

($0 < \mu < |\gamma + p \ln \alpha|$) that, for all $t \geq t_0$ and $\tau \geq t_0$, $t \geq \tau$, the matriciant $X(t, \tau)$ of equations (3.34) satisfies the estimate

$$\|X(t,\tau)y\| \leq Ke^{-\mu(t-\tau)}\|y\|, \qquad y \in \mathbf{R}^n. \tag{3.43}$$

Write the solution $x(t, x_0)$ of (3.33 in a sufficiently small neighborhood of the point x_0 in the form (3.38). Similarly as in the proof of the previous theorem, we can get

$$e^{\mu(t-t_0)}\|x(t,x_0)\| \leq K\|x_0\|(1+K\alpha)^{i(t_0,t)}e^{K\alpha(t-t_0)},$$

i.e.

$$\|x(t,x_0)\| \leq K_1 e^{-(\mu - p\ln(1+Ka) - Ka + \epsilon)(t-t_0)}\|x_0\|$$

for any $\epsilon > 0$. Here $K_1 = K_1(\epsilon)$.Thus, if a is so small that $\mu - Ka - p\ln(1 + Ka) > 0$, then $\|x(t, x_0)\| \to 0$ as $t \to \infty$.

□

Suppose that in system (3.33) the matrices A and B are constant, i.e. let us consider the equations

$$\frac{dx}{dt} = Ax + g(t,x), \qquad t \neq \tau_i,$$
$$\Delta X|_{t=\tau_i} = Bx + I_i(x). \tag{3.44}$$

We can assume that the matrix A is in its real canonical form.

Theorem 39 *Let $\gamma = \max_j \Re\lambda_j(A)$, $\alpha^2 = \max_j \lambda((E + B^T)(E + B))$, and the times τ_i satisfy inequality (3.40). If $\gamma + p\ln\alpha < 0$, the trivial solution of system (3.44) is asymptotically stable as long as the functions $g(t, x)$ and $I_i(x)$ satisfy inequalities (3.37) with a sufficiently small constant a.*

Proof. It follows from the proof of Theorem 38 that the solutions of the first order approximation of equations (3.44), i.e. the solutions of the system with constant matrices

$$\frac{dx}{dt} = Ax, \qquad t \neq \tau_i, \qquad \Delta X|_{t=\tau_i} = Bx, \tag{3.45}$$

are exponentially stable and the matriciant of this system satisfies estimate (3.43). Now by using this and the idea of the proof of the previous theorem, it is easy to complete the proof.

□

The last two theorems still remain valid if, instead of condition (3.40), we assume that the times τ_i satisfy the inequalities

$$0 < \theta_1 \leq \tau_{i+1} - \tau_i \leq \theta_2. \tag{3.46}$$

In this case, inequality (3.41) should be replaced by

$$\gamma + \frac{1}{\theta} \ln \alpha < 0, \tag{3.47}$$

where

$$\theta = \begin{cases} \theta_2, & \text{if} \quad 0 < \alpha < 1, \\ \theta_1, & \text{if} \quad \alpha \geq 1. \end{cases}$$

Now we will study stability of the trivial solution of equations (3.33) with the assumptions that the system of the first order approximation (3.34) is regular, the functions $g(t,x)$ and $I_i(x)$ satisfy

$$\|g(t,x)\| \leq a(t)\|x\|^m, \qquad \|I_i(x)\| \leq \beta_i \|x\|^m, \qquad m > 1, \tag{3.48}$$

where $a(t)$ is a continuous positive function with the zero characteristic exponent, $\beta_i \geq 0$ with $\beta_i e^{-\epsilon \tau_i} \to 0$ for $i \to \infty$ and an arbitrary $\epsilon > 0$. As for the impulse times τ_i, we assume that the finite limit (3.40) exists.

Before proceeding, we prove an auxiliary result which is an analogue of the Bikhari lemma for piecewise continuous functions.

Lemma 13 *Let a nonnegative piecewise continuous function $u(t)$ with the first kind discontinuities at $t = \tau_i$ satisfy, for $t \geq t_0$, the inequality*

$$u(t) \leq C + \int_{t_0}^{t} v(\tau)\Phi(u(\tau))d\tau + \sum_{t_0 < \tau_i < t} \beta_i u(\tau_i), \tag{3.49}$$

where $C \geq 0$, $\beta_i \geq 0$, $v(t)$ is a positive continuous function, $\Phi(u)$ is a positive continuous nondecreasing function for $0 < u < \bar{u}$ ($\bar{u} < \infty$). Then the function $u(t)$ satisfies the estimate

$$u(t) \leq \Psi_i^{-1}\left(\int_{\tau_i}^{t} v(\tau)d\tau\right), \qquad \tau_i < t \leq \tau_{i+1}, \tag{3.50}$$

if

$$\int_{\tau_i}^{t} v(\tau)d\tau < \Psi_i(\bar{u} - 0),$$

where

$$\Psi_i(u) = \int_{c_i}^{u} \frac{du}{\Phi(u)}, \qquad c_i = (1+\beta_i)\Psi_{i-1}^{-1}\left(\int_{\tau_{i-1}}^{\tau_i} v(\tau)d\tau\right),$$
$$\Psi_0(u) = \int_{c}^{u} \frac{du}{\Phi(u)}, \qquad i = 1, 2, \ldots, \tau_0 = t_0.$$

The proof is carried out by induction on every interval $\tau_i < t \leq \tau_{i+1}$ by using continuity of the function $u(t)$ and applying the Bikhari inequality [38].

□

We give two corollaries of this lemma.

1. If $\Phi(u) = u$, we get an analogue of Gronwell-Bellman lemma for piecewise continuous functions. Indeed, in this case

$$\Psi_i(u) = \ln\frac{u}{c_i}, \qquad \Psi_i^{-1}(u) = c_i e^u,$$
$$c_i = (1+\beta_i)c_{i-1}e^{\int_{\tau_{i-1}}^{\tau_i} v(\tau)d\tau},$$

and so it follows from (3.50) that

$$u(t) \leq C \prod_{t_0<\tau_i<t} (1+\beta_i)e^{\int_{t_0}^{t} v(\tau)d\tau}.$$

2. If $\Phi(u) = u^m$, $m > 1$, then for $u(t)$ we have the estimate

$$u(t) \leq \left\{ C^{1-m} \prod_{t_0<\tau_i<t} (1+\beta_i)^{1-m} - (m-1)\int_{t_0}^{t} v(\tau)d\tau \right\}^{\frac{1}{1-m}} \tag{3.51}$$

if

$$C^{1-m} \prod_{t_0<t_i<t} (1+\beta_i)^{1-m} - (m-1)\int_{t_0}^{t} v(\tau)d\tau > 0.$$

Indeed, if $\Phi(u) = u^m$, we have

$$\begin{aligned} \Psi_i(u) &= \frac{1}{1-m}(u^{1-m} - c_i^{1-m}), \\ \Psi_i^{-1}(u) &= (c_i^{1-m} + (1-m)u)^{\frac{1}{1-m}}, \\ c_i &= (1+\beta_i)\left[c_{i-1}^{1-m} + (1-m)\int_{\tau_{i-1}}^{\tau_i} v(\tau)d\tau\right]^{\frac{1}{1-m}}, \end{aligned}$$

and so inequality (3.51) follows.

Theorem 40 *Let the linear part of equations (3.33), i.e. system (3.34), be regular, functions $g(t,x)$ and $I_i(x)$ satisfy inequalities (3.48), and the finite limit (3.40) exists. If all characteristic exponents of solutions of linear system (3.34) are negative, then the trivial solution of system (3.33) is exponentially Lyapunov stable.*

Proof. Let us make in (3.33) the change of variables

$$x = ye^{-\mu(t-t_0)}, \tag{3.52}$$

where $0 < \mu < -\lambda_k$, and λ_k are characteristic exponents of solutions of equations (3.34). As a result of this change of variables, we get the impulsive differential system

$$\frac{dy}{dt} = C(t)y + G(t,y), \qquad t \neq \tau_i, \qquad \Delta y|_{t=\tau_i} = B_i y + J_i(y), \tag{3.53}$$

where

$$C(t) = A(t) + \mu E, \qquad G(t,y) = e^{\mu(t-t_0)} g(t, ye^{-\mu(t-t_0)}),$$
$$J_i(y) = e^{\mu(\tau_i - t_0)} I_i(ye^{-\mu(\tau_i - t_0)}),$$

and

$$\begin{aligned} \|G(t,y)\| &\leq c_1 e^{(\epsilon-(m-1)\mu)(t-t_0)} \|y\|^m, \qquad \epsilon > 0,\ c_1 > 0, \\ \|J_i(y)\| &\leq c_2 e^{(\epsilon-(m-1)\mu)(\tau_i-t_0)} \|y\|^m, \qquad c_2 > 0. \end{aligned}$$

It is clear that the system

$$\frac{dy}{dt} = C(t), \qquad t \neq \tau_i, \qquad \Delta y|_{t=\tau_i} = B_i y \tag{3.54}$$

is regular and its characteristic exponents are negative. By using the Perron theorem for impulsive differential equations, we can assert that there exist such numbers $C_3 \geq 1$ and $C_4 > 0$ that

$$\|Y(t)\| < C_3, \qquad \|Y(t)Y^{-1}(\sigma)\| < C_4 e^{\epsilon(\sigma-t_0)}, \qquad t_0 \leq \sigma \leq t,$$

where $Y(t)$ is the matriciant of system (3.54).

Let $y(t)$, $y(t_0) = y_0$, be an arbitrary solution of system (3.53), for which $\|y_0\|$ is sufficiently small, and $[t_0, t_0 + T]$ be such an interval that $\|y(t)\| < 1$

for $t \in [t_0, t_0 + T]$ (we will see later that $T = \infty$). The solution $y(t)$ can be written as

$$\begin{aligned} y(t) &= Y(t)y_0 + \int_{t_0}^{t} Y(t)Y^{-1}(\sigma)g(\sigma, y(\sigma))d\sigma + \\ &+ \sum_{t_0<\tau_i<t} Y(t)Y^{-1}(\tau_i)J_i(y(\tau_i)), \end{aligned} \tag{3.55}$$

which, by using properties of the matrix $Y(t)$ and the function $G(t, y)$, implies, for $t_0 < t < t_0 + T$, that

$$\begin{aligned} \|y(t)\| &\leq C_3\|y_0\| + \int_{t_0}^{t} c_1C_4e^{(2\epsilon-(m-1)\mu)(\sigma-\tau_0)}\|y(\sigma)\|^m d\sigma + \\ &+ \sum_{t_0<\tau_i<t} c_2C_4e^{(2\epsilon-(m-1)\mu)(\tau_i-t_0)}\|y(\tau_i)\|. \end{aligned}$$

From this and inequality (3.51) it follows that

$$\begin{aligned} y(t) &\leq C_3\|y_0\| \left\{ \prod_{t_0<\tau_i<t} (1 + c_2C_4e^{(2\epsilon-(m-1)\mu)(\tau_i-t_0)})^{1-m} - \right. \\ &\left. - (C_3\|y_0\|)^{m-1}(m-1)\int_{t_0}^{t} c_1C_4e^{(2\epsilon-(m-1)\mu)(\sigma-t_0)}d\sigma \right\}^{\frac{1}{1-m}} \end{aligned} \tag{3.56}$$

If $\epsilon > 0$ is chosen to be sufficiently small and $\|y_0\|$ is also sufficiently small, then, by using (3.52), the statement of the theorem follows directly from (3.56).

□

Remark. This theorem extends the Lyapunov criterion to impulsive differential systems.

Now we will find a sufficient condition for the zero solution of system (3.33) to be stable without the assumption that the linear part of the system is regular.

Let us define a notion of the measure of *irregularity* of a linear impulsive differential system by analogy with that for a system of ordinary differential equations. Define it to be the number

$$\varkappa = \sum_{k=1}^{n} \lambda_k - \lim_{t\to\infty} \frac{1}{t} \left(\int_{t_0}^{t} Sp\, A(\sigma) d\sigma + \sum_{t_0<\tau_i<t} \ln |\det(E + B_i)| \right).$$

Here by λ_k we have denoted the characteristic exponents of solutions of system (3.34). A necessary and sufficient condition for system (3.34) to be regular is the condition $\varkappa = 0$.

Theorem 41 *If, in system (3.34), the functions $g(t,x)$ and $I_i(x)$ satisfy inequalities (3.48) and the inequality*

$$\max_{1\le k\le n} \lambda_k \le -\varkappa(m-1)^{-1} \le 0,$$

holds, where $\varkappa$ is the measure of irregularity of system (3.34), then the zero solution of nonlinear system (3.33) is asymptotically stable.

Proof. By setting in (3.33)

$$x = X(t)e^{-Dt}y, \tag{3.57}$$

where $X(t)$ is the fundamental matrix of the linear system (3.34),

$$D = \mathrm{diag}(\lambda_1 + \mu, \dots, \lambda_n + \mu), \qquad \varkappa(m-1)^{-1} < \mu < -\max_{1\le k\le n} \lambda_k,$$

we get the following impulsive system

$$\frac{dy}{dt} = Dy + G(t,y), \qquad t \ne \tau_i, \qquad \Delta y|_{t=\tau_i} = J_i(y), \tag{3.58}$$

where

$$\begin{aligned} G(t,y) &= e^{Dt}X^{-1}(t)g(g, X(t)e^{-Dt}y), \\ J_i(y) &= e^{D\tau_i}X^{-1}(\tau_i)(E+B_i)^{-1}I_i(X(\tau_i)e^{-D\tau_i}y). \end{aligned}$$

By using Ostrogradsiĭ-Liouville formula for impulsive systems and taking into account that $\varkappa(m-1)^{-1} < \mu < -\max_{1\le k\le n} \lambda_k$, it is easy to see that

$$\begin{aligned} \chi[e^{Dt}X^{-1}(t)] \le \varkappa + \mu, \qquad \chi[X(t)e^{-Dt}] \le -\mu, \\ \|X(t)e^{-Dt}\| \le M < \infty, \qquad t \ge t_0. \end{aligned}$$

It follows from these inequalities that

$$\|G(t,y)\| \le \rho\|y\|^m, \qquad \|J_i(y)\| \le \rho\|y\|^m, \qquad \rho > 0, \qquad m > 1.$$

Thus the conditions of Theorem 40 hold for system (3.58) and it implies that the zero solution of this system is asymptotically stable. Going back to system (3.33), we see that the statement of the theorem is true.

□

This theorem allows to extend the Masser's result [38] to impulsive differential systems.

3.3 Stability in systems with variable times of impulsive effect

Consider differential equations with impulses occurring at the moment when the phase point intersects given hypersurfaces in the extended phase space. We assume that these hypersurfaces are given by equations $t = \tau_i$, $i = 1, 2, \ldots$ and $\tau_i(x) \to \infty$ as $i \to \infty$.

So in this section we will be studying the following impulsive system

$$\frac{dx}{dt} = f(t,x), \qquad t \neq \tau_i(x),$$
$$\Delta x|_{t=\tau_i(x)} = I_i(x). \tag{3.59}$$

As it was pointed out in section 1.3, there is an essential difficulty in studying system (3.59) because of possible beatings of solutions of this system at a surface $t = \tau_i(x)$. Another essential difference between system (3.59) and a system with fixed times of an impulsive effect is that solutions of (3.59), in general, do not depend on initial conditions continuously in such a way that this continuity be uniform on a finite interval. This fact, proved in 1.3, shows that it is not correct to consider Lyapunov stability of solutions of system (3.59) in the usual sense. Hence it becomes necessary to make the notion of the stability of solutions of equations (3.59) more precise.

Definition 1 *A solution $x(t)$ of system (3.59), defined for all $t \geq t_0$, is called stable (Lyapunov stable) if, for an arbitrary $\epsilon > 0$ and $\eta > 0$, there exists such a number $\delta = \delta(\epsilon, \eta)$ that, for any other solution $y(t)$ of system (3.59), $\|x(t_0) - y(t_0)\| < \delta$ implies that $\|x(t) - y(t)\| < \epsilon$ for all $t \geq t_0$ such that $|t - t_i^0| > \eta$, where t_i^0 are the times at which the solution $x(t)$ intersects the surfaces $t = \tau_i(x)$.*

Definition 2 *A solution $x(t)$ is called asymptotically stable if it is stable in the sense of Definition 1 and if there exists such a number $\delta_0 > 0$ that, for any other solution of this system that satisfies the inequality $\|x(t_0) - y(t_0)\| < \delta_0$, the following holds:*

$$\lim_{t \to \infty} \|x(t) - y(t)\| = 0.$$

In most cases, the problem of stability of a certain solution of system (3.59) can be reformulated in terms of stability of the trivial solution of some other impulsive system. But the procedure of this reduction is more complicated comparing with the similar procedure for a system of ordinary differential equations. However in some cases it might be the same.

Let us study stability of a solution $x = x^0(t)$ of system (3.59), which is defined for all $t > t_0$. Denote by τ_i^0 the time at which the solution $x^0(t)$ intersects the surface $t = \tau_i(x)$, i.e. a solution of the equation

$$t = \tau_i(x^0(t)). \tag{3.60}$$

We suppose that there are no beatings of the solution at the surfaces $t = \tau_i(x)$, and so, for every i, equation (3.60) has only one solution. Also denote by $x(t, \tau, x^0(\tau))$ the solution of the system

$$\frac{dx}{dt} = f(t, x),$$

which passes through the point $x^0(\sigma)$, $\sigma \geq t_0$, at $t = \tau$. For all $i = 1$, 2, ..., the function $x(t, \tau_i, x^0(\tau_i^0 + 0))$ coincides with the solution $x^0(t)$ for $\tau_i^0 < t \leq \tau_{i+1}^0$.

For each $i = 1, 2, \ldots$, consider the equation

$$t = \tau_{i+1}(x(t, \tau_i^0, x^0(\tau_i + 0) + y), \qquad \tau_0^0 = t_0. \tag{3.61}$$

For $y = 0$, this equation has the solution $t = \tau_{i+1}^0$. Assume that (3.61) can be solved with respect to t:

$$t = \bar{\tau}_{i+1}(y), \qquad \tau_{i+1}(0) = \tau_{i+1}^0 \tag{3.62}$$

for all y which belong to some neighborhood of the point $y = 0$.

Denote by $\tilde{x}(t, y)$ the function defined by the relation

$$\tilde{x}(t, y) = x(t, \tau_i^0, x^0(\tau_i^0 + 0))$$

for $\bar{\tau}_i(y) < t \leq \bar{\tau}_{i+1}(y)$, and by $\tilde{I}_i(y)$ – the value of the jump of this function at $t = \bar{\tau}_i(y)$. It follows from the definition of the functions $\tilde{x}(t, y)$ and $\tilde{I}_i(y)$ that

$$\tilde{x}(t, 0) = x^0(t), \qquad \tilde{I}_i(0) = I_i(x^0(\tau_i^0)). \tag{3.63}$$

Let us make in (3.59) the change of variables

$$x = y + \tilde{x}(t, y). \tag{3.64}$$

As a result, system (3.59) will become

$$\frac{dy}{dt} = Y(t, y), \qquad t \neq \bar{\tau}_i(y), \qquad \Delta y|_{t=\bar{\tau}_i(y)} = \bar{I}_i(y), \tag{3.65}$$

where

$$Y(t,y) = f(t + y + \tilde{x}(t,y)) - f(t,\tilde{x}(t,y)),$$
$$\bar{I}_i(y) = I_i(y + \tilde{x}(\tau_i(y)), y) - \tilde{I}_i(y).$$

Using equalities (3.63) we see that

$$Y(t,0) = 0, \qquad \bar{I}_i(0) = 0$$

for all $t \geq t_0$, $i = 1, 2, \ldots$, i.e. system (3.65) has the trivial solution. This solution corresponds to the solution $x^0(t)$ of system (3.59) via the change of variables (3.64).

Lyapunov stability of the trivial solution of system (3.65) corresponds to stability of the solution $x^0(t)$. Indeed, let the trivial solution of system (3.65) be Lyapunov stable, i.e. for an arbitrary $\epsilon > 0$ there exists $\delta = \delta(\epsilon) > 0$ such that, for any solution $y(t)$, the condition $\|y(t_0)\| < \delta$ implies that $\|y(t)\| < \epsilon$ for all $t \geq t_0$. It follows from the change of variables (3.64) that the solution $y = 0$ corresponds to a solution $x^0(t)$, and denote by $x(t)$ the solution that corresponds to the solution $y(t)$. Let τ_i' be the times at which the solution $x(t)$ meets the surfaces $t = \tau_i(x)$. Because $\tau_i' \to \tau_i^0$ as $y \to 0$, for an arbitrary $\eta > 0$ there is such $\delta_1 = \delta_1(\eta) > 0$ that if $\|y(t_0)\| < \delta$, then $|\tau_i' - \tau_i^0| < \eta$. Hence, given arbitrary $\epsilon > 0$ and $\eta > 0$, one can find such $\delta = \delta(\epsilon, \eta)$ that the inequality $\|x(t_0) - x^0(t_0)\| < \delta$ would imply that $\|x(t) - x^0(t)\| < \epsilon$ for all $t \geq t_0$, $|t - \tau_i^0| > \eta$.

In the sequel we will be considering stability of the zero solution of an impulsive system. Suppose that in (3.59) the functions $f(t,x)$ and $I_i(x)$ satisfy the conditions

$$f(t,0) = 0, \qquad I_i(0) = 0 \tag{3.66}$$

for all $t \geq t_0$, $i = 1, 2, \ldots$.

We give sufficient conditions for solutions of system (3.59) not to have beatings at the surfaces $t = \tau_i(x)$.

Lemma 14 *Let the function $f(t,x)$ be continuous with respect to t and x (piecewise continuous with respect to t) for $t \geq t_0$, $\|x\| \leq h$, the functions $I_i(x)$, $i = 1, 2, \ldots$, be continuous for $\|x\| \leq h$, and the functions $\tau_i(x)$ satisfy the Lipschitz conditions*

$$|\tau_i(x') - \tau(x'')| \leq N\|x' - x''\| \tag{3.67}$$

for all $i = 1, 2, \ldots$, $\|x'\| \leq h$, $\|x"\| \leq h$, *and the inequality*

$$\tau_i(x) \geq \tau_i(x + I_i(x)). \tag{3.68}$$

Then, if the number h *is sufficiently small, then the integral curve of any solution of system (3.59),* $x(t)$, $\|x(t_0)\| < h$, *which is defined for all* $t \geq t_0$ *and lying in the domain* $\|x(t_0)\| \leq h$, *intersects each surface* $t = \tau_i(x)$ *only once.*

A proof of this lemma follows from the proof of Lemma 5. In this case a contradiction to the assumption that an inequality of type (1.44) holds can be obtained by choosing a sufficiently small neighborhood of the zero solution, i.e. by choosing h sufficiently small.

□

Suppose that system (3.59) can be written in the form

$$\frac{dx}{dt} = A(t)x + g(t,x), \qquad t \neq \tau_i(x),$$
$$\Delta x|_{t=\tau_i(x)} = B_i x + \bar{I}_i(x), \tag{3.69}$$

where $A(t)$ is a bounded matrix, which is continuous for $t \geq t_0$, B_i are constant matrices, the functions $g(t,x)$ and $\bar{I}_i(x)$ are defined for $t \geq t_0$, $\|x\| \leq h$, and

$$\|g(t,x)\| \leq \alpha(t)\|x\|, \qquad \|\bar{I}_i(x)\| \leq \beta_i\|x\| \tag{3.70}$$

for all $t \geq t_0$, x, $\|x\| \leq h$, $i = 1, 2, \ldots$, $\alpha(t) > 0$, $\beta_i > 0$. We suppose that the surface $t = \tau_i(x)$ satisfy conditions (3.67), (3.68) and they are separated from each other uniformly with respect to x, $\|x\| \leq h$, i.e. for some $\theta > 0$,

$$\sup_i(\min_{\|x\|\leq h} \tau_{i+1}(x) - \max_{\|x\|\leq h} \tau_i(x)) \geq \theta. \tag{3.71}$$

Together with system (3.69), let us consider the linear impulsive system

$$\frac{dx}{dt} = A(t)x, \qquad t \neq \tau_i, \qquad \Delta x|_{t=\tau_i} = B_i x, \tag{3.72}$$

where the times τ_i are such that

$$|\tau_i - \tau_i(0)| \leq \bar{\Delta} \tag{3.73}$$

with $\bar{\Delta} = \bar{\Delta}(h)$ being a positive constant, $\bar{\Delta}(h) \to 0$ for $h \to 0$.

Theorem 42 *Let the right-hand sides of equations (3.69) be such that inequalities (3.70), (3.71), and*

$$\tau_i(x) \geq \tau_i((E + B_i)x + \bar{I}_i(x)) \tag{3.74}$$

hold for all x, $\|x\| \leq h$, $i = 1, 2, \ldots$.

If, for any τ_i which satisfy inequality (3.73), solutions of system (3.72) are stable or exponentially stable, then the trivial solution of system (3.69) will be stable or exponentially stable correspondingly if $\alpha(t)$ and β_i satisfy the conditions

$$\int_{t_0}^{\infty} \alpha(t)dt < \infty \qquad \text{and} \qquad \prod_{\tau_i > t_0} (1 + \beta_i) < \infty. \tag{3.75}$$

Proof. Let $x(t)$ be an arbitrary solution of (3.69), which passes through the point x_0, $\|x_0\| \leq h_1 < h$ at $t = t_0$.

Denote by τ_i^0 the times at which this solution meets the surfaces $t = \tau_i(x)$, i.e. the solutions of the equation

$$t = \tau_i(x(t)). \tag{3.76}$$

Because inequality (3.74) is assumed to hold, equation(3.76) has a unique solution for all i, at least for those values of t, for which the solution $x(t)$ does not leave the h-neighborhood of the trivial solution. For these values of t, $x(t)$ also satisfies the system

$$\begin{aligned} \frac{dx}{dt} &= A(t)x + g(t, x), \qquad t \neq \tau_i^0, \\ \Delta x|_{t=\tau_i^0} &= B_i x + \bar{I}_i(x). \end{aligned} \tag{3.77}$$

And so $x(t)$ can be written in the integral form

$$\begin{aligned} x(t) &= X(t, t_0)x_0 + \int_{t_0}^{t} X(t, \sigma)g((\sigma, x(\sigma))d\sigma \\ &\quad + \sum_{t_0 < \tau_i < t} X(t, \tau_i^0)\bar{I}_i(x(\tau_i^0)), \end{aligned} \tag{3.78}$$

where $X(t, t_0)$ is the matriciant of system (3.72) with τ_i replaced by τ_i^0.

If the solutions of equations (3.72) are stable, the matriciant is bounded, for example, by a constant K, and so by using (3.70), it follows from (3.78) that

$$\|x(t)\| \leq K\|x_0\| + \int_{t_0}^{t} K\alpha(\sigma)\|x(\sigma)\|d\sigma + \sum_{t_0 < \tau_i^0 < t} K\beta_i\|x(\tau_i^0)\|,$$

from which, by applying Lemma 1, we get

$$\|x(t)\| \leq K \prod_{t_0 < \tau_i^0 < t} (1 + K\beta_i) e^{\int_{t_0}^{t} K\alpha(\sigma) d\sigma} \|x_0\|. \tag{3.79}$$

If $h_1 < h$ is taken so small that

$$K \prod_{\tau_i^0 > t_0} (1 + K\beta_i) e^{\int_{t_0}^{\infty} K\alpha(\sigma) d\sigma} h_1 \leq h,$$

we are lead to the conclusion that the solution $x(t)$ is defined for all $t \geq t_0$ and it intersects each surface $t = \tau_i(x)$ only once. Since $x(t)$ was an arbitrary solution, using condition (3.75) and estimate (3.79) we see that the first statement of the theorem holds.

If the solutions of equations (3.72) are exponentially stable, then for all $t \geq \sigma$ the matriciant $X(t, \sigma)$ admits the estimate

$$\|X(t, \sigma)\| \leq K e^{-\gamma(t-\sigma)}, \qquad K \geq 1, \qquad \gamma > 0, \tag{3.80}$$

from which, by using (3.78), we have

$$\begin{aligned} \|x(t)\| &\leq K e^{-\gamma(t-t_0)} \|x_0\| + \int_{t_0}^{t} K e^{-\gamma(t-\sigma)} \alpha(\sigma) \|x(\sigma)\| d\sigma + \\ &+ \sum_{t_0 < \tau_i^0 < t} K e^{-\gamma(t-\tau_i^0)} \beta_i \|x(\tau_i^0)\|. \end{aligned}$$

Lemma 1 implies that

$$e^{\gamma(t-t_0)} \|x(t)\| \leq K \prod_{t_0 < \tau_i^0 < t} (1 + K\beta_i) e^{\int_{t_0}^{t} K\alpha(\sigma) d\sigma} \|x_0\|,$$

i.e.

$$\|x(t)\| \leq h e^{-\gamma(t-t_0)}.$$

□

Suppose that the functions $g(t, x)$ and $\bar{I}_i(x)$, instead of inequalities (3.70), satisfy the conditions

$$\|g(t, x)\| \leq a\|x\|, \qquad \|\bar{I}_i(x)\| \leq a\|x\|. \tag{3.81}$$

Theorem 43 *Let the right-hand sides of equations (3.69) be such that inequalities (3.71), (3.74), and (3.81) hold. If, for the matriciant of the system (3.72), estimate (3.80) holds, then the zero solution of system (3.72) is asymptotically stable for sufficiently small values of the constant* a.

Proof. The proof goes along the same lines as the proof of the previous theorem. Indeed, let $x(t)$ be an arbitrary solution of system (3.69), which, at $t = t_0$, passes through a point x_0 which belongs to a small neighborhood of the point $x = 0$. It is possible to find such $h_1 < h$ and such $T \leq \infty$ that this solution lies in the h-neighborhood of the zero solution for $t_0 < t \leq t+T$ and intersects each surface $t = \tau_i(x)$ only once. Let τ_i^0 be a solution of the equations $t = \tau_i(x(t))$. The function $x(t)$ is a solution of system (3.77) for $t_0 < t < t_0 + T$ and hence it can be represented in the integral form (3.78), from which by using inequalities (3.80) and (3.81) we obtain the integral estimate

$$\begin{aligned}\|x(t)\| \;\leq\; & Ke^{-\gamma(t-t_0)}\|x_0\| + Ka\int_{t_0}^{t} e^{-\gamma(t-\tau)}\|x(\sigma)\|d\sigma + \\ & + \sum_{t_0<\tau_i^0<t} Kae^{-\gamma(t-\tau_i^0)}\|x(\tau_i^0)\|.\end{aligned}$$

From this estimate and Lemma 1, it follows that

$$\|x(t)\| \leq Ke^{-(\gamma-Ka)(t-t_0)}(1+Ka)^{i(t_0,t)}\|x_0\|. \tag{3.82}$$

The surfaces $t = \tau_i(x)$ are mutually separated, i.e. they satisfy condition (3.81), and so we have from (3.82)

$$\|x(t)\| \leq Ke^{-\left(\gamma-Ka-\frac{1}{\theta}\ln(1+Ka)\right)(t-t_0)}\|x_0\|. \tag{3.83}$$

If we require that the constant a be so small that

$$\gamma - Ka - \frac{1}{\theta}\ln(1+Ka) > 0, \tag{3.84}$$

and x_0 satisfy the condition $K\|x_0\| < h$, then the solution $x(t)$ will not leave the h-neighborhood of the zero solution for all $t \geq t_0$ and consequently this solution intersects each surface $t = \tau_i(x)$, $i = 1, 2, \ldots$ only once and thus it follows from inequality (3.83) that the zero solution of (3.69) is asymptotically stable.

□

An essential condition used in the proof of the previous theorem was the condition that the matriciant of equations (3.72) satisfies estimate (3.80). In the previous section we gave sufficient conditions for inequality (3.80) to hold. By using these remarks and the reasoning used to prove the last theorem, it is easy to see that the following two theorems hold.

Theorem 44 *Let the largest eigenvalue of the matrix* $\tilde{A}(t) = \frac{1}{2}(A(t) + A^T(t))$ *satisfy the inequality* $\Lambda(t) \leq \gamma$ *for all* $t \geq t_0$, *and let the largest eigenvalues of the matrix* $(E + B_i^T)(E + B_i)$ *be such that* $\Lambda_i^2 \leq \alpha^2$ *for all* $i = 1, 2, \ldots$, *and the limit*

$$\lim_{T\to\infty} \frac{i_x(t, t+T)}{T} = p \tag{3.85}$$

exist and be uniform with respect to $t \geq t_0$ *and* x, $\|x\| \leq h$, *where* $i_x(t, t+T)$ *is the number of points of* $\tau_i(x)$ *in the interval* $[t, t+T]$. *Suppose also that for all* $t \geq t_0$, x, $\|x\| < h$, *and all* $i = 1, 2, \ldots$, *inequalities (3.67), (3.74), and (3.81) hold.*

If $\gamma + p \ln \alpha < 0$ *and the constant* a *in inequalities (3.81) is sufficiently small, then the zero solution of equations (3.69) is asymptotically stable.*

Consider the case when the matrices $A(t)$ and B_i in (3.69) are constant, i.e. we will consider the system

$$\begin{aligned} \frac{dx}{dt} &= Ax + g(t, x), \qquad t \neq \tau_i(x), \\ \Delta x_{t=\tau_i(x)} &= Bx + \bar{I}_i(x). \end{aligned} \tag{3.86}$$

Theorem 45 *Suppose that the functions* $g(t, x)$ *and* $\bar{I}_i(x)$ *satisfy inequalities (3.81) and the surfaces* $t = \tau_i(x)$ *be such that inequality (3.67) holds and*

$$\tau_i(x) \geq \tau_i((E + B)x + \bar{I}_i(x)) \tag{3.87}$$

for all $i = 1, 2, \ldots$, $\|x\| \leq h$. *Let also the finite limit (3.85) exist and* $\gamma = \max_j \Re\lambda_j(A)$, $\alpha^2 = \max_j \lambda_j((E + B)^T(E + B))$. *Then, if the inequality* $\gamma + p \ln \alpha < 0$ *holds, then the zero solution of system (3.86) is asymptotically stable as long as* a *in (3.81) is sufficiently small.*

From this theorem it follows that the trivial solution of the system of ordinary differential equations remains stable if impulsive effects are added to the system.

Theorem 46 *Let in the system*

$$\frac{dx}{dt} = Ax + g(t,x), \qquad t \neq \tau_i(x),$$
$$\Delta x|_{t=\tau_i(x)} = \bar{I}_i(x). \tag{3.88}$$

the functions $g(t,x)$, $\bar{I}_i(x)$, and $\tau_i(x)$ be the same as in equations (3.86) and

$$\tau_i(x) \geq \tau_i(x + \bar{I}_i(x)) \tag{3.89}$$

for all x, $\|x\| \leq h$, $i = 1, 2, \ldots$. If the real parts of all the eigenvalues of the matrix A are negative, then, for small values of a, the zero solution of equations (3.88) is asymptotically stable.

At the end of this section we remark that three last theorems remain true if, instead of the assumption that limit (3.85) exists for the functions $t = \tau_i(x)$, we assume that the inequalities

$$0 < \theta_1 \leq \min_{\|x\| \leq h} \tau_{i+1}(x) - \max_{\|x\| \leq h} \tau_i(x) \leq \theta_2 \tag{3.90}$$

hold for all $i = 1, 2, \ldots$. The condition that the expression $\gamma + p \ln \alpha$ be negative is replaced then by the requirement that the expression $\gamma + \dfrac{1}{\theta} \ln \alpha$ be negative, with $\theta = \theta_1$ if $\alpha \geq 1$, and $\theta = \theta_2$ if $0 < \alpha < 1$.

3.4 Direct Lyapunov method for studying stability of solutions of impulsive systems

In the foregoing sections of this chapter, stability of solutions of impulsive differential systems has been studied by finding appropriate estimates for the matriciant of the corresponding linear system of the first order approximation. As in the theory of ordinary differential equation, stability of solutions of impulsive systems can also be studied by using the so-called direct method due to Lyapunov.

Let a scalar-valued function $V(t,x)$, $V(t,0)=0$, be defined and continuously differentiable in the region

$$Z_0 = \{(t;\, x) | t \geq t_0,\ \|x\| < h_0\}.$$

We say that a function $V(t,x)$ is *positive (negative) constant* in the region Z_0 if, for all $(t,x) \in Z_0$, $V(t,x) \geq 0$ $(V(t,x) \leq 0)$.

A function $V(t,x)$ is called *positive (negative) definite* in the region Z_0 if there exists a scalar valued function $X(x)$, $W(0)=0$, continuous for all $\|x\| < h$ such that

$$V(t,x) \geq W(x) > 0 \qquad (V(t,x) \leq -W(x)) \qquad \text{for} \quad x \neq 0.$$

Consider stability of the zero solution of the impulsive differential system

$$\begin{aligned} \frac{dx}{dt} = f(t,x), \qquad & t \neq \tau_i(x), \qquad \Delta x|_{t=\tau_i(x)} = I_i(x), \\ f(t,0) = 0, \qquad & I_i(0) = 0, \qquad \tau_i(x) < \tau_{i+1}(x) \end{aligned} \tag{3.91}$$

with the assumption that the functions $F(t,x)$, $I_i(x)$ are continuous in the region

$$Z = \{(t;\, x) | t \geq t_0,\ \|x\| \leq h < h_0\},$$

and the functions $\tau_i(x)$ and the number h satisfy the conditions of Lemma 14, which would insure that there are no beatings of solutions of system (3.91) at the surfaces $t = \tau_i(x)$.

Theorem 47 *If there exists a positive definite function $V(t,x)$ satisfying in the region Z the inequalities*

$$\begin{aligned} & \frac{\partial V(t,x)}{\partial t} + \langle grad_x V(t,x),\, f(t,x)\rangle \leq 0, \\ & V(\tau_i(x),\, x + I_i(x)) \leq V(\tau_i(x),\, x), \end{aligned} \tag{3.92}$$

then the trivial solution of system (3.91) is stable. If, instead of the second inequality of (3.92), we have

$$V(\tau_i(x), x + I_i(x)) - V(\tau_i(x), x) \leq -\psi(V(\tau_i(x), x) \qquad (3.93)$$

for all $i = 1, 2, \ldots$, *where* $\psi(s)$ *is a function continuous with respect to* $s \geq 0$, $\psi(0) = 0$, *and* $\psi(s) > 0$ *for* $s > 0$, *then the zero solution of (3.91) is asymptotically stable.*

Proof. Let an arbitrary $\epsilon > 0$ be fixed and let $l = \inf_{t \geq t_0, \epsilon \leq \|x\| < h} V(t, x)$ and a positive number δ be so small that $\sup_{\|x\| < \delta} V(t_0, x) = m < l$. Take an arbitrary solution $x(t)$, $x(t_0) = x_0$, of system (3.91) such that $x_0 \in \mathcal{J}_\delta$, $\mathcal{J}_\delta$ is a ball with the radius δ centered at the point x_0, and consider the function $v(t) = V(t, x(t))$. If we assume that, at a moment t^*, $\|x(t^*)\| = \epsilon$, then $v(t^*) = V(t^*, x(t^*)) \geq l$. Besides, inequalities (3.92) imply that the function $V(t, x)$ is nonincreasing along any solution of system (3.91) that lies in the region Z, so that $v(t^*) \leq v(t_0) = V(t_0, x_0) \leq m < l$. This contradiction proves the first statement of the theorem.

Suppose now that, instead of the second inequality of (3.92), inequality (3.93) holds. We shall prove that the trivial solution is asymptotically stable. To do that it is sufficient to show that $\lim_{t \to \infty} v(t) = 0$. The first inequality of (3.92) and inequality (3.93) imply that the function $v(t)$ is nonincreasing and, since it is bounded from below, the limit $\lim_{t \to \infty} v(t) = \alpha$ exists. Suppose that $\alpha > 0$. Let $c = \min_{\alpha \leq s \leq v(t_0)} \psi(s)$. If the function $x(t)$ intersects the surfaces $t = \tau_i(x)$ at the points $(\tau_i(x_i), x_i)$, then, by (3.93), we have

$$v(\tau_i(x_i) + 0) - v(\tau_i(x_i)) \leq -\psi(v(\tau_i(x_i)))$$

for all $i = 1, 2, \ldots$. Since $\alpha \leq v(\tau_i(x_i)) \leq v(t_0)$, we see that $-\psi(v(\tau_i(x_i))) \leq -c$, and, consequently,

$$v(\tau_i(x_i) + 0) - v(\tau_i(x_i)) \leq -c.$$

Because, by the first inequality of (3.92), the function $v(t)$ is nonincreasing on every interval where it is continuous, we get that $v(\tau_i(x_i) + 0) \geq v(\tau_{i+1}(x_{i+1}))$. From this it follows that, for any naturel k,

$$\begin{aligned} v(\tau_k(x_k) + 0) \;&\leq\; v(\tau_k(x_k) + 0) + \sum_{i=0}^{k-1} (v(\tau_i(x_i) + 0) - \\ &\quad -v(\tau_{i+1}(x_{i+1}))) = v(t_0) + \sum_{i=1}^{k} (v(\tau_i(x_i) + 0) - \\ &\quad -v(\tau_i(x_i))) \leq v(t_0) - kc. \end{aligned}$$

The right-hand side of this inequality becomes negative for large values of k, which contradicts that the function $V(t,x)$ is positive definite. Hence the assumption that $\alpha > 0$ leads to a contradiction.

□

Theorem 48 *Suppose there exists a positive definite function $V(t,x)$, which in the region Z satisfies the inequalities*

$$\frac{\partial V}{\partial t} + \langle grad_x V(t,x)\, f(t,x)\rangle \le -\varphi(V(t,x)), \tag{3.94}$$

$$V(\tau_i(x),\, x + I_i(x)) \le \psi(V(\tau_i(x),\, x)), \qquad i = 1,2,\ldots, \tag{3.95}$$

where $\varphi(s)$, $\psi(s)$ are functions continuous for $s \ge 0$, $\varphi(0) = \psi(0) = 0$, and $\varphi(s) > 0$, $\psi(s) > 0$ for $s > 0$, and let

$$\sup_i(\min_{\|x\|\le h} \tau_{i+1}(x) - \max_{\|x\|\le h} \tau_i(x)) = \theta > 0. \tag{3.96}$$

Then, if the functions $\varphi(s)$ and $\psi(s)$ are such that

$$\int_a^{\psi(a)} \frac{ds}{\psi(s)} \le \theta \tag{3.97}$$

for some $a_0 > 0$ and all $a \in]0, a_0]$, then the zero solution of system (3.91) is stable. If, instead of inequality (3.97), we have

$$\int_a^{\psi(a)} \frac{ds}{\psi(s)} \le \theta - \gamma, \tag{3.98}$$

for some $\gamma > 0$, then the zero solution of system (3.91) will be asymptotically stable.

Proof. Let us fix an arbitrary sufficiently small $\epsilon > 0$ and let $l = \inf_{t \ge t_0, \|x\| \ge \epsilon} V(t,x)$. For this ϵ choose δ to be so small that the inequality $m = \sup_{\|x\|<\delta} V(t_0,x) < l$ hold and let $x(t)$, $x(t_0) = x_0 \in \mathcal{J}_\delta$ be an arbitrary solution of system (3.91) that starts in $\mathcal{J}_\delta$. We will show that this solution will never leave the ball $\mathcal{J}_\epsilon$. Consider the function $v(t) = v(t, x(t))$. To prove the theorem it is sufficient to show that $v(t) < l$ for all $t \ge t_0$. An assumption that $x(t)$ will leave the ball $\mathcal{J}_\epsilon$ without reaching the surface $t = \tau_1(x)$ at a certain moment t^* leads to a contradiction because, on one hand, $v(t^*) = V(t^*, x(t^*)) \ge l$, and on the other hand, the function $v(t)$

is not increasing for $x(t) \in \bar{\mathcal{J}}_h$ and $v(t^*) \le m < l$. So, $x(t)$ intersects the surface $t = \tau_1(x)$, for example, at a point $(\tau_1(x_1), x_1)$. By using inequality (3.95), we have for t, $t_0 \le t \le \tau_1(x_1)$, $v'(t) \le -\varphi(v(t))$, and so

$$-\int_{t_0}^{\tau_1(x_1)} \frac{v'(t)dt}{\varphi(v(t))} \ge \tau_1(x_1) - t_0.$$

By setting in this inequality $v(t) = s$ and using (3.96), we get

$$\int_{v(\tau_1)(x_1)}^{v(t_0)} \frac{ds}{\phi(s)} \ge \tau_1(x_1) - t_0 \ge \theta.$$

By replacing a by $v(\tau_1(x_1))$ in inequality (3.97) and using (3.95), we can write

$$\int_{v(\tau_1(x_1))}^{v(\tau_1(x_1)+0)} \frac{ds}{\varphi(s)} \le \int_{v(\tau_1(x_1))}^{\psi(v(\tau_1(x_1)))} \frac{ds}{\varphi(s)} \le \theta.$$

From these two inequalities it follows that

$$\int_{v(\tau_1(x_1)+0)}^{v(t_0)} \frac{ds}{\varphi(s)} = \int_{v(\tau_1(x_1))}^{v(t_0)} \frac{ds}{\varphi(s)} - \int_{v(\tau_1(x_1))}^{v(\tau_1(x_1)+0)} \frac{ds}{\varphi(s)} \ge 0,$$

and this implies that $v(\tau_1(x_1) + 0) \le v(t_0)$.

To end the proof of the first statement of the theorem it is sufficient to apply induction to get that $v(\tau_i(x_i) + 0) \le v(t_0)$ for all $i = 1, 2, \ldots$.

Now suppose that, instead of (3.97), inequality (3.98) holds and the solution $x(t)$ intersects the surfaces $t = \tau_i(x)$ at the points $(\tau_i(x_i), x_i)$. By inequality (3.95), we have

$$-\int_{\tau_i(x_i)}^{\tau_{i+1}(x_{i+1})} \frac{v'(t)}{\varphi(v(t))} dt \ge \tau_{i+1}(x_{i+1}) - \tau(x_i) \ge \theta.$$

Setting $a = v(\tau_{i+1}(x_{i+1}))$ in (3.98) and using (3.95), we see that

$$\int_{v(\tau_{i+1}(x_{i+1}))}^{v(\tau_{i+1}(x_{i+1})+0)} \frac{ds}{\varphi(s)} \le \int_{v(\tau_{i+1}(x_{i+1}))}^{\psi(v(\tau_{i+1}(x_{i+1})))} \frac{ds}{\varphi(s)} \le \theta - \gamma.$$

From the last two inequalities we get

$$\begin{aligned} \int_{a_{i+1}^+}^{a_i^+} \frac{ds}{\varphi(s)} &= \int_{a_{i+1}}^{a_i^+} \frac{ds}{\varphi(s)} - \int_{a_{i+1}}^{a_{i+1}^+} \frac{ds}{\varphi(s)} \ge \gamma, \\ a_i^+ &= v(\tau_i(x_i) + 0). \end{aligned}$$

So the sequence $\{a_i^+\}$ satisfies the inequality

$$\int_{a_{i+1}^+}^{a_i} \frac{ds}{\varphi(s)} \geq \gamma, \qquad i = 1, 2, \ldots, \tag{3.99}$$

from where it follows that this sequence is decreasing for $i \to \infty$. Let us show that $\lim_{t\to\infty} v(\tau_i(x_i) + 0) = 0$. Suppose the converse, i.e. assume that $\lim_{t\to\infty} v(\tau_i(x_i) + 0) = \alpha > 0$. Let $c = \min_{\alpha \leq s \leq v(t_0)} \varphi(s)$. From (3.99) we get

$$\gamma \leq \int_{a_{i+1}^+}^{a_i^+} \frac{ds}{\varphi(s)} \leq \frac{1}{c}(v(\tau_i(x_i) + 0) - v(\tau_{i+1}(x_{i+1}) + 0)),$$

i.e. $v(\tau_i(x_i) + 0) - v(\tau_{i+1}(x_{i+1}) + 0) \geq \gamma c = \text{const}$ and this contradicts to convergence of the sequence $v(\tau_i(x_i) + 0)$. So $v(\tau_i(x_i) + 0) \to 0$ for $i \to \infty$. To end the proof recall that, by (3.95), $v(t)$ is decreasing on every interval of continuity $]\tau_i(x_i), \tau_{i+1}(x_{i+1})]$, and so $\sup_{\tau_i(x_i) < t < \tau_{i+1}(x_{i+1})} v(t) = v(\tau_i(x_i) + 0)$, which, together with the inequality $v(\tau_i(x_i) + 0) > v(\tau_{i+1}(x_{i+1}) + 0)$ that holds for all i, leads to the inequality $v(t) < v(\tau_i(x_i) + 0)$ for all $t > \tau_i(x_i)$. Consequently it follows from the condition $v(\tau_i(x_i) + 0) \to 0$ for all $i \to \infty$ that $v(t) \to 0$ for $t \to \infty$, and so $\|x(t)\| \to 0$ for $t \to \infty$.

□

Additionally assume that the functions $\tau_i(x)$ are such that, for some $\theta_1 > 0$,

$$\max_{\|x\| \leq h} \tau_i(x) - \min_{\|x\| \leq h} \tau_{i-1}(x) \leq \theta_1 \tag{3.100}$$

for all $i = 1, 2, \ldots$.

Theorem 49 *Let there exist a positive definite function $V(t, x)$ satisfying in the domain Z_0 the conditions*

$$\frac{\partial V(t,x)}{\partial t} + \langle grad_x V(t,x), f(t,x) \rangle \leq \varphi(V(t,x)), \tag{3.101}$$

$$V(\tau_i(x), x + I_i(x)) \leq \psi(V(\tau_i(x), x)), \tag{3.102}$$

for $i = 1, 2, \ldots$, where the functions $\varphi(s)$ and $\psi(s)$ are the same as in Theorem 48. If the functions $\varphi(s)$ and $\psi(s)$ are such that, for some a_0,

$$\int_{\psi(a)}^{a} \frac{ds}{\varphi(s)} \geq \theta_1, \tag{3.103}$$

for all $a \in]0, a_0]$, then the zero solution of system (3.91) is stable; if, instead of (3.102), the inequality

$$\int_{\psi(a)}^{a} \frac{ds}{\varphi(s)} \geq \theta_1 + \gamma \tag{3.104}$$

holds for some $\gamma > 0$, then the zero solution of equations (3.91) is asymptotically stable.

We leave the proof of this theorem since it is similar to the proof of the preceding theorem.

Let us consider a few examples.

1. We shell consider stability of the lower equilibrium point of a pendulum subject to an impulsive effect with the following motion equations

$$\ddot{x} + \sin x = 0, \qquad t \neq \tau_i(x, \dot{x}),$$
$$\Delta x|_{t=\tau_i(x,\dot{x})} = -x + \arccos\left(-\tfrac{\dot{x}^2}{2} + \cos x\right),$$
$$\Delta \dot{x}|_{t=\tau_i(x,\dot{x})} = -\dot{x},$$

which can be written as the system

$$\dot{x} = y, \qquad \dot{y} = -\sin x, \qquad t \neq \tau_i(x, y),$$
$$\Delta x|_{t=\tau_i(x,y)} = -x + \arccos\left(-\tfrac{y^2}{2} + \cos x\right),$$
$$\Delta y|_{t=\tau_i(x,y)} = -y.$$

Let us study stability of the zero solution of this system. Take the total energy of an unperturbed pendulum

$$V(x, y) = 1 - \cos x + \frac{y^2}{2}.$$

to be a Lyapunov function. We find that

$$\frac{dV}{dt} = y \sin x - y \sin x = 0,$$
$$V(x + \Delta x, y + \Delta y) = 1 - \cos\left(\arccos\left(-\frac{y^2}{2} + \cos x\right)\right) =$$
$$= 1 - \cos x + \frac{y^2}{2} = V(x, y).$$

Conditions of Theorem 47 hold regardless of the properties of the surfaces $t = \tau_i(x,y)$, and so the zero solution of this system is stable.

2. Let us study stability of the lower equilibrium of a pendulum with an impulsive effect such that this equilibrium can be made asymptotically stable. Consider the system

$$\dot{x} = y, \qquad \dot{y} = -\sin x, \qquad t \neq \tau_i(x,y),$$
$$\Delta x|_{t=\tau_i(x,y)} = \alpha x + \beta y, \qquad \Delta y|_{t=\tau_i(x,y)} = -\beta x + \alpha y.$$

As in the previous example, take the function $V(x,y)$ to be $V(x,y) = 1 - \cos x + \dfrac{y^2}{2}$. The derivative of this function, calculated by using the system under consideration, identically equals to zero.

Moreover,

$$V(x+\Delta x, y+\Delta y) - V(x,y) = \frac{1}{2}(\alpha^2 + 2\alpha + \beta^2)(x^2+y^2) +$$
$$+ 0(x^2+y^2) = \frac{1}{2}[(\alpha+1)^2 + \beta^2 - 1](x^2+y^2) +$$
$$+ \gamma(x,y)(x^2+y^2),$$

where $\gamma(x,y) \to 0$ for $x^2+y^2 \to 0$.

Let $\frac{1}{2}((\alpha+1)^2 + \beta^2 - 1) = l < 0$, i.e. $(\alpha+1)^2 + \beta^2 < 1$. There exists such $h > 0$ that $|\gamma(x,y)| \leq \epsilon < -l$ as soon as $x^2+y^2 \leq h^2$, and we have

$$V(x+\Delta x, y+\Delta y) - V(x,y) \leq (l+\epsilon)(x^2+y^2).$$

Consequently, if $(\alpha+1)^2+\beta^2 < 1$, then the zero solution of the considered system is asymptotically stable.

3. Consider stability of the zero solution of the system

$$\dot{x} = -y + x^3, \qquad \dot{y} = x + y^3, \qquad t \neq \tau_i(x,y),$$
$$\Delta x|_{t=\tau_i(x,y)} = -\alpha x^3 + \beta y^3,$$
$$\Delta y|_{t=\tau_i(x,y)} = \beta x^3 - \alpha y^3,$$

where $\alpha > 0$, $\beta > 0$, and $\tau_i(x,y) = i + x^2 + y^2$.

It can be checked that solutions of this system satisfy the conditions of Lemma 14 in a sufficiently small neighborhood of the origin and so there are no beatings at the surfaces $t = \tau_i(x, y)$. Set $V(x, y) = x^2 + y^2$. Then

$$\begin{aligned}
\frac{dV}{dt} &= 2x^4 + 2y^4 \le 2V^2(x, y), \\
V(x + \Delta x,\, y + \Delta y) &= x^2 + y^2 - 2\alpha(x^4 + y^4) + 2(x^2 + y^2)xy + \\
&+ (\alpha^2 + \beta^2)(x^6 + y^6) - 4\alpha\beta x^3 y^3 \le \\
&\le V(x, y) - (\alpha - \beta)V^2(x, y) + (\alpha^2 + \beta^2)V^3(x, y).
\end{aligned}$$

It follows from this that we can take $\varphi(s) = 2s^2$, $\psi(s) = s - (\alpha - \beta)s^2 + (\alpha^2 + \beta^2)s^3$. By using the fact that

$$\max_{x^2+y^2 \le h^2} \tau_i(x, y) - \min_{x^2+y^2 \le h^2} \tau_{i-1}(x, y) = 1 + h^2,$$

$$\int_{a-(\alpha-\beta)a^2+(\alpha^2+\beta^2)a^3}^{a} \frac{ds}{2s^2} = \frac{\alpha - \beta - (\alpha^2 - \beta^2)a}{2(1 - (\alpha - \beta)a + (\alpha^2 + \beta^2)a^2)},$$

we see that, for the zero solution of the system under consideration to be asymptotically stable, it is sufficient that $\alpha - \beta > 2$.

Let us find sufficient conditions for the zero solution of system (3.91) to be unstable. In the theorems that give these conditions, we assume that the function $V(t, x)$ exists and has the following properties:

a) the intersection of the region $\Pi = \{(t, x) \in Z \mid V(t, x) > 0\}$, where the function $V(t, x)$ is positive, and a plane $t = \text{const}$ is a nonempty open set adherent to the origin for any $t \ge t_0$;

b) the function $V(t, x)$ is bounded in Π.

Theorem 50 *If there exists a function $V(t, x)$ with properties a) and b) and satisfying in the region Π the conditions*

$$\frac{\partial V(t, x)}{\partial t} + \langle grad_x V(t, x), f(t, x)\rangle \ge 0, \tag{3.105}$$

$$V(\tau_i(x), x + I_i(x)) - V(\tau_i(x), x) \ge \psi(V(\tau_i(x), x)), \tag{3.106}$$

$$i = 1, 2, \ldots,$$

where $\psi(s)$ is a function continuous for $s \ge 0$, $\psi(0) = 0$, $\psi(s) > 0$ for $s > 0$, then the zero solution of system (3.91) is unstable.

Proof. By the conditions of the theorem, in any neighborhood of the point $x = 0$ there exists such a point x_0 that $V(t_0, x_0) > 0$. We will prove that the solution $x(t)$ that starts at the point x_0 will eventually leave the ball T_h. Suppose the converse, i.e. $x(t) \in T_h$ for all $t \geq t_0$. Let $x(t)$ intersect the surfaces $t = \tau_i(x)$ at points $(\tau_i(x_i), x_i)$. Consider the function $v(t) = V(t, x(t))$. By inequalities (3.95), (3.95), $v(t)$ is a nondecreasing function, and hence $v(t) \geq v(t_0) > 0$ for all $t \geq t_0$. This means that $(t, x(t)) \in \Pi$ for all $t \geq t_0$.

Let $c = \min_{v(t_0) \leq s \leq a_0} \psi(s)$, where $a_0 = \sup_{(t,x)\in\Pi} V(t, x)$. It is clear that $c > 0$ and $v(\tau_i(x_i) + 0) - v(\tau_i(x_i)) \geq c$, $i = 1, 2, \ldots$. Because $v'(t) \geq 0$ for $t \neq \tau_i(x_i)$, we have $v(\tau_{i-1}(x_{i-1}) + 0) - v(\tau_i(x_i)) \leq 0$, $i = 1, 2, \ldots$. So, for any natural k, we have that

$$\begin{aligned} v(\tau_k(x_k) + 0) &\geq v(\tau_k(x_k) + 0) + \sum_{i=1}^{k} (v(\tau_{i-1}(x_{i-1}) + 0) - v(\tau_i(x_i))) = \\ &= v(t_0) + \sum_{i=1}^{k} (v(\tau_i(x_i) + 0) - v(\tau_i(x_i))) \geq \\ &\geq v(t_0) + kc. \end{aligned}$$

The right-hand side of the last inequality becomes unbounded as $k \to \infty$, which is a contradiction to $(t, x(t)) \in \Pi$ because the function $V(t, x)$ is bounded in Π. This contradiction shows that the zero solution is unbounded.

□

The next theorem gives sufficient conditions for the zero solution of system (3.91) to be unbounded in the case when this solution of the corresponding differential system without an impulsive effect may be stable or even asymptotically stable.

Theorem 51 *Suppose that there exists a function $V(t, x)$ with properties a) and b) and satisfying in the region Π the inequalities*

$$\frac{\partial V(t,x)}{\partial t} + \langle grad_x V(t, x),\, f(t, x)\rangle \geq -\varphi(V(t, x)), \tag{3.107}$$

$$V(\tau_i(x),\, x + I_i(x)) \geq \psi(V(\tau_i(x), x)), \tag{3.108}$$

$$i = 1, 2, \ldots,$$

where $\varphi(s)$, $\psi(s)$ are continuous functions with $\varphi(0) = \psi(0) = 0$ and $\varphi(s) > 0$, $\psi(s) > 0$ for $s > 0$. Also assume that the functions $\tau_i(x)$ satisfy condition

(3.99). If the functions $\varphi(s)$ and $\psi(s)$ are such that, for some γ,

$$\int_a^{\psi(a)} \frac{ds}{\varphi(s)} \geq \theta_1 + \gamma \tag{3.109}$$

for all $a \in]0, a_0]$, then the trivial solution of system (3.91) is unstable.

Proof. In every arbitrarily small δ-neighborhood of the point $x = 0$ there exists such a point x_0 that $V(t_0, x_0) > 0$. We will prove that the solution that starts at this point will eventually leave the ball $\mathcal{T}_h$. Suppose, conversely, that $x(t) \in \mathcal{T}_h$ does not intersect the surfaces $t = \tau_i(x)$ at points $(\tau_i(x_i), x_i)$ for all $t \geq t_0$.

Let us show that, from the assumption $x(t) \in \mathcal{T}_h$, it will follow that $(t, x(t)) \in \Pi$ for all $t \geq t_0$. Indeed, it is impossible that $(\tau_i(x_i), x_i) \in \Pi$ and $(\tau_i(x_i), x(\tau_i(x_i) + 0)) \notin \Pi$ because, by (3.108),

$$\begin{aligned} V(\tau_i(x_i),\, x(\tau_i(x_i) + 0)) &= V(\tau_i(x_i),\, I_i(x_i)) \geq \\ &\geq \psi(V(\tau_i(x_i),\, x_i)) > 0. \end{aligned}$$

It we assume that the phase point $(t, x(t))$ leaves the region Π, then it necessarily must intersect its boundary. Let t^* be the smallest moment when this occurs. If $\tau_k(x_k) < t \leq \tau_{k+1}(x_{k+1})$, then, denoting $v(t) = V(t, x(t))$, we have that $v(t^*) = 0$ and $v(\tau_k(x_k) + 0) > 0$. It follows from (3.107) that

$$-\int_{\tau_k(x_k)}^{t^*} \frac{v'(t)dt}{\varphi(v(t))} \leq t^* - \tau_k(x_k)$$

and, by using (3.99),

$$\int_{v(t^*)}^{v(\tau_k(x_k)+0)} \frac{ds}{\varphi(s)} \leq t^* - \tau_k(x_k) \leq \theta_1. \tag{3.110}$$

Fix $a' > 0$ such that $0 < \psi(a') \leq v(\tau_k(x_k) + 0)$. By using inequalities (3.108) – (3.110) we get a contradictory chain of the inequalities

$$\theta_1 + \gamma \leq \int_{a'}^{\psi(a')} \frac{ds}{\varphi(s))} \leq \int_0^{v(\tau_k(x_k)+0)} \frac{ds}{\varphi(s)} = \int_{v(t^*)}^{v(\tau_k(x_k)+0)} \frac{ds}{\varphi(s)} \leq \theta_1.$$

Hence, if $x(t) \in \mathcal{T}_h$, then $(t, x(t)) \in \Pi$ for $t \geq t_0$. So it follows from (3.107) that

$$\int_{a_{i+1}}^{a_i^+} \frac{ds}{\varphi(s)} \leq \theta_1, \qquad i = 0, 1, \ldots, \qquad a_i = v(\tau_i(x_i)),$$
$$a_i^+ = v(\tau_i(x_i) + 0).$$

By subtracting this inequality from (3.109) with $a = a_{i+1} = v(\tau_{i+1}(x_{i+1}))$ and using (3.108), we obtain

$$\gamma \leq \int_{a_{i+1}}^{\psi(a_{i+1})} \frac{ds}{\varphi(s)} - \int_{a_{i+1}}^{a_i^+} \frac{ds}{\varphi(s)} \leq \int_{a_{i+1}}^{a_{i+1}^+} \frac{ds}{\varphi(s)} - \int_{a_{i+1}}^{a_i^+} \frac{ds}{\varphi(s)}$$

or

$$\int_{a_i^+}^{a_{i+1}^+} \frac{ds}{\varphi(s)} \geq \gamma, \qquad i = 0, 1, 2, \dots . \tag{3.111}$$

This shows that $\{a_i\}$ is an increasing sequence and it is bounded by a_0 since $(t, x(t)) \in \Pi$.

Let $\min_{v(t_0) \leq s \leq a_0} \varphi(s) = c > 0$. From (3.111) we get

$$\gamma \leq \frac{1}{c} \int_{a_i^+}^{a_{i+1}^+} ds = \frac{1}{c}(v(\tau_{i+1}(x_{i+1}) + 0) - v(\tau_i(x_i) + 0)),$$

or $v(\tau_{i+1}(x_{i+1})+0) - v(\tau_i(x_i)+0) \geq \gamma c$. Hence, for any natural k, $v(\tau_k(x_k)+0) \geq k\gamma c + v(t_0)$, which is a contradiction since the sequence $\{v(\tau_i(x_i) + 0)\}$ is bounded.

□

The following statement can be proved in the same way as Theorem 51.

Theorem 52 *Let there exist a function $V(t, x)$ with properties a) and b), satisfying in the region Π the inequalities*

$$\frac{\partial V(t,x)}{\partial t} + \langle grad_x V(t,x), f(t,x) \rangle \geq \varphi(V(t,x)), \tag{3.112}$$

$$V(\tau_i(x), x + I_i(x)) \geq \psi(V(\tau_i(x), x))), \qquad i = 1, 2, \dots, \tag{3.113}$$

$$i = 1, 2, \dots ,$$

where $\varphi(0) = \psi(0) = 0$, $\varphi(s) > 0$, $\psi(s) > 0$ for $s > 0$. Assume also that the functions $\tau_i(x)$ satisfy condition (3.96). If the functions $\varphi(s)$ and $\psi(s)$ are such that, for some $\gamma > 0$,

$$\int_{\psi(a)}^{a} \frac{ds}{\varphi(s)} \leq \theta - \gamma, \tag{3.114}$$

for $a \in]0, a_0]$, then the zero solution of system (3.91) is unstable.

Chapter 4

Periodic and almost periodic impulsive systems

4.1 Nonhomogeneous linear periodic systems

In this section we state and prove the main conditions for existence of periodic solutions of nonhomogeneous linear periodic impulsive differential systems, i.e. the systems

$$\frac{dx}{dt} = A(t)x + f(t), \qquad t \neq \tau_i, \qquad \Delta x|_{t=\tau_i} = B_i + a_i, \tag{4.1}$$

where $A(t)$ and $f(t)$ are continuous (piecewise continuous) T-periodic matrix and vector function respectively, the constant matrix B_i, constant vectors a_i, and the times τ_i are such that

$$B_{i+p} = B_i, \qquad a_{i+p} = a_i, \qquad \tau_{i+p} = \tau_i + T \tag{4.2}$$

for some natural p, $i \in \mathbb{Z}$. We also assume that $\det(E + B_i) \neq 0$ and $\tau_0 < 0 < \tau_1 < \tau_2 < \ldots < \tau_p < T$.

Let $X(t)$, $X(0) = E$, be the matriciant of the homogeneous system corresponding to (4.1), i.e. a solution of the following matrix Cauchy problem for impulsive systems:

$$\frac{dX}{dt} = A(t)X, \qquad t \neq \tau_i, \qquad \Delta X|_{t=\tau_i} = B_i X, \qquad X(0) = E. \tag{4.3}$$

Any solution $x(t, x_0)$, $x(0, x_0) = x_0$, of system (4.1) can be written as

$$x(t, x_0) = X(t)x_0 + \int_0^t X(t,\tau)f(\tau)d\tau + \sum_{0<\tau_i<t} X(t, \tau_i)a_i. \tag{4.4}$$

Amongst these solutions, that one will be T-periodic, for which x_0 satisfies the equation

$$(E_X(T))x_0 = \int_0^T X(T,\tau)f(\tau)d\tau + \sum_{i=1}^{p} X(T, \tau_i)a_i. \tag{4.5}$$

Suppose that $\det(E - X(T)) \neq 0$. This is equivalent to the condition that the homogeneous system corresponding to (4.1) does not have any nontrivial T-periodic solutions, i.e. the identity is not one of its multipliers. In this case, equation(4.5) has the unique solution

$$x_0 = (E_X(T))^{-1}\left[\int_0^T X(T,\tau)f(\tau)d\tau + \sum_{i=1}^{p} X(T, \tau_i)a_i\right], \tag{4.6}$$

and so system (4.1) has the unique T-periodic solution

$$x^*(t) = X(t)(E - X(T))^{-1}\left(\int_0^T X(T,\tau)f(\tau)d\tau + \sum_{i=1}^{p} X(T,\tau_i)a_i\right) + \\ + \int_0^t X(t,\tau)f(\tau)d\tau + \sum_{0<\tau_i<t} X(t,\tau_i)a_i, \qquad t \geq 0.$$

If we denote

$$G(t,\tau) = \begin{cases} X(t)(E - X(T))^{-1}X^{-1}(\tau), & 0 \leq \tau \leq t \leq T, \\ X(t+T)(E - X(T))^{-1}X^{-1}(\tau), & 0 \leq t < \tau \leq T, \end{cases} \tag{4.7}$$

then the periodic solution can be written as

$$x^*(t) = \int_0^T G(t,\tau)f(\tau)d\tau + \sum_{i=1}^{p} G(t,\tau_i)a_i. \tag{4.8}$$

The function $G(t,\tau)$ is called a *Green's function* of system (4.1). We list some properties of the function $G(t,\tau)$:

1) if $t \neq \tau_i$, then $G(\tau + 0, \tau) - G(\tau,\tau) = E$;

2) $G(0,\tau) = G(T,\tau)$;

3) if $t \neq \tau$, the function $G(t,\tau)$ satisfies the impulsive differential system

$$\frac{dG(t,\tau)}{dt} = A(t)G(t,\tau), \qquad t \neq \tau_i, \qquad \Delta G|_{t=\tau_i} = B_iG(\tau_i,\tau),$$

so that $G(\tau_i + 0, \tau) = (E + B_i)G(\tau_i,\tau)$;

4) $G(t, \tau_i + 0) = G(t,\tau_i)(E + B_i)^{-1}$.

By using properties of the function $X(t)$ (see section 2.1), it is easy to see that the listed properties of the function $G(t,\tau)$ hold. It should also be remarked that properties 1) – 4) determine the Green's function $G(t,\tau)$ uniquely.

Let

$$\max_{t,\tau\in[0,T]} \|G(t,\tau)\| = \frac{K}{T+p}. \tag{4.9}$$

It follows from (4.8) that $x^*(t)$ admits the estimate

$$\|x^*(t)\| \leq \frac{K}{T+p}\left(\int_0^T \|f(t)\|dt + \sum_{i=1}^{p} \|a_i\|\right). \tag{4.10}$$

Hence we have proved the following statement.

Theorem 53 *If the homogeneous system which corresponds to (4.1) does not have nontrivial T-periodic solutions, then, for any T-periodic function $f(t)$ and any periodic sequence a_i ($a_{i+p} = a_i$, $i \in \mathbb{Z}$), system (4.1) has a unique T-periodic solution $x^*(t)$ and this solution satisfies estimate (4.10).*

Now consider the case when the homogeneous system, corresponding to system (4.1), has nontrivial T-periodic solutions.

Initial conditions for such solutions are determined from the equations

$$(E - X(T))x_0 = 0. \tag{4.11}$$

Suppose that the homogeneous system has $k \leq n$ linearly independent solutions. Then the algebraic system (4.11) has exactly k linearly independent solutions, i.e. the rank of the matrix $E - X(T)$ equals to $n - k$.

As shown in section 2.5, the matriciant of the adjoint system

$$\frac{dx}{dt} = -A^T(t)x, \qquad t \neq \tau_i, \qquad \Delta x|_{t=\tau_i} = -(E + B_i^T)^{-1}B_i x \tag{4.12}$$

is the matrix $Y(t) = (X^T(t))^{-1}$, and so the initial conditions for nontrivial T-periodic solutions of the adjoint system must satisfy the condition

$$(E - (X^T(T))^{-1})y_0 = 0 \quad \text{or} \quad (X(T) - E)^T y_0 = 0. \tag{4.13}$$

The rank of the matrix $(X(T) - E)^T$ is equal to the rank of the matrix $X(T) - E$, i.e. $n - k$, and so the system of algebraic equations (4.13) has k linearly independent solutions and hence the adjoint system (4.12) has the corresponding k linearly independent T-periodic solutions.

Theorem 54 *Let the linear homogeneous T-periodic system, which corresponds to equation (4.1), has k linearly independent T-periodic solutions $\varphi_1(t)$, $\varphi_2(t)$, ..., $\varphi_k(t)$ ($1 \leq k \leq n$).*

System (4.1) has T-periodic solutions if and only if the following conditions hold

$$\int_0^T \langle \psi_j(t), f(t)\rangle dt \;+\; \sum_{i=1}^{p}\langle \psi_j(\tau_i), a_i\rangle = 0,$$
$$j = 1, 2, \ldots, k, \tag{4.14}$$

where $\psi_1(t)$, $\psi_2(t)$, ..., $\psi_k(t)$ are linearly independent T-periodic solutions of the adjoint system (4.12). In this case the T-periodic solutions of system (4.1) form a k-parameter family of solutions.

Proof. Let $x(t)$ be a T-periodic solution of the nonhomogeneous system (4.14). Because the solution is T-periodic, it follows that the initial condition $x(0) = x_0$ must satisfy the condition

$$(E - X(T))x_0 = \int_0^T X(T)X^{-1}(\tau)f(\tau)d\tau + \sum_{i=1}^{p} X(t)X^{-1}(\tau_i)a_i. \quad (4.15)$$

Let $\psi(t)$ be a nontrivial T-periodic solution of the adjoint system (4.12). Then $(E - X(T))^T\psi(0) = 0$, and consequently

$$\begin{aligned} 0 &= \langle (E - X(T))^T\varphi(0), x_0\rangle = \langle\psi(0), (E - X(T))x_0\rangle = \\ &= \left\langle X^T(T)\psi(0), \int_0^T X^{-1}(\tau)f(\tau)d\tau + \sum_{i=1}^{p} X^{-1}(\tau_i)a_i \right\rangle = \\ &= \int_0^T \langle (X^T(\tau))^{-1}\psi(0), f(\tau)\rangle d\tau + \sum_{i=1}^{p}\langle (X^T(\tau_i))^{-1}\psi(0), a_i\rangle = \\ &= \int_0^T \langle\psi(\tau), f(\tau)\rangle d\tau + \sum_{i=1}^{p}\langle\psi(\tau_i), a_i\rangle. \end{aligned}$$

This proves the necessity. Let us prove that conditions (4.14) are sufficient. Suppose that equalities (4.14) hold.

If y_0 is an eigen vector of the matrix $X^T(T)$, which corresponds to the multiplier $\rho = 1$, then the solution $y(t)$, $y(0) = y_0$, of the adjoint system (4.12) is T-periodic and

$$y(t) = Y(t)y_0 = \sum_{j=1}^{k} c_j\psi_j(t).$$

Hence we have

$$\begin{aligned} 0 &= \int_0^T \left\langle \sum_{j=1}^{k} c_j\psi_j(t), f(t)\right\rangle dt + \sum_{i=1}^{p}\sum_{j=1}^{k}\langle c_j\psi_j(\tau_i), a_i\rangle \\ &= \int_0^T \langle Y(t)y_0, f(t)\rangle dt + \sum_{i=1}^{p}\langle Y(\tau_i)y_0, a_i\rangle = \\ &= \int_0^T \langle X^T(T)y_0, X^{-1}(t)f(t)\rangle dt + \sum_{i=1}^{p}\langle X^T(T)y_0, X^{-1}(\tau_i)a_i\rangle = \\ &= \left\langle y_0, \int_0^T X(T)X^{-1}(t)f(t)dt + \sum_{i=1}^{p} X(T)X^{-1}(\tau_i)a_i \right\rangle. \end{aligned}$$

Thus the system $(X(T)-E)^T y_0 = 0$ is equivalent to the system

$$
\begin{aligned}
(X(T)-E)y_0 &= 0, \\
\left\langle y_0, \int_0^T X(T)X^{-1}(t)f(t)dt + \sum_{i=1}^{p} X(T)X^{-1}(\tau_i)a_i \right\rangle &= 0,
\end{aligned}
$$

and so the ranks of these systems are the same. And, hence, denoting by b the row-vector

$$
b = \left(\int_0^T X(T)X^{-1}(t)f(t)dt + \sum_{i=1}^{p} X(T)X^{-1}(\tau_i)a_i \right)^T,
$$

we get for the rank of system (4.15)

$$
\begin{aligned}
\operatorname{rank}((X(T)-E)b^T &= \operatorname{rank}\begin{pmatrix} (X(T)-E^T \\ b \end{pmatrix} = \operatorname{rank}(X(T)-E) = \\
&= \operatorname{rank}(X(T)-E) = n-k.
\end{aligned}
$$

By Kronecker-Capelli theorem, system (4.15) which determines initial conditions for T-periodic solutions of nonhomogeneous system (4.1), has exactly k linearly independent solutions.

□

Existence of T-periodic solutions of linear periodic system (4.1) is closely related to this system having bounded solutions. This relation is given by a Massera theorem applied to impulsive systems.

Theorem 55 *If the linear nonhomogeneous T-periodic impulsive system (4.1) has a solution $x^*(t)$, bounded for $t \geq 0$, then this system must have a T-periodic solution.*

Proof. By (4.4), a solution $x^*(t)$ of nonhomogeneous system (4.1), bounded for $t \geq 0$, can be written as

$$
\begin{aligned}
x^*(t) &= X(t)x^*(0) + \int_0^t X(t)X^{-1}(\tau)f(\tau)d\tau + \\
&+ \sum_{0<\tau_i<t} X(t)X^{-1}(\tau_i)a_i. \qquad (4.16)
\end{aligned}
$$

Whence

$$
x^*(T) = X(T)x^*(0) + b, \qquad (4.17)
$$

where

$$b = \int_0^T X(T)X^{-1}(\tau)f(\tau)d\tau + \sum_{i=1}^{p} X(T)X^{-1}(\tau_i)a_i.$$

Since the nonhomogeneous system is periodic, the function $x^*(t+mT)$ is also a solution, and so

$$x^*(mT) = X^m(T)x^*(0) + \sum_{j=0}^{m-1} X^j(T)b \tag{4.18}$$
$$(m = 1, 2, \ldots).$$

Suppose that system (4.1) does not have T-periodic solutions. Then the linear algebraic system

$$(E - X(T))x_0 = b, \tag{4.19}$$

which expresses the periodicity condition, does not have solutions and, in particular, $\det(E - X(T)) = 0$. Whence it follows that the adjoint system $(E - X(T))^T x = 0$ has a nontrivial solution $x = x^0$ such that $\langle b, x^0 \rangle \neq 0$. It is clear that $x = X^T(T)x^0$, and so

$$x^0 = (X^T(T))^k x^0 \qquad (k = 0, 1, 2, \ldots).$$

By using (4.18), we find that

$$\langle x^*(mT), x^0 \rangle = \langle x^*(0), (X^m(T))^T x^0 \rangle + \\ + \sum_{j-0}^{m-1} \langle b, (X^m(T))^T x^0 \rangle,$$

$\langle x^*(0), x^0 \rangle + m\langle b, x^0 \rangle \to \infty$ for $m \to \infty$, which is a contradiction since the solution $x^8(t)$ is bounded.

Thus the conditions of the theorem imply that system (4.19) has solutions and thus there exists at least one T-periodic solution of the nonhomogeneous system (4.1).

□

Corollary. *If the linear nonhomogeneous T-periodic impulsive system (4.1) does not have any T-periodic solutions, then all the solutions of this system are unbounded both for $t \geq 0$ and $t \leq 0$.*

4.2 Nonlinear periodic systems

Now we will establish some sufficient conditions for existence of periodic solutions of the nonlinear impulsive system

$$\frac{dx}{dt} = A(t)x + f(t,x), \qquad t \neq \tau_i, \qquad \Delta x|_{t=\tau_i} = B_i x + I_i(x), \tag{4.20}$$

where $A(t)$ is a continuous T-periodic matrix, the function $f(t,x)$ is continuous with respect to its variables (piecewise continuous with the first kind discontinuities in t at $t = \tau_i$) and T-periodic with respect to t; the matrices B_i, functions $I_i(x)$, and the times τ_i are such that

$$B_{i+p} = B_i, \qquad I_{i+p}(x) = I_i(x), \qquad \tau_{i+p} = \tau_i + T \tag{4.21}$$

for all $i \in \mathbb{Z}$ and some natural p.

To do that, we will use two general theorems on existence of a fixed point of an operator on a complete normed space. We will only give the statements of these theorems since the proofs can be found in many books on functional analysis and its application (see e.g. [57]).

Theorem 56 *Let $\mathcal{B}$ be a Banach space of elements $x, y, \ldots$ with the norms $\|x\|, \|y\|, \ldots$, F – a mapping of the ball $\|x\| \leq h$ in the space $\mathcal{B}$ into the space $\mathcal{B}$, which satisfies the condition*

$$\|F(x) - F(y)\| \leq \rho\|x - y\|, \qquad 0 < \rho < 1. \tag{4.22}$$

Assume that

$$\|F(0)\| \leq h(1 - \rho). \tag{4.23}$$

Then the mapping F has a unique fixed point x_0 such that $F(x_0) = x_0$. Moreover, this point can be found as a limit of the iterations $x_1 = F(0)$, $x_2 = F(x_1)$, $x_3 = F(x_2)$, $\ldots$, satisfying the estimate

$$\|x_0 - y_0\| \leq \frac{\rho^k}{1-\rho}\|F(0)\| \leq \rho^2 h. \tag{4.24}$$

It should be noted that if F maps the ball $\|x\| \leq h$ into itself, then condition (4.23) is redundant.

Theorem 57 *Let F be such a mapping of a closed convex subset S of a Banach space $\mathcal{B}$ into itself that the image $F(S)$ of the set S has the compact closure. Then F has a fixed point $x_0 \in S$.*

We will prove the following statement.

Theorem 58 *Let in system (4.20) the functions $f(t,x)$ and $I_i(x)$, besides the assumptions made, also satisfy the Lipschitz condition in x,*

$$\|f(t,x)-f(t,y)\|+\|I_i(x)+I_i(y)\|\leq N\|x-y\|, \tag{4.25}$$

which is uniform with respect to $0\leq t\leq T$ and $i=1,\ 2,\ \ldots,\ p$.

Suppose that the linear homogeneous T-periodic system

$$\frac{dx}{dt}=A(t)x, \qquad t\neq\tau_i, \qquad \Delta x|_{t=\tau_i}=B_i x \tag{4.26}$$

does not have nontrivial T-periodic solutions. If the Lipschitz constant N is so small that

$$KN<1, \tag{4.27}$$

then system (4.20) has a unique T-periodic solution.

Note that it is not necessary that the functions $f(t,x)$ and $I_i(x)$ be defined and satisfy inequality (4.25) for all $x\in\mathbb{R}^n$. In the course of the proof it will suffice to consider these functions in the set $\|x\|\leq h$, where h is such that $Km\leq h(1-KN)$ and $m=\max\{\|f(t,0)\|_0,\max_i\|I_i(0)\|\}$.

Proof. Introduce the Banach space of piecewise continuous functions (with the first kind discontinuities at $t=\tau_i$) $\varphi(t)$, which are T-periodic, with the norm $\|\varphi(t)\|_0=\max_{t\in[0,T]}\|\varphi(t)\|$. Convergence of a sequence $\varphi_1(t)$, $\varphi_2(t),\ldots$ in $\mathcal{B}$ is equivalent to its usual uniform convergence on the interval $[0,T]$.

Let $\varphi(t)$ be a piecewise continuous function with the first kind discontinuities at $t=\tau_i$, T-periodic, and satisfying the inequality $\|\varphi(t)\|_0\leq h$. By Theorem 53, the equation

$$\frac{dx}{dt}=A(t)x+f(t,\varphi(t)), \qquad t\neq\tau_i, \qquad \Delta x|_{t=-\tau_i}=B_i x+I_i(\varphi(\tau_i)) \tag{4.28}$$

has the unique T-periodic solution

$$x^*(t)=\int_0^T G(t,\tau)f(\tau,\varphi(\tau))d\tau+\sum_{i=1}^{p}G(t,\tau_i)I_i(\varphi(\tau_i)), \tag{4.29}$$

where $G(t,\tau)$ is defined by (4.26). On the set of all such functions $\varphi(t)$, define an operator F by

$$F[\varphi(t)]=\int_0^T G(\tau,\tau)f(\tau,\varphi(\tau))d\tau+\sum_{i=1}^{p}G(t,\tau_i)I_i(\varphi(\tau_i)).$$

If $\varphi_1(t), \varphi_2(t) \in \mathcal{B}$, then

$$\|F\varphi_1(t) - F\varphi_2(t)\| \le \int_0^T G(t,\tau)\|N\|\varphi_1(\tau) - \varphi_2(\tau)\|d\tau + \\ + \sum_{i=1}^{p} \|G(t,\tau_i)\|N\|\varphi_1(\tau_i) - \varphi_2(\tau_i)\|.$$

Whence

$$\|F\varphi_1(t) - F\varphi_2(t)\| \le \frac{KN}{T+p}\left(\int_0^T \|\varphi_1(\tau) - \varphi_2(\tau)\|d\tau + \right. \\ \left. + \sum_{i=1}^{p} \|\varphi_1(\tau_i) - \varphi_2(\tau_i)\|\right), \tag{4.30}$$

where the constant K is determined according to (4.1).

From the last inequality we get

$$\|F\varphi_1(t) - F\varphi_2(t)\|_0 \le KN\|\varphi_1(t) - \varphi_2(t)\|_0. \tag{4.31}$$

Moreover, if

$$m = \max\{\|f(t,0)\|_0, \max_i \|I_i(0)\|\}, \tag{4.32}$$

then

$$\|F(0)\|_0 \le Km. \tag{4.33}$$

Choose h such that $Km \le h(1-KN)$. Then the mapping F, by estimates (4.30) and (4.31), and condition (4.27), satisfies the conditions of Theorem 56 and, consequently, it has a fixed point $F\varphi^*(t) = \varphi^*(t)$ satisfying the equation

$$\varphi^*(t) = \int_0^T G(t,\tau)f(\tau,\varphi^*(t))dt + \sum_{i=1}^{p} G(t,\tau_i)I_i(\varphi^*(\tau_i)), \tag{4.34}$$

i.e. a unique T-periodic solution of system (4.20).

□

The function $\varphi^*(t)$ can be found by taking the limit of the uniformly convergent sequences of T-periodic functions $\varphi_m(t)$:

$$\varphi_0(t) \equiv 0, \quad \varphi_{m+1}(t) = \int_0^T G(t,\tau)f(\tau,\varphi_m(\tau))d\tau + \tag{4.35}$$

$$+\sum_{i=1}^{p} G(t,\tau_i)I_i(\varphi_m(\tau_i)), \quad m = 0,1,2,\ldots. \tag{4.36}$$

Besides, the following estimate holds:

$$\|\varphi^*(t) - \varphi_m(t)\|_0 \leq (KN)^m h. \tag{4.37}$$

Theorem 59 *Let in system (4.20) the functions $f(t, x)$ and $I_i(x)$ be continuous for $\|x\| \leq h$, $0 \leq t \leq T$, and satisfy the condition*

$$K\|f(t,x)\| \leq h, \quad K\|I_i(x) \leq h, \quad 0 \leq t \leq T, \\ i = 1, 2, \ldots, p, \quad \|x\| \leq h. \tag{4.38}$$

If the homogeneous system (4.26) does not have nontrivial T-periodic solutions, then system (4.20) has at least one T-periodic solution.

Proof. Let $\mathbb{B}$ be the Banach space of piecewise continuous (with the first kind discontinuities at $t = \tau_i$) T-periodic functions, introduced in the proof of the preceding theorem. As before we define the mapping $F : S_h \to \mathbb{B}$, where $S_h = \{\varphi(t) \in \mathbb{B} \mid \|\varphi(t)\|_0 \leq h\}$ by formula (4.30). To prove the theorem it will suffice to show that the mapping F has a fixed point, $F[\varphi^*(t)] = \varphi^*(t)$.

By using (4.30) and (4.28), it follows that

$$\|F\varphi(t)\| \leq \frac{K}{T+p}\left(\int_0^T \|f(\tau, \varphi(\tau))d\tau + \sum_{i=1}^{p} \|I_i(\varphi(\tau_i))\|\right). \tag{4.39}$$

If $\varphi(t) \in S_h$, then, by (4.38), $\|F\varphi(t)\|_0 \leq h$, i.e. $F\varphi(t) \in S_h$. Hence F maps the ball S_h into itself. Besides, by inequality (4.29),

$$\|F\varphi_1(t) - F\varphi_2(t)\|_0 \leq \frac{K}{T+p}\left[\int_0^T \|f(\tau, \varphi_1(\tau)) - f(\tau, \varphi_2(\tau))\|d\tau + \right. \\ \left. + \sum_{i=1}^{p} \|I_i(\varphi_1(\tau_i)) - I_i(\varphi_2(\tau_i))\|\right].$$

Because the functions $f(t, x)$ and $I_i(x)$ are continuous, it is evident that if $\|\varphi_1(t) - \varphi_2(t)\|_0 \to 0$, then $\|F\varphi_1(t) - F\varphi_2(t)\|_0 \to 0$. So the mapping is continuous.

To end the proof by applying the Schauder theorem, it remains to show that the set $F(S)$ in $\mathbb{B}$ has the compact closure, i.e. any sequence $\varphi_1(t)$, $\varphi_2(t), \ldots$ has a uniformly convergent subsequence.

Take an arbitrary sequence $\{\varphi_j(t)\}$, $j = 1, 2, \ldots$, $\varphi_j(t) \in F(S_h)$. This sequence is uniformly bounded and equicontinuous on the interval $[0, \tau_1]$ and so, by Arzela theorem, there is a subsequence $\{\varphi_j^{(1)}\}$, uniformly convergent on $[0, \tau_1]$. Consider the subsequence $\{\varphi_j^{(1)}\}$ on the interval $(\tau_1, \tau_2]$. On this interval, the subsequence $\{\varphi_2^{(1)}\}$ is uniformly bounded and equicontinuous, and so it has a subsequence $\{\varphi_j^{(2)}\}$ uniformly convergent on the interval $[0, \tau_2]$. Continuing this process for the intervals $(\tau_2 \tau_3], \ldots (\tau_p, T]$, we see that the sequence $\{\varphi_j(t)\}$ has a subsequence $\{\varphi_i^{(p+1)}\}$, which will converge uniformly on the interval $[0, T]$. This means that $F(S_h)$ has the compact closure in $\mathbb{B}$ and, by Theorem 57, we conclude that F has a fixed point in S_h, i.e. system (4.20) has at least one T-periodic solution.

□

Consider the problem of existence and approximate calculation of periodic solutions of a weakly nonlinear impulsive system with the assumption that this system is subjected to an impulsive effect when the phase point intersects the hypersurfaces $t = \tau_i(x)$. Such a system can be written as

$$\frac{dx}{dt} = A(t)x + f(t, x), \qquad t \neq \tau_i(x), \qquad \Delta x|_{t=\tau_i(x)} = B_i x + I_i(x), \tag{4.40}$$

where $A(t)$, $f(t, x)$ are matrix and vector functions, continuous and T-periodic in t, the constant matrices B_i, the functions $I_i(x)$ and $\tau_i(x)$ are such that

$$B_{i+p} = B_i, \qquad I_{i+p}(x) = I_i(x), \qquad \tau_{i+p} = \tau_i(x) + T \tag{4.41}$$

for all $i \in \mathbf{Z}$ and some natural p. We also assume that the functions $f(t, x)$, $I_i(x)$, and $\tau_i(x)$ satisfy the Lipschitz condition in x uniformly with respect to $t \in \mathbf{R}$ and $i \in \mathbf{Z}$, i.e.

$$\begin{aligned} \|f(t, x') - f(t, x'')\| &\leq L\|x' - x''\|, \\ \|I_i(x') - I_i(x'')\| &\leq L\|x' - x''\|, \qquad (4.42) \\ \|\tau_i(x') - \tau(x'')\| &\leq N\|x' - x''\| \qquad (4.43) \end{aligned}$$

for all x', x'', $\|x'\| \leq h$, $\|x''\| \leq h$, with L, N being Lipschitz constants.

To insure that there are no beatings of solutions of equations (4.20) at the surfaces $t = \tau_i(x)$, we require to hold the following condition for all $i \in \mathbf{Z}$, $\|x\| \leq h$,

$$\tau_i(x) \geq \tau_i((E + B_i)x + I_i(x)) \tag{4.44}$$

Suppose that system (4.40) has a unique T-periodic solution $x = x^*(t)$ in the region $\|x\| \leq T$ and let us denote by τ_i the times at which the integral curve of this solution intersects the surfaces $t = \tau_i(x)$. Then $x = x^8(t)$ is a T-periodic solution of the system

$$\frac{dx}{dt} = A(t)x + f(t,x), \qquad t \neq \tau_i^0, \qquad \Delta x|_{t=\tau_i^0} = B_i x + I_i(x). \tag{4.45}$$

If we additionally assume that the linear part of system (4.45), i.e. the system

$$\frac{dx}{dt} = A(t)x, \qquad t \neq \tau_i^0, \qquad \Delta x|_{t=\tau_i^0} = B_i x \tag{4.46}$$

does not have nontrivial T-periodic solutions, then, by the theorems of the preceding section, the solution $x = x^*(t)$, if the Lipschitz constants L and N are small, can be found as a limit of a uniformly convergent sequence of the periodic functions

$$\varphi_0(t) \equiv 0, \quad \varphi_{m+1}(t) = \int_0^T G(t,\tau)f(\tau,\varphi_m(\tau))d\tau + \\ + \sum_{i=1}^{p} G(t,\tau_i^0)I_i(\varphi_m(\tau_i^0)), \quad m = 0,1,2,\ldots. \tag{4.47}$$

Here $G(t,\tau)$ is a Green's function for the periodic solution, and is given by equality (4.7).

To find periodic solutions of equations (4.40), we will do the following. Fix p points y^1, y^2, ..., y^p, $\|y^j\| < h$, $j = 1, 2, \ldots, p$ and construct the sequence of periodic functions

$$\begin{aligned} x_{m+1}(t,y^1,\ldots,y^p) &= \int_0^T G(t,\tau)f(\tau,x^m(\tau,y^1,\ldots,y^p))d\tau + \\ &+ \sum_{i=1}^{p} G(t,\tau_i(y^i))I_i(y^i), \ m = 0,1,2,\ldots, \\ x_0(t,y^1,\ldots,y^p) &\equiv 0. \end{aligned} \tag{4.48}$$

If such a sequence of periodic functions converges uniformly,

$$\lim_{m\to\infty} x_m(t,y^1,\ldots,y^p) = x_\infty(t,y^1,\ldots,y^p),$$

then the limit function $x_\infty(t,y^1,y^2,\ldots,y^p)$ is a periodic solution of the system

$$\begin{aligned} \frac{dx}{dt} &= A(t)x + f(t,x), \qquad t \neq \tau_i(y^i), \\ \Delta x|_{t=\tau_i(y^i)} &= B_i x + I_i(y^i), \qquad y^i = y^{i+p}. \end{aligned} \tag{4.49}$$

If we choose y^j so that

$$y^i = x_\infty(\tau_i(y^i), y^1, \ldots, y^p), \qquad i = 1, 2, \ldots, p, \tag{4.50}$$

then the function $x_\infty(t, y^1, \ldots y^p)$ will be exactly the needed T-periodic solution of system (4.40).

Hence, to prove that system (4.40) has a periodic solution, we need to resolve two problems: to prove that the sequence of functions (4.48) converges uniformly and to show that system (4.50) has a solution.

It follows from the proof of Theorem 58 that if inequalities (4.6) – (4.8) hold, then the sequence of periodic functions (4.48) converges uniformly to a limit function $x_\infty(t, y^1, \ldots, y^p)$ for any fixed y^1, y^2, ..., y^p, $\|y^j\| < h$, $j = \overline{1,p}$. Let us establish some properties of the functions $x_m(t, y^1, \ldots y^p)$ and $x_\infty(t, y^1, \ldots y^p)$, depended on the points y^1, ..., y^p. The following two lemmas hold.

Lemma 15 *There exists such a positive constant $L_\infty = L_\infty(L, N)$, $L_\infty \to 0$ for $L \to 0$, and $N \to 0$ such that, for any y^1, ..., y^p, $\|y^i\| < h$, and z^1, ..., z^p, $\|x^i\| < h$, $i = 1$, ..., p, the inequality*

$$\|x_m(t, y^1, \ldots, y^p) - x_m(t, z^1, \ldots, z^p) \leq L_\infty \sum_{i=1}^{p} \|y^i - z^i\|$$

holds for all $\overline{\tau}_i < t \leq \underline{\tau}_{i+1}$, $m = 1, 2, \ldots,$ where

$$\overline{\tau}_i = \max(\tau_i(y^i), \ , \tau_i(z^i)), \qquad \underline{\tau}_i = \min(\tau_i(y^i), \ , \tau_i(z^i)).$$

Lemma 16 *The functions $x_\infty(\tau_i(y^i), y^1, \ldots, y^p)$, $i = 1, \ldots, p$ satisfy the Lipschitz condition with respect to every variable $y^1, \ldots, y^p$ with the Lipschitz constant $L_0 = L_0(L, N)$, $L_0 \to 0$ for $L \to 0$ and $N \to 0$.*

These lemmas can be proved in the same way as Lemmas 4 and in [140] or as Lemmas 19 and 20 in the succeeding section.

Now we take up the problem of solvability of system (4.50). By Lemma 16, the functions $x_\infty(\tau_i(y^i), y^1, \ldots y^p)$, $i = 1$, ..., p, satisfy the Lipschitz conditions in y^i, $i = 1$, ..., p, with the Lipschitz constant $L_0 = L_0(L, N)$. If we assume that the constants L and N are so small that $pL_0(L, N) < 1$, then equations (4.50) have a unique solution in the region $\|y\| \leq h$, which can be found by a simple iteration of determining approximations to this solution by the formulas

$$y^i_{k+1} = x_\infty(\tau_i(y^i_k), y^1_k, \ldots, y^p_k), \qquad i = 1, 2, \ldots, p, \quad k = 0, 1, \ldots,$$

and taking, for the initial approximation, the function $y_0^i = 0$, $i = 1, \ldots, p$.

It is not possible, in general, to find the functions $x_\infty(\tau_i(y^i), y^1, \ldots, y^p)$, so to find the points of intersection of the periodic solution and the surfaces $t = \tau_i(x)$, we need to solve the equations

$$y^i = x_m(\tau_i(y^i), \ , y^1, \ldots, y^p).$$

If $pL_0(L, N) < 1$, these equations have a unique solution in the region $\|y\| < h$, $y^1 = y_m^1$, $y^2 = y_m^2$, ..., $y^p = y_m^p$, and the difference between this solution and a solution of the exact equations (4.50) can be found as follows

$$\begin{aligned}
\|y_*^i - y_m^i\| &= \|x_\infty(\tau_i(y_*^i), y_*^1, \ldots, y_*^p) - x_m(\tau_i(y_m^i), y_m^1, \ldots, y_m^p)\| \leq \\
&\leq \|x_\infty(\tau_i(y_*^i), y_*^1, \ldots, y_*^p) - x_\infty(\tau_i(y_m^i), y_m^1, \ldots, y_m^p)\| + \\
&\quad + \|x_\infty(\tau_i(y_m^i), y_m^1, \ldots, y_m^p) - x_m(\tau(y_m^i), y_m^1, \ldots, y_m^p)\| \leq \\
&\leq L_0 \sum_{i=1}^{p} \|y_*^i - y_m^i\| + (LN)^m h, \quad i = 1, \ldots, p,
\end{aligned}$$

i.e.

$$\|y_*^i - y_m^i\| \leq \frac{(LN)^m ph}{1 - pL}, \qquad i = 1, \ldots, p. \tag{4.51}$$

Formulating the obtained results, we get the following statement.

Theorem 60 *Let the functions defining the T-periodic system (4.40) be such that inequalities (4.42) – (4.44) hold. If the linear T-periodic system*

$$\frac{dx}{dt} = A(t)x, \qquad t \neq \tau_i(0), \qquad \Delta x|_{t=\tau_i(0)} = B_i x$$

does not have nonzero T-periodic solutions, then, for sufficiently small Lipschitz constants L and N, system (4.40) has a unique T-periodic solution.

4.3 Numerical-Analytical method for finding periodic solutions

A numerical-analytical method for finding periodic solutions of a certain class of systems of ordinary differential equations as a limit of a uniformly convergent sequence of periodic functions has been suggested and developed in [131, 146].

We will use the ideas of the numerical-analytical method to study periodic nonlinear impulsive equations

$$\frac{dx}{dt} = f(t,x), \qquad t \neq \tau_i \qquad \Delta x|_{t=\tau_i} = I_i(x). \tag{4.52}$$

Here $x \in D \subset \mathbf{R}^n$, D is a closed bounded region of the Euclidean space $\mathbf{R}^n$, the function $f(t,x)$ is continuous (piecewise continuous with first kind discontinuities at $t = \tau_i$), $I_i(x)$ are continuous functions, defined for

$$(t,x) \in \mathbf{R} \times \mathbf{D} = (-\infty, \infty) \times \mathbf{D}. \tag{4.53}$$

We assume that system (4.52) is T-periodic in t. This means that the function $f(t,x)$ is T-periodic in t uniformly with respect to $x \in D$, the functions $I_i(x)$ and the times τ_i are such that

$$I_{i+p}(x) = I_i(x), \qquad \tau_{i+p} = \tau_i + T \tag{4.54}$$

for all $i \in \mathbf{Z}$, $x \in D$, and some natural p.

We also assume that the functions $f(t,x)$ and $I_i(x)$ satisfy the Lipschitz condition in $x \in D$ uniformly with respect to $t \in \mathbf{R}$, $i \in \mathbf{Z}$,

$$\begin{aligned} \|f(t,x') - f(t,x'')\| &\leq K_1\|x' - x''\|, \\ \|I_i(x') - I_i(x'')\| &\leq K_2\|x' - x''\| \end{aligned} \tag{4.55}$$

with the Lipschitz constants K_1 and K_2.

Let

$$\max_{\substack{t\in[0,T] \\ x\in D}} \|f(t,x)\| \leq M, \qquad \max_{\substack{1\leq i\leq p \\ x\in D}} \|I_i(x)\| \leq M. \tag{4.56}$$

We will call a d-neighborhood of a point $x_0 \in \mathbf{R}^n$ the set of the points $x \in \mathbf{R}^n$ such that $\|x - x_0\| < d$.

For system (4.52) and the region D, we take a subset D_1 of $\mathbf{R}^n$, the points of which are contained in the set D together with their $\frac{MT}{2}\left(1 + \frac{4p}{T}\right)$-neighborhood.

The idea of the subsequent investigations is based on the following lemma.

Lemma 17 *Let in the system*

$$\frac{dx}{dt} = f(t) - \mu, \qquad t \neq \tau_i, \qquad \Delta x|_{t=\tau_i} = I_i \tag{4.57}$$

the function $f(t)$ be T-periodic, the constant vectors I_i and the sequence of times τ_i be such that $I_{i+p} = I_i$ and $\tau_{i+p} = \tau_i + T$. Then there exists a constant vector μ, for which the solution that passes for $t = 0$ through a given point x_0 is T-periodic.

Proof. Indeed, by integrating equations (4.57), we see that the vector μ can be taken to be

$$\mu = \frac{1}{T}\left[\int_0^T f(t)dt + \sum_{i=1}^{p} I_i\right]. \tag{4.58}$$

□

Denote by $\overline{f(t)}$ the integral mean of the function $f(t)$ taken over the interval $[0, T]$, i.e.

$$\overline{f(t)} = \frac{1}{T}\int_0^T f(\sigma)d\sigma.$$

In the sequel we will need the following statement.

Lemma 18 *Let $f(t)$ be a function, continuous (piecewise continuous with first kind discontinuities) on the interval $[0, T]$. For all $t \in [0, T]$, the inequality*

$$\left\|\int_0^T [f(\sigma) - \overline{f(t)}]d\sigma\right\| \leq 2t\left(1 - \frac{t}{T}\right) \max_{t\in[0,T]} \|f(t)\| \tag{4.59}$$

holds.

Proof. Because

$$\begin{aligned}
\int_0^t \left[f(\sigma) - \frac{1}{T}\int_0^T f(\sigma)d\sigma\right] d\sigma &= \int_0^t f(\sigma)d\sigma - \frac{t}{T}\int_0^T f(\sigma)d\sigma = \\
&= \left(1 - \frac{t}{T}\right)\int_0^t f(\sigma)d\sigma - \frac{t}{T}\int_t^T f(\sigma)d\sigma,
\end{aligned}$$

we have that

$$\begin{aligned}
\left\|\int_0^t [f(\sigma) - \overline{f(t)}]d\sigma\right\| &\leq \left(1 - \frac{t}{T}\right)\int_0^t \|f(\sigma)\|d\sigma + \frac{t}{T}\int_t^T \|f(\sigma)\|d\sigma \\
&\leq 2t\left(1 - \frac{t}{T}\right) \max_{t\in[0,T]} \|f(t)\|.
\end{aligned}$$

□

Let the region D and the functions $f(t,x)$, $I_i(x)$, besides the assumptions made before, satisfy the following:

1) there exists a nonempty closed set D_0, contained together with its $\frac{MT}{2}\left(1+\frac{4p}{T}\right)$-neighborhood in the region D;

2) the Lipschitz constants K_1 and K_2 are such that

$$\frac{K_1T}{3} + pK_2\left(2+\frac{K_1T}{3}\right) < 1. \tag{4.60}$$

Now we begin the construction of periodic solutions of system (4.52). Suppose that system (4.52) has a T-periodic solution and the point $x_0 \in D$, through which the solution passes at $t = 0$, is given. An algorithm of constructing such a solution is given by the following theorem.

Theorem 61 *If system (4.52), satisfying the foregoing conditions, has a T-periodic solution $x = \varphi(t)$, which passes through a point $x_0 \in D_0$ at $t = 0$, then this solution is a limit of a uniformly convergent sequence of the periodic functions*

$$\varphi(t) = \lim_{m\to\infty} x_m(t,x_0), \tag{4.61}$$

defined on the period $t \in [0,T]$ by the formula

$$\begin{aligned} x_{m+1}(t,x_0) &= x_0 + \int_0^t [f(\sigma,x_m(\sigma,x_0)) - \overline{f(\sigma,x_m(\sigma,x_0))}]d\sigma + \\ &\quad + \sum_{0<\tau_i<t} I_i(x_m(\tau_i,x_0)) - t\overline{I(x_m(\tau_i,x_0))}, \qquad (4.62) \\ x_0(t,x_0) &\equiv x_0, \quad m = 0,1,2,\ldots \end{aligned}$$

Here

$$\overline{f(t,x(t))} = \frac{1}{T}\int_0^T f(\sigma,x(\sigma))d\sigma,$$

$$\overline{I(x(\tau_i,x_0))} = \frac{1}{T}\sum_{i=1}^{p} I_i(x(\tau_i,x_0)).$$

Proof. Every periodic function $x_m(t, x_0)$ defined on the period by (4.62) admits, according to Lemma 18, the estimate

$$\|x_m(t, x_0) - x_0\| \leq M\alpha(t) + 2pM, \tag{4.63}$$

where $\alpha(t) = 2t\left(1 - \frac{t}{T}\right) \leq \frac{T}{2}$. Whence it follows that every function $x_m(t, x_0)$ is defined for any natural m and $x_m(t, x_0) \in D$ for all $-\infty < t < \infty$ (because it is periodic) and $x_0 \in D_0$.

To prove that the sequence of functions (4.62) converges, make an estimate for $x_{m+1}(t, x_0) - x_m(t, x_0)$. For $m = 0$, by (4.63), we have

$$\|x_1(t, x_0) - x_0\| \leq M\alpha(t) + 2pM \leq \frac{MT}{2}\left(1 + \frac{4p}{T}\right). \tag{4.64}$$

If $m = 1$, then

$$\|x_2(t, x_0) - x_1(t, x_0)\| \leq K_1\left(\frac{MT}{3} + 2pT\right)\alpha(t) +$$

$$+2pK_2\frac{MT}{2}\left(1 + \frac{4p}{T}\right) = N_1\alpha(t) + M_1 \leq \frac{N_1T}{2} + M_1, \tag{4.65}$$

where

$$N_1 = K_1\left(\frac{T}{3} + 2pM\right), \qquad M_1 = pK_2MT\left(1 + \frac{4p}{T}\right).$$

If we assume that

$$\|x_m(t, x_0) - x_{m-1}(t, x_0)\| \leq N_{m-1}\alpha(t) + M_{m-1} \leq \frac{N_{n-1}T}{2} + M_{m-1},$$

then, by using (4.62), we find that

$$\|x_{m+1}(t, x_0) - x_m(t, x_0)\| \leq K_1\left(\frac{N_{m-1}T}{T} + M_{m-1}\right)\alpha(t) +$$

$$+2pK_2\left(\frac{N_{n-1}T}{2} + M_{m-1}\right).$$

By using induction, we prove that, for an arbitrary natural m,

$$\|x_{m+1}(t, x_0) - x_m(t, x_0)\| \leq N_m\alpha(t) + M_m \leq \frac{T}{2}N_m + M_m, \tag{4.66}$$

where the constants N_m and M_m satisfy the recurrent relations

$$N_{m+1} = \frac{K_1 T}{3} n_m + K_1 M_m, \qquad M_{m+1} = pK_2 T N_m + 2pK_2 M_m,$$
$$N_0 = M, \qquad M_0 = 2pM. \tag{4.67}$$

Whence it can be seen that for the sequence of functions $x_m(t, x_0)$ to be uniformly convergent, it is sufficient that all the solution of difference equations (4.67) go to zero for $m \to \infty$. For this, it is necessary and sufficient that the eigen values of the matrix

$$\begin{pmatrix} \frac{K_1 T}{3} & K_1 \\ pK_2 T & 2pK_2 \end{pmatrix} \tag{4.68}$$

of these equations satisfy the inequalities $|\lambda_1| < 1$ and $|\lambda_2| < 1$.

As known (see e.g. [38]), the roots of the polynomial $F(\lambda) = \lambda^2 + a\lambda + b$, where a and b are real numbers, lie inside the circle $|\lambda| < 1$ in the complex plane if

$$-1 + |a| < b < 1. \tag{4.69}$$

The characteristic polynomial of matrix (4.68) is

$$F(\lambda) = \lambda^2 - \left(\frac{K_1 T}{3} + 2pK_2\right)\lambda - \frac{pK_1 K_2 T}{3}.$$

For this polynomial, conditions (4.69) become

$$-1 + \frac{K_1 T}{3} + 2pTK_2 < -\frac{pK_1 K_2 T}{3} < 1, \tag{4.70}$$

which are equivalent to inequality (4.60).

Hence, by condition (4.60), the eigen values of matrix (4.68) satisfy $|\lambda_1| < 1$ and $|\lambda_2| < 1$, and this implies that the sequence of the functions $x_m(t, x_0)$ converges uniformly on the interval $[0, T]$, and, since the functions are periodic, on the whole real axis.

We set

$$\lim_{m \to \infty} x_m(t, x_0) = x_\infty(t, x_0)$$

and show that

$$\varphi(t) = x_\infty(t, x_0). \tag{4.71}$$

By passing to the limit for $m \to \infty$, we see that $x_\infty(t, x_0)$ is a periodic solution of the equation

$$\begin{aligned} x(t, x_0) &= x_0 + \int_0^t [f(\sigma, x(\sigma, x_0)) - \overline{f(\sigma, x(\sigma, x_0))}]d\sigma + \\ &+ \sum_{0<\tau_i<t} I_i(x(\tau_i, x_0)) - t\overline{I(x(\tau_i, x_0))}. \end{aligned} \tag{4.72}$$

Moreover, by Lemma 4.53, inequality (4.60), and the estimate

$$\begin{aligned} \|x_{m+k}(t, x_0) - x_m(t, x_0)\| &\leq \sum_{i=0}^{k-1} \|x_{m+i+1}(t, x_0) - x_{m+i}(t, x_0)\| \leq \\ &\leq \sum_{i=0}^{k-1} M_{m+i+1} \leq \sum_{i=0}^{k-1} q^{m+i} \frac{MT}{2}\left(1 + \frac{4p}{T}\right), \end{aligned} \tag{4.73}$$

where q is a positive eigen value of matrix (4.68), we get the inequalities

$$\|x_\infty(t, x_0) - x_0\| \leq \frac{MT}{2}\left(1 + \frac{4p}{T}\right), \tag{4.74}$$

$$\|x_\infty(t, x_0) - x_m(t, x_0)\| \leq q^m (1-q)^{-1} \frac{MT}{2}\left(1 + \frac{4p}{T}\right). \tag{4.75}$$

By the conditions of the theorem, the function $x = \varphi(t)$ is a periodic solution of equations (4.52) and so it satisfies the equation

$$x(t) = x_0 + \int_0^t f(\sigma, x(\sigma))d\sigma + \sum_{0<\tau_i<t} I_i(x(\tau_i)), \qquad t \geq 0 \tag{4.76}$$

and has the property

$$\frac{1}{T}\int_0^T f(t, \varphi(t))dt + \frac{1}{T}\sum_{i=1}^{p} I_i(\varphi(\tau_i)) = 0. \tag{4.77}$$

It follows from relations (4.72), (4.76), and (4.77) that $\varphi(t)$, similarly to $x_\infty(t, x_0)$, is a T-periodic solution of equation (4.72). So to finish the proof of the theorem, it is sufficient to show that equation (4.72) can not have two distinct T-periodic solutions. Suppose the converse. Let $x(t, x_0)$

and $y(t, x_0)$ be two distinct solutions of equation (4.72). Then the difference $x(t, x_0) - y(t, x_0)$, for $0 \leq t \leq T$, can be estimated as

$$x(t,x_0) - y(t,x_0) \leq \left(1 - \frac{t}{T}\right) \int_0^t K_1 \|x(\sigma,x_0) - y(\sigma,x_0)\| d\sigma +$$
$$+ \frac{t}{T} \int_0^T K_1 \|x(\sigma,x_0) - y(\sigma,x_0)\| d\sigma +$$
$$+ \sum_{0<\tau_i<t} K_2 \|x(\tau_i,x_0) - y(\tau_i,x_0)\| + \frac{t}{T} \sum_{i=1}^{p} K_2 \|x(\tau_i,x_0) - y(\tau_i,x_0)\|.$$

By setting $\|x(t,x_0) - y(t,x_0)\| = r(t)$, the latter inequality can be rewritten as

$$r(t) \leq \left(1 - \frac{t}{T}\right) \int_0^t K_1 r(\sigma) d\sigma + \frac{t}{T} \int_0^T K_1 r(\sigma) d\sigma +$$
$$+ \sum_{0<\tau_i<t} K_2 r(\tau_i) + \frac{t}{T} \sum_{i=1}^{p} K_2 r(\tau_i). \tag{4.78}$$

Let

$$\max_{t \in [0,T]} r(t) = r_0 \geq 0.$$

By integrating this inequality at the mth step, we get

$$r(t) \geq N_m \alpha(t) + M_m, \tag{4.79}$$

where the constants N_m and M_m are given as solutions of the system of the difference equations

$$\begin{pmatrix} N_{m+1} \\ M_{m+1} \end{pmatrix} = \begin{pmatrix} \frac{K_1 T}{3} & K_1 \\ pK_2 T & 2pK_2 \end{pmatrix} \begin{pmatrix} N_m \\ M_m \end{pmatrix} \tag{4.80}$$

with the initial conditions $N_0 = 0$, $M_0 = r_0$, i.e.

$$\begin{pmatrix} N_m \\ M_m \end{pmatrix} = \begin{pmatrix} \frac{K_1 T}{3} & K_1 \\ pK_2 T & 2pK_2 \end{pmatrix}^m \begin{pmatrix} 0 \\ r_0 \end{pmatrix}. \tag{4.81}$$

Whence it is easy to see that if inequality (4.60) holds, then $N_m \to 0$ and $M_m \to 0$ as $m \to 0$. It follows from (4.79) that $r(t) \equiv 0$, i.e. $x(t, x_0) = y(t, x_0)$, which finishes the proof.

□

We now consider the problem of existence of periodic solutions of system (4.52).

By the theorem just proved, finding a periodic solution of system (4.52) is reduced to calculating the functions $x_m(t, x_0)$ if we know that such a solution exists and the point x_0, through which it passes at $t = 0$, is known.

If we know the functions $x_m(t, x_0)$, the problem of existence of periodic solutions can be solved in the following manner.

Denote by $\Delta(x_0)$ the expression

$$\Delta(x_0) = \frac{1}{T}\int_0^T f(\sigma, x_\infty(\sigma, x_0))d\sigma + +\frac{1}{T}\sum_{i=1}^{p} I_i(x_\infty(\tau_i, x_0)), \tag{4.82}$$

where $x_\infty(t, x_0) = \lim_{m\to\infty} x_m(t, x_0)$ and $x_m(t, x_0)$ are given by (4.62). Because $x_\infty(t, x_0)$ satisfies the relation

$$\begin{aligned} x_\infty(t, x_0) &= x_0 + \int_0^t [f(\sigma, x_\infty(\sigma, x_0)) - \overline{f(\sigma, x_\infty(\sigma, x_0))}]d\sigma + \\ &+ \sum_{0<\tau_i<t} I_i(x_\infty(\tau_i, x_0)) - \overline{tI(x_\infty(\tau_i, x_0))}, \end{aligned}$$

the function $x_\infty(t, x_0)$ is a periodic solution of system (4.52) for $\Delta(x_0) = 0$.

Hence the existence of periodic solutions of system (4.52) is related to existence of zeros of the function $\Delta(x_0)$. However, to find the function $\Delta(x_0)$ is practically impossible, so there is the problem: how by using the functions

$$\Delta_m(x_0) = \frac{1}{T}\int_0^T f(\sigma, x_m(\sigma, x_0))d\sigma + \frac{1}{T}\sum_{i=1}^{p} I_i(x_m(\tau_i, x_0)) \tag{4.83}$$

to make a conclusion on whether the function $\Delta(x_0)$ has zeros, i.e. system (4.52) has periodic solutions? This problem can be solved by using the following theorem.

Theorem 62 *Let for impulsive differential system (4.52) the following conditions hold: a) for some integer $m \geq 0$, the mapping $\Delta_m : D_0 \to \mathbf{R}^n$ has an isolated singular point, i.e. the index of $\Delta_m(x_0) = 0$ is nonzero; b) there exists a closed convex region $\bar{D}_1 \subseteq D_0$ containing one singular point x^0, such that the inequality*

$$\inf_{x\in\Gamma_{D_1}} \|\Delta_m(x)\| > \frac{q^m}{1-q}\left(K_1 + \frac{K_2 p}{T}\right)\frac{MT}{2}\left(1 + \frac{4p}{T}\right), \tag{4.84}$$

where q is a positive eigen value of matrix (4.68), holds on its boundary. Then system (4.52) has a T-periodic solution $x = \varphi(t)$ with $\varphi(0) \in D_0$.

Proof. The index of an isolated singular point of a continuous mapping $\Delta_m(x)$ equals to the characteristic of the vector field which this map generates on a sufficiently small sphere S^{n-1} centered at x^0. Because there is only one singular point in D_1 and D_1 is homeomorphic to a ball in $\mathbf{R}^n$, the characteristic of the vector field Δ_m on the sphere S^{n-1} equals to its characteristic on Γ_{D_1} The vector fields $\Delta_m(x)$ and $\Delta(x)$ are *homotopic* in Γ_{D_1}. This follows since there is a family of the vector fields, depended on a parameter θ, $0 \leq \theta \leq 1$, everywhere continuous in Γ_{D_1},

$$V(\theta, x_0) = \Delta_m(x_0) + \theta(\Delta(x_0) - \Delta_m(x_0)),$$

connecting the vector fields $V(0, x_0) = \Delta_m(x_0)$ and $V(1, x_0) = \Delta(x_0)$ and not equal to zero for any value of the parameter. Indeed, by using estimate (4.75), we have

$$\begin{aligned}||\Delta(x_0) - \Delta_m(x_0)|| &\leq \frac{1}{T}\int_0^T K_1||x_\infty(\sigma, x_0) - x_m(\sigma, x_0)||d\sigma + \\ &+\frac{1}{T}\sum_{i=1}^{p} K_2||x_\infty(\tau_i, x_0) - x_m(\tau_i, x_0)|| \leq \\ &\leq q^m(1-q)^{-1}\left(K_1 + \frac{pK_2}{T}\right)\frac{MT}{2}\left(1 + \frac{4p}{T}\right). \qquad (4.85)\end{aligned}$$

Taking into account (4.74) we see that, for all $x \in \Gamma_{D_1}$,

$$||V(\theta, x)|| \geq ||\Delta_m(x)|| - ||\Delta_m(x) - \Delta(x)|| > 0.$$

Because the characteristics of the fields homotopic on a compact set are equal, the characteristic of the field $\Delta(x)$ on Γ_{D_1} equals to the index of the singular point in the region D_1, on the boundary of which we have inequality (4.74).

□

We can always calculate the index if system (4.52) is two dimensional. If the dimension is greater than two, to calculate the index becomes more difficult. However, there are a number of criteria that allow to decide in this case whether the index is different from zero or not.

In particular, if the function $\Delta_m(x)$ is continuously differentiable in a neighborhood of the point x^0 and $\left.\frac{\partial \Delta_m(x)}{\partial x}\right|_{x=x^8} \neq 0$, then the index of the point x^0 is not equal to zero.

There is a certain arbitrariness in the choice of the region D_1 such that inequality (4.84) hold on its boundary. In particular, for impulsive periodic systems in the standard form

$$\frac{dx}{dt} = \epsilon f(t,x), \qquad t \neq \tau_i, \qquad \Delta x|_{t=\tau_i} = \epsilon I_i(x), \tag{4.86}$$

where ϵ is a small positive parameter, the region D_1 can be chosen to be any sufficiently small ball centered at the singular point.

Thus the problem of existence of T-periodic solutions of system (4.86) is settled by the following statement.

Theorem 63 *Let the functions $f(t,x)$, $I_i(x)$ and the times τ_i, which define system (4.86), be as in equations (4.52). If the averaged system*

$$\frac{dy}{dt} = \epsilon f_0(y) \equiv \frac{\epsilon}{T}\left[\int_0^T f(t,y)dt + \sum_{i=1}^{p} I_i(y)\right]$$

has an isolated stationary point y_0, $f(y_0) = 0$, and the index of the function $f(y)$ is not equal to zero at the point y_0, then system (4.85), for sufficiently small values of the parameter ϵ, has a T-periodic solution $x = \varphi(t,\epsilon)$, $\lim_{\epsilon\to 0}\varphi(t,\epsilon) = 0$.

This theorem substantiates the averaging principle for periodic impulsive systems in general form.

Note that if system (4.52) is one-dimensional, i.e. x is a scalar quantity, then Theorem 63 can be strengthened by not requiring that the singular point be isolated.

Theorem 64 *Let, for the scalar T-periodic impulsive equation*

$$\frac{dx}{dt} = f(t,x), \qquad t \neq \tau_i, \qquad \Delta x|_{t=\tau_i} = I_i(x), \tag{4.87}$$

where the functions $f(t,x)$ and $I_i(x)$ are defined for all $t \in \mathbf{R}$, $x \in [a,b]$, the function $\Delta_m(x)$, defined for some m by (4.74), satisfy the inequalities

$$\begin{aligned} \min_{a+h\le x\le b-h} \Delta_m(x) &\le -\frac{q^m}{1-q}\left(K_1 + \frac{K_2 p}{T}\right)h, \\ \max_{a+h\le x\le b-h} \Delta_m(x) &\ge \frac{q^m}{1-q}\left(K_1 + \frac{K_2 p}{T}\right)h, \end{aligned} \tag{4.88}$$

where $h = \dfrac{MT}{2}\left(1 + \dfrac{4p}{T}\right), \quad b - a > 2h.$

Then equation (4.87) has at least one periodic solution $x = \varphi(t)$, *for which* $a + h \leq \varphi(0) \leq b - h$.

Proof. Let x_1 and x_2 be such points of the segment $[a+h, b-h]$ that

$$\Delta_m(x_1) = \min_{a+h \leq x \leq b-h} \Delta_m(x), \qquad \Delta_m(x_2) = \max_{a+h \leq x \leq b-h} \Delta_m(x).$$

By using inequalities (4.85) and (4.87), we have

$$\begin{aligned} \Delta(x_1) &= \Delta_m(x_1) + [\Delta(x_1) - \Delta_m(x_1)] \leq 0, \\ \Delta(x_2) &= \Delta_m(x_2) + [\Delta(x_2) - \Delta_m(x_2)] \geq 0. \end{aligned}$$

Because the function $\Delta(x)$ is continuous, it follows from these two inequalities that there exists a point $x^0 \in [x_1, x_2]$ such that $\Delta(x^0) = 0$. This is sufficient for existence of a periodic solution of impulsive equation (4.77). □

Now, for impulsive systems we will study the questions of existence of a periodic solution and constructing its approximation with the assumption that the impulsive effect in the system occurs when the phase point intersects hypersurfaces in the extended phase space.

Consider the system of equations

$$\frac{dx}{dt} = f(t,x), \qquad t \neq \tau_i(x), \qquad \Delta x|_{t=\tau_i(x)} = I_i(x), \tag{4.89}$$

where $f(t,x)$ is a function continuous in x (piecewise continuous in t), and $f(x,t)$, $I_i(x)$, $\tau_i(x)$ satisfy the Lipschitz condition

$$\begin{aligned} \|f(t,x') - f(t,x'')\| &\leq K_1\|x' - x''\|, \\ \|I_i(x') - I_i(x'')\| &\leq K_2\|x' - x''\|, \\ |\tau_i(x') - \tau_i(x'')| &\leq N\|x' - x''\| \end{aligned} \tag{4.90}$$

uniformly with respect to $t \in \mathbf{R}$, $i \in \mathbf{Z}$ for all x', $x'' \in D$, where D is a closed bounded domain in the Euclidean space $\mathbf{R}^n$.

We also assume that system (4.89) is T-periodic in t, i.e. the function $f(t,x)$ is T-periodic in t, and the functions $I_i(x)$ and $\tau(x)$ are such that, for some natural p,

$$I_{i+p}(x) = I_i(x), \qquad \tau_{i+p}(x) = \tau_i(x) + T \tag{4.91}$$

for all $x \in D$, $i \in \mathbf{Z}$.

Because in system (4.89) there may beatings of solutions occurring at the surfaces $t = t_i(x)$, to exclude this possibility we require that the functions $t_i(x)$ and $I_i(x)$ satisfy the inequalities

$$\tau_i(x) \geq \tau_i(x + I_i(x)) \tag{4.92}$$

for all $x \in D$ and the separating condition for the surfaces $t =_i (x)$, i.e.

$$\min_{x\in D} \tau_{i+1}(x) - \max_{x\in D} \tau_i(x) > 0. \tag{4.93}$$

It should be noted that even if the conditions that exclude the beatings of the solutions at the surfaces $t = \tau_i(x)$ hold, still systems (4.89) and (4.52) are essentially different. The difference is that the iteration procedure used for equations (4.52) can not be directly transferred to this case because now, at every step, the function $x_m(t,x_0)$ will have discontinuity points, which are in general different from the discontinuity points of the function $x_{m-1}(t,x_0)$. And so the functions $x_m(t,x_0)$ do not converge uniformly with respect to $t \in [0,T]$.

To construct an iterative process for finding T-periodic solutions of equations (4.89) we do as follows. Fix p points $y_i \in D$, $i = \overline{1,p}$ and construct the sequence of T-periodic functions defined on the period $[0,T]$ by the formulas

$$\begin{aligned} x_{m+1}(t,x_0,y^1,\ldots,y^p) = x_0 + \int_0^t [f(\sigma, x_m(\sigma,x_0,y^1,\ldots,y^p)) - \\ -\overline{f(\sigma, x_m(\sigma,y^1,\ldots,y^p))}]d\sigma + \sum_{0<\tau_i(y^i)<t} I_i(y^i) - t\overline{I(y^i)}, \end{aligned} \tag{4.94}$$

where $\overline{I(y^i)} = \dfrac{1}{T}\sum_{i=1}^p I_i(y^i)$.

If this sequence of functions converges uniformly for $t \in [0,T]$, then the limit function

$$x_\infty(t,x_0,y^1,\ldots,y^p) = \lim_{m\to\infty} x_m(t,x_0,y^1,\ldots,y^p) \tag{4.95}$$

is a T-periodic solution of the system

$$\frac{dx}{dt} = f(t,x) - \bar{\Delta}(x, y^1, \ldots, y^p), \qquad t \neq \tau_i(y^i),$$
$$\Delta x|_{t=\tau_i(y^i)} = I_i(y^i). \tag{4.96}$$

Here

$$\bar{\Delta}(x_0, y^1, \ldots, y^p) = \overline{f(t, x_\infty(t, x_0, y^1, \ldots, y^p)} + \frac{1}{T}\sum_{i=1}^{p} I_i(y^i),$$
$$y^{i+p} = y^i. \tag{4.97}$$

If x_0, y^1, ..., y^p are chosen so that

$$\bar{\Delta}(x, y^1, \ldots, y^p) = 0,$$
$$y^i = x_\infty(\tau_i(y^i), x_0, y^1, \ldots, y^p), \qquad i = \overline{1,p}, \tag{4.98}$$

then the limit function $x_\infty(t, x_0, y^1, \ldots y^p)$ will be the needed periodic solution of system (4.89).

Hence, the question of existence of T-periodic solutions of equation (4.89) can be reduced to finding conditions that would make the convergence of the sequence of functions (4.94) uniform and equations (4.98) solvable.

It follows from the proof of Theorem 61 that if the functions $f(t,x)$ and $I_i(x)$ satisfy inequalities (4.90) and conditions 1) and 2) hold, then, for any $x_0 \in D_1$ and $y^1, \ldots, y^p \in D$, the sequence of functions (4.94) converges uniformly to a limit T-periodic function $x_\infty(t, x_0, y^1, \ldots y^p)$. Let us establish some properties of the functions $x_m(t, x_0, y^1, \ldots y^p)$ and $x_\infty(t, x_0, y^1, \ldots y^p)$.

Lemma 19 *There exists such a positive constant $K' = K'(K_1, K_2)$ that, for any y^1, ..., y^p and z^1, ..., $z^p \in D$, the following inequality holds,*

$$\|x_m(t, x_0, y^1, \ldots, y^p) - x_m(t, x_0, z^1, \ldots, z^p)\| \leq K' \sum_{i=1}^{p} \|y^i - z^i\| \tag{4.99}$$

for all $\overline{\tau}_i < t < \underline{\tau}_{i+1}$, $m = 1, 2, \ldots,$ where

$$\overline{\tau}_i = \max(\tau_i(y^i), \ , \tau_i(z^i)), \qquad \underline{\tau}_i = \min(\tau_i(y^i), \ , \tau_i(z^i)).$$

Proof. For $m = 0$ we have from relations (4.94)

$$\|x_1(t, x_0, y^1, \ldots, y^p) - x_1(t, x_0, z^1, \ldots, z^p)\| \leq 2K_2 \sum_{i=1}^{p} \|y^i - z^i\|$$

for all t, $\overline{\tau}_i < t \leq \underline{\tau}_{i+1}$.

If $m = 1$, then, for $\overline{\tau}_i < t \leq \underline{\tau}_{i+1}$,

$$\|x_2(t, x_0, y^1, \ldots, y^p) - x_2(t, x_0, z^1, \ldots, z^p)\| \leq$$
$$\leq 2K_2[K_1\alpha(t) + 1] \sum_{i=1}^{p} \|y^i - z^i\|.$$

By using induction, we see that for all $m = 0, 1, 2, \ldots$,

$$\overline{\tau}_i < t \leq \underline{\tau}_{i+1},$$

$$\|x_{m+1}(t, x_0, y^1, \ldots, y^p) - x_{m+1}(t, x_0, z^1, \ldots, z^p)\| \leq$$
$$\leq 2K_2[K'_m\alpha(t) + 1] \sum_{i=1}^{p} \|y^i - z^i\|.$$

where the constants K'_m satisfy the recurrence relation

$$K'_{m+1} = K_1 \left(\frac{K'_m T}{3} + 1 \right), \qquad K'_0 = 0.$$

Inequality (4.60) implies that the sequence K'_m is uniformly bounded:

$$K'_m \leq \frac{K_1}{1 - \frac{K_1 T}{3}}.$$

So, for all $m = 0, 1, 2, \ldots$ and $\overline{\tau}_i < t \leq \underline{\tau}_{i+1}$,

$$\|x_{m+1}(t, x_0, y^1, \ldots, y^p) - x_{m+1}(t, x_0, z^1, \ldots, z^p)\| \leq$$
$$\leq 2K_2 \left[\frac{K_1}{1 - \frac{K_1 T}{3}} \alpha(t) + 1 \right] \sum_{i=1}^{p} \|y^i - z^i\|.$$

To end the proof, we can set

$$K' = 2K_2 \left(1 + \frac{K_1 T}{2} \cdot \frac{1}{1 - \frac{K_1 T}{3}} \right).$$

□

It follows from Lemma 19 that the functions $x_m(\tau(y^i), x_0, y^1, \dots, y^p)$ satisfy the Lipschitz conditions with respect to y^i. Indeed, let y^i, $z^i \in D$ and $\tau_i(y^i) < \tau_i(z^i)$. Then

$$\begin{aligned} &\|x_m(\tau_i(y^i), x_0, y^1, \dots, y^p) - x_m(\tau_i(z^i), x_0, y^1, \dots, z^i, \dots, y^p)\| \leq \\ &\leq K'\|y^i - z^i\| + \|x_m(\tau_i(y^i), x_0, y^1, \dots, z^i, \dots, y^p) - \\ &- x_m(\tau_i(z^i), x_0, y^1, \dots, z^i, \dots, y^p)\|. \end{aligned} \tag{4.100}$$

Because the function $x(t, x_0, y^1, \dots, z^i, \dots y^p)$ is continuous in the interval $(\tau_i(y^i), \tau_i(z^i))$, it is easy to get from (4.94) the estimate

$$\begin{aligned} &\|x_m(\tau_i(y^i), x_0, y^1, \dots, z^i, \dots, y^p) - x_m(\tau_i(z^i), x_0, y^1, \dots, z^i, \dots, y^p)\| \leq \\ &\leq \left(2 + \frac{p}{T}\right) M|\tau_i(y^i) - \tau_i(z^i)| \leq \left(2 + \frac{p}{T}\right) MN\|y^i - z^i\|. \end{aligned}$$

Finally we have

$$\begin{aligned} &\|x_m(\tau_i(y^i), x_0, y^1, \dots, z^i, \dots, y^p) - x_m(\tau_i(z^i), x_0, y^1, \dots, z^i, \dots, y^p)\| \leq \\ &\leq \left[K' + \left(2 + \frac{p}{T}\right) MN\right] \|y^i - z^i\| \end{aligned} \tag{4.101}$$

for all $m = 1, 2, \dots$. By using inequality (4.101), we get the following statement.

Lemma 20 *The function $x_\infty(\tau_i(y^i), x_0, y^1, \dots, y^p)$ satisfies the Lipschitz condition with respect to y^i with the Lipschitz constant*

$$N' = K' + \left(2 + \frac{p}{T}\right) MN. \tag{4.102}$$

Proof. Indeed, since the sequence of functions $x_m(t, x_0, y^1, \dots, y^p)$ converges uniformly to the limit function $x_\infty(t, x_0, y^1, \dots y^p)$ and the functions $x_m(\tau_i(y^i), x_0, y^1, \dots y^p)$ satisfy the Lipschitz condition with respect to y^i with the same constant N', the limit function will also satisfy the Lipschitz condition with this constant.

□

If the Lipschitz constant $N' = N'(K_1, K_2, N)$ is less than one, then equations (4.98) can be solved for y^i, $y^i = y^i(x_0)$, $i = \overline{1,p}$, and so the problem of existence of T-periodic solutions of system (4.89) is reduced to finding isolated zeros of the function $\bar{\Delta}_m(x_0, y_m^1(x_0), \dots y_m^p(x_0))$, where $y_m^i(x_0)$, $i = 1, \dots, p$, are solutions of the equations

$$y^i = x_m(\tau_i(y^i), x_0, y^1, \dots, y^p), \qquad i = 1, \dots, p. \tag{4.103}$$

Sufficient conditions for existence of T-periodic solutions of equations (4.89) are given by the following theorem.

Theorem 65 *Let the functions $f(t,x)$, $I_i(x)$, and $\tau_i(x)$, which define the T-periodic system (4.89), satisfy inequalities (4.90) – (4.93), conditions 1) and 2) hold with K_1, K_2, and N being such that $N'(K_1, K_2, N) < 1$.*

If

a) for an integer $m \geq 0$ the mapping $\bar{\Delta}_m(x_0, y_m^1(x_0), \dots y_m^p(x_0)) : D_0 \to \mathbf{R}^n$ has an isolated singular point with the nonzero index;

b) there exists a closed convex region $\bar{D} \subseteq D_0$ containing only this singular point and such that the inequality

$$\inf_{x_0 \in \Gamma_{\bar{D}}} \|\bar{\Delta}_m(x_0, y_m(x_0), \dots, y_m^p(x_0))\| >$$
$$> \frac{q^m}{1-q}\left(K_1 + \frac{K_2 p}{T}\right) \times \frac{MT}{2}\left(1 + \frac{4p}{T}\right), \tag{4.104}$$

holds on its boundary.

Then system (4.89) has a T-periodic solution $x = \varphi(t)$, $\varphi(0) \in D_0$ and this solution can be found as the limit of a uniformly convergent sequence of the T-periodic functions defined on the period by relation (4.94)

A proof of this theorem is similar to the proof of Theorem 62.

□

In particular it follows from Theorem 65 that, for T-periodic impulsive differential systems in the standard form

$$\frac{dx}{dt} = \epsilon f(t,x), \qquad t \neq \bar{\tau}_i + \epsilon\tau_i(x), \qquad \Delta x|_{t=\bar{\tau}_i+\epsilon\tau_i(x)} = \epsilon I_i(x), \tag{4.105}$$

where the functions $f(t,x)$, $I_i(x)$, $\tau_i(x)$ are as in equations (4.89), $\bar{\tau}_i(x)$ is a sequence of constants satisfying $\bar{\tau}_{i+p} = \bar{\tau}_i + T$, and the functions $\tau_i(x)$ are subject to the conditions

$$\tau_i(x) \geq \tau_i(x + \epsilon I_i(x))$$

for all $x \in D$, the following statement holds.

Theorem 66 *For all sufficiently small ϵ, for system (4.105) to have a T-periodic solution which continuously depends on ϵ in a neighborhood of the point $\epsilon = 0$ it is necessary that the averaged system*

$$\frac{dy}{dt} = \epsilon f_0(y) \equiv \frac{\epsilon}{T}\left[\int_0^T f(t,y)dt + \sum_{i=1}^{p} I_i(y)\right]$$

have a stationary point $y = y_0 \in D$ and sufficient that this stationary point be isolated with the nonzero index.

Theorem 66 substantiates the averaging principle for impulsive equations in the standard form.

4.4 Almost periodic sequences

In this section we give main properties of almost periodic sequences, which will be used for studying almost periodic impulsive differential systems.

Let $\{x_i\}$ be a sequence in $\mathbf{R}^n$, defined for $i \in \mathbf{Z}$, where $\mathbf{Z}$ is the set of integers. An integer p is called an *ϵ-almost period* of a sequence $\{x_i\}$ if, for any $i \in \mathbf{Z}$,

$$\|x_{i+p} - x_i\| < \epsilon. \tag{4.106}$$

If a sequence $\{x_i\}$ is p-periodic, i.e. $x_{i+p} = x_i$ for all $i \in \mathbf{Z}$, then, for any $\epsilon > 0$, the numbers np, $n \in \mathbf{Z}$, are ϵ-almost periods of this sequence. It is easy to see that if p is an ϵ-almost period of a sequence $\{x_i\}$, then $-p$ is also an ϵ-almost period of this sequence; if p and q are two ϵ-almost periods of $\{x_i\}$, then $p + q$ and $p - q$ are 2ϵ-almost periods of this sequence.

Definition. *A sequence $\{x_i\}$ is called almost periodic if for any $\epsilon > 0$ there exists a relatively dense set of its ϵ-periods, i.e. there exists such a natural number $N = N(\epsilon)$ that, for an arbitrary $k \in \mathbf{Z}$, there is at least one number p in the segment $[k, k+N]$, for which inequality (4.106) holds for all $i \in \mathbf{Z}$.*

Theorem 67 *An almost periodic sequence $\{x_i\}$ is bounded.*

Proof. Let $\epsilon > 0$ and $k \in \mathbf{Z}$ be fixed. There exists such a natural number N that there is an ϵ-almost period p in the interval $[-k, -k+N]$. Because $-k \leq p \leq -k + N$, we have that $0 \leq p + k \leq N$ and

$$\|x_k\| \leq \|x_k - x_{k+p}\| + \|x_{k+p}\| < \epsilon + \max_{0 \leq i \leq N} \|x_i\| = \epsilon + M.$$

Because k is an arbitrary integer, the theorem is proved.

□

Theorem 68 *Let $y = f(x)$ be a function uniformly continuous in the ball $\|x\| \leq h$ and $\{x_i\}$ be an almost periodic sequence such that $\|x_i\| \leq h$ for all $i \in \mathbf{Z}$. Then the sequence $\{y_i\}$, $y_i = f(x_i)$, is almost periodic.*

Proof. For an arbitrary $\epsilon > 0$ there exists such $\delta > 0$ that

$$\|f(x') - f(x'')\| < \epsilon$$

if

$$\|x' - x''\| < \delta, \qquad \|x'\| \leq h, \qquad \|x''\| \leq h.$$

If p is a δ-almost period of the sequence $\{x_i\}$, then

$$\|y_{i+p} - y_i\| = \|f(x_{i+p}) - f(x_i)\| < \epsilon$$

for all $i \in \mathbf{Z}$. Hence p is an ϵ-almost period of the sequence $\{y_i\}$.

□

Consider the set of all bounded sequences $\{x_i\}$, $i \in \mathbf{Z}$, $x_i \in \mathbf{R}^n$. By introducing in this set the norm

$$\|x_i\|_0 = \sup_{i \in \mathbf{Z}} \|x_i\|,$$

we get a linear normed space. It is easy to see that this space is complete. Convergence in this space of a sequence $\{x_i^k\}$, $i \in \mathbf{Z}$, $k \in \mathbf{N}$, to a sequence $\{x_i^0\}$ just means the convergence with respect to the index k, which is uniform with respect to $i \in \mathbf{Z}$.

Using the norm introduced in the space of bounded sequences, we can define convergence in the space of almost periodic sequences.

A sequence of almost periodic sequences $\{x_i^m\}$, $i \in \mathbf{Z}$, $m \in \mathbf{N}$, converges to a sequence $\{y_i\}$ if $\lim_{m\to\infty} \|x_i^m - y_i\|_0 = 0$.

Theorem 69 *Let, for every $m \in \mathbf{N}$, the sequence $\{x_i^m\}$, $i \in \mathbf{Z}$ be almost periodic. If the sequences $\{x_i^m\}$ converges to a sequence $\{y_i\}$ as $m \to \infty$, then $\{y_i\}$ is almost periodic.*

Proof. Fix an arbitrary $\epsilon > 0$. Because $\|x_i^m - y_i\|_0 \to 0$ as $m \to \infty$, there is such $N = N(\epsilon)$ that

$$\|x_i^N - y_i\| < \frac{\epsilon}{3}$$

for all $i \in \mathbf{Z}$. Let p be an $\frac{\epsilon}{3}$-period of the sequence $\{x_i^N\}$. By using the previous inequality, we obtain

$$\begin{aligned}\|y_{i+p} - y_i\| &\le \|y_{i+p} - x_{i+p}^N\| + \|x_{i+p}^N - x_i^N\| + \\ &+\|x_i^N - y_i\| < \frac{\epsilon}{3} + \frac{\epsilon}{3} + \frac{\epsilon}{3} = \epsilon.\end{aligned}$$

Because $\frac{\epsilon}{3}$-periods p are relatively dense, the limit sequence $\{y_i\}$ is almost periodic.

□

Theorem 70 *A sequence $\{x_i\}$ is almost periodic if and only if for any sequence of integers $\{m_i\}$ there exists such a subsequence $\{m_{k_j}\}$ that the sequence $\{x_{i+m_{k_j}}\}$ converges for $j \to \infty$ uniformly with respect to $i \in \mathbf{Z}$.*

Proof. The proof follows [55].

Necessity. Let a sequence $\{x_i\}$ be almost periodic, $\{m_k\}$ be a subsequence of integers, and $\epsilon > 0$. There exists such $N = N(\epsilon)$ that there is an ϵ-almost period p_k of the sequence $\{x_i\}$ in the interval $[m_k - N, m_k]$. Because $m_k - N < p_k \leq m_k$, we have $0 \leq m_k < N$.

Let $q_k = m_k - p_k$. The sequence $\{q_k\}$ can take only a finite number of values, so there exists q such that $q_k = q$ for an infinite number of indices k, k_j^1. Because $\|x_{i+m_k} - x_{i+q_k}\| = \|x_{i+m_k} - x_{i+m_k-p_k}\| < \epsilon$, we see that $\|x_{i+m_{k_j^1}} - x_{i+q}\| < \epsilon$ for any $i \in \mathbf{Z}$. Fix a monotone decreasing to zero sequence of positive numbers $\{\epsilon_j\}$. Choose from $\{x_{i+m_k}\}$ a subsequence $x_{m_{k_j^1}}\}$, for which $\|x_{i+m_{k_j^1}} - x_{i+q^1}\| < \epsilon_1$. From this subsequence choose a subsequence $\{x_{i+m_{k_j^2}}\}$ such that $\|x_{i+m_{k_j^2}} - x_{j+q^2}\| < \epsilon_2$. Continuing in this manner we get a subsequence $\{x_{i+m_{k_j^r}}\}$ with $\|x_{i+m_{k_j^r}} - x_{i+q^r}\| < \epsilon_r$. Take the diagonal sequence $\{x_{i+m_{k_j^j}}\}$ and show that it converges as $j \to \infty$ uniformly with respect to i. Let $\epsilon > 0$ and N' be such that $\epsilon_{N'} < \frac{\epsilon}{2}$. Then

$$\begin{aligned}\|x_{i+m_{k_s^r}} - x_{i+m_{k_s^s}}\| &\leq \|x_{i+m_{k_r^r}} - x_{i+q^{N'}}\| + \|x_{i+q^{N'}} - x_{i+m_{k_s^s}}\| < \\ &< \epsilon_{N'} + \epsilon_{N'} < \epsilon,\end{aligned}$$

for r, $s \geq N'$ because $\{m_{k_j^r}\}_j$ and $\{m_{k_j^s}\}_j$ are subsequences of the sequence $\{m_{k_j}^{N'}\}_j$. Hence $\|x_{i+m_{k_r^r}} - x_{i+m_{k_s^s}}\| < \epsilon$, so that the sequence $\{x_{i+m_{k_j^j}}\}$ is Cauchy and so it converges uniformly with respect to $i \in \mathbf{Z}$.

Sufficiency. Suppose that the condition of the theorem holds and the sequence $\{x_i\}$ is not almost periodic. Then there exists such $\epsilon_0 > 0$ that, for any natural N, there is N consecutive integers not equal to any ϵ_0-almost period. Let L_N be such a set of the consecutive integers, m_1 be arbitrary, and m_2 be such that $m_1 - m_2$ lie in L_1. For example, if m is a number from L_1, then we can take $m_2 = m_1 - m$. To make the notations the same, denote $L_1 = L_{\nu_1}$ and choose such $\nu_2 > |m_1 - m_2|$ and m_3 that $m_3 - m_1$ and $m_3 - m_2$ be in L_{ν_2}. Such a choice is possible as can be seen from the following considerations.

Let l, $l+1$, ..., $l + \nu_2 - 1$ be numbers from L_{ν_2} and $m_2 \leq m_1$. Take $m_3 = l + m_1$ such that $m_3 - m_1 \in L_{\nu_2}$, $m_3 - m_2 = m_3 - m_1 + m_1 - m_2 =$

$l + m_1 - m_2 < l + \nu_2$, and $m_3 - m_2 \geq l$, so $m_3 - m_2 \in L_{\nu_2}$. In the general case, choose $\nu_i > \max_{1 \leq \mu < \nu \leq i} |m_\mu - m_\nu|$ and then m_{i+1} so that the numbers $m_{i+1} - m_\nu$ would belong to L_{ν_i} for $1 \leq \mu \leq i$. The number m_{i+1} is chosen in the same way, namely, $m_{i+1} = l + m_j$, where $m_j = \max_{1 \leq \mu \leq i} m_\mu$. For the sequence $\{m_i\}$ we have

$$\sup_{i \in \mathbf{Z}} \|x_{i+m_r} - x_{i+m_s}\| = \sup_{i \in \mathbf{Z}} \|x_{i+m_r-m_s} - x_i\|.$$

But $m_r - m_s \in L_{\nu_{r-1}}$, $r \geq s$, and so it is not an ϵ_0-period (it was assumed that $L_{N'}$ does not contain any ϵ_0-almost period). Whence it follows that there exists an index i, for which $\|x_{i+m_r-m_s} - x_i\| \geq \epsilon_0$, which means that $\sup_{i \in \mathbf{Z}} \|x_{i+m_r} - x_{i+m_s}\| \geq \epsilon_0$ or $\|x_{i+m_r} - x_{i+m_s}\| \geq \epsilon_0$. However it follows from the assumptions that the sequence $\{m_i\}$ contains such a subsequence $\{m_{k_j}\}$ that the sequence $\{x_{i+m_{k_j}}\}_j$ converges uniformly with respect to $i \in \mathbf{Z}$. This means that there exists such an index j_0 that $\|x_{i+m_{k_j}} - x_{i+m_{k_l}}\| < \epsilon_0$. This contradiction finishes the proof.

□

The proved theorem implies the following important statements.

Theorem 71 *If sequences $\{x_i\}$ and $\{y_i\}$ are almost periodic, then the sequence $\{x_i + y_i\}$ is also almost periodic.*

Proof. Let the sequences $\{x_i\}$ and $\{y_i\}$ be almost periodic. From an arbitrary sequence of integers $\{m_k\}$ choose a subsequence $m_{k'_j}$ such that the sequence $\{x_{i+m_{k'_j}}\}$ converges uniformly with respect to $i \in \mathbf{Z}$. From the sequence $\{m_{k'_j}\}$ choose a subsequence $\{m_{k_j}\}$, for which the sequence $\{y_{i+m_{k_j}}\}_j$ converges uniformly with respect to $i \in \mathbf{Z}$. Whence it follows that there exists a subsequence $\{m_{k_j}\}$ of the sequence $\{m_k\}$ such that the sequence $\{x_{i+m_{k_j}} + y_{i+m_{k_j}}\}_j$ converges uniformly with respect to $i \in \mathbf{Z}$. By the previous theorem, it is sufficient that the sequence $\{x_i + y_i\}$ be almost periodic.

□

Theorem 72 *If $\{x_i\}$ is an almost periodic sequence and $\{\alpha_i\}$ is an almost periodic sequence of real numbers, then the sequence $\{\alpha_i x_i\}$ is also almost periodic.*

Proof. In any sequence of integers $\{m_k\}$ there exists a subsequence $\{m_{k_j}\}$ such that the sequences $\{x_{i+m_{k_j}}\}$ and $\{\alpha_{i+m_{k_j}}\}$ converge uniformly with respect to $i \in \mathbf{Z}$. This means that the sequence $\{\alpha_{i+m_{k_j}} \cdot x_{i+m_{k_j}}\}$ converges uniformly with respect to $i \in \mathbf{Z}$, and this implies, by Theorem 70, that the sequence $\{\alpha_i x_i\}$ is almost periodic.

□

From the preceding theorems we conclude that the set of all almost periodic sequences $\{x_i\}$, $i \in \mathbf{Z}$, $x_i \in \mathbf{R}^n$, forms a linear space. In this linear space we introduce a norm by setting

$$\|x_i\|_0 = \sup_{i \in \mathbf{Z}} \|x_i\|.$$

Then this linear space becomes a complete linear space, i.e. a Banach space.

We will establish one more property of almost periodic sequences.

Theorem 73 *For given two almost periodic sequences $\{x_i\}$ and $\{y_i\}$, and any $\epsilon > 0$ there exists a relatively dense set of their common ϵ-almost periods.*

Proof. Fix an arbitrary $\epsilon > 0$. There exist integers $N_1 = N_1(\epsilon)$ and $N_2 = N_2(\epsilon)$ such that, in the intervals $[k, k + N_1]$, $[k, k + N_2]$, there are at least one $\frac{\epsilon}{2}$-almost period of the sequences $\{x_i\}$ and $\{y_i\}$ respectively. Let $N_3 = N_3(\epsilon) = \max\{N_1, N_2\}$. For any $k \in \mathbf{Z}$, in the interval $[k, k + N_3]$ there exist at least one $\frac{\epsilon}{2}$-almost period p_1 of the sequence $\{x_i\}$ and at least one $\frac{\epsilon}{2}$-almost period p_2 of the sequence $\{y_i\}$. Because $|p_1 - p_2| \leq N_3$, the difference $p_2 - p_1$ can take only a finite number of values regardless of which set of the consecutive N_3 numbers that contain p_1 and p_2 is taken. We will say that two pairs of the numbers (p_1, p_2) and (p_1', p_2') are equivalent if $|p_1 - p_2| = |p_1' - p_2'|$. Because $|p_1 - p_2|$ takes only a finite number of values, the number of equivalence classes of this relation is finite.

Choose a system of representatives for these classes, (p_1^r, p_2^r), $r = 1, 2, \ldots, s$. Set $N_4 = N_4(\epsilon) = \max_r |p_1^r|$, $N = N_3 + 2N_4$, and show that between any two numbers from the interval $[k, k + N_4]$ there exists an ϵ-almost period common to the sequences $\{x_i\}$ and $\{y_i\}$. Let k be an integer, p_1, p_2 – $\frac{\epsilon}{2}$-almost periods, which belong to the interval $[k + N_4, k + N_4 + N_3]$, (p_1^r, p_2^r) be a representative of the class containing the pair (p_1, p_2) such that $|p_1 - p_2| = |p_1^r - p_2^r|$. We have either $p_1^r - p_2^r = p_1 - p_2$ or $p_1^r - p_2^r = p_2 - p_1$, and so $p_1^r - p_1 = p_2^r - p_2 = -p$ or $p_1^r + p_1 = p_2^r + p_2 = p$. It follows from the inequality $|p_1^r| \leq N_4$ that $k < p_1 + p_1^r \leq k + N_3 + 2N_4 = k + N$ and

$k < p_1 - p_1^\tau \leq k + N_3 + 2N_4 = k + N$, i.e. in both cases, $k < p \leq k + N$. The number p defined in this way, is a common ϵ-almost period of both sequences.

Indeed

$$\begin{aligned}
\|x_{i+p} - x_i\| &= \|x_{i+p_1 \pm p^\tau{}_1} - x_i\| \leq \|x_{i+p_1 \pm p^\tau{}_1} - x_{i+p_1}\| + \|x_{i+p_1} - x_i\| < \\
&< \frac{\epsilon}{2} + \frac{\epsilon}{2} = \epsilon, \\
\|y_{i+p} - y_i\| &= \|y_{i+p_2 \pm p^\tau{}_2} - y_i\| \leq \|y_{i+p_2 \pm p^\tau{}_2} - y_{i+p_2}\| + +\|y_{i+p_2} - y_i\| < \\
&< \frac{\epsilon}{2} + \frac{\epsilon}{2} = \epsilon.
\end{aligned}$$

□

Let us prove that, for an almost periodic sequence, there exists a finite mean value.

Theorem 74 *For every almost periodic sequence $\{x_i\}$ there exists the finite mean value*

$$M(\{x_i\}) = \lim_{n\to\infty} \frac{1}{n} \sum_{j=i}^{i+n-1} x_j. \tag{4.107}$$

which is uniform with respect to $i \in \mathbf{Z}$.

Proof. First, we find an estimate of the value

$$\sum_{j=i}^{i+n-1} x_j - \sum_{j=0}^{n-1} x_j.$$

Because the sequence $\{x_i\}$ is almost periodic, for the number $\frac{\epsilon}{4} > 0$, there exists such a number $N = N(\epsilon)$ that, amongst the integers from the interval $[i, i + N]$, one can find an $\frac{\epsilon}{4}$-almost period p, i.e. an integer p such that

$$\|x_{i+p} - x_i\| < \frac{\epsilon}{4} \tag{4.108}$$

for all $i \in \mathbf{Z}$.

Having chosen N and p, write

$$\sum_{j=i}^{i+n-1} x_j - \sum_{j=0}^{n-1} x_j = \sum_{j=p}^{p+n-1} x_j - \sum_{j=0}^{n-1} x_j + \sum_{j=\nu}^{p-1} x_j - \sum_{j=\nu+n}^{p+n-1} x_j.$$

By using inequality (4.108) and that $0 \le p_1 - i \le N$, we get

$$\left\| \sum_{j=i}^{i+n-1} x_j - \sum_{j=0}^{n-1} x_j \right\| \le \sum_{j=0}^{n-1} \|x_{j+p} - x_j\| + \sum_{j=i}^{p-1} \|x_j\| + \sum_{j=i+n}^{p+n-1} \|x_j\| \le$$

$$\le \frac{\epsilon}{4} \cdot n + 2N \sup_{i\in\mathbf{Z}} \|x_i\|. \tag{4.109}$$

Let us prove that the sequence $\{y_k\}$,

$$y_k = \frac{1}{k} \sum_{j=0}^{k-1} x_j,$$

has a limit for $k \to \infty$. We use the Cauchy criterion. For any two natural numbers m and n we have

$$\left\| \frac{1}{m} \sum_{j=0}^{m-1} x_j - \frac{1}{n} \sum_{j=0}^{n-1} x_j \right\| \le \frac{1}{mn} \left\| n \sum_{j=0}^{m-1} x_j - \sum_{j=0}^{mn-1} x_j \right\| +$$

$$+ \frac{1}{mn} \left\| \sum_{j=0}^{mn-1} x_j - m \sum_{j=0}^{n-1} x_j \right\| \le \frac{1}{mn} \sum_{\nu=1}^{n} \left\| \sum_{j=0}^{m-1} x_j - \sum_{j=(\nu-1)m}^{\nu m-1} x_j \right\| +$$

$$+ \frac{1}{mn} \sum_{\nu=1}^{m} \left\| \sum_{j=(\nu-1)n}^{\nu n-1} x_j - \sum_{j=0}^{n-1} x_j \right\|.$$

Whence, by using estimate (4.109), we have

$$\left\| \frac{1}{m} \sum_{j=0}^{m-1} x_j - \frac{1}{n} \sum_{j=0}^{n-1} x_j \right\| \le \frac{1}{mn} \cdot n \left(\frac{\epsilon}{4} m + 2N \sup_{i\in\mathbf{Z}} \|x_i\| \right) +$$

$$+ \frac{1}{mn} \cdot m \left(\frac{\epsilon}{4} n + 2N \sup_{i\in\mathbf{Z}} \|x_i\| \right) =$$

$$= \frac{\epsilon}{2} + 2N \left(\frac{1}{m} + \frac{1}{n} \right) \sup_{i\in\mathbf{Z}} \|x_i\|.$$

Choosing N_0 so large that the inequality

$$N_0 > \frac{8N}{\epsilon} \sup_{i\in\mathbf{Z}} \|x_i\|$$

holds, we see that, for all $m > N_0$, $n > N_0$,

$$\left\| \frac{1}{m} \sum_{j=0}^{m-1} x_j - \frac{1}{n} \sum_{j=0}^{n-1} x_j \right\| < \epsilon.$$

The Cauchy criterion implies that the sequence $\{y_i\}$ converges, i.e. the limit

$$\lim_{n\to\infty} \frac{1}{n} \sum_{j=0}^{n-1} x_j = x^0. \tag{4.110}$$

exists. Finally, we will show that limit (4.107) exists. Note that it follows from inequality (4.109) that

$$\left\| \frac{1}{k} \sum_{j=(i-1)k}^{ik-1} x_j - \frac{1}{k} \sum_{j=0}^{k-1} x_j \right\| \leq \frac{\epsilon}{4} + \frac{2N}{k} \sup_{i\in\mathbf{Z}} \|x_i\|,$$

for $i = 1, 2, \ldots$. Consequently, for the expression

$$\frac{1}{n} \sum_{k=1}^{n} \left[\frac{1}{i} \sum_{j=(k-1)i}^{ki-1} x_j - \frac{1}{i} \sum_{j=0}^{i-1} x_j \right] = \frac{1}{ni} \sum_{j=0}^{ni-1} x_j - \frac{1}{i} \sum_{j=0}^{i-1} x_j,$$

we also have

$$\left\| \frac{1}{ni} \sum_{j=0}^{ni-1} x_j - \frac{1}{i} \sum_{j=0}^{i-1} x_j \right\| \leq \frac{\epsilon}{4} + \frac{2N}{i} \sup_{i\in\mathbf{Z}} \|x_j\|.$$

By passing in this inequality to the limit for $n \to \infty$, we get

$$\left\| x^0 - \frac{1}{i} \sum_{j=0}^{i-1} x_j \right\| \leq \frac{\epsilon}{4} + \frac{2N}{i} \sup_{i\in\mathbf{Z}} \|x_j\|. \tag{4.111}$$

It follows from inequalities (4.109) and (4.110) that

$$\left\| x^0 - \frac{1}{n} \sum_{j=i}^{i+n-1} x_j \right\| \leq \frac{\epsilon}{2} + \frac{4N}{n} \sup_{i\in\mathbf{Z}} \|x_j\| < \epsilon$$

for

$$n > \frac{8N}{\epsilon} \sup_{i\in\mathbf{Z}} \|x_j\|.$$

And this means that

$$\lim_{n\to\infty} \frac{1}{n} \sum_{j=1}^{i+n-1} x_i = M(\{x_i\}) \equiv x^0.$$

□

Let $\{\tau_i\}$ be a sequence of real numbers, indexed by the set of integers $\mathbf{Z}$, such that $\tau_i \to -\infty$ for $i \to -\infty$ and $\tau_i \to +\infty$ for $i \to +\infty$. We also assume that this sequence is strictly increasing, i.e. $\tau_{i+1} > \tau_i$ for all $i \in \mathbf{Z}$. As before, we denote by $i(t, t+Y)$ the number of terms of the sequence $\{\tau_i\}$, which belong to the interval $(t, t+T]$ and assume that

$$\overline{\lim_{T\to\infty}} \frac{i(t,t+T)}{T} = p < \infty \tag{4.112}$$

uniformly in $t \in \mathbf{R}$.

Lemma 21 *The uniform in $t \in \mathbf{R}$ boundedness of the upper limit (4.112) is equivalent to the following statement: there exist such number $l > 0$ and a natural number q that any interval of the time axis of length l contains no more than q terms of the sequence $\{\tau_i\}$.*

Proof. Since the upper limit (4.112) is bounded, it follows that, for the number $q_1 = p + 1$, there is such l that

$$\frac{i(t,t+l)}{l} \le q_1$$

for all $t \in \mathbf{R}$. Consequently, any interval of the time axis of length l contains no more than $q = [lq_1] + 1$ points of the sequence $\{\tau_i\}$. Here $[(\cdot)]$ means the integer part of the number $(\cdot)$.

Let there exist such numbers $l > 0$ and natural q that any interval of the time axis of length l contains no more than q points of the sequence $\{\tau_i\}$. Write T as $T = ml + \tau$ for some natural m and $0 \le \tau < l$. Then

$$\frac{i(t,t+T)}{T} \le \frac{i(t,t+l) + i(t+l,t+2l) + \ldots + i(t+ml,t+T)}{ml},$$

and, consequently,

$$\overline{\lim_{T\to\infty}} \frac{i(t,t+T)}{T} \le \lim_{m\to\infty} \frac{(m+1)q}{ml} = \frac{q}{l}.$$

□

If the sequence $\{\tau_i\}$ satisfies the condition

$$\tau_{i+1} - \tau_i \geq \theta > 0 \tag{4.113}$$

for all $i \in \mathbf{Z}$ and some θ, then any interval of the time axis of length θ does not contain more than one point of τ_i, and so,

$$\overline{\lim_{T\to\infty}} \frac{i(t, t+T)}{T} \leq \frac{1}{\theta}.$$

If the moments τ_i are such that the sequence $\{\tau_{i+1} - \tau_i\}$ is p-periodic, i.e. $\tau_{i+p} = \tau_i + \omega$, then in this case the limit

$$\lim_{T\to\infty} \frac{i(t, t+T)}{T} = \frac{p}{\omega} \tag{4.114}$$

exists.

Consider one more class of the sequences $\{\tau_i\}$, for which there exists a finite uniform in $t \in \mathbf{R}$ limit of type (4.114).

Lemma 22 *Let the numbers τ_i, $i \in \mathbf{Z}$, be such that the sequence $\{\bar{\tau}_i\}$, $\bar{\tau}_i = \tau_{i+1} - \tau_i$, $i \in \mathbf{Z}$, is almost periodic. Then there exists the limit*

$$\lim_{T\to\infty} \frac{i(t, t+T)}{T} = p, \tag{4.115}$$

uniform with respect to $t \in \mathbf{R}$.

Proof. First of all we show that the limit

$$\lim_{n\to\infty} \frac{\tau_n}{n} = \frac{1}{p}$$

exists and is different from zero. Indeed, without loss of generality, we can assume that $\tau_{-1} < 0$ and $\tau_0 \geq 0$. Then

$$\tau_n = \tau_0 + \sum_{j=0}^{n-1} \bar{\tau}_j$$

and thus

$$\frac{\tau_n}{n} = \frac{\tau_0}{n} + \frac{1}{n}\sum_{j=0}^{n-1} \bar{\tau}_j.$$

Because the sequence $\{\bar{\tau}_i\}$ is almost periodic, by Theorem 74, there exists a finite mean value, i.e. there exists the finite limit

$$\lim_{n\to\infty}\frac{1}{n}\sum_{j=0}^{n-1}\bar{\tau}_j$$

and this limit is nonzero because the terms of the sequence $\{\bar{\tau}_i\}$ are positive. Thus, there exists the nonzero limit

$$\lim_{n\to\infty}\frac{\tau_n}{n}=\lim_{n\to\infty}\left(\frac{\tau_0}{n}+\frac{1}{n}\sum_{j=0}^{n-1}\bar{\tau}_j\right)=\frac{1}{p}$$

and so the limit

$$\lim_{n\to\infty}\frac{i(0,\tau_n)}{\tau_n}=p$$

also exists. Now it is easy to prove that

$$\lim_{T\to\infty}\frac{i(0,T)}{T}=p. \tag{4.116}$$

Indeed,

$$\frac{T}{i(0,T)}=\frac{\tau_i+\theta_i}{i}$$

for some natural i, $0\le\theta_i\le\sup_{i\in\mathbf{Z}}\bar{\tau}_i$. Hence,

$$\frac{T}{i(0,T)}-\frac{\tau_i}{i}=o\left(\frac{1}{i}\right).$$

Whence equality (4.116) follows.

In the same way as it was shown in the proof of Theorem 74, the property of a sequence $\{\bar{\tau}_i\}$ implies that

$$\left|\sum_{j=\nu}^{\nu+i-1}\bar{\tau}_j-\sum_{j=0}^{i-1}\bar{\tau}_j\right|\le\frac{\epsilon i}{4}+2N\sup_{j\in\mathbf{Z}}\bar{\tau}_j. \tag{4.117}$$

By using this inequality, we can write

$$\left|\frac{1}{i}\sum_{j=(\nu-1)i}^{\nu i-1}\bar{\tau}_j-\frac{1}{i}\sum_{j=0}^{i-1}\bar{\tau}_j\right|\le\frac{\epsilon}{4}+\frac{2N}{i}\sup_{j\in\mathbf{Z}}\bar{\tau}_j$$

for $\nu = 1, 2, \ldots$. Consequently, for the arithmetic mean

$$\frac{1}{n}\sum_{\nu=1}^{n}\left[\frac{1}{i}\sum_{j=(\nu-1)i}^{\nu i-1}\bar{\tau}_j - \frac{1}{i}\sum_{j=0}^{i-1}\bar{\tau}_j\right] = \frac{1}{ni}\sum_{j=0}^{ni-1}\bar{\tau}_j - \frac{1}{i}\sum_{j=0}^{i-1}\bar{\tau}_j,$$

we also have the estimate

$$\left|\frac{1}{ni}\sum_{j=0}^{ni-1}\bar{\tau}_j - \frac{1}{i}\sum_{j=0}^{i-1}\bar{\tau}_j\right| \le \frac{\epsilon}{4} + \frac{2N}{i}\sup_{j\in\mathbf{Z}}\bar{\tau}_j. \tag{4.118}$$

By using inequalities (4.117) and (4.118), we get

$$\left|\frac{1}{i}\sum_{j=\nu}^{\nu+i-1}\bar{\tau}_j - \frac{1}{p}\right| < \frac{\epsilon}{2} + \frac{4N}{i}\sup_{j\in\mathbf{Z}}\bar{\tau}_j,$$

i.e.

$$\lim_{n\to\infty}\frac{\tau_{i+n} - \tau_i}{n} = \frac{1}{p} \tag{4.119}$$

uniformly with respect to $i \in \mathbf{Z}$.

Let the interval $[t, t+T]$ contain k terms of the sequence $\{\tau_i\}$, $\tau_{\nu+1}, \tau_{\nu+2}, \ldots, \tau_{\nu+k}$. Then

$$\frac{T}{i(t, t+T)} = \frac{T}{k} = \frac{\tau_{\nu+k} - \tau_\nu + \theta_k}{k}, \qquad |\theta_k| \le \sup_{j\in\mathbf{Z}}\bar{\tau}_i.$$

By (4.119), it follows from this relation that

$$\lim_{T\to\infty}\frac{i(t, t+T)}{T} = p$$

for all $t \in \mathbf{R}$.

□

We will give some more properties of the sequences $\{\tau_i\}$, for which $\{\bar{\tau}_i\}$ is an almost periodic sequence.

Set, for any integers i and j, $\tau_i^j = \tau_{i+j} - \tau_i$ and consider the sequences $\{\tau_i^j\}$, $i \in \mathbf{Z}$, $j \in \mathbf{Z}$. It is easy to verify that the numbers τ_i^j satisfy

$$\tau_{i+k}^j - \tau_i^j = \tau_{i+j}^k - \tau_i^k, \qquad \tau_i^j - \tau_i^k = \tau_{i+k}^{j-k}. \tag{4.120}$$

The family of the sequences $\{\tau_i^j\}$ will be called *equipotentially almost periodic* if for an arbitrary $\epsilon > 0$ there exists a relatively dense set of ϵ-almost periods, that are common to all the sequences $\{\tau_i^j\}$.

Let $\epsilon > 0$ and Γ_ϵ be a set of real numbers, for which there exists at least one such number k that

$$|\tau_i^k - r| < \epsilon, \qquad i \in \mathbf{Z}. \tag{4.121}$$

Denote by $\mathbf{P}_r$ the set of the numbers p satisfying inequality (4.121) for fixed ϵ and r, and by $\mathbf{P}_\epsilon$ – the union of the sets $\mathbf{P}_r$, taken for all $r \in \Gamma_\epsilon$, i.e. $\mathbf{P}_\epsilon = \cup_{r\in\Gamma_\epsilon}\mathbf{P}_r$.

For the sequences $\{\tau_i^j\}$ defined above, the following three lemmas hold [55].

Lemma 23 *If p is an almost period, common to all the sequences $\{\tau_i^j\}$, $j = 0, \pm 1, \ldots$, then $\tau_0^p \in \Gamma_\epsilon$ and $p \in \mathbf{P}_\epsilon$. If $p \in \mathbf{P}_\epsilon$, then p is a 2ϵ-almost period, common to all the sequences $\{\tau_i^j\}$.*

To see that the lemma holds, it is sufficient to use the first inequality from (4.120).

□

Lemma 24 *Let $\gamma \subset \Gamma_\epsilon$, $\Gamma \neq \emptyset$, and $\mathbf{P}$ be such a subset of the union of $\mathbf{P}_r$, $r \in \Gamma$, that $\mathbf{P} \cap \mathbf{P}_r \neq \emptyset$ for all $r \in \Gamma$. Then the set Γ is dense if and only if the set $\mathbf{P}$ is relatively dense.*

Proof. The integers from $\mathbf{P}$ can be arranged so that the sequence $\{p_k\}$ is increasing. Because the sequence $\{\tau_i\}$ is increasing, so will be the sequence $\mathbf{P}' = \{\tau_0^{p_k}\}$. It can be directly verified that, if one of the sets $\mathbf{P}$, $\mathbf{P}'$ is relatively dense, so will be the other. Consequently, it will be sufficient to show that $\mathbf{P}'$ is relatively dense if and only if $\mathbf{P}$ is relatively dense. It will be proved if we show that

1) $\lim_{k\to-\infty} p_k = -\infty$ if and only if $\lim_{k\to-\infty} \tau_0^{p_k} = -\infty$;

2) $\lim_{k\to+\infty} p_k = +\infty$ if and only if $\lim_k \to +\infty\tau_0^{p_k} = +\infty$;

3) the sequence $\{p_{k+1} - p_k\}$ is bounded if and only if the sequence $\{\tau_0^{p_{k+1}} - \tau_0^{p_k}\}$ is bounded.

Statements 1) and 2) follow directly from properties of the sequences $\{\tau_i\}$, $\{p_k\}$, $\{\tau_0^{p_k}\}$.

It remains to prove statement 3). By using the first and the second relation of (4.120), we get

$$\begin{aligned} |(\tau_0^{p_{k+1}} - \tau_0^{p_k}) - \tau^{p_{k+1}-p_k}| &= |(\tau_0^{p_{k+1}} - \tau_0^{p_k}) - \tau_{-p_k+p_k}^{p_{k+1}-p_k}| = \\ &= |(\tau_0^{p_{k+1}} - \tau_0^{p_k}) - (\tau_{-p_k}^{p_{k+1}} - \tau_{-p_k}^{p_k})| \leq \\ &\leq |\tau_{-p_k}^{p_k} - \tau_0^{p_k}| + |\tau_{-p_k}^{p_{k+1}} - \tau_0^{p_{k+1}}|. \end{aligned}$$

By applying Lemma 29, we find that

$$|(\tau_0^{p_{k+1}} - \tau_0^{p_k}) - \tau_0^{p_{k+1}-p_k}| < 2\epsilon + 2\epsilon,$$

and, because $\{\tau_i\}$ is almost periodic, statement 3) follows.

□

From the preceding lemmas, we have

Lemma 25 *The following statements are equivalent:*

a) the sequences $\{\tau_i^j\}$, $i \in \mathbf{Z}$, $j = 0, \pm 1, \pm 2, \ldots$, are equipotentially almost periodic;

b) the set $\mathbf{P}$ is relatively dense for any $\epsilon > 0$;

c) the set Γ_ϵ is relatively dense for any $\epsilon > 0$.

We will give two examples of the sequences $\{\tau_i\}$, for which the sequences $\{\tau_i^j\}$ are equipotentially almost periodic.

Let the sequence $\{\tau_i^1\}$, $\tau_i^1 = \tau_{i+1} - \tau_i$, be periodic with the period p, i.e. τ_i^1 for all $i \in \mathbf{Z}$ and $\tau_0^p = T$. From the equality $\tau_{i+p}^1 = \tau_i^1$, it follows that $\tau_{i+1+p} - \tau_{i+1} = \tau_{i+p} - \tau_i$, and thus $\tau_{i+1}^p = \tau_i^p$ so that $\tau_i^p = T$, $i \in \mathbf{Z}$. Because $\tau_i^{mp} = \sum_{j=0}^{m-1} \tau_{i+jp}^p$ for $m > 0$ and $\tau_i^{mp} = -\sum_{j=0}^{m} \tau_{i+jp}^p$ for $m < 0$, it follows that $\tau_i^{mp} = mT$ for all $i \in \mathbf{Z}$, $m \in \mathbf{Z}$. By using the first relation of (4.120), we get

$$\tau_{i+p}^j - \tau_i^j = \tau_{i+j}^p - \tau_i^p = 0,$$

and so the sequences $\{\tau_i^j\}$, $i \in \mathbf{Z}$, $j = 0, \pm 1, \pm 2, \ldots$, are periodic with the period p, and even more so they are equipotentially almost periodic.

Let $\{\alpha_i\}$ be an almost periodic sequence of reals with $\sup_{i \in \mathbf{Z}} |\alpha_i| = \alpha < \frac{T}{2}$, $T > 0$, and let $\{\tau_i\} = \{iT + \alpha_i\}$. Then $\tau_{i+1} - \tau_i \geq T - 2\alpha > 0$ and, consequently, the sequence $\{\tau_i\}$ is strictly increasing with $\lim_{t\to-\infty}\tau_i = -\infty$, $\lim_{i\to+\infty}\tau_i = \infty$.

Let p be an $\frac{\epsilon}{2}$-almost period of the sequence $\{\alpha_i\}$. For any i, $j \in \mathbf{Z}$ we have

$$|\tau_{i+p}^j - \tau_i^j| = |(\alpha_{i+j+p} - \alpha_{i+j}) - (\alpha_{i+p} - \alpha_i)| < \frac{\epsilon}{2} + \frac{\epsilon}{2} = \epsilon,$$

which means that the family of the sequences $\{\tau_i^j\}$, $i \in \mathbf{Z}$, $j = 0, \pm 1, \ldots$, is equipotentially almost periodic.

Lemma 26 *If the family of sequences $\{\tau_i^j\}$ is equipotentially almost periodic, then for any $i > 0$ there exists such a natural number q that in any interval of the real axis of length l there are not more than q terms of the sequence $\{\tau_i\}$.*

Proof. Set $\epsilon = 1$. By Lemma 25, there exists such a number $r \in \Gamma_\epsilon$ that $r - 1 > l$. From the inequalities $r > 1$ and $|\tau_0^k - r| < 1$ it follows that $\tau_0^k > 0$ and hence $k \geq 1$. Set $q = k$ and let $(a, a+l)$ be an interval of the real axis and τ_{i_0} – the smallest term of the sequence $\{\tau_i\}$, which belongs to this interval. Inequality $|\tau_{i_0}^k - r| < 1$ implies that

$$\tau_{i_0+k} > \tau_{i_0} + r - 1 > \tau_{i_0} + l > a + l.$$

□

Lemma 27 *Suppose that the sequences $\{\tau_i^j\}$, $i \in \mathbf{Z}$, $j = 0, \pm 1, \ldots$, are equipotentially almost periodic. Then for any $\epsilon > 0$ there exists such $l = l(\epsilon) > 0$ that for any interval $\mathcal{J}$ of the real axis of length l there are such a subinterval $I \subset \mathcal{J}$ of length ϵ and an integer q that*

$$|\tau_i^q - r| < \epsilon, \qquad i \in \mathbf{Z}, \qquad \mathbf{r} \in \mathbf{I}. \tag{4.122}$$

Proof. By Lemma 25, the set $\Gamma_{\epsilon/2}$ is relatively dense. Consequently there exists $l_1 = l_1(\epsilon) > 0$ such that for any interval $\mathcal{J}_1$ of length l_1 there are such $r_1 \in \mathcal{J}_1$ and integer q that

$$|\tau_i^q - r_1| < \frac{\epsilon}{2}, \qquad i \in \mathbf{Z}. \tag{4.123}$$

Let $l = l_1 + \epsilon$ and $\mathcal{J} = (a, a+l)$ be an interval of the real axis of length l. For the interval $\mathcal{J}_1 = \left(a + \frac{\epsilon}{2}, a + \frac{\epsilon}{2} + l_1\right)$, which has the length l_1, there exist numbers $r_1 \in \mathcal{J}_1$ and q satisfying condition (4.123). Take $J = \left(r_1 - \frac{\epsilon}{2}, r_1 + \frac{\epsilon}{2}\right)$. Then $J \subset \mathcal{J}$ and, for $r \in J$, we have

$$|\tau_i^q - r| \leq |\tau_i^q - r_1| + |r_1 - r| < \epsilon, \qquad i \in \mathbf{Z}.$$

□

Lemma 28 *Let a sequence $\{\tau_i\}$ be such that the sequences $\{\tau_i^j\}$, $j = 0, \pm 1, \ldots$, are equipotentially almost periodic. Then for any $\epsilon > 0$ there is such a number $l > 0$ that for any η, $0 < \eta < \epsilon$ and any interval $\mathcal{J}$ of length l there exist such integers q and m that $m\eta \in \mathcal{J}$ and $|\tau_i^q - m\eta| < \epsilon$ for all $i \in \mathbf{Z}$.*

Proof. Let the number $l = l(\epsilon)$ be as given by Lemma 27, $\mathcal{J}$ – an interval of length l, and j and q be chosen for the interval $\mathcal{J}$ as in Lemma 27. There exists such a number $r_0 \in J$ that $r_0 + \epsilon \in J$. If r_0 is not a multiple of η, then, by adding a number smaller than η, we get $r = m\eta$, a multiple of η, and clearly $m\eta \in J$.

□

We give two more statements on the properties of the sequences $\{\tau_i^j\}$ and their relation to the Bohr almost periodic functions. For a proof of these statements see [55].

Lemma 29 *Let the sequences $\{\tau_i^j\}$, $j = 0, \pm 1, \pm 2, \ldots$, be equipotentially almost periodic and a function $\Phi(t)$ be Bohr almost periodic. Then for any $\epsilon > 0$ there exists such $l = l(\epsilon) > 0$ that for any interval $\mathcal{J}$ of length l there are such $r \in \mathcal{J}$ and an integer q that*

$$|\tau_i^q - r| < \epsilon, \qquad |\Phi(t + r) - \Phi(t)| < \epsilon$$

for all integers i and all $t \in \mathbf{R}$.

Lemma 30 *Let $\Phi(t)$ be an Bohr almost periodic function. If a sequence $\{\tau_i\}$ is such that the sequences $\{\tau_i^j\}$, $j \in \mathbf{Z}$, are equipotentially almost periodic, then the sequence of numbers $\{\Phi(\tau_i)\}$ is almost periodic.*

As an example of a sequence $\{\tau_I\}$, for which the sequences $\{\tau_i^j\}$ are equipotentially almost periodic, consider the sequence

$$\tau_i = i + a_i, \qquad a_i = \frac{1}{4}|\sin i - \sin i\sqrt{2}|.$$

By Lemma 30, the sequence $\{a_i\}$ is almost periodic. For the sequence $\{\tau_i\}$, $\tau_i = i + a_i$, we have $\tau_{i+1} - \tau_i \geq 1 - 2\sup_i a_i > 0$ and so, the sequence $\{\tau_i\}$ is strictly increasing with $\tau_i \to -\infty$ for $i \to -\infty$ and $\tau_i \to +\infty$ for $i \to +\infty$. If p is an $\frac{\epsilon}{2}$-almost period of the sequence $\{a_i\}$, then, for any integers i and j,

$$|\tau_{i+p}^j - \tau_i^j| = |a_{i+j+p} - a_{i+p} + a_i| < \epsilon,$$

i.e. the sequences $\{\tau_i^j\}$, $j = 0, \pm 1, \pm 2, \ldots$, are equipotentially almost periodic.

In the sequel, we will see that the property of a sequence $\{\tau_i^1\}$ to be separated from zero, i.e. $\inf_i \tau_i^1 = \theta > 0$, is important for studying impulsive differential equations.

Now we construct an example of a sequence $\{\tau_i\}$, for which all the sequences $\{\tau_i^j\}$, $j = 0, \pm 1, \pm 2, \ldots$, are equipotentially almost periodic and $\inf_i \tau_i^1 = 0$.

Consider the sequence $\{a_i\}$ defined as above. It is known [65] that for any natural n there are such integers i and m that

$$|i(1 - \sqrt{2}) - 2m\pi| < \frac{2\pi}{n}.$$

Because the function $\sin t$ is continuous and periodic, the difference $\sin i - \sin i\sqrt{2}$ can be arbitrarily small and, so, the sequence $\tau_i^1\}$, $\tau_i = i + a_i$, is not separated from zero. Set $\tau_{2i} = i + a_i$, $\tau_{2i-1} = i - a_i$. Let us show that, for thus defined sequence $\{\tau_i\}$, the sequences $\{\tau_i^j\}$, $j = 0, \pm 1, \pm 2, \ldots$, are equipotentially almost periodic.

Indeed, consider four cases:

a) $i = 2p$, $j = 2r + 1$, $\quad p, r = 0, \pm 1, \pm 2, \ldots$,

$$\begin{aligned} \tau_{i+2q}^j - \tau_i^j &= \tau_{i+j+2q} - \tau_{i+2q} - \tau_{i+q} + \tau_i = \\ &= p + r + q - a_{p+r+q} - p - r + a_{p+r} - p - q - a_{p+q} + p + a_i = \\ &= a_{p+r} - a_{p+r+q} + a_p - a_{p+q}. \end{aligned}$$

In the same way we find for other cases that

b) $i = 2p$, $j = 2r$,

$$\tau_{i+2q}^j - \tau_i^j = a_{p+r+q} - a_{p+q} - a_{p+r} + a_p;$$

c) $i = 2p+1$, $j = 2r$,

$$\tau^j_{i+2q} - \tau^j_i = a_{p+r+q} - a_{p+r} - a_{p+q} + a_p;$$

d) $i = 2p+1$, $j = 2r+1$,

$$\tau^j_{i+2q} - \tau^j_i = a_{p+r+q} - a_{p+r} - a_{p+q} + a_p;$$

It follows from these relations that, if q is an $\frac{\epsilon}{2}$-almost period of the sequence $\{\tau_i\}$, then, for any i and j, the inequality

$$|\tau^j_{i+2q} - \tau^j_i| < |a_{p+q+r} - a_{p+r}| + |a_{p+q} - a_p| < \epsilon,$$

holds, i.e. $2q$ is an ϵ-almost period of the sequences $\{\tau^j_i\}$, $j = 0, \pm 1, \pm 2, \dots$. For the constructed sequence $\{\tau_i\}$, $\inf_i \tau^1_i = 0$.

4.5 Almost periodic functions

Consider a linear impulsive differential system

$$\frac{dx}{dt} = Ax + f(t), \qquad t \neq \tau_i, \qquad \Delta x|_{t=\tau_i} = a_i,$$

where A is a constant $n \times n$ matrix with $\Re\lambda_j(A) < 0$, $f(t)$ is a Bohr almost periodic function, $\{a_i\}$ is an almost periodic sequence of points of $\mathbf{R}^n$, and $\{\tau_i\}$ is a sequence of real numbers such that the sequence $\{\bar{\tau}_i\}$ is almost periodic. With the assumption that $\sup_{i\in\mathbf{Z}} \bar{\tau}_i = \theta > 0$, this system has a unique solution

$$x^*(t) = \int_{-\infty}^{t} e^{A(t-\sigma)} f(\sigma) d\sigma + \sum_{\tau_i < t} e^{A(t-\tau_i)} a_i$$

bounded on the whole real axis. Unlike the case of ordinary differential equations, the solution $x^*(t)$ is not a Bohr almost periodic function because $x^*(t)$ is a discontinuous function.

In order to be able to consider in the sequel almost periodic impulsive differential systems and to say that they have almost periodic solutions, we enlarge the class of almost periodic functions in the way so that if, in a considered differential system, the function $f(t)$ belongs to this class, then the function $x^*(t)$ will also belong to this class.

Let a function $\varphi(t) = (\varphi_1(t), \ldots, \varphi_n(t))$ be defined on the real axis $\mathbf{R}$, piecewise continuous with first kind discontinuities at the points of a fixed sequence $\{\tau_i\}$. We assume that the sequence $\{\tau_i\}$ is such that the derived sequence $\{\tau_i^j\}$, $j = 0, \pm 1, \pm 2, \ldots$, is equipotentially almost periodic.

We call a function $\varphi(t)$ *almost periodic* if:

a) for any $\epsilon > 0$ there exists a positive number $\delta = \delta(\epsilon)$ such that if the points t' and t'' belong to the same interval of continuity and $|t' - t''| < \delta$, then $\|\varphi(t') - \varphi(t'')\| < \epsilon$;

b) for any $\epsilon > 0$ there exists a relatively dense set Γ of ϵ-almost periods such that if $\tau \in \Gamma$, then $\|\varphi(t + \tau) - \varphi(t)\| < \epsilon$ for all $t \in \mathbf{R}$ which satisfy the condition $|t - \tau_i| > \epsilon$, $i = 0, \pm 1, \pm 2, \ldots$

This definition of an almost periodic function apparently coincides with one of the definitions of almost periodicity in the space of generalized functions [74].

It is clear that a periodic piecewise continuous function with first kind discontinuities is an almost periodic function in the sense of the preceding definition.

In the sequel, almost periodicity will be understood, if not stated otherwise, in the above defined sense.

Now consider one more example of an almost periodic function.

Let $\{\mu_i\}$ be an almost periodic sequence of vectors in $\mathbf{R}^n$ and the sequence $\{\tau_i\}$ be such that the family of the sequences $\{\tau_i^j\}$ is equipotentially almost periodic. Let us show that the step function $\varphi(t) = \mu_i$, $\tau_i \le t < \tau_{i+1}$, is almost periodic. To do this, it is sufficient to check that condition b) of the definition holds. By Lemma 27, for any $\epsilon_1 > 0$ there exists a relatively dense set of real numbers, Γ, for the elements of which there exists at least one integer p satisfying

$$|\tau_i^p - \tau| < \frac{\epsilon_1}{2}, \qquad i = 0, \pm 1, \pm 2, \ldots. \tag{4.124}$$

By Lemma 24, the set of all such p, taken for all $\tau \in \Gamma$, is also relatively dense. Denote it by $\mathbf{P}$. Let the density index of the set $\mathbf{P}$ be equal to N_1, i.e. between any integers k and $k + N_1$ there exists at least one integer p such that inequality (4.124) holds. Denote by N_2 the density index of the set Q of all $\frac{\epsilon}{2}$-almost periodic sequences $\{\mu_i\}$. We will assume that $\epsilon_1 < \epsilon_2$. If $N_3 = \max(N_1, N_2)$, then there are elements of the sets $\mathbf{P}$ and $\mathbf{Q}$ between any integers k and $k + N_3$. It is clear that, for such elements $p \in \mathbf{P}$ and $q \in Q$, $|p - q| \le N_3$. Because the difference $p - q$ can take only a finite number of values, we will say that two pairs (p, q) and (p', q') are equivalent if $|p - q| = |p' - q'|$ regardless of where an interval of the length N_3 is taken on the real axis. It is clear that there is only a finite number of equivalence classes. Choose for these classes a system of representatives, (p^r, q^r), $r = \overline{1, s}$. Set $N_4 = \max_r |p^r|$, $N = N_3 + 2N_4$. Let n_0 be an integer, p and q be elements of the sets $\mathbf{P}$ and $\mathbf{Q}$, lying between $n_0 + N_4$ and $n_0 + N_4 + N_3$, (p^r, q^r) – a representative of the class containing the pair (p, q) such that $|p - q| = |p^r - q^r|$. We have either $p^r - q^r = p - q$ or $p^r - q^r = q - p$, which means that $p^r - p = q^r - q = -\nu$ or $p^r + p = q^r + q = \nu$. From the inequality $|p^r| \le N_4$, it follows that $n_0 < p + p^r \le n_0 + N_3 + 2N_4 = n_0 + N$ and $n_0 < p - p^r \le n_0 + N_3 + 2N_4 = n_0 + N_4$, i.e. in both cases, $n_0 < \nu < n_0 + N$. If r', r'' are elements of the set Γ such that

$$|\tau_i^p - r'| < \frac{\epsilon_1}{2}, |\tau_i^{p^r} - r''| < \frac{\epsilon_1}{2}, , i = 0, \pm 1, \pm 2, \ldots,$$

then, by using the relation $\tau_i^j - \tau_i^k = \tau_{i+k}^{j-k}$ for $r = r' \pm r''$, we find that

$$|\tau_i^{p^r+p} - r| = |\tau_i^{p^r} + \tau_{i+p}^{p^r} - (r' + r'')| < \frac{\epsilon_1}{2} + \frac{\epsilon_1}{2} = \epsilon_1, \quad (4.125)$$

$$|\tau_i^{p^r-p} - r| = |\tau_{i-p}^{p^r} + \tau_{i-p}^{p} - (r' - r'')| < \frac{\epsilon_1}{2} + \frac{\epsilon_1}{2} = \epsilon_1, \quad (4.126)$$

By Lemma 24, the set of the numbers r, which satisfy (4.125) and (4.126) for some p and p^r, and the set $\mathbf{P}$ are relatively dense sets. For the sequence $\{\mu_i\}$, we find that

$$|\mu_{i+q^r\pm q} - \mu_i| \le |\mu_{i+q^r\pm q} - \mu_{i+q^r}| + +|\mu_{i+q^r} - \mu_i| < \\ < \frac{\epsilon}{2} + \frac{\epsilon}{2} = \epsilon.$$

We have shown that there exist relatively dense sets of real numbers, Γ, and of integers such that

$$|\tau_k^q - r| < \epsilon_1, \qquad |\mu_{k+q} - \mu_k| < \epsilon, \qquad \epsilon_1 < \epsilon, \quad k = 0, \pm 1, \ldots. \quad (4.127)$$

If $\tau_i + \epsilon < t < \tau_{j+1} - \epsilon$, then $\tau_j + r + \epsilon < t + r < \tau_{j+1} + r - \epsilon$ and, by (4.127), $\tau_{i+q} < t + r < \tau_{i+q+1}$. Consequently, the relation $\|\varphi(t+r) - \varphi(t)\| = \|\mu_{i+q} - \mu_i\| < \epsilon$ for any $t \in \mathbf{R}$ satisfying the inequalities $|t - \tau_i| > \epsilon$, $i = 0$, $\pm 1, \ldots$. This finishes the proof that the function $\varphi(t)$ is almost periodic.

We give the main properties of almost periodic functions.

Theorem 75 *An almost periodic function is bounded on the real axis.*

Proof. Let $\varphi(t)$ be an almost periodic function and $l = l(1)$ be the density index of the set Γ_1,

$$M = \max_{0 \le t \le l} \|\varphi(t)\|, \qquad \|\varphi(t') - \varphi(t'')\| \le M_1$$

if $|t' - t''| \le 1$ and t', t'' belong to the same interval of continuity of the function $\varphi(t)$. By the definition of almost periodicity, for any $t \in \mathbf{R}$, $|t - \tau_i| > 1$, there exists 1-almost period r such that $t + r \in [0, l]$ and $\|\varphi(t+r) - \varphi(t)\| < 1$. Whence it follows that $\|\varphi(t)\| < M + M_1 + 1$ for any $t \in \mathbf{R}$.

□

Theorem 76 *If $\varphi(t)$ is an almost periodic function, then for any $\epsilon > 0$ there exists a relatively dense set of intervals of a fixed length γ, $0 < \gamma < \epsilon$, which consist of ϵ-almost periods of the function $\varphi(t)$.*

Proof. Let l be the density index of the set Γ of $\frac{\epsilon}{2}$-almost periods of the function $\varphi(t)$ and the number $\frac{\gamma}{2} = \delta(\frac{\epsilon}{2})$ be determined by using the uniform continuity of this function, i.e. if t' and t'' belong to the same interval of continuity of the function $\varphi(t)$ and $|t' - t''| < \frac{\gamma}{2}$, then $\|\varphi(t') - \varphi(t'')\| < \frac{\epsilon}{2}$. We can assume that $\gamma < \frac{\epsilon}{2}$. Set $L = l + \gamma$ and consider an arbitrary interval $[a, a + L]$. By the definition of an almost period, there exists an $\frac{\epsilon}{2}$-almost period $r \in [a - \frac{\gamma}{2}, a + \frac{\gamma}{2} = l]$. Then $[r - \frac{\gamma}{2}, r + \frac{\gamma}{2}] \subset [a, a + L]$. Let ξ be an arbitrary number from the interval $[r - \frac{\gamma}{2}, r + \frac{\gamma}{2}]$. By using inequality $|\xi - r| < \delta$ and denoting $t' = t - r + \xi$, we find, for all $t \in \mathbf{R}$ such that $|t - \tau_i| > \epsilon$, that $|t' - \tau_i| > \frac{\epsilon}{2}$ and

$$\|\varphi(t + \xi) - \varphi(t)\| \leq \|\varphi(t' + r) - \varphi(t')\| + \\ + \|\varphi(t') - \varphi(t)\| < \frac{\epsilon}{2} + \frac{\epsilon}{2} = \epsilon.$$

□

This theorem implies that for a given $\epsilon > 0$ there exists a relatively dense set of ϵ-almost periods of the function $\varphi(t)$, which are multiples of γ.

Theorem 77 *Let $\varphi(t)$ be an almost periodic function with the values in the set $E \subset \mathbf{R}^n$. If $F(y)$ is a uniformly continuous function defined on the set E, then the function $F(\varphi(t))$ is almost periodic.*

Theorem 4.126 can be proved in the same way as the similar theorem for Bohr almost periodic functions.

Lemma 31 *For any two almost periodic functions with discontinuities at the points of the same sequence, for any $\epsilon > 0$ there exists a relatively dense set of their common ϵ-almost periods.*

Proof. Let $\varphi_1(t)$ and $\varphi_2(t)$ be two almost periodic functions with the same sequence of points of discontinuity. By the corollary to Theorem 76, there exist such numbers l_1 and l_2 that every interval $[a, a + l_1]$ and $[a, a + l_2]$ will contain the corresponding $\frac{\epsilon}{4}$-almost periods r_1 and r_2, which are multiples of γ, $0 < \gamma < \frac{\epsilon}{4}$. If we set $l = \max(l_1, l_2)$, then in each segment $[a, a+l]$ there exists a pair of $\frac{\epsilon}{4}$-almost periods $r_1 = n'\gamma$ and $r_2 = n''\gamma$ with integer n' and n''. Because $r_1 - r_2 = (n' - n'')\gamma = n\gamma$ and $|n\gamma| \le l$, we see that $n\gamma$ can take only a finite number of values. Let these values be $n_1\gamma, n_2\gamma, \dots, n_p\gamma$ with the pairs of almost periods $(r_1^1, r_2^1), (r_1^2, r_2^2), \dots, (r_1^p, r_2^p)$ as their representatives, i.e. such that $r_1^s - r_2^s = n_s\gamma$, $s = \overline{1,p}$. Set $\max_s |r_1^s| = T$. Let $[a, a+l+2T]$ be an arbitrary segment of length $l + 2T$. In the segment $[a+T, a+l+T]$ take two $\frac{\epsilon}{4}$-almost periods of the functions $\varphi_1(t)$ and $\varphi_2(t)$, $r_1 = n'\gamma$ and $r_2 = n''\gamma$, and let $r_1 - r_2 = n_s\gamma = r_1^s - r_2^s$. Whence we get

$$r = r_1 - r_1^s > r_2 - r_2^s, \qquad r \in [a,\ a+l+2T]. \tag{4.128}$$

The set of all the numbers given by relation (4.128), Γ, is relatively dense and consists of multiples of γ. Let us show that there is a subset $\Gamma_0 \subset \Gamma$, relatively dense in $\mathbf{R}$, such that, for all $r \in \Gamma_0$, $|t + r - \tau_i| > \frac{\epsilon}{2}$ if $|t - \tau_i| > \epsilon$, $i = 0, \pm 1, \dots$.

Let $l = l(\epsilon)$ be the density index of the set Γ, $l = l'(\epsilon)$ is the density index of the set Γ' of real numbers, multiples of γ, defined by Lemma 27 for $\frac{\epsilon}{4}$. Clearly, γ can be chosen to be sufficiently small so that the inequalities $l < +\infty$, $l' < +\infty$ hold. Set $l'' = \max(l, l')$. Then for any interval $[a, a+l'']$ there exist such integers m, m', and q that $m\gamma, m'\gamma \in [a, a+l'']$ and

$$|t_k^q - m\gamma| < \frac{\epsilon}{4}, \quad \|\varphi_j(t + m'\gamma) - \varphi_j(t)\| < \frac{\epsilon}{2}, \quad j = 1, 2,\ k = 0, \pm 1, \dots. \tag{4.129}$$

The differences $m - m'$ can take only a finite number of values, for example, n_s, $s = \overline{1,p}$. For every n_s, we fix triples (m_s, m_s', q) and consider them as representatives of the class determined by the number n_s. Set $\lambda = \max_{1 \le s \le p} |m_s'\gamma|$. Let $\mathcal{I} = [a, a+l''+2\lambda]$ be an interval of length $l'' + 2\lambda$. For the subinterval $\mathcal{I}' = [a+\lambda, a+\lambda+l'']$, $\mathcal{I}' \subset \mathcal{I}$, of length l'', we can find the integers m', m, q satisfying (4.128) and the condition $m\gamma, m'\gamma \in \mathcal{I}$. Let $m\gamma - m'\gamma = n_s\gamma$, i.e. $m\gamma - m'\gamma = m_s\gamma - m_s'\gamma$, and so $m - m_s = m' - m_s'$. Set $r = (m - m_s)\gamma$, $h = q - q_s$. Clearly, $r \in \mathcal{I}$. By using the second formula of (4.120), we get for any integer k that

$$\begin{aligned}
|\tau_k^h - r| &= |\tau_k^{q-q_s} - r| = |\tau_{k-q-q_s}^{q-q_s} - r| = \\
&= |\tau_{k-q_s}^{q} - \tau_{k-q_s}^{q_s} - m\gamma + m_s\gamma| \le \\
&\le |\tau_{k-q_s}^{q} - m\gamma| + |\tau_{k-q_s}^{q_s} - m_s\gamma| < \frac{\epsilon}{4} + \frac{\epsilon}{4} = \frac{\epsilon}{2}.
\end{aligned}$$

Let $|t - \tau_i| > \epsilon$ and, for definiteness sake, $\tau_k + \epsilon < t < \tau_{k+1} - \epsilon$. Then $\tau_k + r + \epsilon < t + r < \tau_{k+1} + r - \epsilon$. Because $|\tau_i^h - r| < \frac{\epsilon}{2}$ for $t = 0, \pm 1, \pm 2, \ldots$, we have that $\tau_i^h - \frac{\epsilon}{2} < r < \tau_i^h + \frac{\epsilon}{2}$ and, consequently,

$$\tau_{k+h} - \tau_k + \tau_k + \epsilon - \frac{\epsilon}{2} < t + r < \tau_{k+1} + \tau_{k+h+1} - -\tau_{k+1} + \frac{\epsilon}{2} - \epsilon,$$

$$\tau_{k+h} + \frac{\epsilon}{2} < t + r < \tau_{k+h+1} - \frac{\epsilon}{2},$$

i.e.

$$|t + r - \tau_i| > \frac{\epsilon}{2}, \qquad i = 0, \pm 1, \pm 2, \ldots.$$

Hence the set Γ_0 is nonempty and relatively dense in $\mathbf{R}$. Whence it follows for $j = 1, 2$, $r \in \Gamma_0$ that

$$\begin{aligned}\|\varphi_j(t + r) - \varphi_i(t)\| &= \|\varphi_j(t + (m' - m'_s)\gamma) - \varphi(t)\| \leq \\ &\leq \|\varphi_j(t + (m' - m'_s)\gamma) - \varphi_j(t + m'\gamma)\| + \|\varphi_j(t) - \varphi_j(t + m'\gamma)\| < \\ &< \frac{\epsilon}{2} + \frac{\epsilon}{2} = \epsilon.\end{aligned}$$

□

Theorem 78 *The sum of two almost periodic functions which have a common sequence of points of discontinuities is also an almost periodic function.*

Theorem 79 *The scalar product of two almost periodic functions which have a common sequence of points of discontinuity is also an almost periodic function.*

Theorem 80 *The quotient* $\dfrac{\varphi(t)}{\psi(t)}$ *of two almost periodic scalar functions is an almost periodic function if*

$$\inf_{t \in \mathbf{R}} \|\psi(t)\| > 0.$$

Theorems 78 – 80 can be proved, by using Lemma 31, similarly to the corresponding statements for Bohr almost periodic functions.

4.6 Almost periodic differential systems

Let a system of ordinary differential equations be given,

$$\frac{dx}{dt} = A(t)x, \tag{4.130}$$

where $A(t)$ is a square $n \times n$ matrix, the elements of which are continuous real-valued functions of a real variable t.

Denote by $U(t,s)$ a matriciant of system (4.130). The following statement holds [55].

Lemma 32 *If* $\sup \|A(t)\| = m < +$, *then*

$$\|U(t,s) - E\| < e^{m|t-s|} - 1, \qquad \|U(t,s)\| < e^{m|t-s|}, \qquad t,\, s \in \mathbf{R}.$$

We will prove that the following lemmas hold.

Lemma 33 *Let the matrix* $A(t)$ *be uniformly continuous and bounded on* $\mathbf{R}$. *Then for any* $\epsilon > 0$ *and a fixed* $\theta > 0$ *there exists such* $\delta = \delta(\epsilon) > 0$ *that* $\|U(t', s') - U(t,s)\| < \epsilon$ *if* $|t - t'| < \delta$, $|s - s'| < \delta$, *and* $|t - s| \leq \theta$.

Proof. Set $t' = t + \alpha$, $s' = s + \beta$. Let $\epsilon > 0$ be arbitrary and $\delta = \delta(\epsilon)$, $0 < \delta < 1$, be such that

$$\begin{aligned} \|A(t+r) - A(t)\| &< \frac{\epsilon}{2\theta e^{m(\theta+2)}}, \\ e^{m(|\alpha|+|\beta|)} - 1 &< \frac{\epsilon}{2e^{m\theta}} \end{aligned}$$

for $t \in \mathbf{R}$, $|\alpha| < \delta$, $|\beta| < \delta$.

Denote $V(t,s,\alpha,\beta) = U(t+\alpha, s+\beta)$. By Lemma 32,

$$\|V(s,s,\alpha,\beta)\| < e^{m|\alpha-\beta|} - 1 < \frac{\epsilon}{2e^{m\theta}}.$$

It is easy to check that

$$\frac{\partial V}{\partial t} = A(t)V + (A(t+\alpha) - A(t))U(t+\alpha,\, s+\beta),$$

and so

$$\begin{aligned} V(t,s,\alpha,\beta) = {} & U(t,s)V(s,s,\alpha,\beta) + \\ & + \int_s^t U(t,u)(A(u+r) - A(u))U(u+\alpha,\, s+\beta)du. \end{aligned}$$

Whence,

$$\|V(t,s,\alpha,\beta)\| < \frac{\epsilon e^{m|t-s|}}{2e^{m\theta}} + \frac{\epsilon\theta e^{m(t-s+|\alpha|+\beta|)}}{2\theta e^{m(\theta+2)}} < \epsilon.$$

□

Lemma 34 *Let $A(t)$ be an Bohr almost periodic matrix. If r is an $\frac{\epsilon}{\theta}e^{-m\theta}$-almost period of the matrix $A(t)$, then*

$$\|U(t+r, s+r) - U(t,s)\| < \epsilon \qquad \textit{for} \qquad 0 \le t-s \le \theta.$$

Proof. If we denote $V(t,s) = U(t+r, s+r) - U(t,s)$, then

$$\frac{\partial V}{\partial t} = A(t)V + (A(t+r) - A(t))U(t+r, s+r).$$

Because $Y(s,s) = 0$,

$$V(t,s) = \int_s^t U(t,u)(A(u+r) - A(u))U(u+r, s+r)du.$$

By taking an estimate of this integral, we get

$$\|V(t,s)\| < e^{m|t-s|}\frac{\epsilon(t-s)}{\theta e^{m\theta}} < \epsilon.$$

□

Lemma 35 *Let $U(t,s)$ be the fundamental matrix of solutions of system (4.130) in which the matrix $A(t)$ is Bohr almost periodic, $f(t)$ is an almost periodic function, the sequences of the vectors $\{I_i\}$ and the matrices $\{B_i\}$ are almost periodic, and the sequence of times $\{\tau_i\}$ is such that the sequences $\{\tau_i^j\}$, $j = 0, \pm1, \pm2, \ldots$, are positive and equipotentially almost periodic. Then for any $\epsilon > 0$ and any $\theta > 0$ there exist such a real number ν, $0 < \nu < \epsilon$, and relatively dense sets of real numbers, Γ, and integers $\mathbf{Q}$ that the following relations hold:*

1) $\|U(t+r, s+r) - U(t,s)\| < \epsilon, \qquad 0 \le t-s \le \theta,$
2) $\|f(t+r) - f(t)\| < \epsilon, \quad t \in \mathbf{R},\ |t-\tau_i| > \epsilon,\ i = 0, \pm1, \ldots,$
3) $\|I_{k+q} - I_k\| < \epsilon,$
4) $\|B_{k+q} - B_k\| < \epsilon,$
5) $|\tau_k^q - r| < \nu, \quad k = 0, \pm1, \pm2, \ldots,\ r \in \Gamma,\ q \in \mathbf{Q}.$

Proof.

1. By using corollary of Theorem 75 and Corollary 2 of [38], p.371, it can be checked similarly to Lemma 31 that there exists a relatively dense set of almost periods τ, multiples of γ, $0 < \gamma < \epsilon$, and common to the function $f(t)$ and the matrix $A(t)$, such that

$$\|A(t+\tau) - A(t)\| < \frac{\epsilon}{2\theta} e^{-m\theta}, \qquad \|f(t+\tau) - f(t)\| < \frac{\epsilon}{2},$$
$$\tau = n\gamma, \qquad |t - \tau_i| > \epsilon, \qquad i = 0, \pm 1, \ldots.$$

2. As in the proof of Lemma 31, we get that there exist relatively dense sets of real numbers, Γ, and integers, $\mathbf{P}$, such that

$$\|A(t+r) - A(t)\| < \frac{\epsilon}{2\theta} e^{-m\theta}, \qquad \|f(t+r) - f(t)\| < \frac{\epsilon}{2},$$
$$|t - \tau_k| > \epsilon, \quad |\tau_k^p - r| < \frac{\nu}{2}, \quad 0 < \nu < \epsilon, \quad k = 0, \pm 1, \ldots.$$

3. As known, there exists a relatively dense set of common almost periods of the sequences $\{I_i\}$ and $\{B_i\}$. Hence there exists such a natural number N that the interval between any two adjacent multiples of N contains integers p and q, for which

$$|\tau_k^p - m\gamma| < \frac{\nu}{2}, \qquad \|I_{k+q} - I_k\| < \frac{\epsilon}{2}, \qquad \|B_{k+q} - B_k\| < \frac{\epsilon}{2}. \tag{4.131}$$

The difference $p - q$ can take only a finite number of values, say n_i, $i = \overline{1, r}$. To every n_i there corresponds a pair of integers (p^i, q^i) satisfying (4.131). Fix such a pair to be a representative of the class corresponding to n_i. Set $m = \max_i |p^i|$. Let $l+1$, $l+2$, ..., $l+N+2M$ be arbitrary $N+2M$ integers. Amongst N integers $l+M+1, \ldots, l+M+N$ there exists a pair (p,q) satisfying (4.131). Let $p - q = n_j$. Then $p - p^j = q - q^j = \mu$. It is clear that μ is one of the numbers $l+1, l+2, \ldots, l+N+2M$. For any integer k, we have

$$\|I_{k+\mu} - I_k\| \le \|I_{k+p^j-p} - I_{k+p^j}\| + \|I_{k+p} - I_k\| - \frac{\epsilon}{2} + \frac{\epsilon}{2} = \epsilon.$$

In the same way we see that $\|B_{k+\mu} - B_k\| < \epsilon$. Let r and τ_i be almost periods, multiples of γ, which correspond, by the definition of almost periodicity, to two pairs (p,q), (p^i, q^i). Denoting $r_\mu = r - r_j$, we have

$$\begin{aligned} \|A(t+r_\mu) - A(t)\| &\le \|A(t+r_\mu) - A(t-r_j)\| + \|A(t-r_j) - A(t)\| < \\ &< \frac{\epsilon}{2\theta} e^{-m\theta} + \frac{\epsilon}{2\theta} e^{-m\theta} = \frac{\epsilon}{\theta} e^{-m\theta}. \end{aligned} \tag{4.132}$$

Because $\tau_i^j - \tau_i^k = \tau_{i+k}^{j-k}$,

$$|\tau_k^{\mu} - r_{\mu}| \leq |\tau_{k-p^j}^{p} - r| + |\tau_{k-p^j}^{p^j} - \tau_j| < \frac{\nu}{2} + \frac{\nu}{2} = \nu.$$

By noting that $|\tau_k^p - r| < \gamma < \frac{\epsilon}{2}$ and $|t - \tau_i| > \epsilon$, we have

$$|t + r - \tau_i| > \frac{\epsilon}{2}, \qquad i = 0, \pm 1, \ldots.$$

Indeed let, for definiteness sake, $\tau_i + \epsilon < t < \tau_{i+1} - \epsilon$. Then $\tau_i + r + \epsilon < t + r < \tau_{i+1} - \epsilon + r$. But $\tau_{i+p} < \tau_i + r + \nu$ and $\tau_{i+1} + r - \nu < \tau_{i+p+1}$, and so either $\tau_{i+p} + \frac{\epsilon}{2} < t + r < \tau_{i+p+1} - \frac{\epsilon}{2}$ or $|t + r - \tau_i| > \frac{\epsilon}{2}$. By taking into account these inequalities, we find that

$$\begin{aligned} \|f(t + r_{\mu}) - f(t)\| &\leq \|f(t + r_{\mu}) - f(t + r)\| + \|f(t + r) - f(t)\| < \\ &< \frac{\epsilon}{2} + \frac{\epsilon}{2} = \epsilon. \end{aligned}$$

To finish the proof it remains to note that estimate 1 follows from Lemma 34 □

Denote by $X(t, s)$ the Cauchy matrix for the linear impulsive differential system

$$\frac{dx}{dt} = A(t)x, \qquad t \neq t_i, \qquad \Delta x|_{t=t_i} = B_i x, \tag{4.133}$$

where $A(t)$ is a continuous $n \times n$ matrix, B_i, $i = 0, \pm 1, \ldots$, are constant square matrices of dimension n, the sequence $\{\tau_i\}$ is strictly increasing. For $\tau_k \leq t < \tau_{k+1}$, $\tau_{m-1} \leq s < \tau_m$, $s < t$,

$$X(t, s) = U(t, \tau_k) \prod_{s \leq \tau_i < t} (E + B_i) U(\tau_i, \tau_{i-1})(E + B_m) U(\tau_m, s). \tag{4.134}$$

Lemma 36 *Let the matrix $A(t)$ be Bohr almost periodic, the sequence $\{B_i\}$ be almost periodic, and the sequences $\{\tau_i^j\}$ be equipotentially almost periodic. If the Cauchy matrix $X(t, s)$ satisfies the inequality*

$$\|X(t, s)\| \leq Ce^{-\alpha(t-s)}, \qquad t \geq s,$$

where C and α are positive real numbers, then the diagonal of the matrix $X(t, s)$ is almost periodic, i.e. for any $\epsilon > 0$, $t, s \in \mathbf{R}$, $|t - \tau + i| > \epsilon$,

$|s - \tau_i| > \epsilon$, $i = 0, \pm 1, \ldots$ *there exists a relatively dense set of almost periods,* Γ, *such that, for* $r \in \Gamma$, *we have*

$$\|X(t+r, s+r) - X(t,s)\| < \epsilon \Gamma e^{-\frac{\alpha}{2}(t-s)},$$

where Γ *is a positive constant.*

Proof. Because

$$\begin{aligned}\frac{\partial X}{\partial t} &= A(t)X(t+r, s+r) + (A(t+r) - \\ &\quad -A(t))X(t+r, s+r), \qquad t \neq \tau_i', \\ \Delta X(t+r, s+r)|_{t=\tau_i'} &= B_i X(\tau_i' + r, s+r) + \\ &\quad +(B_{i+q} - B_i)X(\tau_i + r, s+r),\end{aligned}$$

where $\tau_i' = \tau_i - r$, the numbers r and q are defined in the conditions of Lemma 35, we have

$$\begin{aligned}X(t+r, s+r) &= X(t,r) + \\ &+ \int_s^t X(t,u)(A(u+r) - A(u))X(u+r, s+r)du + \\ &+ \sum_{s \leq \tau_i' \leq t} X(t, \tau_i')(B_{i+q} - B_i)X(\tau_i' + r, s+r).\end{aligned}$$

As in the proof of Lemma 35, it can be checked that $|t - \tau_i'| > \epsilon$ implies that $\tau_{i+q}' < t + r < \tau_{i+q+1}'$. Further we have

$$\begin{aligned}&\|X(t+r, s+r) - X(t,s)\| \leq \\ &\leq \int_s^t \|X(t,u)\| \, \|A(u+r) - A(u)\| \, \|X(u+r, s+r)\| du + \\ &\quad + \sum_{s < \tau_i' < t} \|X(t, \tau_i')\| \, \|B_{i+q} - B_i\| \, \|X(\tau_i' + r, s+r)\| \leq \\ &\leq \int_s^t \epsilon C e^{-\alpha(t-s)} du + \sum_{s < \tau_i' < t} \epsilon C e^{-\alpha(t-s)} = \\ &= \frac{\epsilon C}{\alpha} e^{-\alpha(t-s)}(t-s) + i(s,t)\epsilon C e^{-\alpha(t-s)},\end{aligned}$$

where $i(s,t)$ is the number of the points τ_i' in the interval (s,t). Let N be the number defined in Lemma 26. Then

$$\begin{aligned}&\|X(t+r, s+r) - X(t,s)\| \leq \\ &\leq \epsilon \left(\frac{2C}{\alpha^2} + \frac{2NC}{\alpha} \right) e^{-\frac{\alpha}{2}(t-s)} = \epsilon \Gamma e^{-\frac{\alpha}{2}(t-s)}.\end{aligned}$$

□

Consider the impulsive differential system

$$\frac{dx}{dt} = A(t)x + f(t), \qquad t \neq \tau_i,$$
$$\Delta x|_{t=\tau_i} = B_i x + I_i, \tag{4.135}$$

where the matrices $A(t)$, B_i, $i = 0, \pm 1, \pm 2, \ldots$, the sequences $\{\tau_i\}$, $\{I_i\}$, and the function $f(t)$ satisfy the conditions of Lemmas 35 and 36. Denote

$$\gamma = \sup_t \Lambda(t), \qquad \alpha^2 = \sup_t \Lambda_i^2,$$

where $\Lambda(t)$ is the greatest eigen value of the matrix $\frac{1}{2}(A(t)+A^T(t))$, Λ_i is the greatest eigen value of the matrix $(E+B_i)^T(E+B_i)$, ^{T} is the conjugation sign. By Lemma 22, there exists a finite limit

$$\lim_{t\to\infty} \frac{i(t_0, t_0+t)}{t} = p,$$

which is uniform for $t_0 \in \mathbf{R}$.

By Theorem 33, we have

$$\|X(t,s)\| \leq K e^{(\epsilon+\gamma+p\ln\alpha)(t-s)},$$

where $\epsilon > 0$ is arbitrary, $K = K(\epsilon) \geq 1$. Let $\beta(\epsilon) = \epsilon + \gamma + p\ln\alpha$.

Theorem 81 *If system (4.135) satisfies all the listed above conditions and $\gamma + p\ln\alpha < 0$, then this system admits a unique almost periodic solution and this solution is asymptotically stable.*

Proof. Let us show that an almost periodic solution is given by the expression

$$x_0(t) = \int_{-\infty}^{t} X(t,s)f(s)ds + \sum_{t_i<t} X(t,t_i)I_i. \tag{4.136}$$

First we check that the integral expression in the right-hand side of (4.136) makes sense. Indeed, choose $\epsilon > 0$ to be so small that $\beta = \beta(\epsilon) < 0$, fix it and denote

$$M = \sup_t \|f(t)\| + \sup_t \|I_i\|.$$

Then

$$\begin{aligned}\|x_0(t)\| &= \int_{-\infty}^{t} \|X(t,s)\|\,\|f(s)\|ds + \sum_{t_i<t} \|X(t,\tau_i)\|\,\|I_i\| \le \\ &\le \int_{-\infty}^{t} Ke^{\beta(t-s)}Mds + \sum_{\tau_i<t} KMe^{\beta(t-\tau_i)} < \\ &< KM\left(\frac{1}{-\beta} + \frac{e^{(\frac{1}{N}-1)\beta}}{1-e^{\beta\frac{1}{N}}}\right).\end{aligned}$$

Substituting into (4.135) we see that $x_0(t)$ is a solution of this system.

Take an ϵ-almost period defined in Lemmas 35 and 36. By using Lemma 26, we find that

$$\begin{aligned}\|x_0(t+r) - x_0(t)\| &\le \int_{-\infty}^{t} \|X(t+r,\, s+r)f(s+r) - X(t,s)f(s)\|ds + \\ &+ \sum_{\tau_i<t} \|X(t+r,\, \tau_{i+q})I_{i+q} - X(t,\tau_i)I_i\| < \Gamma_1(\epsilon)\epsilon,\end{aligned}$$

where $\Gamma_1(\epsilon)$ is a bounded positive function of ϵ.

Because any solution $x(t)$ of system (4.135) can be represented as

$$x(t) = X(t,t_0)x_0 + \int_{t_0}^{t} X(t,s)f(s)ds + \sum_{t_0\le\tau_i<t} X(t,\tau_i)I_i, \tag{4.137}$$

for any two distinct solutions of this system, $\varphi(t)$ and $\psi(t)$, we have

$$\|\varphi(t) - \psi(t)\| \le Ke^{\beta(t-t_0)}\|\varphi(t_0) - \psi(t_0)\|,$$

whence it follows that a solution of system (4.135) is asymptotically stable and unique.

□

Now let us study a weakly nonlinear impulsive system given by

$$\begin{aligned}\frac{dx}{dt} &= A(t)x + f(t,x), \qquad t \ne t_i, \\ \Delta x|_{t=\tau_i} &= B_i x + I_i(x),\end{aligned} \tag{4.138}$$

where the matrices $A(t)$, B_i, $i = 0, \pm1, \ldots$, are the same as in system (4.135), the sequences $\{\tau_i^j\}$, $j = 0, \pm1, \pm2, \ldots$, are equipotentially almost periodic and $\inf_i \bar{\tau}_i^j = \theta > 0$. Let the function $f(t,x)$ be almost periodic in t and the sequence $\{I_i(x)\}$ – almost periodic in i and uniform in the region

$$t \in \mathbf{R}, \qquad \|x\| < h, \qquad i = 0, \pm 1, \ldots. \tag{4.139}$$

Assume that the functions $f(t,x)$ and $I_i(x)$ satisfy the Lipschitz condition

$$\|f(t,x) - f(t,y)\| + \|I_i(x) - I_i(y)\| \leq L\|x - y\| \tag{4.140}$$

and are uniform and bounded in the region (4.139),

$$\sup_{\substack{-\infty<t<+\infty \\ \|x\|<h}} \|f(t,x)\| + \sup_{\substack{-\infty<t<+\infty \\ \|x\|<h}} \|I_i(x)\| = \mathcal{H} < +\infty.$$

Let us prove the following.

Lemma 37 *If $\varphi(t)$ is an almost periodic function and* $\inf_i \bar{\tau}_i^j = \theta > 0$, *then $\{\varphi(t_i)\}$ is an almost periodic sequence.*

Proof. Let $0 < 3\epsilon < \theta$. Construct a sequence $\{\tau_i'\}$ satisfying the condition $\tau_i - \tau_i' = 2\epsilon$, $i = 0, \pm 1, \pm 2, \ldots$. For ϵ_1 choose the numbers r and q according to Lemma 35 such that $\|\varphi(t+r) - \varphi(t)\| < \epsilon_1$ and $|\tau'^q_k - r| < \nu$, $0 < \nu < \epsilon_1$, for all $|t - \tau_i'| > \epsilon_1$, $t \in \mathbf{R}$, $k = 0, \pm 1, \pm 2, \ldots$, $i = 0, \pm 1, \pm 2, \ldots$. Because $-\nu < \tau_{k+q} - \tau_k - r < \nu$ and $\tau_k' + r = \tau_k - 2\epsilon_1 + r$, we see that $0 < 2\epsilon_1 - \nu < \tau_{k+q} - \tau_k' - r < 2\epsilon_1 + \nu < 3\epsilon_1$. Thus, if $\|\varphi(t') - \varphi(t'')\| < o(3\epsilon_1)$ for t', t'' belonging to the same interval of continuity with $|t' - t''| < 3\epsilon_1$, then, assuming that $2o(3\epsilon_1) + \epsilon_1 < \epsilon < \theta$, we find that

$$\begin{aligned}\|\varphi(\tau_{k+q}) - \varphi(\tau_k)\| \leq \|\varphi(\tau_{k+q}) - \varphi(\tau_k' + r)\| + \|\varphi(\tau_k' + r) - \\ -\varphi(\tau_k')\| + \|\varphi(\tau_k') - \varphi(\tau_k)\| < 2o(3\epsilon_1) + \epsilon_1 < \epsilon.\end{aligned}$$

□

Let $\epsilon > 0$ be such that $\beta = \beta(\epsilon) < 0$ and

$$a = \frac{1}{-\beta} + \frac{e^{(\frac{1}{N}-1)\beta}}{1 - e^{\beta\frac{1}{N}}},$$

where N is the natural number given by Lemma 26. The following theorem holds.

Theorem 82 *If system (4.138) satisfies all the conditions stated above and*

$$1) \quad KHa < h,$$
$$2) \quad KLa < 1,$$
$$3) \quad \beta + KL + \frac{1}{N}\ln(1 + KL) < 0,$$

then this system has a unique asymptotically stable almost periodic solution.

Proof. Let $\mathfrak{R}$ be the space of all almost periodic functions with discontinuities at the points of the same sequence $\{\tau_i\}$. If $\varphi(t) \in \mathfrak{R}$, then we take $\|\varphi(t)\|_0 = \sup_t \|\varphi(t)\|$ to be the norm of this function in $\mathfrak{R}$. We consider a subset of $\mathfrak{R}$, $\mathbf{D}$ of all almost periodic functions $\varphi(t)$ such that $\|\varphi(t)\|_0 < h$ and define an operator T on $\mathfrak{R}$ as follows: if $\varphi(t) \in \mathfrak{R}$, then

$$T(\varphi(t)) = \int_{-\infty}^{t} X(t,s)f(s,\varphi(s))ds + \sum_{\tau_i < t} X(t,\tau_i)I_i(\varphi(\tau_i)). \tag{4.141}$$

Let us show that $T(\mathbf{D}) \subseteq \mathbf{D}$. Indeed, if $\|\varphi(t)\|_0 < h$, then

$$\begin{aligned} \|T(\varphi(t)) &\le \int_{-\infty}^{t} \|X(t,s)\| \, \|f(s,\varphi(s))\|ds + \sum_{\tau_i<t} \|X(t,\tau_i)\| \, \|I_i(\varphi(\tau_i))\| \le \\ &\le \int_{-\infty}^{t} Ke^{\beta(t-s)}Hds + \sum_{\tau_i<t} KHe^{\beta(t-\tau_i)} \le KHa < h. \end{aligned}$$

Moreover, if the function $\varphi(t)$ is almost periodic, then, by Lemma 37, the sequence $\{\varphi(t_i)\}$ is almost periodic and, by using the method of finding common almost periods, it is possible to show that the sequence $\{I_i(\varphi(t_i))\}$ is almost periodic. Theorem 77 imp;lies that the function $f(t,\varphi(t))$ is almost periodic. Whence, by using Lemmas 35 and 36 we find that, if $\varphi(t) \in \mathbf{D}$, then there exists a relatively dense set Γ of ϵ-almost periods of the function $\varphi(t)$ such that, for $r \in \Gamma$, $t \in \mathbf{R}$, $|t - \tau_i| > \epsilon$, $i = 0, \pm 1, \ldots$, the following inequality holds,

$$\begin{aligned} &\|T(\varphi(t+r)) - T(\varphi(t))\| \le \\ &\le \int_{-\infty}^{t} \|X(t+r,s+r)f(s+r,\varphi(s+r)) - X(t,s)f(s,\varphi(s))\|ds + \\ &\quad + \sum_{\tau_i<t} \|X(t+r,\tau_{i+q})I_{i+q}(\varphi(\tau_{i+q})) - X(t,\tau_i)I_i(\varphi(\tau_i))\| \le \Gamma_2(\epsilon)\epsilon, \end{aligned}$$

where $\Gamma_2(\epsilon)$ is a certain bounded function of ϵ. Hence we have proved that $T(\mathbf{D}) \subseteq \mathbf{D}$. If $\varphi,\ \psi \in \mathbf{D}$, then

$$
\begin{aligned}
\|T(\varphi(t)) - T(\psi(t))\| \ \leq\ & \int_q -\infty^t \|X(t,s)\|\, \|f(s,\varphi(s)) - f(s,\psi(s))\| ds + \\
& + \sum_{\tau_i<t} \|X(t,\tau_i)\|\, \|I_i(\varphi(\tau_i)) - I_i(\psi(\tau_i))\| \leq \\
\leq\ & KLa\|\varphi(t) - \psi(t)\|_0,
\end{aligned}
$$

whence, by condition 2) of the theorem, it follows that T is a contraction operator in $\mathbf{D}$, and so there exists a unique almost periodic solution of system (4.138).

Now, by using the integral form of system (4.138),

$$
\begin{aligned}
x(t) = X(t,t_0)x_0 + \int_{t_0}^{t} X(t,s)f(s,x(s))ds + \\
+ \sum_{t_0 \leq \tau_i < t} X(t,\tau_i)I_i(x(\tau_i)),
\end{aligned}
\tag{4.142}
$$

we find that any two solutions $\varphi(t)$ and $\psi(t)$ of this system satisfy the inequality

$$
\|\varphi(t) - \psi(t)\| \leq K\|\varphi(t_0) - \psi(t_0)\| e^{(\beta + KL + \frac{1}{N}\ln(1+KL))(t-t_0)}. \tag{4.143}
$$

Indeed, it follows from (4.142) that

$$
\begin{aligned}
\|\varphi(t) - \psi(t)\| \leq & \|X(t,t_0)\|\, \|\varphi(t_0) - \psi(t_0)\| + \\
& + \int_{t_0}^{t} \|X(t,s)\|\, \|f(s,\varphi(s)) - f(s,\psi(s))\| ds + \\
& + \sum_{\tau_i<t} \|X(t,\tau_i)\|\, \|I_i(\varphi(\tau_i)) - I_i(\psi(\tau_i))\| \leq \\
\leq & Ke^{\beta(t-t_0)}\|\varphi(t_0) - \psi(t_0)\| + \int_{t_0}^{t} Ke^{\beta(t-s)} L\|\varphi(s) - \psi(s)\| ds + \\
& + \sum_{\tau_i<t} KLe^{\beta(t-\tau_i)}\|\varphi(\tau_i) - \psi(\tau_i)\|.
\end{aligned}
$$

By multiplying both sides of the last inequality by $e^{-\beta t}$ and denoting $v(t) = \|\varphi(t) - \psi(t)\| e^{-\beta t}$, we find that

$$
v(t) \leq Kv(t_0) + \int_{t_0}^{t} KLv(s)ds + \sum_{t_i<t} KLv(t_i).
$$

By applying to this relation the analogue of the Gronwell–Bellman lemma, we get

$$v(t) \leq Kv(t_0) \prod_{t_0 \leq \tau_i < t} (1 + KL)e^{KL(t-t_0)}.$$

or

$$\|\varphi(t) - \psi(t)\| \leq K\|\varphi(t_0) - \psi(t_0)\| \prod_{t_0 \leq \tau_i < t} (1 + KL)e^{(\beta+KL)(t-t_0)}.$$

It can be seen from Lemma 26 that inequality (4.143) holds and so, by condition 3) of the theorem, solutions of system (4.138) are asymptotically stable.

□

4.7 Homogeneous linear periodic systems

We will be considering a system of impulsive ordinary differential equations in the form

$$\frac{dx}{dt} = A(t)x + f(t), \qquad t \neq \tau_i,$$
$$\Delta x|_{t=\tau_i} = B_i x + I_i, \tag{4.144}$$

where $x \in \mathbb{R}^n$, $A(t)$ is a continuous T-periodic $n \times n$-matrix, B_i, $i = 0, \pm 1$, ..., are constant $n \times n$ matrices, the sequence of times $\{\tau_i\}$ satisfies the condition $\tau_{i+p} = \tau_i + T$, $i = 0, \pm 1, \ldots$. The functions $f(t)$ and the sequence $\{I_i\}$ are almost periodic.

In this section, for system (4.144), we will prove an analogue of the theorem on almost periodicity of a bounded solution, which holds for systems of ordinary differential equations [26], and give sufficient conditions for existence of almost periodic solutions of nonhomogeneous linear systems. We also consider the question of asymptotic stability of almost periodic solutions.

Let $X(t,s)$ be the Cauchy matrix (4.134) of the homogeneous system, corresponding to equations (4.144). Denote

$$X(t) = X(t,0), \qquad \Lambda = \frac{1}{T}\mathrm{Ln}X(T), \qquad \Phi(t) = X(t)e^{-\Lambda t}.$$

In Section 2.7 it was shown that $\Phi(t)$ is a nonsingular matrix, which is periodic with the period T, piecewise continuous with discontinuities at the points τ_i, $i = 0, \pm 1, \pm 2, \ldots$.

Let us apply the change of coordinates $x = \Phi(t)y$ to system (4.144). Then this system becomes

$$\frac{dy}{dt} = \Lambda y + g(t), \qquad t \neq \tau_i,$$
$$\Delta y|_{t=\tau_i} = \mathcal{I}_i, \tag{4.145}$$

where $g(t) = \Phi^{-1}(t)f(t)$, $\mathcal{I}_i \Phi^{-1}(\tau_i + 0)I_i$.

Using properties of the matrix $\Phi(t)$ it is easy to check that the function $g(t)$ and the sequence $\{\mathcal{I}_i\}$ are periodic and the question of boundedness, almost periodicity, and asymptotic characteristics of system (4.144) is reduced to that of system (4.145).

It should be noted that if the matrix $A(t)$ does not depend on time, then the Cauchy matrix $X(t,s)$ and, consequently, the matrix $\Phi(t)$ can be found explicitly. Indeed, if $A(t) = A$, then

$$X(t) = e^{A(t-\tau_j)} \prod_{0<\tau_i<t} (E + B_i)e^{A(\tau_i-\tau_{i-1})}(E + B_k)e^{A\tau_k},$$

if $\tau_j < \tau \leq \tau_{j+1}$, $\tau_{k-1} < 0 \leq \tau_k$, $J > K$, and

$$X(T) = e^{A(T-\tau_{k+p+1})} \prod_{i=k}^{k+p-1} (E + B_i)e^{A(\tau_i-\tau_{i-1})}(E + B_k)e^{A\tau_k}.$$

Let $f(t)$ be an almost periodic scalar-valued function, $\{a_i\}$ – an almost periodic number sequence, the sequence $\{\tau_i\}$ be such that $\tau_{i+p} = \tau_i + T$, $i = 0$, ± 1, ± 2,.... To make the exposition more clear, we restate the corollary of Lemma 4.133 as

Lemma 38 *For any real number $\epsilon > 0$ there exist such a real number ν, $0 < \nu < \epsilon$, and relatively dense sets of real numbers τ and integers, $\mathbf{Q}$, that the following relations hold:*

a) $|f(t+\tau) - f(t)| < \epsilon$, $t \in \mathbf{R}$, $|t - \tau_i| > \epsilon$, $i = 0, \pm 1, \ldots;$

b) $|a_{i+q} - a_i| < \epsilon$, $i = 0, \pm 1, \pm 2, \ldots;$

c) $|\tau_k^q - \tau| < \nu$, $k = 0, \pm 1, \ldots, \tau \in \tau$, $q \in \mathbf{Q}$.

Lemma 39 *If the conditions of Lemma 38 hold and the sum*

$$F(t) = \int_0^t f(s)ds + \sum_{\substack{0<\tau_i<t \\ (t<\tau_i<0)}} a_i$$

is bounded, then $F(t)$ is almost periodic.

Proof. Let $F(t)$ be bounded and $m = \inf_t(F(t)$, $M = \sup_t F(t)$. Assume that $M > m$ (the case of $M = m$ is trivial).

1. Let us show that, for the function $F(t)$ defined on the real axis $\mathbf{R}$, there exists a relatively dense set of points (x_1, x_2), which gives up to ϵ the variation of the function $F(t)$, equal to $M - m$.

By the definition of the upper and lower bounds, there exists such a pair of points (s_1, s_2) that

$$F(s_1) < m + \frac{\epsilon}{16}, \qquad F(s_2) > M - \frac{\epsilon}{16}.$$

Let $|s_1 - \tau_i| > \epsilon_1$, $|s_2 - \tau_i| > \epsilon_1$, $i = 0, \pm 1, \ldots$. Denote

$$d = |s_1 - s_2|, \qquad s = \min(s_1, s_2), \qquad M_1 = \max_t |f(t)|,$$

$$\epsilon_2 = \min\left(\epsilon_1, \frac{\epsilon}{8d}, \frac{\epsilon}{16\left(\frac{d}{T}+1\right)pM_1}, \frac{\epsilon}{8\left(\frac{d}{T}+1\right)p}\right).$$

By using Lemma 38, choose an almost period $r = r(\epsilon_2)$ for the function $f(t)$. By this lemma, the points $s_j + r = x_j$, $j = 1, 2$, form a relatively dense set and, hence, there exists a number $l = l(\epsilon_2) > 0$ such that any segment $[a, a+l]$ contains a point $s + r$. Thus every segment of length $l + d$ contains a pair of points (x_1, x_2) and so these pairs form a relatively dense set. As in the proof of Lemma 35, it can be verified that $\tau_i + \epsilon < s < \tau_{i+1} - \epsilon$ implies $\tau_{i+q} < s + r < \tau_{i+q+1}$. Further, we have

$$F(x_2) - F(x_1) = \left(F(s_2) + \int_{s_2}^{s_2+r} f(s)ds + \sum_{s_2<\tau_i<s_2+r} a_i\right) -$$
$$-\left(F(s_1) + \int_{s_1}^{s_1+r} f(s)ds + \sum_{s_1<\tau_i<s_1+r} a_i\right) = F(s_2) - F(s_1) +$$
$$+\int_{s_1}^{s_2} (f(s+r) - f(s))ds + \sum_{s_1<\tau_i<s_2} (a_{i+q} - a_i),$$

and so

$$F(x_2) - F(x_1) > \left(M - \frac{\epsilon}{16}\right) - \left(m + \frac{\epsilon}{16}\right) - \frac{\epsilon d}{8d} -$$
$$-\frac{2M_1\epsilon\left(\frac{d}{T}+1\right)p}{16\left(\frac{d}{T}+1\right)pM_1} - \frac{\epsilon\left(\frac{d}{T}+1\right)p}{8\left(\frac{d}{T}+1\right)pM_1} = M - m - 4\cdot\frac{\epsilon}{8} =$$
$$= M - m - \frac{\epsilon}{2}.$$

Thus

$$(M - F(x_2)) + (F(x_1) - m) < \frac{\epsilon}{2}.$$

Because the numbers $M - F(x_2)$ and $F(x_!) - m$ are nonnegative, it follows from the latter inequality that

$$M - F(x_2) < \frac{\epsilon}{2}, \qquad F(x_1) - m < \frac{\epsilon}{2},$$

i.e. the relatively dense set of the pairs (x_1, x_2) defines the variation of the function $F(t)$ up to ϵ.

2. Denote

$$\epsilon_3 = \min\left(\frac{\epsilon}{3L}, \frac{\epsilon}{6\left(\frac{L}{2T}+1\right)p}, \frac{\epsilon}{12M_1\left(\frac{L}{2T}+1\right)p}\right),$$

and let $r = r(\epsilon_3)$ be chosen according to Lemma 38. We use properties of the pairs (x_1, x_2) to get an upper and lower estimates of the difference $F(t+r) - F(t)$ for $|t - \tau_i| > \epsilon$. To find a lower estimate, choose a point x_1 in the segment $\left[t - \frac{L}{2}, t + \frac{L}{2}\right]$ such that

$$F(x_1) < m + \frac{\epsilon}{2},$$

with $|t - x_1| \leq \frac{L}{2}$. We get

$$\begin{aligned}
F(t+r) - F(t) &= \left(F(x_1+r) + \int_{x_1+r}^{t+r} f(s)ds + \sum_{x_1+r<\tau_i<t+r} a_i\right) - \\
&- \left(F(x_1) + \int_{x_1}^{t} f(s)ds + \sum_{x_1<\tau_i<t} a_i\right) = F(x_1+r) - F(x_1) + \\
&+ \int_{x_1}^{t} (f(s+r) - f(s))ds + \sum_{x_1<\tau_i<t} (a_{i+q} - a_i) > m - \left(m + \frac{\epsilon}{2}\right) - \\
&- \frac{\epsilon}{3L} \cdot \frac{L}{2} - \frac{\epsilon 2M_1}{12\left(\frac{L}{2T}+1\right)pM_1}\left(\frac{L}{2T}+1\right)p - \\
&- \frac{\epsilon}{6\left(\frac{L}{2T}+1\right)p}\left(\frac{L}{2T}+1\right)p = m - \left(m + \frac{\epsilon}{2}\right) - 3 \cdot \frac{\epsilon}{6} = \\
&= -\epsilon.
\end{aligned} \tag{4.146}$$

In the same way, by taking a point x_2 from the segment $\left[t - \frac{L}{2}, t + \frac{L}{2}\right]$ and using that $|t - x_2| \leq \frac{L}{2}$, we wind

$$F(t+r) - F(t) = F(x_2 + r) - F(x_2) + \int_{x_2}^{t} (f(s+r) -$$

$$-f(s))ds + \sum_{x_2 < \tau_i < t} (a_{i+q} - a_i) < M - \left(M - \frac{\epsilon}{2}\right) + \frac{\epsilon}{3L} \cdot \frac{L}{2} +$$

$$+ \frac{\epsilon 2M_1 \left(\frac{L}{2T} + 1\right) p}{12 \left(\frac{L}{2T} + 1\right) pM_1} + + \frac{\epsilon \left(\frac{L}{2T} + 1\right) p}{6 \left(\frac{L}{2T} + 1\right) p} = \epsilon. \tag{4.147}$$

□

By using Lemma 35, we can prove the following theorem in a way similar to Lemma 39.

Theorem 83 *Let $f(t)$ be a scalar-valued almost periodic function, $\{a_i\}$ – an almost periodic number sequence, $\{\tau_i\}$ be such that the sequences $\{\tau_i^j\}$, $j = 0, \pm 1, \ldots$, are equipotentially almost periodic and* $\inf_i \tau_i^1 = \theta > 0$. *Then if the sum*

$$F(t) = \int_0^t f(s)ds + \sum_{\substack{0 < \tau_i < t \\ (t < \tau_i < 0)}} a_i$$

is bounded, then it is also almost periodic.

Corollary. *The integral*

$$F(t) = \int_0^t f(s)ds$$

of an almost periodic function $f(t)$ is an almost periodic function if and only if it is bounded.

It should be noted that Lemma 39 and Theorem 83 are true in the case of a vector space.

Lemma 40 *Let a scalar impulsive differential equation be given,*

$$\frac{dx}{dt} = \lambda x + f(t), \qquad t \neq \tau_i,$$
$$\Delta x|_{t=\tau_i} = a_i, \tag{4.148}$$

where λ is a complex number, $f(t)$ is an almost periodic function, $\{a_i\}$ is an almost periodic sequence, and the sequence $\{\tau_i\}$ satisfies the condition $\tau_{i+p} = \tau_i + T$, $i = 0, \pm 1, \pm 2, \ldots$, p is a natural number. If equation (4.145) has a bounded solution, then this solution is almost periodic.

Proof. A general solution of equation (4.148) can be written as

$$x(t) = e^{\lambda t}\left(x_0 + \int_0^t e^{-\lambda s} f(s)ds \overset{+}{(-)} \sum_{\substack{0<\tau_i<t \\ (t<\tau_i<0)}} e^{-\lambda \tau_i} a_i \right), \tag{4.149}$$

1. Let $\Re\lambda = \alpha > 0$. Then $|e^{\lambda t}| \to \infty$ as $t \to \infty$. Consequently, for the solution (4.148) to be bounded, we need to set

$$x_0 = -\int_0^\infty e^{-\lambda s} f(s)ds - \sum_{\tau_i>0} e^{-\lambda\tau_i} a_i.$$

It is easy to verify that the integral and the sum in the formula converge.

By substituting this value of x_0 into (4.149), we find that

$$x(t) = -\int_t^\infty e^{\lambda(t-s)} f(s)ds - \sum_{\tau_i>t} e^{\lambda(t-\tau_i)} a_i.$$

If $r = r(\epsilon)$ is an almost ϵ-period, given by Lemma 35, then

$$|x(t+r) - x(t)| \le \int_t^\infty e^{\lambda(t-s)} |f(s+r) - f(s)| ds + \\ + \sum_{\tau_i>t} e^{\lambda(t-\tau_i)} |a_{i+q} - a_i| < \Gamma(\epsilon)\epsilon,$$

where $\Gamma(\epsilon)$ is a bounded function of ϵ. Hence $x(t)$ is an almost periodic function.

2. Let $\Re\lambda = \alpha < 0$. Then $|e^{\lambda t}| \to \infty$ as $t \to \infty$. As in case **1**, we find that a bounded solution, written as

$$x(t) = \int_{-\infty}^t e^{\lambda(t-s)}(s)ds + \sum_{\tau_i<t} e^{\lambda(t-\tau_i)} a_i,$$

is an almost periodic function.

3. Let $\Re\lambda = 0$, $\lambda = i\nu$, and a bounded solution

$$x(t) = e^{-i\nu t}\left(x_0 + \int_0^t e^{-i\nu s} f(s)ds \overset{+}{\underset{(-)}{}} \sum_{\substack{0<\tau_k<t \\ (t<\tau_k<0)}} e^{-i\nu\tau_k} a_k\right).$$

exist. Then the sum

$$\int_0^t e^{-i\nu s} f(s)ds \overset{+}{\underset{(-)}{}} \sum_{\substack{0<\tau_k<t \\ (t<\tau_k<0)}} e^{-i\nu\tau_k} a_k \tag{4.150}$$

is bounded and, since the function $e^{-i\nu s}f(s)$ and the sequence $\{e^{-i\nu\tau_k}a_k\}$ are almost periodic, by the lemma, the sum (4.150) is an almost periodic function. In this case, all the solutions of equation (4.148) are almost periodic.

□

Theorem 84 *If differential system (4.144) has a bounded solution, then this solution is almost periodic.*

Proof. It was remarked at the beginning of the paragraph that the question of almost periodicity of a bounded solution of system (4.144) is equivalent to the question of almost periodicity of a bounded solution of system (4.145) which is obtained from system (4.144) by the periodic transformation $x = \Phi(t)y$.

As known, there exists a nonsingular transformation $y = Sz$, where S is a constant matrix, which transforms system (4.145) into a system with an upper triangular coefficient matrix. Hence, without loss of generality, we can assume that system (4.145) has the form

$$\begin{aligned}
\frac{dy_1}{dt} &= \lambda_1 y_1 + b_{12}y_2 + \ldots + b_{1n}y_n + g_1(t), \\
&\cdot \quad \cdot \quad \cdot \quad \cdot \quad \cdot \quad \cdot \quad \cdot \quad \cdot \\
\frac{dy_{n-1}}{dt} &= \ldots \lambda_k y_{n-1} + b_{n-1}y_n + g_{n-1}(t), \\
\frac{dy_n}{dt} &= \ldots \lambda_k y_n + g_n(t), \\
\Delta y_1|_{t=\tau_i} &= Q_{i1}, \ldots, \Delta y_n|_{t=\tau_i} = Q_{in}.
\end{aligned} \tag{4.151}$$

By solving system (4.151) from bottom to top and using Lemma 39, we find that the bounded solution of this system is almost periodic. Now we use the transformation $\Phi^{-1}(t)$.

□

Theorem 85 *Let system (4.144) satisfy all the stated above conditions and, moreover, let the matrix* $\Lambda = \frac{1}{T}\operatorname{Ln} X(T)$ *not have the eigenvalues, the real part of which is zero. Then system (4.144) has a unique almost periodic solution. If, additionally, the eigenvalues of the matrix* Λ *have only negative real parts, then the almost periodic solution is asymptotically stable.*

Proof. Without loss of generality, we can assume that $\Lambda = \operatorname{diag}(P, N)$, where P and N are square matrices of order m and $n-m$ such that

$$\Re\lambda_j(P) > 0, \qquad j = \overline{1,m}, \qquad \Re\lambda_j(N) < 0, \qquad j = \overline{m+1,n}.$$

Set

$$G(t) = \begin{cases} -\ \operatorname{diag}(e^{Pt}, 0), & \text{for}\, t < 0, \\ \ \ \operatorname{diag}(0, e^{Nt}), & \text{for}\, t > 0. \end{cases}$$

In the sequel we will use the following known properties of the matrix $G(t)$:

1) $G(0+) - G(0-) = E_n$, where E_n is the $n \times n$-unit matrix;

2) $\|G(t)\| \le Ce^{-\alpha|t|}$, where C and α are positive constants;

3) $\dfrac{\partial G(t)}{dt} = \Lambda G(t)$ for $t \neq 0$.

Let us show that

$$y_0(t) = \int_{-\infty}^{\infty} G(t-s)g(s)ds + \sum_{i=-\infty}^{\infty} G(t-\tau_i)Q_i \tag{4.152}$$

is a solution of system (4.145). Indeed, because

$$\begin{aligned} \|y_0(t)\| &\le \int_{-\infty}^{\infty} \|G(t-s)\|\,\|g(s)\|ds + \sum_{i=-\infty}^{\infty} \|G(-\tau_i)\|\,\|Q_i\| \le \\ &\le C\max\left(\frac{2}{\alpha}, \frac{2}{1-e^{-\alpha\nu}}\right)(\sup_i \|g(t)\| + \sup_i \|Q_i\|), \end{aligned}$$

where $\gamma = \min_{1\leq i\leq p} \tau_i^1$, the right-hand side of expression (4.152) is defined. For $t \neq \tau_i$, we have

$$\begin{aligned}
\frac{dy_0}{dt} &= (G(0+)-G(0-))g(t) + \int_{-\infty}^{t} \Lambda G(t-s)g(s)ds + \\
&+ \int_{t}^{\infty} \Lambda G(t-s)g(s)ds + \Lambda \sum_{i=-\infty}^{\infty} G(t-\tau_i)Q_i = \Lambda y_0 + g(t), \\
\Delta y_0|_{t=\tau_j} &= \sum_{i=-\infty}^{\infty} G(\tau_j - \tau_i +)q_i - \sum_{i=-\infty}^{\infty} G(\tau_j-\tau_i)q_i = Q_j.
\end{aligned}$$

This means that $y_0(t)$ is a solution of system (4.145). Let $r = r(s)$ be an ϵ-almost period chosen by Lemma (4.144). Then

$$\begin{aligned}
y_0(t+r) - y_0(t) = \int_{-\infty}^{\infty} G(t+r-s)g(s)ds + \sum_{i=-\infty}^{\infty} G(t+r-\tau_i)Q_i - \\
- \int_{-\infty}^{\infty} G(t-s)g(s)ds - \sum_{i=-\infty}^{\infty} G(t-\tau_i)Q_i = \\
= \int_{-\infty}^{\infty} G(t-s)(g(s+r)-g(s))ds + \sum_{-\infty}^{\infty} G(t-\tau_i)(Q_{i+q}-Q_i)
\end{aligned}$$

and

$$\begin{aligned}
\|y_0(t+r) - y_0\| &\leq \int_{-\infty}^{\infty} Ce^{-\alpha|t-s|}\|g(s+r)-g(s)\|ds + \\
&+ \sum_{i=-\infty}^{\infty} Ce^{-\alpha|t-\tau_i|}\|Q_{i+q}-Q_i\| < \Gamma(\epsilon)\epsilon,
\end{aligned}$$

where $\Gamma(\epsilon)$ is a bounded function of ϵ.

Suppose that $y_1(t)$ is an almost periodic solution of system (4.145), not equal to $y_0(t)$. Then the difference $y_1(t) - y_0(t)$ is an almost periodic solution of the homogeneous system

$$\frac{dy}{dt} = \Lambda y, \tag{4.153}$$

which has only the zero solution. This contradiction proves that an almost periodic solution is unique.

Going back to system (4.144), we see that a unique almost periodic solution equals to

$$x_0(t) = \int_{-\infty}^{\infty} \Phi(t)G(t-s)\Phi^{-1}(s)ds + \\ + \sum_{i=-\infty}^{\infty} \Phi(t)G(t-\tau_i)\Phi^{-1}(\tau_i)I_i. \quad (4.154)$$

It follows from this relation that

$$\|x_0(t)\| \leq K(\sup_i \|f(t)\| + \sup_i \|I_i\|),$$

where K is a positive constant.

If all the real parts of the matrix Λ have negative real parts, then, since the difference of two solutions of system (4.145) is a solution of system (4.153), solutions of system (4.145), and hence of (4.144), are asymptotically stable.

□

Consider the quasilinear impulsive system

$$\frac{dx}{dt} = A(t)x + f(t,x), \quad t \neq \tau_i, \\ \Delta x|_{t=\tau_i} = B_i x + I_i(x), \quad (4.155)$$

where the matrices $A(t)$, B_i, $i = 0, \pm 1, \ldots$, and the sequence $\{\tau_i\}$ satisfy the conditions of Theorem 85, the function $f(t,x)$ is almost periodic in t, and the sequence $\{I_i(x)\}$ is almost periodic in i uniformly with respect to x on every compact set. Suppose that the Lipschitz condition

$$\|f(t,x) - f(t,y)\| + \|I_i(x) - I_i(y)\| \leq L\|x - y\| \quad (4.156)$$

holds uniformly with respect to $t \in \mathbf{R}$ and $i = 0, \pm 1, \ldots$.

Theorem 86 *If $\Re\lambda_j(\Lambda) \neq 0$, $j = \overline{1,n}$, then, for a sufficiently small Lipschitz constant L, system (4.155) has a unique almost periodic solution. If $\Re\lambda_j(\Lambda) < 0$, $j = \overline{1,n}$, and L is sufficiently small, then the periodic solution is asymptotically stable.*

Proof. Let $\mathfrak{N}$ be the space of all almost periodic functions with discontinuities at the points of the same sequence $\{\tau_i\}$.

Define an operator T on $\mathfrak{N}$ by

$$T(\varphi(t)) = \int_{-\infty}^{\infty} \Phi(t)G(t-s)\Phi^{-1}(s)f(s,\varphi(s))ds + \\ + \sum_{i=-\infty}^{\infty} \Phi(t)G(t-\tau_i)\Phi^{-1}(\tau_i)I_i(\varphi(\tau_i)),$$

$\varphi(t) \in \mathfrak{N}$. By using Lemmas 35 and 36, and properties of the matrix $G(t)$, it is possible to show similarly as in Theorem 82 that the operator T is defined on $\mathfrak{N}$ and $T(\mathfrak{N}) \subseteq \mathfrak{N}$. Besides, if

$$2MCL\left(\frac{1}{\alpha} + \frac{1}{1-e^{-\gamma\alpha}}\right) < 1,$$

where $M = \sup_{s,t\in\mathbf{R}} \|\Phi(t)\|\|\Phi^{-1}(s)\|$, then the operator T is a contraction and, consequently, there exists a unique almost periodic solution of system (4.144).

If $\Re\lambda_j(\Lambda) < 0$, $j = \overline{1,n}$, then the Cauchy matrix $X(t,s)$ of the corresponding homogeneous linear system (4.155) is subject to the estimate

$$\|X(t,s)\| \le Ke^{\beta(t-s)} \qquad \text{for} \qquad t \ge s,$$

where $K \ge 1$, $\beta < 0$ are constants.

By applying the integral form of (4.156) to system (4.155), we can find that any two solutions of this system satisfy the following inequality

$$\|\varphi(t) - \psi(t)\| \le Ke^{\beta(t-t_0)}\|\varphi(t_0) - \psi(t_0)\| + \\ + \int_{t_0}^{t} Ke^{\beta(t-s)}L\|\varphi(s) - \psi(s)\|ds + \sum_{t_0<\tau_i<t} KLe^{\beta(t-\tau_i)}\|\varphi(\tau_i) - \psi(\tau_i)\|.$$

Applying the argument used in the proof of inequalities (4.143), we get

$$\|\varphi(t) - \psi(t)\| \le Ke^{p}e^{\left(\beta+KL+\frac{p}{T}\ln(1+KL)\right)(t-t_0)}\|\varphi(t_0) - \psi(t_0)\|,$$

whence the asymptotic stability of solution $\varphi(t)$ follows if the constant L is sufficiently small.

□

We remark that all the results of sections 4.6 and 4.7 hold if we assume that the matrix $A(t)$ is piecewise continuous with first kind discontinuities at the points of the sequence $\{\tau_i\}$.

Chapter 5

Integral sets of impulsive systems

5.1 Bounded solutions of nonhomogeneous linear systems

Let us study the question of existence of solutions, bounded for all $t \in \mathbf{R}$, of the nonhomogeneous linear impulsive system

$$\frac{dx}{dt} = A(t)x + f(t), \qquad t \neq \tau_i, \qquad \Delta x|_{t=\tau_i} = B_i x + a_i, \tag{5.1}$$

where $A(t)$ and $f(t)$ are matrix and vector functions correspondingly, continuous (piecewise continuous with first kind discontinuities at $t = \tau_i$), bounded for all $t \in \mathbf{R}$; B_i and a_i are constant vector and matrix, uniformly bounded with respect to $i \in \mathbf{Z}$, $\det(E + B_i) \neq 0$. The sequence of the times τ_i is indexed by integers such that $\tau_i \to -\infty$ as $i \to -\infty$ and $\tau_i \to +\infty$ as $i \to +\infty$. Besides, we assume that

$$\overline{\lim_{T\to\infty}} \frac{i(t, t+T)}{T} = p < \infty \tag{5.2}$$

uniformly with respect to $t \in \mathbf{R}$. As follows from Lemma 21, this condition is equivalent to the condition that it is possible to choose such numbers l and natural q that any interval of the time axis of length l contains no more than q points of the sequence $\{\tau_i\}$.

Together with system (5.1), we also consider its homogeneous part, i.e. the equations

$$\frac{dx}{dt} = A(t)x, \qquad t \neq \tau_i, \qquad \Delta x|_{t=\tau_i} = B_i x, \tag{5.3}$$

and suppose that this system is hyperbolic. Without infringing the generality, we assume that the matrices $A(t)$ and B are given in the block diagonal form,

$$A(t) = \begin{pmatrix} A_+(t) & 0 \\ 0 & A_-(t) \end{pmatrix}, \qquad B_i = \begin{pmatrix} B_i^+ & 0 \\ 0 & B_I^- \end{pmatrix}$$

and, if $\Phi_+(t, \tau)$ and $\Phi_-(t, \tau)$ are the corresponding matriciants of the following linear systems

$$\frac{dx_1}{dt} = A_+(t)x_1, \qquad t \neq \tau_i, \qquad \Delta x_1|_{t=\tau_i} = B_i^+ x_1 \tag{5.4}$$

and

$$\frac{dx_2}{dt} = A_-(t)x_2, \qquad t \neq \tau_i, \qquad \Delta x_2|_{t=\tau_i} = B_i^- x_2, \tag{5.5}$$

then

$$\begin{aligned}\|\Phi_+(t,\tau)\| &\le Ke^{-\gamma(t-\tau)} \text{ for } t \ge \tau, &(5.6)\\ \|\Phi_-(t,\tau)\| &\le Ke^{\gamma(t-\tau)} \text{ for } t \le \tau. &(5.7)\end{aligned}$$

The following theorem holds.

Theorem 87 *If the linear impulsive system (5.3) is hyperbolic, then, for every function $f(t)$, bounded on the whole axis, and a bounded sequence $\{a_i\}$, system (5.1) has a unique solution $x^*(t)$, bounded for all $t \in \mathbf{R}$, and there is such a positive constant C that*

$$\|x^*(t)\| \le C(\sup_{t\in\mathbf{R}} \|f(t)\| + \sup_{t\in\mathbf{Z}} \|a_i\|). \tag{5.8}$$

Proof.
Denote by $G(t,\tau)$ the matrix

$$G(t,\tau) = \begin{cases} \text{diag}(\Phi_+(t,\tau), 0) & \text{for } t > \tau, \\ \text{diag}(0, \Phi_-(t,\tau)) & \text{for } t < \tau. \end{cases} \tag{5.9}$$

By inequalities (5.6) and (5.6), the function $G(t,\tau)$ satisfies the estimate

$$\|G(t,\tau)\| \le Ke^{-\gamma|t-\tau|} \tag{5.10}$$

for all $t, \tau \in \mathbf{R}$.

By using the matrix $G(t,\tau)$, define the function

$$x^*(t) = \int_{-\infty}^{\infty} G(t,\tau)f(\tau)d\tau + \sum_{i=-\infty}^{\infty} G(t,\tau_i)a_i. \tag{5.11}$$

The right-hand side of (5.11) is defined since inequalities (5.2) and (5.10) imply that the integral and the sum in (5.11) converge uniformly. Indeed, from (5.11) we have

$$\begin{aligned}\|x^*(t)\| &\le \int_{-\infty}^{\infty} Ke^{-\gamma(t-\tau)}\|f(\tau)\|d\tau + \sum_{i=-\infty}^{\infty} Ke^{-\gamma|t-\tau_i|}\|a_i\| \le \\ &\le \frac{2K}{\gamma}\sup_{t\in\mathbf{R}}\|f(t)\| + K\sup\|a_i\| \sum_{i=-\infty}^{\infty} e^{-\gamma|t-\tau_i|}.\end{aligned}$$

Because by (5.1), any interval of the t-axis of length l does not contain more that q points of the sequence $\{\tau_i\}$, for $j \geq i$ we have

$$j - i + 1 \leq q\left(\left[\frac{\tau_j - \tau_i}{l}\right] + 1\right) \leq q\left(\frac{\tau_j - \tau_i}{l} + 1\right),$$

and, consequently,

$$\tau_j - \tau_i \leq l\left(\frac{1}{q} - 1\right) + \frac{l}{q}(j - i). \tag{5.12}$$

Thus

$$\sum_{i=-\infty}^{\infty} e^{-\gamma|t-\tau_i|} = \sum_{\tau_i \leq t} e^{-\gamma(t-\tau_i)} + \sum_{\tau_i > t} e^{-\gamma(\tau_i - t)} \leq \sum_{\tau_i \leq \tau_j} e^{-\gamma(\tau_j - \tau_i)} +$$

$$+ \sum_{\tau_i > \tau_j} e^{-\gamma(\tau_i - \tau_{j+1})} \leq 2e^{\gamma l\left(1 - \frac{1}{q}\right)} \sum_{m=0}^{\infty} e^{-\frac{\gamma l m}{q}} = \frac{2e^{\gamma l\left(1-\frac{1}{q}\right)}}{1 - e^{-\frac{\gamma l}{q}}}.$$

Hence, the final estimate will be

$$\|x^*(t)\| \leq C(\sup_{t \in \mathbf{R}} \|f(t)\| + \sup_{t \in \mathbf{Z}} \|a_i\|), \tag{5.13}$$

where

$$C = \max\left(\frac{2K}{\gamma}, \frac{2Ke^{\gamma l\left(1-\frac{1}{q}\right)}}{1 - e^{-\frac{\gamma l}{q}}}\right). \tag{5.14}$$

Let us show that the function $x^*(t)$, given by (5.11), is a solution of system (5.1). Indeed, write $x^*(t)$ as

$$\begin{aligned} x^*(t) \;=\; & \int_{-\infty}^{t} G(t,\tau)f(\tau)d\tau + \int_{t}^{\infty} G(t,\tau)f(\tau)d\tau + \\ & + \sum_{\tau_i < t} G(t,\tau_i)a_i + \sum_{\tau_i > t} G(t,\tau_i)a_i. \end{aligned} \tag{5.15}$$

By differentiating $x^*(t)$ for $t \neq \tau_i$, we get

$$\begin{aligned} \frac{dx^*}{dt} \;=\; & \int_{-\infty}^{t} \frac{dG(t,\tau)}{dt} f(\tau)d\tau + \sum_{\tau_i < t} \frac{dG(t,\tau_i)}{dt} a_i + \\ & + \int_{t}^{\infty} \frac{dG(t,\tau)}{dt} f(\tau)d\tau + \sum_{\tau_i > t} \frac{dG(t,\tau_i)}{dt} a_i + \\ & + (G(t,t-0) - G(t,t+0))f(t) = A(t)x^*(t) + f(t) \end{aligned}$$

since

$$\frac{dG(t,\tau)}{dt} = A(t)G(t,\tau), \qquad t \neq \tau, \qquad t \neq \tau_i,$$

and, for $t = \tau$, $t \neq \tau_i$, $G(t, t-0) - G(t, t+0) = E$. It can also be seen from (5.15) that, for $t = \tau_i$,

$$x^*(\tau_j + 0) - x^*(\tau_j) = \int_{-\infty}^{\infty} [G(\tau_j + 0, \tau) - G(\tau_j, \tau)] f(\tau) d\tau +$$

$$+ \sum_{i=-\infty}^{\infty} [G(\tau_j + 0, \tau_i) - G(\tau_j, \tau_i)] a_i + a_j = B_j x^*(\tau_j) + a_j.$$

So, the function $x^*(t)$ is a solution of system (5.1), bounded on the whole axis. Uniqueness of such a solution follows from the fact that the difference of two solutions bounded on the whole axis is a solution of the homogeneous system (5.3), which is bounded on the whole axis. However, system (5.3), since it is hyperbolic, admits only the trivial solution as a solution bounded on the whole axis.

Note that if, in system (5.1), $A(t) = A_+(t)$, $B_i = B_i^+$, then the matrix $G(t,\tau)$, in this case, has the form $G(t,\tau) = \Phi_+(t,\tau)$ and the unique solution $x^*(t)$, bounded on the whole axis, can be written as

$$x^*(t) = \int_{-\infty}^{t} \Phi_+(t,\tau) f(\tau) d\tau + \sum_{\tau_i < t} \Phi_+(t,\tau_i) a_i. \tag{5.16}$$

This solution, by inequality (5.6), is also asymptotically stable.

Consider the particular case of system (5.1),

$$\frac{dx}{dt} = Ax + f(t), \qquad t \neq \tau_i, \qquad \Delta x|_{t=\tau_i} = Bx + a_i. \tag{5.17}$$

Here A and B are constant matrices, $\det(E+B) \neq 0$, the vectors $f(t)$ and a_i are the same as in system (5.1), and, concerning the times τ_i, we assume that the finite limit

$$\lim_{T\to\infty} \frac{i(t, t+T)}{T} = p. \tag{5.18}$$

exists and is uniform with respect to $t \in \mathbb{R}$.

Theorem 88 *Let, in system (5.17), the matrices A and B commute. If the real parts of the matrix*

$$\Lambda = A + p\,\mathrm{Ln}(E+B) \tag{5.19}$$

are not equal to zero, then system (5.17) has a unique solution, bounded on the whole axis. It is asymptotically stable if the real parts of all eigenvalues are negative.

Proof. By using a nonsingular matrix S, write the matrix Λ in the form

$$\Lambda = S^{-1}\mathrm{diag}(\Lambda^{+}, \Lambda^{-})S, \tag{5.20}$$

where Λ^{+} is a square matrix, the eigenvalues of which have positive eigenvalues, and the eigenvalues of the matrix Λ^{-} are negative. Define the matrix $G(t,\tau)$ by the relations

$$G(t,\tau) = \begin{cases} -S^{-1}\mathrm{diag}(e^{\Lambda_{+}(t-\tau)}, 0)S(E+B)^{-p(t-\tau)+i(t,\tau)}, & t<\tau \\ S^{-1}\mathrm{diag}(0, e^{\Lambda_{-}(t-\tau)})S(E+B)^{-p(t-\tau)+i(t,\tau)}, & t>\tau, \end{cases} \tag{5.21}$$

If $\tau_i < t < \tau_{i+1}$, $\tau_j < \tau < \tau_{j+1}$, $t \neq \tau$, the matrix $G(t,\tau)$ satisfies the relation

$$\frac{dG(t,\tau)}{dt} = AG(t,\tau), \tag{5.22}$$

if $t = \tau_i$, $\tau \neq \tau_i$, then

$$G(\tau_i + 0, \tau) - G(\tau_i - 0, \tau) = BG(\tau_i - 0, \tau), \tag{5.23}$$

and, at $t = \tau$, $\tau \neq \tau_i$, the function $G(t,\tau)$ has the jump

$$G(\tau + 0, \tau) - G(\tau - 0, \tau) = E.$$

We will show that (5.22) holds.

Let $t > \tau$. Then

$$\begin{aligned} \frac{dG(t,\tau)}{dt} &= S^{-1}\mathrm{diag}(0, e^{\Lambda_{-}(t-\tau)})[\mathrm{diag}(0,\Lambda_{-})S - \\ &\quad -\mathrm{Sp}\,\mathrm{Ln}(E+B)](E+B)^{-p(t-\tau)+i(t,\tau)}. \end{aligned}$$

Since

$$\mathrm{diag}(0, e^{\Lambda_{-}(t-\tau)})\mathrm{diag}(0,\Lambda_{-}) = \mathrm{diag}(0, e^{\Lambda_{-}(t-\tau)})\mathrm{diag}(\Lambda_{+},\Lambda_{-}),$$

we have that

$$\begin{aligned} \frac{dG(t,\tau)}{dt} &= S^{-1}\mathrm{diag}(0, e^{\Lambda_{-}(t-\tau)})S[A- \\ &\quad -p\,\mathrm{Ln}(E+B)](E+B)^{-p(t-\tau)+i(t,\tau)} = AG(t,\tau), \end{aligned}$$

because the matrix A commutes with the matrix $S^{-1}\operatorname{diag}(0, e^{\Lambda_-(t-\tau)})$, for the matrices A and Λ commute and, hence, the matrices $e^{\Lambda t}$ and A also commute. The identity

$$e^{\Lambda t}A = Ae^{\Lambda t}$$

is equivalent to

$$\operatorname{diag}(e^{\Lambda_+ t}, e^{\Lambda_- t})SAS^{-1} = SAS^{-1}\operatorname{diag}(e^{\Lambda_+ t}, e^{\Lambda_- t}),$$

from which it follows that

$$\operatorname{diag}(0, e^{\Lambda_- t})SAS^{-1} = SAS^{-1}\operatorname{diag}(0, e^{\Lambda_- t})$$

or

$$S^{-1}\operatorname{diag}(0, e^{\Lambda_- t})SA = AS^{-1}\operatorname{diag}(0, e^{\Lambda_- t})S.$$

Similarly we can show that (5.22) holds for $t < \tau$, $t \neq \tau_i$, $\tau \neq \tau_i$. Property (5.23) of the matrix $G(t, \tau)$ can be verified directly by using that every matrix $S^{-1}\operatorname{diag}(e^{\Lambda_+ t}, 0)S$ and $S^{-1}\operatorname{diag}(0, e^{\Lambda_- t})S$ commutes with the matrix B since the matrix B commutes with the matrix Λ.

By the assumption, the real parts of the matrix Λ are nonzero, the times τ_i are such that the finite limit (5.18) exists, and so there exist positive numbers K and γ such that

$$\|G(t, \tau)\| \leq Ke^{-\gamma|t-\tau|}, \qquad t, \tau \in \mathbf{R}. \tag{5.24}$$

This can be seen from

$$\begin{aligned} \|e^{\Lambda_+(t-\tau)}\| &\leq K_1 e^{\gamma_1(t-\tau)}, \qquad t \leq \tau, \\ \|e^{\Lambda_-(t-\tau)}\| &\leq K_2 e^{-\gamma_2(t-\tau)}, \qquad t \geq \tau, \end{aligned}$$

where $0 < \gamma_1 < \min_j \Re\lambda_j(\Lambda_+)$, $0 < \gamma_2 < \min_j[-\Re\lambda_j(\Lambda_-)]$, K_1, K_2 are certain positive constants and

$$\|(E + B)^{-p(t-\tau)+i(t,\tau)}\| \leq K_3(\epsilon)e^{\epsilon\|\operatorname{Ln}(E+B)\|\,|t-\tau|}$$

with $\epsilon > 0$, $K_3(\epsilon)$ being a positive constant.

By using the matrix $G(t, \tau)$, we can define the function

$$x^*(t) = \int_{-\infty}^{\infty} G(t, \tau)f(\tau)d\tau + \sum_{i=-\infty}^{\infty} G(t, \tau_i)a_i. \tag{5.25}$$

The right-hand side of (5.25) is defined since, by (5.18) and (5.24), the integral and the sum converge for all $t \in \mathbf{R}$ and this convergence is uniform on every finite interval of the real axis.

Indeed,

$$\left\| \int_{-\infty}^{\infty} G(t,\tau) f(\tau) d\tau \right\| \leq \int_{-\infty}^{\infty} K e^{-\gamma|t-\tau|} \|f(\tau)\| d\tau \leq \frac{2K}{\gamma} \sup_{t \in \mathbf{R}} \|f(t)\|. \quad (5.26)$$

By (5.18), there is such a number $l > 0$ that

$$\left| \frac{i(t, t+l)}{l} - p \right| < 1$$

for all $t \in \mathbf{R}$, i.e. any time interval of length l does not contain more than $(p+1)l$ points of the sequence $\{\tau_i\}$ and so

$$\begin{aligned} \left\| \sum_{i=-\infty}^{\infty} G(t,\tau) I_i \right\| &\leq \sum_{i=-\infty}^{\infty} K e^{-\gamma|t-\tau|} \|I_i\| \leq \\ &\leq \sum_{m=-\infty}^{\infty} \sum_{t+ml<\tau_i<t+(m+l)l} K e^{-\gamma|t-\tau_i|} \|I_i\| \quad (5.27) \\ &\leq \frac{2K(p+1)l}{1-e^{-\gamma l}} \sup \|I_i\|. \end{aligned}$$

By writing $x^*(t)$ as

$$\begin{aligned} x^*(t) &= \int_{-\infty}^{t} G(t,\tau) f(\tau) d\tau + \int_{t}^{\infty} G(t,\tau) f(\tau) d\tau + \\ &+ \sum_{\tau_i<t} G(t,\tau_i) I_i + \sum_{\tau_i>t} G(t,\tau_i) I_i \end{aligned}$$

and taking the formal derivative with respect to t for $t \neq \tau_i$, we get, by using (5.22), that

$$\frac{dx^*}{dt} = Ax^* + f(t).$$

The differentiation is legitimate since the improper integrals and sums, obtained after the formal differentiation, converge uniformly on every finite interval.

For $t = \tau_i$, by using (5.23), we have

$$\Delta x^*|_{t=\tau_i} = Bx^*(\tau_i) + I_i,$$

i.e. $x^*(t)$ is a solution of system (5.17). Boundedness of $x^*(t)$ follows from (5.26) and (5.27).

Uniqueness of the solution $x^*(t)$ of system (5.17), which is bounded for all $t \in \mathbf{R}$, follows because the difference of such two solutions is a solution of the system

$$\frac{dx}{dt} = Ax, \qquad t \neq \tau_i; \qquad \Delta x|_{t=\tau_i} = Bx,$$

bounded on the whole real axis. This system, however, with the assumptions made, has only the zero solution.

We will show that if the real parts of the matrix Λ are negative, then the solution $x^*(t)$ is asymptotically stable.

Let $x = \varphi(t)$ be an arbitrary solution of system (5.17). Then the difference $z(t) = \varphi(t) - x^*(t)$ is a solution of the system

$$\frac{dz}{dt} = Az, \qquad t \neq \tau_i; \qquad \Delta z|_{t=\tau_i} = Bz \tag{5.28}$$

and, for $t \geq 0$, it can be written as

$$z(t) = e^{At}(E + B)^{i(0,t)} z_0$$

or

$$z(t) = e^{\Lambda t}(E + B)^{pt\left(\frac{i(0,t)}{pt} - 1\right)} z_0.$$

Whence it can be seen from (5.28) that if $\Re\lambda_j(\Lambda) < 0$, then all solutions of equations (5.28) approach to zero as $t \to \infty$ and this implies asymptotic stability of the solution $x = x^*(t)$.

5.2 Existence of bounded solutions of nonlinear systems

Using results of the previous section we will study the problem of existence of a solution, bounded on the whole axis, of a nonlinear impulsive system.

Consider the system

$$\frac{dx}{dt} = A(t)x + f(t,x), \qquad t \neq \tau_i, \qquad \Delta x|_{t=\tau_i} = B_i x + I_i(x), \tag{5.29}$$

where the matrices $A(t)$, B_i, and the times τ_i are the same as in equations (5.1), the function $f(t,x)$ is defined for all $t \in \mathbf{R}$, $x \in \mathbf{R}^n$, piecewise continuous with respect to t with first kind discontinuities at $t = \tau_i$, and continuous with respect to x. We also assume that the functions $f(t,x)$ and $I_i(x)$ are bounded for $x = 0$, i.e.

$$\|f(t,0)\| + \|I_i(0)\| \leq M \tag{5.30}$$

for all $t \in \mathbf{R}$, $i \in \mathbf{Z}$, and that they satisfy the Lipschitz condition in x with the Lipschitz constant N:

$$\|f(t,x) - f(t,y)\| + \|I_i(x) - I_i(y)\| \leq N\|x - y\| \tag{5.31}$$

for all $t \in \mathbf{R}$, $i \in \mathbf{Z}$, $x, y \in \mathbf{R}^n$.

Theorem 89 *Let system (5.29) satisfy the conditions stated above and its linear part, i.e. system (5.3) be hyperbolic. Then, for sufficiently small values of the Lipschitz constant N, system (5.29) has a unique solution, bounded for all $t \in \mathbf{R}$. This solution will be asymptotically stable if the solutions of system (5.3) are asymptotically stable.*

Proof. Consider a sequence of functions defined by the recurrence relation

$$\begin{aligned} x_{m+1}(t) &= \int_{-\infty}^{\infty} G(t,\tau) f(\tau, x_m(\tau)) d\tau + \sum_{i=-\infty}^{\infty} G(t,\tau_i) I_i(x_m(\tau_i)), \\ & x_0(t) \equiv 0, \qquad m = 0,1,2,\ldots, \end{aligned} \tag{5.32}$$

where $G(t,\tau)$ is given by (5.37). Let us prove that the functions $x_m(t)$ are uniformly bounded for all $t \in \mathbf{R}$. Indeed, the function $x_0(t)$ is bounded for all $t \in \mathbf{R}$. Suppose that the functions $x_j(t)$, $j = 1, 2, \ldots, m$ are bounded. Then, for the function $x_{m+1}(t)$, we have

$$\begin{aligned}
&\|x_{m+1}(t)\| \leq \|x_{m+1}(t) - x_1(t)\| + \|x_1(t)\| \leq \\
&\leq \left\| \int_{-\infty}^{\infty} G(t,\tau)(f(\tau, x_m(\tau)) - f(\tau, 0))d\tau \right\| + \\
&+ \left\| \int_{-\infty}^{\infty} G(t,\tau) f(\tau, 0) d\tau \right\| + \\
&+ \left\| \sum_{i=-\infty}^{\infty} G(t,\tau_i)(I_i(x_m(\tau_i)) - I_i(0)) \right\| + \left\| \sum_{i=-\infty}^{\infty} G(t,\tau_i) I_i(0) \right\|.
\end{aligned}$$

Whence, by inequalities (5.10), (5.12), and (5.30), we get

$$\begin{aligned}
\|x_{m+1}(t)\| &\leq \frac{2K}{\gamma} N \sup_{t\in\mathbf{R}} \|x_m(t)\| + \frac{2K}{\gamma} M + \\
&+ \frac{2K e^{\gamma l \left(1 - \frac{1}{q}\right)}}{1 - e^{-\frac{\gamma l}{q}}} (N \sup_{t \in \mathbf{Z}} \|x_m(\tau_i)\| + M \leq \\
&\leq 2C(N \sup_{t\in\mathbf{R}} \|x_m(t)\| + M),
\end{aligned}$$

and consequently

$$\sup_{t\in\mathbf{R}} \|x_{m+1}(t)\| \leq 2CN \sup_{t\in\mathbf{R}} \|x_m(t)\| + 2CM. \tag{5.33}$$

If the Lipschitz constant is so small that

$$2CN < 1, \tag{5.34}$$

then it follows from (5.33) that

$$\sup_{t\in\mathbf{R}} \|x_m(t)\| \leq \frac{2CM}{1 - 2CN} \tag{5.35}$$

for all $m = 1, 2, \ldots$, i.e. the sequence of functions (5.32) is uniformly bounded.

By differentiating $x_{m+1}(t)$ with respect to t with $t \neq \tau_i$ and calculating the difference $x_{m+1}(\tau_i + 0) - x_{m+1}(\tau_i)$, we see that the function $x_{m+1}(t)$ is a unique solution of the system

$$\frac{dx}{dt} = A(t)x + f(t, x_m(t)), \qquad t \neq \tau_i,$$
$$\Delta x|_{t=\tau_i} = B_i x + I_i(x_m(\tau_i)), \tag{5.36}$$

which is bounded for all $t \in \mathbf{R}$.

Let us show that the convergence of the sequence $\{x_m(t)\}$ is uniform. To do that, estimate the difference $x_{m+1}(t) - x_m(t)$. We have

$$\begin{aligned} \|x_{m+1}(t) - x_m(t)\| &\leq KN \int_{-\infty}^{\infty} e^{-\gamma|t-\tau|}\|x_m(\tau) - x_{m-1}(\tau)\|d\tau + \\ &+KN \sum_{i=-\infty}^{\infty} e^{-\gamma|t-\tau_i|}\|x_m(\tau_i) - x_{m-1}(\tau_i)\| \leq \\ &\leq 2NC \sup_{t\in\mathbf{R}} \|x_m(t) - x_{m-1}(t)\|. \end{aligned}$$

Hence,

$$\sup_{t\in\mathbf{R}} \|x_{m+1}(t) - x_m(t)\| \leq 2NC \sup_{t\in\mathbf{R}} \|x_m(t) - x_{m-1}(t)\| \tag{5.37}$$

for all $m = 1, 2, \ldots$. Whence it follows that, for all $m = 1, 2, \ldots$, the following inequality holds

$$\sup_{t\in\mathbf{R}} \|x_{m+1}(t) - x_m(t)\| \leq (2NC)^m \cdot CM. \tag{5.38}$$

This, by (5.34), implies that the functions $x_m(t)$ converge uniformly.

Set

$$\lim_{m\to\infty} x_m(t) = x^*(t).$$

From inequality (5.32) we have

$$\sup_{t\in\mathbf{R}} \|x^*(t)\| \leq \frac{2MC}{1 - 2NC}. \tag{5.39}$$

Since the functions $f(t, x)$ and $I_i(x)$ are continuous with respect to x, by passing to the limit as $m \to \infty$, we see that the limit function $x^*(t)$ satisfy the relation

$$x^*(t) = \int_{-\infty}^{\infty} G(t, \tau)f(\tau, x^*(\tau))d\tau + \sum_{i=-\infty}^{\infty} G(t, \tau_i)I_i(x^*(\tau_i)). \tag{5.40}$$

If we differentiate $x^*(t)$ with respect to t for $t \neq \tau_i$ and calculate the value of the jump of the function $x^*(t)$ at $t\tau_i$, we see that $x^*(t)$ is a solution of system (5.29). It can be seen that $x^*(t)$ is a unique solution of system (5.29), bounded for all $t \in \mathbf{R}$, since each function $x_{m+1}(t)$ is a unique solution of (5.36), bounded for all $t \in \mathbf{R}$.

To end the proof, we need to show that the solution $x^*(t)$ is asymptotically stable if such are the solutions of system (5.31).

With these assumptions, the matrix $G(t,\tau)$ is

$$G(t,\tau) = \Phi(t,\tau), \tag{5.41}$$

where $\Phi(t,\tau)$, $\Phi(\tau,\tau) = E$, is the matriciant of system (5.31) and, for all $t \geq \tau$, is subject to the estimate

$$\|\Phi(t,\tau)\| \leq Ke^{-\gamma(t-\tau)}, \qquad t \geq \tau, \tag{5.42}$$

for some positive K and γ since the system is hyperbolic.

The solution $x^*(t)$ itself is a uniform limit of the sequence of functions

$$x_{m+1}(t) = \int_{-\infty}^{\infty} \Phi(t,\tau)f(\tau, x_m(\tau))d\tau + \sum_{\tau_i<t} \Phi(t,\tau_i)I_i(x_m(\tau_i)) \tag{5.43}$$

and satisfies the relation

$$x_{m+1}(t) = \int_{-\infty}^{t} \Phi(t,\tau)f(\tau, x^*(\tau))d\tau + \sum_{\tau_i<t} \Phi(t,\tau_i)I_i(x^*(\tau_i)). \tag{5.44}$$

Let $x(t,x_0)$, $x(0,x_0) = x_0$, be an arbitrary solution of system (5.29), which starts, for $t = 0$, at the point x_0 that lies in a sufficiently small neighborhood of the point $x^*(0) = X_0^*$. Then, for $t > 0$, the difference $x(t,x_0) - x^*(t)$ can be written as

$$\begin{aligned} x(t,x_0) - x^*(t) &= \Phi(t,0)(x_0 - x_0^*) + \\ &+ \int_0^t \Phi(t,\tau)[f(\tau, x(\tau,x_0)) - f(\tau, x^*(\tau))]d\tau + \\ &+ \sum_{0<\tau_i<t} \Phi(t,\tau_i)[I_i(x(\tau_i,x_0)) - I_i(x^*(\tau_i))]. \end{aligned}$$

This relation together with inequalities (5.31) and (5.42) imply that

$$\|x(t,x_0) - x^*(t)\| \le Ke^{-\gamma t}\|x_0 - x_0^*\| + \\ + \int_0^t Ke^{-\gamma(t-\tau)} N\|x(\tau,x_0) - x^*(\tau)\| d\tau + \\ + \sum_{0<\tau_i<t} Ke^{-\gamma(t-\tau_i)} N\|x(\tau_i,x_0) - x^*(\tau_i)\|$$

or

$$u(t) \le K\|x_0 - x_0^*\| + \int_0^t KNu(\tau)d\tau + \sum_{0<\tau_i<t} KNu(\tau_i), \tag{5.45}$$

where

$$u(t) = e^{\gamma t}\|x(t,x_0) - x^*(t)\|.$$

According to the analogue of the Gronwell-Bellman lemma, we get from (5.45)

$$u(t) \le K\|x_0 - x_0^*\| e^{KNt}(1 + KN)^{i(0,t)}. \tag{5.46}$$

Because

$$i(0,t) \le q + \frac{q}{l}t,$$

we see that

$$u(t) \le K_1 e^{\left(KN + \frac{q}{l}\ln(1+KN)\right)t}\|x_0 - x_0^*\|,$$

where $K_1 = K(1 + KN)^q$ and so, finally,

$$\|x(t,x_0) - x^*(t)\| \le K_1 e^{-\gamma - N_1)t}\|x_0 - x_0^*\| \tag{5.47}$$

for all $t \ge 0$, $N_1 = KN + \frac{q}{l}\ln(1 + KN)$.

If the constant N is required to be so small that not only (5.34) holds but also

$$KN + \frac{q}{l}\ln(1 + KN) < \gamma, \tag{5.48}$$

then, by (5.47), $\|x(t,x_0) - x^*(t)\| \to 0$ for $t \to \infty$, i.e. the solution $x^*(t)$ is asymptotically stable.

5.3 Integral sets of systems with hyperbolic linear part

In this section we will give sufficient conditions for existence of integral sets of nonlinear impulsive differential equations and study the behavior of solutions lying in the integral set and in its neighborhood.

Consider the system

$$\frac{dz}{dt} = A(t)z + f(t,z), \qquad t \neq \tau_i, \qquad \Delta z|_{t=\tau_i} = B_i z + I_i(z) \tag{5.49}$$

and assume that its linear part, i.e. the system

$$\frac{dz}{dt} = A(t)z, \qquad t \neq \tau_i, \qquad \Delta z|_{t=\tau_i} = B_i z \tag{5.50}$$

is hyperbolic. In this case, without infringing the generality, we can assume that the matrices $A(t)$ and $B(t)$ are in the block diagonal form, $A(t) = \mathrm{diag}(A_+(t)), A_-(t)), B_i = \mathrm{diag}(B_i^+, B_i^-)$ and, hence, system (5.49) is written in the form

$$\frac{dx}{dt} = A_+(t)x + F(t,x,y), \qquad \frac{dy}{dt} = A_-(t)y + G(t,x,y), \qquad t \neq \tau_i$$
$$\Delta x|_{t=\tau_i} = B_i^+ x + I_i^{(1)}(x,y), \qquad \Delta y|_{t=\tau_i} = B_i^- y + I_i^{(2)}(x,y). \tag{5.51}$$

Here

$$\mathrm{col}(x,y) = z, \qquad \mathrm{col}(F,G) = f,$$
$$\mathrm{col}(I_i^{(1)}, I_i^{(2)}) = i_i, \qquad x \in \mathbf{R}^k, \qquad y \in \mathbf{R}^{n-k}, \qquad 0 \leq k \leq n.$$

Suppose that the functions $F(t,x,y), G(t,x,y)$ are continuous in t (piecewise continuous with first kind discontinuities at $t = \tau_i$). The functions $F(t,x,y)$, $G(t,x,y)$, $U_i^{(1)}(x,y)$, $I_i^{(2)}(x,y)$ are assumed to be defined for all $t \in \mathbf{R}$, $x \in \mathbf{R}^n$, $y \in \mathbf{R}^{n-k}$ and satisfy the conditions

$$\|F(t,x',y') - F(t,x'',y'')\| + \|G(t,x',y') - G(t,x'',y'')\| \leq$$
$$\leq N(\|x' - x''\| + \|y' - y''\|), \tag{5.52}$$
$$\|I_i^{(1)}(x',y') - I_i^{(1)}(x'' - y'')\| + \|I_i^{(2)}(x',y') - I_i^{(2)}(x'' - y'')\| \leq$$
$$\leq N(\|x' - x''\| + \|y' - y''\|), \tag{5.53}$$
$$\|F(t,0,0)\| + \|G(t,0,0)\| + \|I_i^{(1)}(0,0)\| + \|I_i^{(2)}(0,0)\| \leq M,$$

for all $t \in \mathbf{R}$, $x \in \mathbf{R}^k$, $y \in \mathbf{R}^{n-k}$.

Because system (5.50) is hyperbolic, the matriciants $\Phi_+(t,\tau)$, $\Phi_-(t,\tau)$ of the system

$$\frac{dx}{dt} = A_+(t)x, \qquad t \neq \tau_i, \qquad \frac{dy}{dt} = A_-(t)y, \qquad t \neq \tau_i,$$
$$\Delta x|_{t=\tau_i} = B_i^+ x, \qquad \Delta y|_{t=\tau_i} = B_i^- y. \tag{5.54}$$

are subject to estimates (5.6), (5.7).

Let us prove

Lemma 41 *Let in the system*

$$\frac{dx}{dt} = A_+(t)x + F_0(t,x), \qquad \frac{dy}{dt} = A_-(t)y + G_0(t,x), t \neq \tau_i$$
$$\Delta x|_{t=\tau_i} = B_i^+ x + \tilde{I}_i^{(1)}(x), \qquad \Delta y|_{t=\tau_i} = B_i^- y + \tilde{I}_i^{(2)}(x). \tag{5.55}$$

the functions $F_0(t,x)$, $G_0(t,x)$, $\tilde{I}_i^{(1)}(x)$, $\tilde{I}_i^{(2)}(x)$ be defined for all $t \in \mathbf{R}$, $x \in \mathbf{R}^k$, piecewise continuous with respect to t with first kind discontinuities at $t = \tau_i$ and satisfy the Lipschitz conditions in x:

$$\|F_0(t,x') - F_0(t,x'')\| + \|G_0(t,x') - G_0(t,x'')\| +$$
$$\|\tilde{I}_i^{(1)}(x') - \tilde{I}_i^{(1)}(x'')\| + \|\tilde{I}_i^{(2)}(x') - \tilde{I}_i^{(2)}(x'')\| \leq N\|x' - x''\| \tag{5.56}$$

for all $t \in \mathbf{R}$, $i \in \mathbf{Z}$, $x', x'' \in \mathbf{R}^k$.

We also assume that

$$\|F_0(t,0)\| + \|G_0(t,0)\| + \|\tilde{I}_i^{(1)}(0)\| + \|\tilde{I}_i^{(2)}(0)\| \leq M \tag{5.57}$$

for all $t \in \mathbf{R}$, $i \in \mathbf{Z}$, and the times τ_i satisfy condition (5.2).

Then, for sufficiently small values of the Lipschitz constant N, system (5.55) has an integral set

$$y = u^0(t,x), \tag{5.58}$$

where the function $u^0(t,x)$ is piecewise continuous with respect to t with first kind discontinuities at $t = \tau_i$ and such that

$$\|u^0(t,x') - u^0(t,x'')\| \leq K^2 N(1+KN)^2 a_1 \|x' - x''\| \tag{5.59}$$

for all $t \in \mathbf{R}$, $x', x'' \in \mathbf{R}^k$.

For any two solutions $(x_1(t), y_1(t))$, $(x_2(t), y_2(t))$ that belong to the integral set, we have the following inequality

$$\|x_1(t) - x_2(t)\| + \|y_1(t) - y_2(t)\| \leq K(1 + KN)^2(1 + KNa) \times \times \|x_1(t_0) - x_2(t_0)\| e^{-\gamma_1 (t-t_0)}, \quad t \geq t_0, \quad (5.60)$$

where γ_1 and a are positive constants with the values given below.

Proof. By the conditions of the lemma, every solution of the system

$$\frac{dx}{dt} = A_+(t)X + F_1(t, x) \quad (5.61)$$

is defined for all $t \in \mathbf{R}$ and, for any pair (t_0, x_0), $t_0 \in \mathbf{R}$, $x_0 \in \mathbf{R}^k$, a solution that passes through the point x_0 at $t = t_0$ is unique.

Any integral curve of the impulsive system

$$\frac{dx}{dt} = A_+(t)X + F_1(t, x) \qquad t \neq \tau_i, \qquad \Delta\, x|_{t=\tau_i} = B_i^+ x + \tilde{I}_i^{(1)}(x) \quad (5.62)$$

consists of parts of integral curves of system (5.61). For sufficiently small values of the Lipschitz constant N, the mapping $x \to (E + B_i)x + \tilde{I}_i^{(1)}(x)$ is bijective for any $i \in \mathbf{Z}$ and hence the solution $x_t(t_0, x_0)$, $x_{t_0}(t_0, x_0) = x_0$, of system (5.62) is unique for any $t_0 \in \mathbf{R}$ and $x_0 \in \mathbf{R}^n$. Moreover, existence of the finite upper limit (5.50) implies that the solution is bounded on the entire time axis $\mathbf{R}$.

Let $x_t(t_0, x_0)$, $X_{t_0}(t_0, x_0) = x_0$, be an arbitrary solution of system (5.62). Substitute $x_t(t_0, x_0)$ into the second system of equations (5.55),

$$\frac{dy}{dt} = A_-(t)y + G_1(t, x_t(t_0, x_0)), \qquad t \neq \tau_i$$
$$\Delta y|_{t=\tau_i} = B_i^- y + \tilde{I}_i^{(2)}(x_{\tau_i}(t_0, x_0)). \quad (5.63)$$

The family of solutions of system (5.63),

$$y_t(t_0, x_0) = -\int_t^\infty \Phi_-(t, \tau) G_1(\tau, x_\tau(t_0, x_0)) d\tau - - \sum_{\tau - I > t} \Phi_-(t, \tau_i) \tilde{I}_i^{(2)}(x_{\tau_i}(t_0, x_0)), \quad (5.64)$$

which depends on t_0 and x_0 considered as parameters, covers the set $\Gamma_0(t)$:

$$y = u(t, x) = -\int_t^\infty \Phi_-(t, \tau) G_1(\tau, x_\tau(t, x)) d\tau - - \sum_{\tau_i > t} \Phi_-(t, \tau_i) \tilde{I}_i^{(2)}(x_{\tau_i}(t, x)). \quad (5.65)$$

We will show that $\Gamma_0(t)$ is an integral set and inequalities (5.59), (5.60) hold. Indeed, if $y_0 = u(t_0, x_0)$, then, by using that $x_\tau(t, x_t(t_0, x_0)) = x_\tau(t_0, x_0)$, we have

$$\begin{aligned} y_t(t_0, x_0) &= u(t, x_t(t_0, x_0)) = \\ &= -\int_t^\infty \Phi_-(t,\tau)G_1(\tau, x_\tau(t, x_t(t_0, x_0)))d\tau - \\ &- \sum_{\tau>t} \Phi_-(t,\tau_i)\tilde{I}_i^{(2)}(x_{\tau_i}(t, x_t(t_0, x_0))) = \\ &= \int_t^\infty \Phi_-(t,\tau)G_1(\tau, x_\tau(t_0, x_0))d\tau - \\ &- \sum_{\tau_i>t} \Phi_-(t,\tau_i)\tilde{I}_i^{(2)}(x_{\tau_i}(t_0, x_0)), \end{aligned}$$

i.e. $y = u(t, x_t(t_0, x_0))$ is a solution of system (5.63). This means that the set $\Gamma_0(t)$ is entirely filled with integral curves of system (5.55) and hence it is an integral set of this system.

Let now $x_t(t_0, x')$, $x_{t_0}(t_0, x') = x'$, and $x_t(t_0, x'')$, $x_{t_0}(t_0, x'') = x''$, be two arbitrary solutions of system (5.62). The difference of these solutions, for $t \geq t_0$, can be written as

$$\begin{aligned} x_t(t_0, x') - x_t(t_0, x'') &= \Phi(t, t_0)(x' - x'') + \\ &+ \int_{t_0}^t \Phi_+(t,\tau)(F(\tau, x_\tau(t_0, x')) - F(\tau, x_\tau(t_0, x'')))d\tau + \\ &+ \sum_{t_0<\tau_i<t} \Phi_+(t,\tau_i)(\tilde{I}_i^{(1)}(x_{\tau_i}(t_0, x')) - \tilde{I}_i^{(1)}(x_{\tau_i}(t_0, x''))), \end{aligned}$$

whence, by inequalities (5.34) and (5.56), it follows that

$$\begin{aligned} e^{\gamma(t-t_0)}\|x_t(t_0, x') - x_t(t_0, x'')\| &\leq K\|x'' - x'\| + \\ &= \int_{t_0}^t KNe^{\gamma(\tau-t_0)}\|x_\tau(t_0, x') - x_\tau(t_0, x'')\|d\tau + \\ &+ \sum_{t_0<\tau_i<t} KNe^{\gamma(\tau_i-t_0)}\|x_{\tau_i}(t_0, x') - x_{\tau_i}(t_0, x'')\|. \end{aligned}$$

By using this inequality and an analogue of the Gronwell-Bellman lemma, we get

$$\begin{aligned} e^{\gamma(t-t_0)}\|x_t(t_0, x') - x_t(t_0, x'')\| &\leq \\ &\leq Ke^{KN(t-t_0)}(1 + KN)^{i(t_0,t)}\|x' - x''\|. \end{aligned} \tag{5.66}$$

Whence by applying (5.30), we finally get

$$\|x_t(t_0, x') - x_t(t_0, x'')\| \le K_1 e{-}\gamma_1(t - t_0)\|x' - x''\|. \tag{5.67}$$

for all $t \ge t_0$, $x', x'' \in \mathbf{R}^k$, where

$$K_1 = K(1 + KN)^q, \qquad \gamma_1 = \gamma - KN - \frac{q}{l}\ln(1 + KN).$$

Assume now that N is so small that $\gamma_1 > 0$. Then it readily follows from (5.65) that

$$\|u(t, x') - u(t, x'')\| \le \Big(\int_t^\infty Ke^{\gamma(t-\tau)}NK_1e^{-\gamma_1(\tau-t)}d\tau +$$

$$+\sum_{\tau>t} Ke^{\gamma(t-\tau_i)}NK_1e^{-\gamma_1(\tau_i-t)}\Big)\|x' - x''\| \le$$

$$\le KNK_1\left(\frac{1}{\gamma+\gamma_1} + \frac{e^{(\gamma+\gamma_1)l\left(1-\frac{1}{q}\right)}}{1 - e^{-(\gamma+\gamma_1)\frac{l}{q}}}\right)\|x' - x''\|.$$

For any two solutions of system (5.55), which lie on the integral set $y = u(t, x)$, estimate (5.67) holds and

$$\|y_t(t_0, y_0') - y_t(t_0, y_0'')\| \le$$

$$\le \int_t^\infty Ke^{\gamma(t-\tau)}N\|x_\tau(t_0, x_0') - x_\tau(t_0, x_0'')\|d\tau +$$

$$+\sum_{\tau_i>t} Ke^{\gamma(t-\tau_i)}N\|x_{\tau_i}(t_0, x_0') - x_{\tau_i}(t_0, x_0'')\| \le$$

$$\le KNK_1\left(\frac{1}{\gamma+\gamma_1} + \frac{e^{(\gamma+\gamma_1)l\left(1-\frac{1}{q}\right)}}{1 - e^{-(\gamma+\gamma_1)\frac{l}{q}}}\right)e^{-\gamma_1(t-t_0)}\|x_0' - x_0''\|, \tag{5.68}$$

where $y_0' = u(t_0, x_0')$, $y'' = u(t_0, x_0'')$.

□

Also note that system (5.55) has a unique solution $(x^*(t), y^*(t))$, which is bounded for all $t \in \mathbf{R}$. Indeed, by Theorem 89, for sufficiently small N, equations (5.62) have a unique solution $x^*(t)$, bounded for all $t \in \mathbf{R}$, which is a limit of the uniformly convergent sequence of functions

$$\begin{aligned} x_{m+1}(t) &= \int_{-\infty}^{t} \Phi_+(t,\tau)F_1(\tau, x_m(\tau))d\tau + \\ &+ \sum_{\tau_i<t} \Phi_+(t,\tau_i)\tilde{I}_i^{(1)}(x_m(\tau_i)). \end{aligned} \tag{5.69}$$

The system

$$\frac{dy}{dt} = A_-(t)y + G_1(t, x^*(t)), \qquad t \neq \tau_i,$$
$$\Delta y|_{t=\tau_i} = B_- y + \tilde{I}_i^{(2)}(x^*(\tau_i))$$

has the unique solution

$$\begin{aligned} y_*(t) &= -\int_{t}^{\infty} \Phi_-(t,\tau)G_1(\tau, x^*(\tau))d\tau - \\ &- \sum_{\tau_i>t} \Phi_-(t,\tau_i)I_i^{(2)}(x^*(\tau_i)), \end{aligned} \tag{5.70}$$

which is bounded for all $t \in \mathbf{R}$, so that $(x^*(t), y^*(t))$ is a unique solution of system (5.55), bounded for all $t \in \mathbf{R}$.

Clearly, this solution lies in the integral set $\Gamma_0(t)$, so that, for all $t \in \mathbf{R}$,

$$y^*(t) = u(t, x^*(t)). \tag{5.71}$$

It follows from inequalities (5.67) and (5.68) that all solutions of equations (5.55) that lie in the set Γ_0 eventually approach the solution $(x^*(t), y^*(t))$, i.e. we have

$$\begin{aligned} &\|x(t) - x^*(t)\| + \|y(t) - y^*(t)\| \leq \\ &\leq K_1 \left[1 + KN \left(\frac{1}{\gamma+\gamma_1} + \frac{e^{(\gamma+\gamma_1)l\left(1-\frac{1}{q}\right)}}{1 - e^{-(\gamma+\gamma_1)\frac{l}{q}}} \right) \right] \times \\ &\times \|x(t_0) - x^*(t_0)\| e^{-\gamma_1(t-t_0)} \end{aligned}$$

for any solution $(x(t), y(t))$ that satisfy the condition $y(t_0) = u(t_0, x(t_0))$.

To show that the integral sets of equations (5.51) exist, we additionally assume at the beginning that

$$F(t,0,0) = G(t,0,0) = I_i^{(1)}(0,0) = I_i^{(2)}(0,0) = 0. \tag{5.72}$$

We look for the integral set Γ_+ of system (5.51) as a limit of the sequence of sets $\Gamma_+^{(m)}$,

$$\Gamma_+ = \lim_{m\to\infty} \Gamma_+^{(m)}; \qquad \Gamma_+^{(m)} : , y = u_+^{(m)}(t,x), \qquad m = 0,1,\ldots, \tag{5.73}$$

each of which is an integral set of the system

$$\begin{aligned} &\frac{dx}{dt} = A_+(t)x + F(t,x,u_+^{(m-1)}(t,x)), \\ t \neq \tau_i, \quad & \\ &\frac{dy}{dt} = A_-(t)x + G(t,x,u_+^{(m-1)}(t,x)), \\ &\Delta x|_{t=\tau_i} = B_i^+ x + I_i^{(1)}(x,u^{(m-1)}(\tau_i,x)), \\ &\Delta y|_{t=\tau_i} = B_i^- x + I_i^{(2)}(x,u^{(m-1)}(\tau_i,x)). \end{aligned} \tag{5.74}$$

Take the set $y \equiv 0$ as $\Gamma_+^{(0)}$ and, by using Lemma 41, define the set $\Gamma_+^{(1)}$ as follows:

$$\begin{aligned} y = u_+^{(1)}(t,x) \quad &= \quad -\int_t^{\infty} \Phi_-(t,\tau)G(\tau,x_\tau^{(1)}(t,x,0))d\tau - \\ &- \sum_{\tau_i>t} \Phi_-(t,\tau_i)I_i^{(2)}(x_{\tau_i}^{(1)}(t,x),0), \end{aligned} \tag{5.75}$$

where $x_\tau^{(1)}(t_0,x_0)$ is a general solution of the system

$$\begin{aligned} \frac{dx}{dt} &= A_+(t)x + F(t,x,0), \qquad t \neq \tau_i, \\ \Delta x|_{t=\tau_i} &= B_i^+ x + I_i^{(1)}(x,0). \end{aligned} \tag{5.76}$$

If the sets $\Gamma_+^{(1)}, \Gamma_+^{(2)}, \ldots, \Gamma_+^{(m-1)}$ are found, by using Lemma 5.49, we define the set $\Gamma_+^{(m)}$ as follows:

$$\begin{aligned} y &= u_+^{(m)}(t,x) = \\ &= -\int_t^{-\infty} \Phi_-(t,\tau)G(\tau,x_\tau^{(m)}(t,x),u_+^{(m-1)}(\tau,x_\tau^{(m)}(t,x)))d\tau - \\ &- \sum_{\tau_i>t} \Phi_-(t,\tau_i)I_i^{(2)}(x_\tau^{(m)}(t,x),u_+^{(m-1)}(\tau_i,x_{\tau_i}^{(m)}(t,x))), \end{aligned} \tag{5.77}$$

where $x_i^{(m)}$ is a general solution of the system

$$\frac{dx}{dt} = A_+(t)x + F(t, x, u_+^{(m-1)}(t, x)), \qquad t \neq \tau_i,$$
$$\Delta x|_{t=\tau_i} = B_i^+ x + I_i^{(1)}(x, u_+^{(m-1)}(\tau_i, x)). \tag{5.78}$$

To prove that the sequence of functions $u_+^{(m)}(t, x)$ converges to a limit function that defines an integral set of system (5.51), we establish some properties of these functions and $x_t^m(t_0, x)$.

Lemma 42 *For an arbitrary number σ, $0 < \sigma < \gamma$, one can find such a number $N_0 > 0$ that, for all $0 < N \leq N_0$, the functions $u_+^{(m)}(t, x)$ and $x_t^{(m)}(t_0, x)$ satisfy the inequalities*

$$\|u_+^{(m)}(t, x') - u_+^{(m)}(t, x'')\| \leq 2K^2Na\|x' - x''\|, \tag{5.79}$$
$$\|x_t^{(m)}(t_0, x') - x_t^{(m)}(t_0, x'')\| \leq K_0 e^{-\sigma(t-t_0)}\|x' - x''\| \tag{5.80}$$

for all $-\infty < t_0 < \infty$, $m = 0, 1, 2, \ldots$, where

$$K_0 = K(1 + KN(1 + 2K^2Na))^q,$$
$$a = \left(\frac{1}{\gamma + \sigma} + \frac{e^{(\gamma+\sigma)l\left(1-\frac{1}{q}\right)}}{1 - e^{-(\gamma+\sigma)\frac{l}{q}}}\right).$$

Proof. Fix an arbitrary number σ, $0 < \sigma < \gamma$, and choose N_0 such that, for all $0 < N \leq N_0$, the two inequalities hold simultaneously:

$$KN(1 + 2K^2Na) + \frac{q}{l}\ln(1 + KN(1 + 2K^2Na)) \leq \gamma - \sigma,$$
$$(1 + 2K^2Na)(1 + KN(1 + 2K^2Na))^q \leq 2. \tag{5.81}$$

Because every function $x_i^{(m)}(t_0, x_0)$ can be written as

$$x_t^{(m)}(t, x_0) = \Phi_+(t, t_0)x_0 +$$
$$+ \int_{t_0}^{t} \Phi_+(t, \tau)F(\tau, x_\tau^{(m)}(t_0, x_0), u_+^{(m-1)}(\tau, x_\tau^{(m)}(t_0, x_0)))d\tau +$$
$$+ \sum_{t_0 < \tau_i < t} \Phi_+(t, \tau_i)I_i^{(1)}(x_{\tau_i}^{(m)}(t_0, x_0), u_+^{(m-1)}(\tau_i, x_{\tau_i}^{(m)}(t_0, x_0))), \tag{5.82}$$

by using inequalities (5.52), (5.54) for the case when $m = 1$, we have

$$\begin{aligned}\|x_t^{(1)}(t_0,x') - x_t^{(1)}(t_0,x'')\| &\le Ke^{-\gamma(t-t_0)}\|x' - x''\| + \\ &+ \int_{t_0}^{t} Ke^{-\gamma(t-\tau)}N\|x_\tau^{(1)}(t_0,x') - x_\tau^{(1)}(t_0,x'')\| + \\ &+ \sum_{t_0<\tau_i<t} Ke^{-\gamma(t-\tau_i)}N(\|x_{\tau_i}^{(1)}(t_0,x') - x_{\tau_i}^{(1)}(t_0,x'')\|).\end{aligned}$$

by applying the analogue of the Gronwell-Bellman lemma to the function

$$\|x_t^{(1)}(t_0,x') - x_t^{(1)}(t_0,x'')\|e^{\gamma(t-t_0)},$$

we get

$$\begin{aligned}&\|x_t^{(1)}(t_0,x') - x_t^{(1)}(t_0,x'')\|e^{\gamma(t-t_0)} \le \\ &\le Ke^{KN(t-t_0)}(1+KN)^{i(t_0,t)}\|x' - x''\|,\end{aligned}$$

or

$$\begin{aligned}&\|x_t^{(1)}(t_0,x') - x_t^{(1)}(t_0,x'')\|e^{\gamma(t-t_0)} \le \\ &\le K(1+KN)^q e^{-(\gamma-N_1)(t-t_0)}\|x' - x''\|,\end{aligned}$$

where $N_1 = KN + \frac{q}{l}\ln(1+KN)$, i.e. (5.80) holds when $m = 1$.

By (5.75), we get for the difference

$$\begin{aligned}&\|u_+^{(1)}(t,x') - u_+^{(1)}(t,x'')\| \le \\ &\le \int_t^{\infty} Ke^{\gamma(t-\tau)}N\|x_\tau^{(1)}(t,x') - x_\tau^{(1)}(t,x'')\|d\tau + \\ &+ \sum_{\tau_i>t} Ke^{\gamma(t-\tau_i)}N\|x_{\tau_i}^{(1)}(t,x') - x_{\tau_i}^{(1)}(t,x'')\| \le \\ &\le K^2N(1+KN)^q a\|x' - x''\|,\end{aligned}$$

which, together with the second inequality of (5.81), shows that inequality (5.79) is true for $m = 1$.

Assume that inequalities (5.79) and (5.80) hold for $m = 1, 2, \ldots, k$.

Then, for $m = k+1$, we get from (5.82) for $t \geq t_0$ that

$$\|x_t^{(k+1)}(t_0,x') - x_t^{(k+1)}(t_0,x'')\| \leq Ke^{-\gamma(t-t_0)}\|x' - x''\| +$$
$$+ \int_{t_0}^{t} Ke^{-\gamma(t-\tau)} N(\|x_\tau^{(k+1)}(t_0,x') - x_\tau^{(k+1)}(t_0,x'')\| +$$
$$+\|u_+^{(k)}(\tau, x_\tau^{(k+1)}(t_0,x')) - u_+^{(k)}(\tau, x_\tau^{(k=1)}(t_0,x''))\|)d\tau +$$
$$+ \sum_{t_0<\tau_i<t} Ke^{-\gamma(t-\tau_i)} N(\|x_{\tau_i}^{(k+1)}(t_0,x') - x_{\tau_i}^{(k+1)}(t_0,x'')\| +$$
$$+\|u_+^{(k)}(\tau_i, x_{\tau_i}^{(k+1)}(t_0,x')) - \|u_+^{(k)}(\tau_i, x_{\tau_i}^{(k+1)}(t_0,x''))\|) \leq$$
$$\leq K(1 + KN(1+2K^2Na))^q \times e^{-(\gamma - N_2)(t-t_0)}\|x' - x''\|,$$

where $N_2 = KN(1+2K^2Na) + \frac{q}{l}\ln(1 + KN(1+2K^2Na))$. The latter inequality can be written by using the first inequality of (5.81) as

$$\|x_t^{(k+1)}(t_0,x') - x_t^{(k+1)}(t_0,x'')\| \leq Ke^{-\sigma(t-t_0)}\|x' - x''\|,$$

which shows that inequality (5.80) is valid if $m = k+1$ for all $t \geq t_0$, and hence it holds for all $m = 1, 2, \ldots, t \geq t_0$.

For $m = k+1$, we have from (5.77) that

$$\|u_+^{(k+1)}(t,x') - u_+(k+1)(t,x'')\| \leq$$
$$\leq \int_t^{\infty} Ke^{\gamma(t-t_0)} N\left[\|x_\tau^{(k+1)}(t,x') - x_\tau^{(k+1)}(t,x'')\| +\right.$$
$$+\|u_+^{(k)}(\tau, x_\tau^{(k+1)}(t,x')) - u_+^{(k)}(\tau, x_\tau^{(k+1)}(t,x''))\|\Big] +$$
$$+ \sum_{\tau_i>t} Ke^{\gamma(t-\tau_i)} N\left[\|x_{\tau_i}^{(k+1)}(t,x') - x_{\tau_i}^{(k+1)}(t,x'')\| +\right.$$
$$+\|u_+^{(k)}(\tau_i, x_{\tau_i}^{(k+1)}(t,x')) - +\|u_+^{(k)}(\tau_i, x_{\tau_i}^{(k+1)}(t,x''))\|\Big].$$

Whence, by using inequality (5.79) for $m = k$ and (5.80) for $m = k+1$, we get the following estimate

$$\|u_+^{(k+1)}(t,x') - u_+^{(k+1)}(t,x'')\| \leq KN(1+2K^2Na)K_0\|x' - x''\|.$$

which, in virtue of the second inequality of (5.81), leads to inequality (5.79) for $m = k+1$, and hence, by induction, proves it for all $m = 1, 2, \ldots$.

□

In the sequel we will need the following statement which is of interest by itself.

Lemma 43 *Let a nonnegative function $u(t)$, piecewise continuous for $t \geq t_0$ satisfy the inequality*

$$u(t) \leq \alpha + \int_{t_0}^{t} [\beta e^{\gamma_1(\tau - t_0)} + \delta u(\tau)]d\tau + \tag{5.83}$$

$$+ \sum_{t_0 < \tau_i < t} [\beta e^{\gamma_1(\tau_i - t_0)} + \delta u(\tau_i)], \tag{5.84}$$

where $\alpha \geq 0$, $\beta \geq 0$, $\gamma_1 > \delta > 0$, τ_i are the points of first kind discontinuity of the function $u(t)$. Then, for $\tau_i < t \leq \tau_{i+1}$, $u(t)$ satisfies the estimate

$$u(t) \leq \alpha(1+\delta)^i e^{\delta(t-t_0)} + \frac{\beta}{\gamma_1 - \delta}[\,1+$$

$$+(1+\gamma_1)\sum_{\nu=0}^{i-1}(1+\delta)^{i-\nu-1}e^{-(\gamma_1-\delta)(t-\tau_{\nu+1})}\,]e^{\gamma_1(t-t_0)}. \tag{5.85}$$

Moreover, if the times τ_i satisfy the property that there exist such numbers $l > 0$ and natural q that any interval of the time axis of length l contains no more than q points of the sequence $\{\delta_i\}$, then there exists such $\delta^* < \gamma_1$ that, for all $0 < \delta < \delta^*$,

$$u(t) \leq \alpha(1+\delta)^q e^{\left(\delta + \frac{q}{l}\ln(1+\delta)\right)(t-t_0)} +$$

$$+\frac{\beta}{\gamma_1 - \delta}\left[1 + (1+\gamma_1)\frac{e^{-(\gamma_1-\delta)l\left(\frac{1}{q}-1\right)}}{1-(1+\delta)e^{-(\gamma_1-\delta)\frac{l}{q}}}\right]e^{\gamma_1(t-t_0)}. \tag{5.86}$$

Proof. Inequality (5.83) can be written on $[t_0, \tau_1]$ as

$$u(t) \leq \alpha + \int_{t_0}^{t} [\beta e^{\gamma_1(\tau - t_0)} + \delta u(\tau)]d\tau. \tag{5.87}$$

By integrating (5.87) we see that, for all $t_0 \leq t \leq \tau_1$,

$$u(t) \leq \alpha e^{\delta(t-t_0)} + \frac{\beta}{\gamma_1 - \delta}(1 - e^{-(\gamma_1-\delta)(t-t_0)})e^{\gamma_1(t-t_0)}. \tag{5.88}$$

If $t \in]\tau_i, \tau_{i+1}]$, $i = 1, 2, \ldots$, and

$$u(t) \leq \alpha_i + \int_{tau_i}^{t} [\beta e^{\gamma_1(\tau - t_0)} + \delta u(\tau)]d\tau,$$

then, by integrating this inequality, we see that

$$u(t) \le \alpha_i e^{\delta(t-tau_i)} + \frac{\beta}{\gamma_1 - \delta}(1 - e^{-(\gamma_1-\delta)(t-tau_i)})e^{\gamma_1(t-t_0)} \tag{5.89}$$

for all $t \in]\tau_i, \tau_{i+1}]$. By using (5.83), (5.88), (5.89) and induction, it is easy to show that

$$\begin{aligned} u(t) \le \alpha(1+\delta)^i e^{\delta(t-t_0)} + \beta\Big\{\sum_{\nu=0}^{i-1}(1+ \\ +\frac{1+\delta}{\gamma_1-\delta}(1-e^{-(\gamma_1-\delta)(\tau_{\nu+1}-\tau_\nu)}))e^{-(\gamma_1-\tau)(t-\tau_{\nu+1})}\Big\} + \\ +\frac{1}{\gamma_1-\delta}(1-e^{-(\gamma-\delta)(t-\tau_i)})\Big\}e^{\gamma_1(t-t_0)} \end{aligned} \tag{5.90}$$

for any $i = 1, 2, \ldots$ and all $t \in]\tau_i, \tau_{i+1}]$. Note that in (5.90) as well as in (5.88) and (5.89), it is not necessary that $\gamma_1 > \delta$. These inequalities hold for any $\delta > 0$ and $\gamma_1 > 0$, in particular, for $\delta = \gamma_1$ we get from (5.90) that

$$\begin{aligned} u(t) \;\; \le \;\; & \alpha(1+\gamma_1)^i e^{\gamma(t-t_0)} + \beta\Big\{\sum_{\nu=0}^{i-1}(1+\gamma_1)^{i-\nu-1}[1+ \\ & +(1+\gamma_1)(\tau_{\nu+1}-\tau_\nu)] + (t-\tau_i)\Big\}e^{-\gamma(t-t_0)}. \end{aligned} \tag{5.91}$$

If $\gamma_1 > \delta > 0$, then by neglecting negative termes in (5.90), we obtain inequality (5.85).

Let the times τ_i satisfy the conditions stated in the lemma. Then the number of the points of the sequence τ_i, which lie in the interval $[t_0, t]$, $i(t_0, t)$, is estimated as follows:

$$i(t_0, t) \le q + \frac{q}{l}(t - t_0)$$

and, for the difference $\tau_j - \tau_i$, we have

$$\tau_i - \tau_\nu \ge l\left(\frac{1}{q} - 1\right) + \frac{l}{q}(i - \nu)$$

for all $i \ge \nu \ge 1$. Hence, if $t \in]\tau_i, \tau_{i+1}]$, then

$$(1+\delta)^i \le (1+\delta)^q e^{\frac{q}{l}\ln(1+\delta)(t-t_0)} \tag{5.92}$$

and

$$\sum_{\nu=0}^{i-1}(1+\delta)^{i-\nu-1}e^{-(\gamma_1-\delta)(t-\tau_{\nu+1})} \leq$$

$$\leq e^{-(\gamma_1-\delta)l\left(\frac{1}{q}-1\right)} \frac{1-(1+\delta)^i e^{-i(\gamma_1-\delta)\frac{l}{q}}}{1-(1+\delta)e^{-(\gamma_1-\delta)\frac{l}{q}}} \tag{5.93}$$

Let a root of the equation $(1+\delta)e^{-(\gamma_1-\delta)\frac{1}{q}} = 1$ be denoted by δ^*. It follows from inequality (5.93) that if $\delta \in]0, \delta^*[$, then, for any natural i and any $t \in]\tau_i, \tau_{i+1}]$, we have

$$\sum_{\nu=0}^{i-1}(1+\delta)^{i-\nu-1}e^{-(\gamma_1-\delta)(t-\tau_{\nu+1})} \leq \frac{e^{-(\gamma_1-\delta)l\left(\frac{1}{q}-1\right)}}{1-(1+\delta)e^{-(\gamma_1-\delta)\frac{l}{q}}}. \tag{5.94}$$

Estimates (5.92) and (5.93) allow, by using inequality (5.85), to get estimate (5.86), which holds for all $t \geq t_0$ and any $0 < \delta < \delta^*$.

□

Corollary. *If conditions of the preceding lemma hold and the times τ_i are such that*

$$\tau_{i+1} - \tau_i \geq \theta$$

for some positive θ, then, for any $0 < \delta < \delta^*$, where δ^* is a root of the equation $(1+\delta)e^{-(\gamma_1-\delta)\theta} = 1$, the inequality

$$u(t) \leq \alpha(1+\delta)e^{\left(\delta+\frac{1}{\theta}\ln(1+\delta)\right)(t-t_0)} +$$
$$+\frac{\beta}{\gamma_1-\delta}\left[1+\frac{1+\gamma_1}{1-(1-\delta)e^{-(\gamma_1-\delta)\theta}}\right]e^{\gamma_1(t-t_0)}. \tag{5.95}$$

holds for all $t \geq t_0$.

Lemma 44 *For a given $0 < \rho < 1$ there exists such a number $N^0 \leq N_0$ that, for all N, $0 < N \leq N^0$, the functions $u^{(m)}(t,x)$ satisfy the inequalities*

$$\|x_t^{(m)}(t_0,x) - x_t^{(m-1)}(t_0,x)\| \leq 2K\rho^{m-1}e^{-\sigma(t-t_0)}\|x\|, \tag{5.96}$$

$$\|u_+^{(m)}(t,x) - u_+^{(m-1)}(t,x)\| \leq 2\rho^{m-1}K^2Na\|x\| \tag{5.97}$$

for any $t \geq t_0$ and $x \in \mathbf{R}^k$.

Proof. For $m = 1$ we have

$$\|u_+^{(1)}(t,x) - u_+^{(0)}(t,x)\| \leq \|u_+^{(1)}(t,x)\| \leq 2K^2Na\|x\|$$

$$\|x_t^{(1)}(t_0,x) - x_t^{(0)}(t_0,x)\| \leq \|x_t^{(1)}(t_0,x)\| \leq K_0e^{-\sigma(t-t_0)}\|x\|,$$

i.e. inequalities (5.83), (5.85) hold. Suppose now that

$$\|x_t^{(m)}(t_0,x) - x_t^{(m-1)}(t_0,x)\| \leq KL_{m-1}e^{-\sigma(t-t_0)}\|x\| \tag{5.98}$$

$$\|u_+^{(m)}(t,x) - u_+^{(m-1)}(t,x)\| \leq \alpha_{m-1}\|x\|, \tag{5.99}$$

$$L_0 = K_0, \qquad \alpha_0 = 2K^2Na.$$

Then we have

$$\|x_t^{(m+1)}(t_0,x) - x_t^{(m)}(t_0,x)\| \leq \int_{t_0}^{t} Ke^{-\gamma(t-\tau)}N[(1+$$
$$+2K^2Na)\|x_\tau^{(m+1)}(t_0,x) - x_\tau^{(m)}(t_0,x)\| +$$
$$+\alpha_{m-1}\|x_\tau^{(m)}(t_0,x)\|]d\tau + \sum_{t_0<\tau_i<t} Ke^{-\gamma(t-\tau_i)}N[(1+$$
$$+2K^2Na)\|x_{\tau_i}^{(m+1)}(t_0,x) - x_{\tau_i}^{(m)}(t_0,x)\| + \alpha_m\|x_{\tau_i}^{(m)}(t_0,x)\|.$$

Hence the function $v(t) = e^{\gamma(t-t_0)}\|x_t^{(m+1)}(t_0,x) - x_t^m(t_0,x)\|$ satisfies the inequality

$$v(t) \leq \int_{t_0}^{t} [KN(1+2K^2Na)v(\tau)+$$
$$+KN\alpha_{m-1}K_0e^{(\gamma-\sigma)(\tau-t_0)}\|x\|]\,d\tau +$$
$$+\sum_{t_0<\tau_i<t} [KN(1+2K^2Na)v(\tau_i) + KN\alpha_{m-1}e^{(\gamma-\sigma)(\tau_i-t_0)}\|x\|],$$

hence it satisfies (5.83).

Thus, by the previous lemma, we have

$$\|x_t^{(m+1)}(t_0,x) - x_t^{(m)}(t_0,x)\| \le \beta b e^{-\sigma(t-t_0)}\|x\|,$$

where

$$\beta = KNK_0\alpha_{m-1}, \qquad \delta = KN((1+2K^2Na),$$
$$b = \frac{1}{\gamma-\delta-\sigma}\left[1+(1+\gamma-\sigma)\frac{e^{(\gamma-\sigma-\delta)l}}{1-(1+\delta)e^{-(\gamma-\delta-\sigma)\frac{l}{q}}}\right].$$

Thus, if we assume that inequality (5.99) is true, then we obtain, for the function $x_t^{(m+1)}(t_0,x) - x_t^{(m)}(t_0,x)$, the following estimate:

$$\|x_t^{(m+1)}(t_0,x) - x_t^{(m)}(t_0,x)\| \le KL_m e^{-\sigma(t-t_0)}\|x\| \tag{5.100}$$

with the constant

$$L_m = NK_0 b\alpha_{m-1}. \tag{5.101}$$

Now let us estimate the difference $u^{(m+1)}(t,x) - u^{(m)}(t,x)$. We have

$$\begin{aligned}
&\|u_+^{(m+1)}(t,x) - u_+^{(m)}(t,x)\| \le \\
&\le \int_t^\infty Ke^{\gamma(t-\tau)}N\big[\|x_\tau^{(m+1)}(t,x) - x_\tau^{(m)}(t,x)\| + \\
&\quad +\|u_+^{(m)}(\tau, x_\tau^{(m-1)}(t,x)) - u_+^{(m-1)}(\tau, x_\tau^{(m)}(t,x)\|\big]d\tau + \\
&\quad +\sum_{\tau_i>t} ke^{\gamma(t-\tau_i)}N\big[\|x_{\tau_i}^{(m+1)}(t,x) - x_{\tau_i}^{(m)}(t,x)\| + \\
&\quad +\|u_+^{(m)}(\tau_i, x_{\tau_i}^{(m_1)}(t,x)) - u_+^{(m-1)}(\tau_i, x_{\tau_i}^{(m)}(t,x))\|\big] \le \\
&\le \int_t^\infty Ke^{\gamma(t-\tau)}N\big[(1+2K^2Na)\|x_\tau^{(m+1)}(t,x) - x_\tau^{(m)}(t,x)\| + \\
&\quad +\alpha_{m-1}\|x_\tau^{(m)}(t,x)\| + \\
&\quad +\sum_{\tau_i>t} Ke^{\gamma(t-\tau_i}N\big[(1+2K^2Na)\|x_{\tau_i}^{(m+1)}(t,x) - x_{\tau_i}(m)(t,x)\| + \\
&\quad +\alpha_{m-1}\|x_{\tau_i}^{(m)}(t,x)\|\big].
\end{aligned}$$

Whence, by using estimates (5.80) and (5.100), we get the inequality

$$\|u_+^{(m+1)}(t,x) - u_+^{(m)}(t,x)\| \le KN\big[K(1+2K^2Na)\alpha_m + K_0\alpha_{m-1}\big]\|x\|,$$

i.e. an inequality of the form

$$\|u_+^{(m+1)}(t,x) - u_+^{(m)}(t,x)\| \le \alpha_m \|x\| \tag{5.102}$$

with the constant α_m equal

$$\begin{aligned} \alpha_m &= KN\big[K(1+2K^2Na)L_m + K_0\alpha_{m-1}]a = \\ &= KNK_0[1+K^2N(1+2K^2Na)b\big]a\alpha_{m-1}. \end{aligned} \tag{5.103}$$

Choose a number $N^0 \le N_0$ such that, for all N, $0 < N \le N^0$, we have

$$NKK_0\big[1+K^2N(1+2K^2Na)b\big]a \le \rho. \tag{5.104}$$

Then it follows from the previous inequality that

$$\alpha_m \le \rho^m \alpha_0 = 2K^2Na\rho^m$$

and (5.105)

$$\alpha_m \le 2K_0K^2N^2ab\rho^{m-1} \le 2\rho^m.$$

Thus, by using estimates (5.100) and (5.102) and applying induction, we see that inequalities (5.96) and (5.97) are fulfilled.

□

The established inequalities (5.96) and (5.97) imply that the sequence of functions $x_t^{(m)}(t_0,x)$ and $u_t^{(m)}(t,x)$ converge with the convergence of the sequence $\{x_t^{(m)}(t_0,x)\}$, as $m\to\infty$, being uniform on the set $-\infty < t_0 < t < \infty$, $\|x\| \le r$ and the convergence of the sequence $\{u_t^{(m)}(t,x)\}$ being uniform for all $t \in \mathbf{R}$ and $\|x\| \le r$, where r is a positive constant. Set

$$x_t(t,x) = \lim_{m\to\infty} x_t^{(m)}(t,x), \qquad u_+(t,x) = \lim_{m\to\infty} u_+^{(m)}(t,x).$$

As a limit of a uniformly convergent sequence of piecewise continuous functions, the function $u_+(t,x)$ is piecewise continuous with first kind discontinuities being at points of the sequence $\{\tau_i\}$. Since the functions $u_+^{(m)}(t,x)$ and $x_t^{(m)}(t_0,x)$ satisfy inequalities (5.79) and (5.80), the limit functions $u(t,x)$ and $x_t(t_0,x)$ also satisfy these inequalities, i.e.

$$\|u_+(t,x') - u_+(t,x'')\| \le 2K^2Na\|x'-x''\| \tag{5.106}$$

$$\|x_t(t_0,x') - x_t(t_0,x'')\| \le K_0e^{-\sigma(t-t_0)}\|x'-x''\|, \qquad t \ge t_0, \tag{5.107}$$

for all $x', x'' \in \mathbf{R}^k$.

Because the functions $G(t,x,y)$, $F(t,x,y)$, $I_i^{(1)}(x,y)$, $I_i^{(2)}(x,y)$, $u(t,x)$ are continuous with respect to x, y, we pass to the limit in equalities (5.77) and (5.82) by setting $m \to \infty$. It can be seen that the functions $x_t(t_0,x_0)$ and $u_t(t,x)$ satisfy the relations

$$\begin{aligned} x_t(t_0,x_0) = {} & \Phi_+(t,t_0)x_0 + \\ & + \int_{t_0}^{t} \Phi_+(t,\tau)F(\tau, x_\tau(t_0,x_0),\ , u_+(\tau, x_\tau(t_0,x_0)))d\tau + \\ & + \sum_{t_0<\tau_i<t} \Phi_+(t,\tau_i)I_i^{(1)}(x_{\tau_i}(t_0,x_0)),\ , U_+(\tau_i, x_{\tau_i}(t_0,x_0)), \end{aligned} \tag{5.108}$$

$$\begin{aligned} u_+(t,x) = {} & -\int_t^{\infty} \Phi_-(t,\tau)G(\tau, x_\tau(t,x),\ , u_+(\tau, x_\tau(t,x))) - \\ & - \sum_{\tau_i>t} \Phi_-(t,\tau_i))I_i^{(2)}(x_{\tau-i}(t,x),\ , u_+(\tau_i, x_{\tau_i}(t,x))). \end{aligned} \tag{5.109}$$

Let us show that the set $\Gamma_+ : y = u_+(t,x)$ is an integral set of system (5.51). To do this, it will suffice to show that $(x_t(t_0,x_0), u_+(t, x_t(t_0,x_0))$ is a solution of system (5.51). Denote

$$x(t) = x_t((t_0,x_0), y(t) = u_+(t, x_t(t_0,x_0))).$$

By differentiating relation (5.108) with respect to $t \neq \tau_i$, we have in the introduced notations

$$\frac{dx}{dt} = A_+(t)x + F(t,x,y),$$

and, for $t = \tau_i$,

$$x(\tau_i + 0) - x(\tau_i) = B_i^+ x(\tau_i) + I_i^{(1)}(x(\tau_i), y(\tau_i)).$$

If we replace in (5.109) x by $x_t(t_0,x_0)$ and use $x_\tau(t, x_t(t_0,x_0)) = x_\tau(t_0,x_0)$, we see that the function $y(t) = u(t, x_t(t_0,x_0))$ satisfies the relation

$$\begin{aligned} y(t) \quad = {} & -\int_t^{\infty} \Phi_-(t,\tau)G(\tau, x(\tau), y(\tau))d\tau - \\ & - \sum \tau_i > t\Phi_-(t,\tau_i)I_i^{(2)}(x(\tau_i), y(\tau_i)). \end{aligned}$$

By differentiating this equality with respect to t, $t \neq \tau_i$, we obtain

$$\frac{dy}{dt} = A_-(t)y + G(t,x,y),$$

and, if $t = \tau_i$,

$$y(\tau_i + 0) - y(\tau_i) = B_i^- y(\tau_i) + I_i^{(2)}(x(\tau_i), y(\tau_i)).$$

Hence $(x_t(t_0, x_0), u_+(t, x_t(t_0, x_0)))$ is a solution of system (5.51) for any $t_0 \in \mathbf{R}$, $x \in \mathbf{R}^k$, which means that the function $y = u_+(t, x)$ defines an integral set of this system, Γ_+.

On the integral set Γ_+, system (5.51) is reduced to

$$\begin{aligned} \frac{dx}{dt} &= A_+(t)x + F(t, x, u_+(t, x)), \qquad t \neq \tau_i, \\ \Delta x|_{t=\tau_i} & B_i^+ x + I_i^{(1)}(x, u_+(\tau_i, x)). \end{aligned} \tag{5.110}$$

Thus, we have proved the following theorem.

Theorem 90 *Let inequalities (5.52), (5.53) and relations (5.2) and (5.72) hold for system (5.51), the linear part of which is hyperbolic. Then there exists such a positive number N^0 that, for all $0 < N \leq N^0$, system (5.51) has the integral set Γ_+: $y = u_+(t, x)$, where $u_+(t, x)$ is a function, piecewise continuous with respect to t with first kind discontinuities at $t = \tau_i$ and satisfying inequality (5.106). System (5.51) is reduced to system (5.110) on the integral set Γ_+.*

Let us study the behavior of solutions of equations (5.51) on the integral set Γ_+.

For a solution $x_t(t_0, x)$ that lies in the integral set Γ_+, inequality (5.107) holds. It follows from this inequality that, for any $x_0 \in \mathbf{R}^k$, a solution $x_t(t_0, x_0)$ that lies in the integral set approaches the origin as time increases and is subject to the estimate

$$\|x_t(t_0, x_0)\| \leq K_0 e^{-\sigma(t - t_0)} \|x\|. \tag{5.111}$$

In the same way as it was proved that the integral set $\Gamma_+ : y = u_+(t, x)$ of equations (5.51) exists, we can prove that these equations have the integral set $\Gamma_- : x = u_-(t, y)$. It can be found by taking a limit of the sequence of sets $\bar{\Gamma}_-^{(m)}$,

$$\Gamma^{(m)} = \lim_{m \to \infty} \Gamma_-^{(m)}; \qquad \Gamma_-^{(m)} = u_-^{(m)}(t, y), \qquad m = 0, 1, \ldots,$$

each of which is an integral set of the system

$$\begin{aligned}
\frac{dx}{dt} &= A_+(t)x + F(t, u_-^{(m-1)}(t,y), y), \\
& \qquad\qquad\qquad\qquad\qquad\qquad t \neq \tau_i, \\
\frac{dy}{dt} &= A_-(t)x + G(t, u_-^{(m-1)}(t,y), y), \\
\Delta x|_{t=\tau_i} &= B_i^+ x + I_i(1)(u_-^{(m-1)}(\tau_i, y), y), \\
\Delta y|_{t=\tau_i} &= B_i^- y + I_i(2)(u_-^{(m-1)}(\tau_i, y), y).
\end{aligned} \tag{5.112}$$

The set $\bar{\Gamma}_-^{(m)}$ is given by the formula

$$\begin{aligned}
\Gamma_-^{(m)} : x = u_-^{(m)}(t,y) = \\
&= \int_{-\infty}^{t} \Phi_+(t,\tau) F(\tau, u^{(m-1)}(\tau, y_\tau^{(m)}(t,\tau)), y_\tau^{(m)}(t,y))) d\tau + \\
&+ \sum_{\tau_i < t} \Phi_+(t,\tau_i) I_i^{(2)}(u^{(m-1)}(\tau_i, y_{\tau_i}^{(m)}(t,y)), y_{\tau_i}^{(m)}(t,y))),
\end{aligned} \tag{5.113}$$

where $y_t^{(m)}(t_0, y_0)$ is a general solution of the system

$$\begin{aligned}
\frac{dy}{dt} &= A_-(t)y + G(t, u_-^{(m-1)}(t,y), y), \qquad t \neq \tau_i, \\
\Delta y|_{t=\tau_i} &= B_i^- y + I_i^{(2)}(u_-^{(m-1)}(\tau_i, y), y).
\end{aligned} \tag{5.114}$$

For sufficiently small values of the Lipschitz constant N, the sequences of functions $y_t^{(m)}(t_0, y_0)$, $u_-^{(m-1)}(t,y)$ converge uniformly and the limit function $u_-(t,y) = \lim_{m\to\infty} u_-^{(m)}(t,y)$ defines for equation (5.51) the integral set Γ_t^- : $x = u_-(t,y)$.

The functions $u_-(t,y)$ and $y_t(t_0,y) = \lim_{t\to\infty} y_t^{(m)}(t_0,y)$ satisfy the inequalities

$$\|u_-(t,y') - u_-(t,y'')\| \leq 2K^2 N a \|y' - y''\|, \tag{5.115}$$

$$\|y_t(t_0,y') - y_t(t_0,y'')\| \leq K_0 e^{\sigma(t-t_0)} \|y' - y''\| \tag{5.116}$$

and the relations

$$\begin{aligned} y_t(t_0, y_0) &= \Phi_-(t, t_0)y_0 + \\ &+ \int_{t_0}^{t} \Phi_-(t, \tau)G(\tau, u_-(\tau, y_\tau(t_0, y_0)), y_\tau(t_0, y_0))d\tau + \\ &+ \sum_{t_0<\tau_i<t} \Phi_-(t, \tau_i)I_i^{(2)}(u_-(\tau_i, y_{\tau_i}(t_0, y_0)), y_{\tau_i}(t_0, y_0)), \\ u_-(t, y) &= \int_{-\infty}^{t} \Phi_+(t, \tau)F(\tau, u_-(\tau, y_\tau(t, y)), y_\tau(t, y))d\tau + \\ &+ \sum_{t_0<\tau_i<t} \Phi_+(t, \tau_i)I_i^{(1)}(u_-(\tau_i, y_{\tau_i}(t, y)), y_{\tau_i}(t, y)). \end{aligned}$$

Restricted to the integral set Γ_-^m, system (5.51) is equivalent to the following impulsive equations

$$\begin{aligned} \frac{dx}{dt} &= A_-(t)y + G(t, u_-(t, y), y), \qquad t \neq \tau_i \\ \Delta y|_{t=\tau_i} &= B_i^- y + I_i^{(2)}(u_-(\tau_i, y), y). \end{aligned} \tag{5.117}$$

For each solution $y_t(t_0, y_0)$ of these equations, the following estimate is true,

$$\|y_t(t_0, x_0)\| \leq K_0 e^{\sigma(t-t_0)} \|y_0\|, \qquad t \leq t_0, \qquad y_0 \in \mathbf{R}^{n-k}. \tag{5.118}$$

Now we can make the following statement.

Theorem 91 *Let the functions, which define system (5.51) with a hyperbolic linear part, satisfy inequalities (5.52) and relations (5.72) and let there exist the finite limit (5.2). Then there exists such a positive number N^0 that, for all $0 \leq N \leq N^0$, system (5.51) has an integral set $\Gamma_t^- : x = u_-(t, y)$, where $u_-(t, y)$ is a function piecewise continuous with respect to t with first kind discontinuities at $t = \tau_i$ and satisfying inequality (5.115). Restricted to the integral set Γ_-^m, system (5.51) is reduced to equations (5.117), solutions of which satisfy inequality (5.118) for $t \leq t_0$.*

Consider the general case of system (5.51), i.e. when the functions $F(t, x, y), G(t, x, y), I_i^{(1)}(x, y), I_i^{(2)}(x, y)$ do not take the zero value at the origin. However, we will assume that these functions are bounded for $x = y = 0$, i.e. they satisfy inequality (5.53). In this case, the integral sets Γ_i^+ and Γ_i^- of system (5.51) exist as well.

Theorem 92 *Let system (5.51), the linear pert of which is hyperbolic, be such that inequalities (5.52), (5.53) and relation (5.2) hold.*

For a sufficiently small Lipschitz constant N there exist integral sets

$$\Gamma_+ : y = u_+(t,x) \text{ and } \Gamma_- : x = u_-(t,y),$$

where $u_+(t,x)$ and $u_-(t,x)$ are $(n-k)$- and k-dimensional functions respectively, piecewise continuous with respect to t with first kind discontinuities at $t = \tau_i$ and satisfying the inequalities

$$\|u_+(t,0)\| \leq M_0, \qquad \|u_-(t,0)\| \leq M_0, \tag{5.119}$$

$$\|u_+(t,x') - u_+(t,x'')\| \leq \beta N\|x' - x''\|, \tag{5.120}$$

$$\|u_-(t,y') - u_-(t,y'')\| \leq \beta N\|y' - y''\|, \tag{5.121}$$

where M_0, β are some positive constants.

For any solution $(x_t(t_0,x_0), u_+(t, x_t(t_0,x_0)))$, lying in the set Γ_+^+, and any solution $(u_-(t, y_t(t_0,y_0)), y_t(t_0,y_0))$, lying in the set Γ_-^-, the following estimates hold,

$$\begin{aligned}&\|x_t(t_0,x_0) - x^*(t)\| + u_+(t, x_t(t_0,x_0)) - y^*(t)\| \leq \\ &\leq K^* e^{-\sigma(t-t_0)}\left[\|x_0 - x^*(t_0)\| + \|u_+(t_0,x_0) - y^*(t_0)\|\right], \\ &t \geq t_0,\end{aligned} \tag{5.122}$$

$$\begin{aligned}&\|u_-(t, y_t(t_0,y_0)) - x^*(t)\| + \|y_t(t_0,y_0) - y^*(t)\| \leq \\ &\leq K^* e^{\sigma(t-t_0)}\left[\|u_-(t_0,y_0) - x^*(t_0)\| + \|y_0 - y^*(t_0)\|\right], \\ &t \leq t_0,\end{aligned} \tag{5.123}$$

where K^*, $0 < \sigma < \gamma$, $\sigma = \sigma(N)$ are positive constants and $(x^*(t), y^*(t))$ are solutions of equations (5.51), bounded on the whole axis.

Proof. By Theorem 5.29, system (5.51) has a unique solution $x = x^*(t)$, $y = y^*(t)$, bounded on the whole axis, if the Lipschitz constant N is sufficiently small.

Let us make in (5.51) the change of variables

$$x = \eta + x^*(t), \qquad y = \xi + y^*(t). \tag{5.124}$$

Then system (5.51) becomes

$$\begin{aligned}\frac{d\eta}{dt} &= A_+(t)\eta + F_1(t,\eta,\xi), \\ \frac{d\xi}{dt} &= A_-(t)\xi + G_1(t,\eta,\xi), \ ,t \neq \tau_i, \\ \Delta\eta_{t=\tau_i} &= B_i^+\eta + I_i^{(3)}(\eta,\xi), \qquad \Delta\xi_{t=\tau_i} = B_i^-\xi + I_i^{(4)}(\eta,\xi),\end{aligned} \tag{5.125}$$

where

$$\begin{aligned}
F_1(t,\eta,\xi) &= F(t,\eta+x^*(t),\,\xi+y^*(t)) - F(t,x^*(t),\,y^*(t)),\\
G_1(t,\eta,\xi) &= G(t,\eta+x^*(t),\,\xi+y^*(t)) - G(t,x^*(t),\,y^*(t)),\\
I_i^{(3)}(\eta,\xi) &= I_i^{(1)}(\eta+x^*(\tau_i),\,\xi+y^*(\tau_i)) - I_i^{(1)}(x^*(\tau_i),y^*(\tau_i)),\\
I_i^{(4)}(\eta,\xi) &= I_i^{(2)}(\eta+x^*(\tau_i),\,\xi+y^*(\tau_i)) - I_i^{(2)}(x^*(\tau_i),y^*(\tau_i)),
\end{aligned}$$

with

$$F_1(t,0,0) = G_1(t,0,0) = I_i^{(3)}(0,0) == I_i^{(4)}(0,0) = 0$$

and the functions $F_1(t,\eta,\xi)$, $G_1(t,\eta,\xi)$, $I_i^{(3)}(\eta,\xi)$, $I_i^{(4)}(\eta,\xi)$ satisfying the inequalities of type (5.52) with the same Lipschitz constant N. By Theorems 91 and 92, we see that if the value of N is sufficiently small, then system (5.125) has integral surfaces $\tilde{\Gamma}_+^+ : \xi = \tilde{u}_+(t,\eta)$ and $\tilde{\Gamma}_-^- : \eta = u_-(i,\xi)$ which have the following properties:

$$\begin{aligned}
&\tilde{u}_+(t,0) = 0, \qquad \tilde{u}_-(t,0) = 0,\\
&\|\tilde{u}_+(t,\eta') - \tilde{u}_+(t,\eta'')\| \le \beta N\|\eta' - \eta''\|,\\
&\|\tilde{u}_-(t,\xi') - \tilde{u}_-(t,\xi'')\| \le \beta N\|\xi' - \xi''\|,
\end{aligned} \tag{5.126}$$

where β is determined only by the linear part of equations (5.125).

Any solution $(\eta(t),\xi(t))$ of equations (5.125) with the initial conditions (t_0,η_0,ξ_0) that satisfy $\xi_0 = \tilde{u}_+(t_0,\eta_0)$ is subject to the inequality

$$\|\eta(t)\| + \|\xi(t)\| \le K^* e^{-\sigma(t-t_0)}[\|\xi(t_0)\| + \|\eta(t_0)\|], \qquad t \ge t_0, \tag{5.127}$$

and any solution $(\eta(t)\xi(t))$, the initial conditions of which satisfy $\eta_0 = u_-(t_0,\xi_0)$, is subject to

$$\|\eta(t)\| + \|\xi(t)\| \le K^* e^{\sigma(t-t_0)}[\|\xi(t_0)\| + \|\eta(t_0)\|] \tag{5.128}$$

for $t \le t_0$.

Set

$$\begin{aligned}
u_+(t,x) &= y^*(t) + \tilde{u}_+(t,x - x^*(t)),\\
u_-(t,y) &= x^*(t) + \tilde{u}_-(t,y - y^*(t)).
\end{aligned}$$

Using the change of variables (5.124) and relations (5.53), (5.126) – (5.128) we see that the functions $u_+(t,x)$ and $x = u_-(t,x)$ satisfy conditions of the theorem and a solution of equations (5.51), which lies in the integral sets $\Gamma_+ : y = u_+(t,x)$ and $\Gamma_- : x = u_-(t,y)$ satisfies estimate (5.122).

□

At the end of this section we will prove that the functions $u_+(t,x)$ and $u_-(t,y)$ that define the integral sets Γ_+ and Γ_- are T-periodic in t if system (5.51) is T-periodic, i.e. if the vector functions $F(t,x,y)$, $G(t,x,y)$ and the matrix functions $A_+(t)$ and $A_-(t)$ are T-periodic and the equalities

$$B^+_{i+p} = B^-_i, \qquad B^-_{i+p} = B^-_i,$$
$$I^{(1)}_{i+p}(x,y) = I_i(1)(x,y), \qquad I^{(2)}_{i+p}(x,y) = I_i(2)(x,y),$$
$$\tau_{i+p} = \tau_i + T \tag{5.129}$$

hold for some natural p.

Let us show that the function $u_+(t,x)$ is T-periodic. It will be sufficient to show that the functions $u^m_+(t,x)$ are T-periodic in t because the sequence $\{u^m_+(t,x)\}$ converges to $u_+(t,x)$ uniformly for $m \to \infty$.

For $m = 1$, the solution $x^{(1)}_t(t_0, x_0)$ of the T-periodic system (5.76) satisfies the identity $x^{(1)}_{t+T}(t_0 + T, x_0) = x^{(1)}_t(t_0, x_0)$, hence it follows from (5.75) that

$$\begin{aligned}
u^{(1)}(t+T,x) &= -\int_{t+T}^{\infty} \Phi_-(t+T,\tau)G(\tau, x^{(1)}_\tau(t+T,x),0)d\tau - \\
&- \sum_{\tau_i > t+T} \Phi_-(t+T,\tau_i)I^{(2)}_i(x^{(1)}_{\tau_i}(t+T,x),0) = \\
&= -\int_{t}^{\infty} \Phi_-(t+T,\tau+T)G(\tau+T, x^{(1)}_{\tau+T}(t+T,x),0)d\tau - \\
&- \sum_{\tau_i > t} \Phi_-(t+T,\tau_j+T)I^{(2)}_{i+p}(x^{(1)}_{\tau_j+T}(t+T,x),0) = \\
&= -\int_{t}^{\infty} \Phi_-(t,\tau)G(\tau, x^{(1)}_\tau(t,x),0)d\tau - \\
&- \sum_{\tau_j > t} \Phi_-(t,\tau_j)I^{(2)}_j(x_{\tau_j}(t,x),0) = u^{(1)}(t,x),
\end{aligned}$$

i.e. the function $u^{(1)}_+(t,x)$ is T-periodic in t. Assume that the functions $u^{(2)}_+$, $\ldots$, $u^{(m-1)}_+$ are also T-periodic in t. Then system (5.78) is T-periodic and the solution $x^{(m)}_t(t_0,x_0)$ of this system satisfies the identity $x^{(m)}_{t+T}(t_0+T,x_0) =$

$x_t^{(m)}(t_0, x_0)$. By using this identity, relations (5.129), and the equality $\Phi_-(t+T, \tau+T) = \Phi_-(t, \tau)$, we readily see from (5.77) that the function $u_+^{(m)}(t,x)$ is T-periodic. By induction we conclude that all the functions $u_+^{(m)}(t,x)$, $m = 1, 2, \ldots$, are T-periodic in t and, hence, such is the limit function $u_+(t,x)$. In the same way we can prove that the function $u_-(t,y)$ is T-periodic.

5.4 Integral sets of a certain class of discontinuous dynamical systems

In the preceding sections we studied the problem of existence of integral sets of differential equations with an impulsive effect that occurs at fixed times. Now we will study the question of existence on integral sets of a certain class of linear and weakly nonlinear differential systems that undergo an impulsive effect when the phase point meets a given set in the phase space.

Consider the system

$$\frac{d\varphi}{dt} = \omega, \qquad \frac{dx}{dt} = A(\varphi)x + f(\varphi), \qquad \varphi \notin \Gamma,$$
$$\Delta|_{\varphi \in \Gamma} = I(\varphi), \tag{5.130}$$

where $x = (x_1, x_2, , x_n)$, $\varphi = (\varphi_1, \varphi_2, \ldots, \varphi_m)$, $\omega = (\omega_1, \omega_2, \ldots \omega_m)$ is a vector with positive coordinates, $f(\varphi)$ and $I(\varphi)$ are functions continuous (piecewise continuous with first kind discontinuities on the set Γ), 2π-periodic in each of their variables φ_j, $j = \overline{1, m}$, $A(\varphi)$ is a continuous square matrix, 2π-periodic with respect to each variable φ_j.

We will regard the point $\varphi = (\varphi_1, \ldots, \varphi_m)$ as a point of an m-dimensional torus Γ^m so that the domain of the functions $A(\varphi)$, $f(\varphi)$, and $I(\varphi)$ will be the torus Γ^m. We assume that the set Γ is a subset of the torus Γ^m, which is a manifold of dimension $m - 1$ defined by an equation $\Phi(\varphi) = 0$ for some continuous scalar function $\Phi(\varphi)$, 2π-periodic with respect to φ.

Denote by $t = t_i(\varphi)$ solutions of the equation

$$\Phi(\varphi + \omega t) = 0. \tag{5.131}$$

Let the function $\Phi(\varphi)$ be such that equation (5.131) has solutions $t = t_i(\varphi)$ since otherwise system (5.130) would not be an impulsive system but an ordinary dynamical system.

Lemma 45 *For any solution $t = t_i(\varphi)$ of equation (5.131), the equality*

$$t_i(\varphi - \omega t) - t_i(\varphi) = t \tag{5.132}$$

holds for all $\varphi \in \Gamma^m$, $t \in \mathbf{R}$.

Proof. If $t_i(\varphi)$ is a solution of (5.131), then, for all $\varphi \in \Gamma^m$, we have $\Phi(\varphi + \omega t_i(\varphi)) = 0$. For any $t \in \mathbf{R}$, the point $\varphi - \omega t$ belongs to Γ^m. Thus, by replacing φ by $\varphi - \omega t$, we get $\Phi(\varphi - \omega t + \omega t_i(\varphi - \omega t)) = 0$. Whence it follows that, for some j and all $\varphi \in \Gamma^m$, $t \in \mathbf{R}$,

$$t_i(\varphi - \omega t) - t = t_j(\varphi). \tag{5.133}$$

If $t = 0$, it follows from (5.133) that $t_i(\varphi) = t_j(\varphi)$ for any $\varphi \in \Gamma^m$ and, consequently, $i = j$, which proves the lemma.

□

Let $G(t, \tau, \varphi)$, $G(\tau, \tau, \varphi) = E$, be a fundamental matrix of the system

$$\frac{dx}{dt} = A(\varphi + \omega t)x. \tag{5.134}$$

It can be readily verified that $G(t, \tau, \varphi)$ satisfies the equalities

$$\begin{gathered} G(t, \tau, \varphi + 2\pi) = G(t, \tau, \varphi), \\ G(t, t + \tau, \varphi - \omega t) = G(0, \tau, \varphi), \end{gathered} \tag{5.135}$$

for all $t, \tau \in \mathbf{R}$ and $\varphi \in \Gamma_m$.

Let the matrix $G(t, \tau, \varphi)$ and the functions $t_i(\varphi)$ be such that the functions

$$\begin{aligned} x_t(\varphi) \;\; &= \;\; \int_{-\infty}^{t} G(t, \tau, \varphi) f(\varphi + \omega\tau) d\tau + \\ &\quad + \sum_{t_i(\varphi) < t} G(t, t_i(\varphi), \varphi) I(\varphi + \omega t_i(\varphi)), \end{aligned} \tag{5.136}$$

which depend on φ as a parameter, are defined for all $t \in \mathbf{R}$ and uniformly bounded. Sufficient conditions for convergence of the integral and the sum in (5.136) will be given later.

Set $x_t(\varphi) = u(\varphi + \omega t)$ and replace in (5.136) φ by $\varphi - \omega t$. Then, by using (5.132) and (5.135), we get

$$\begin{aligned} u(\varphi) &= \int_{-\infty}^{t} G(t, \tau, \varphi - \omega t) f(\varphi + \omega(\tau - t)) d\tau + \\ &\quad + \sum_{t_i(\varphi - \omega t) < t} G(t, t_i(\varphi - \omega t), \varphi - \omega t) I(\varphi + \omega(t_i(\varphi - \omega t) - t)) = \\ &= \int_{-\infty}^{0} G(0, \tau, \varphi) f(\varphi + \omega\tau) d\tau + \\ &\quad + \sum_{t_i(\varphi) < 0} G(0, t_i(\varphi), \varphi) I(\varphi + \omega t_i(\varphi)). \end{aligned} \tag{5.137}$$

With the assumption that the integral and the sum in (5.136) converge uniformly, we see that the function $u(\varphi)$ defines the invariant set of system (5.130),

$$x = u(\varphi), \qquad u(\varphi + 2\pi) = u(\varphi). \tag{5.138}$$

Indeed, a direct verification shows that $x_i(\varphi) = u(\varphi + \omega t)$ satisfies the equation

$$\frac{dx}{dt} = A(\varphi + \omega t)x + f(\varphi + \omega t)$$

if $t \neq t_i(\varphi)$, i.e. if $\varphi + \omega t \notin \Gamma$, and the jump condition

$$\Delta x = u(\varphi + \omega(t_i + 0)) - u(\varphi + \omega(t_i - 0)) = I(\varphi + \omega t_i(\varphi))$$

if $t = t_i(\varphi)$, which means that $(x_i(\varphi), \varphi + \omega t)$ is a solution of system (5.130).

Note that, for the integral and the sum in (5.136) to be convergent, it is sufficient that the function $G(t, \tau, \varphi)$ satisfy the inequality

$$\|G(t, \tau, \varphi)\| \leq K e^{-j(t-\tau)} \tag{5.139}$$

for all $t, \tau \in \mathbf{R}$, $t \geq \tau$, $\varphi \in \Gamma^m$, some positive K and γ, and solutions of equation (5.131) be such that

$$t_{i+1}(\varphi) - t_i(\varphi) \geq \theta > 0 \tag{5.140}$$

for all $i \in \mathbf{Z}$, $\varphi \in \Gamma^m$, and some $\theta > 0$.

Using these conditions we have from (5.136) that

$$\|x_t(\varphi)\| \leq \frac{K}{\gamma} \max_{\varphi \in \Gamma_m} \|f(\varphi)\| + \frac{K}{1 - e^{-\gamma\theta}} \max_{\varphi \in \Gamma_m} \|I(\varphi)\| \tag{5.141}$$

for all $t \in \mathbf{R}$ and $\varphi \in \Gamma^m$. So we have the following statement.

Theorem 93 *Let in system (5.130) the 2π-periodic functions $f(\varphi)$ and $I(\varphi)$ be continuous on Γ^m (piecewise continuous with first kind discontinuities at $\varphi \in \Gamma$), the matrix $A(\varphi)$ be continuous on Γ^m and 2π-periodic.*

If the matrix $G(t, \tau, \varphi)$ satisfies estimate (5.139) and the functions $t_i(\varphi)$ satisfy inequality (5.140), then system (5.130) has an asymptotically stable invariant set

$$x = u(\varphi), \qquad u(\varphi + 2\pi) = u(\varphi).$$

where $u(\varphi)$ is a function piecewise continuous with first kind discontinuities on Γ. Also, there is such a positive constant C which does not depend on the function $f(\varphi)$ and $I(\varphi)$, that

$$\|u(\varphi)\| \leq C \max\{\max_{\varphi \in \Gamma^m} \|f(\varphi)\|, \max_{\varphi \in \Gamma^m} \|I(\varphi)\|\}. \tag{5.142}$$

Proof. If conditions of the theorem hold, the invariant set $x = u(\varphi)$ is defined by the function $u(\varphi)$ from relation (5.137). Estimate (5.142) follows from (5.141) if we set

$$C = \max\left\{\frac{K}{\gamma}, , \frac{k}{1 - e^{-\gamma\theta}}\right\}.$$

The invariant set is asymptotically stable since, by inequality (5.138), any solution $(x = x_t(x_0), \varphi_t = \varphi_0 + \omega t)$ of the initial equation is attracted to the solution $(u(\varphi_0 + \omega t), \varphi_0 + \omega t)$ that lies in the invariant set.

□

Consider the case when the matrix $G(t, \tau, \varphi)$ and the roots of equation (5.138) satisfy inequalities (5.139) and (5.140) respectively. Denote by $\Lambda(\varphi)$ the greatest eigen value of the symmetric matrix $A^*(\varphi) = \frac{1}{2}(A(\varphi) + A'(\varphi))$, where $A'(\varphi)$ is a transpose of the matrix $A(\varphi)$. By the Wazhevky inequality, for any solution of the system

$$\frac{dx}{dt} = A(\varphi + \omega t)x, \tag{5.143}$$

we have

$$\begin{aligned} \|x(t)\| &\le e^{\int_\tau^t \Lambda(\varphi+\omega\sigma)d\sigma}\|x(\tau)\|, \qquad t \ge \tau, \\ &\|x^2\| = \langle x, x\rangle. \end{aligned} \tag{5.144}$$

Hence, if the limit

$$\lim_{T\to\infty}\frac{1}{T}\int_t^{t+T}\Lambda(\varphi + \omega\sigma)d\sigma = \lambda \tag{5.145}$$

exists and is uniform with respect to $t \in \mathbf{R}$ and $\lambda < 0$, then the matrix $G(t, \tau, \varphi)$ satisfies inequality (5.139).

If the coordinates of the vector ω are rationally independent, i.e. $\langle k, \omega\rangle = 0$ with $k = (k_1, \dots, k_m)$ being a vector with integer coordinates is possible only for the zero vector k, then trajectories of the equation $\varphi = \omega$ are uniformly distributed on the torus Γ^m and the time average of the function $\Lambda(\varphi)$ equals to the space average of this function. Hence, if

$$\frac{1}{(2\pi)^m}\int_0^{2\pi}\cdots\int_0^{2\pi}\Lambda(\varphi)d\varphi_1\dots d\varphi_m < 0, \tag{5.146}$$

then the matrix $G(t,\tau,\varphi)$ satisfies inequality (5.139).

Suppose the function $\Phi(\varphi)$ that defines the set Γ is given by $\Phi(\varphi) = \Phi_0(\langle a,\varphi\rangle)$, where $\Phi_0(s)$ is a function 2π-periodic in s with a finite number of zeros p in the period, $a = (a_1, \dots a_m)$ is a vector with positive integer components. Let us show that, with such assumptions, solutions of equation (5.131) satisfy condition (5.140). Indeed, let $s_1, \dots, s_p$ be roots of the equation $\Phi_0(s) = 0$, which belong to the interval $[0, 2\pi]$. To every root s_j there corresponds a family of solutions of equation (5.131),

$$t = t_k^j(\varphi) = \frac{-\langle a,\varphi\rangle + s_j + 2k\pi}{\langle a,\omega\rangle}. \tag{5.147}$$

It is clear that the set of solutions $\{t_k^j(\varphi)\}$, $k \in \mathbf{Z}$, $j = \overline{1,p}$ can be indexed by a single index in such a way that $t_i(\varphi) \to \infty$ as $i \to +\infty$ and $t_i(\varphi) \to -\infty$ as $i \to -\infty$, and the equalities

$$\begin{aligned} t_{i+p}(\varphi) - t_i(\varphi) &= \frac{2\pi}{\langle a,\omega\rangle}, \\ t_{i+1}(\varphi) - t_i(\varphi) &= \frac{s_{j+1} - s_j}{\langle a,\omega\rangle} \end{aligned} \tag{5.148}$$

hold for all $i \in \mathbf{Z}$, $\varphi \in \Gamma^m$, and some $j = \overline{1,p}$. It follows from (5.148) that $t_{i+1}(\varphi) - t_i(\varphi) \geq \theta > 0$ if we take θ to be

$$\frac{1}{\langle a,\omega\rangle} \min_{1\leq j\leq p} (s_{j+1} - s_j).$$

Now we will give sufficient conditions for existence of integral sets of weakly nonlinear impulsive differential system

$$\begin{aligned} \frac{d\varphi}{dt} = \omega, \qquad \frac{dx}{dt} &= A(\varphi)x + f(\varphi,x), \qquad \varphi \notin \Gamma, \\ \Delta|_{\varphi\in\Gamma} &= I(\varphi,x). \end{aligned} \tag{5.149}$$

Here the matrix $A(\varphi)$ and the set Γ are as in system (5.130), the functions $f(\varphi,x)$ and $I(\varphi,x)$ are defined for all $\varphi \in \Gamma^m$, $x \in \mathbf{R}^n$, are continuous (piecewise continuous with first kind discontinuities at the set Γ), 2π-periodic in φ, and satisfy the Lipschitz conditions with respect to x uniformly in $\varphi \in \Gamma^m$,

$$\|f(\varphi,x') - f(\varphi,x'')\| + \|I(\varphi,x') - I(\varphi,x') - I(\varphi,x'')\| \leq N\|x' - x''\| \tag{5.150}$$

for all $x', x'' \in \mathbf{R}^n$.

We have the following statement.

Theorem 94 *If the matrix $G(t,\tau,\varphi)$ satisfies inequality (5.139) and the solutions $t_i(\varphi)$ of equation (5.131) are such that inequality (5.140) holds, then, for a sufficiently small value of the Lipschitz constant N, system (5.149) has an asymptotically stable invariant set*

$$x = u(\varphi), \qquad u(\varphi + 2\pi) = u(\varphi),$$

where $u(\varphi)$ is a function piecewise continuous with first kind discontinuities at $\varphi \in \Gamma$ and

$$\Delta|_{\varphi_t(\varphi)\in\Gamma} = I(\varphi_t(\varphi), \ , u(\varphi_t(\varphi)))|_{\varphi_t(\varphi)\in\Gamma}, \qquad \varphi_t(\varphi) = \varphi + \omega t. \tag{5.151}$$

Proof. As in section 5.3, we look for an integral set of system (5.149) as a limit of the sequence of sets

$$\mathfrak{M}^{(k)} : x = u^{(k)}(\varphi), \quad \varphi \in \Gamma^m, \quad k = 1,2,\ldots, \quad u^0(\varphi) \equiv 0, \tag{5.152}$$

each of which is an integral set of the system

$$\frac{d\varphi}{dt} = \omega, \qquad \frac{dx}{dt} = A(\varphi)x + f(\varphi, u^{(k-1)}(\varphi)), \qquad \varphi \notin \Gamma,$$
$$\Delta|_{\varphi\in\Gamma} = I(\varphi, u^{(k-1)}(\varphi)). \tag{5.153}$$

By Theorem 93, system (5.153) has the integral set

$$x = u^{(k)}(\varphi) = \int_{-\infty}^{0} G(0,\tau,\varphi) f(\varphi + \omega\tau, u^{(k-1)}(\varphi + \omega\tau))d\tau +$$
$$+ \sum_{t_i(\varphi)<0} G(0, t_i(\varphi), \varphi) I(\varphi + \omega t_i(\varphi), u(k-1)(\varphi + \omega t_i(\varphi))). \tag{5.154}$$

for every $k = 1,2,\ldots$. Assuming that the Lipschitz constant is so small that $2NC < 1$, by (5.154) using (5.139), (5.140), (5.142), (5.150), we have

$$\|u(1)(\varphi)\|_0 \equiv \max_{\varphi\in\Gamma^m} \|u(1)(\varphi)\| \leq 2C \max\{f(\varphi,0)\|_0, \ , \|I(\varphi,0)\|_0]$$
$$\|u^{(k)}(\varphi)\|_0 \leq \|u^{(k)}(\varphi) - u^{(1)}(\varphi)\|_0 + \|u^{(1)}(\varphi)\|_0 \leq$$
$$\leq \frac{1}{1-2NC}\|u^{(1)}(\varphi)\|_0 \leq \tag{5.155}$$
$$\leq 2\frac{C}{1-2NC}\max\{\|f(\varphi,0)\|_0, \|I(\varphi,0)\|_0\}. \tag{5.156}$$

Moreover,

$$\|u^{(k+1)}(\varphi) - u^{(k)}(\varphi)\|_0 \leq 2NC\|u^{(k)}(\varphi) - u^{(k-1)}(\varphi)\|_0. \tag{5.157}$$

If $2NC < 1$, estimates (5.155) and (5.157) imply that the sequence of functions $u^{(k)}(\varphi)$ converges uniformly. Let

$$u(\varphi) = \lim_{k\to\infty} u^{(k)}(\varphi). \tag{5.158}$$

Since the convergence of the sequence $u^{(k)}(\varphi)$ is uniform, the limit function $u(\varphi)$ is 2π-periodic, piecewise smooth with first kind discontinuities at the set Γ.

Passing to a limit in (5.154) for $k \to \infty$ we see that the function $u(\varphi)$ satisfies the equality

$$\begin{aligned} u(\varphi) &= \int_{-\infty}^{0} G(0,\tau,\varphi) f(\varphi+\omega\tau, u(\varphi+\omega\tau))d\tau + \\ &+ \sum_{t_i(\varphi)<0} G(0,t_i(\varphi),\varphi) I(\varphi+\omega t_i(\varphi), u(\varphi+\omega t_i(\varphi))). \end{aligned} \tag{5.159}$$

Whence it easy to see that $x = u(\varphi_y(\varphi))$ satisfies the equality

$$\frac{dx}{dt} = A(\varphi+\omega t)x + f(\varphi+\omega t,\, x),$$

for $\varphi_t(\varphi) \neq \Gamma$, i.e $t \neq t_i(\varphi)$, and the condition

$$x(t+0) - x(t-0) = I(\varphi_t(\varphi),\, x(t)).$$

for $\varphi_t(\varphi) \in \Gamma$, i.e. $t = t_i(\varphi)$. Hence, $(x(t), \varphi_t(\varphi))$ is a solution of system (5.149). This means that $x = u(\varphi)$ defines an integral set of system (5.149).

To complete the proof, it remains to show that the invariant set is asymptotically stable.

Let $(y(t), \varphi + \omega t)$ be an arbitrary solution of equations (5.149) and $(x(t), \varphi+\omega t)$ – a solution of these equations, lying in the integral set. Because

$$\begin{aligned} y(t) - x(t) &= y(t) - u(\varphi+\omega t) = G(t,0,\varphi)(y(0) - u(\varphi)) + \\ &+ \int_0^t G(t,\tau,\varphi)\big[f(\varphi+\omega\tau, y(\tau)) - f(\varphi+\omega\tau, u(\varphi+\omega\tau))\big]d\tau + \\ &+ \sum_{0<t_i(\varphi)<t} G(t,t_i(\varphi),\varphi)\big[I(\varphi+\omega t_i(\varphi), y(t_i(\varphi))) - \\ &- I(\varphi+\omega t_i(\varphi), u(\varphi+\omega t_i(\varphi)))\big], \end{aligned}$$

by inequalities (5.139) and (5.150), we have

$$\begin{aligned}\|y(t)-u(\varphi+\omega t)\| &\leq Ke^{-\gamma t}\|y(0)-u(\varphi)\|+\\ &+KN\int_0^t e^{-\gamma(t-\tau)}\|y(\tau)-u(\varphi+\omega\tau)\|d\tau+\\ &+KN\sum_{0<t_i(\varphi)<t} e^{-\gamma(t-t_i(\varphi))}\|y(t_i(\varphi))-u(\varphi+\omega t_i(\varphi))\|.\end{aligned}$$

Whence, by Lemma 1, we have for the function $\|y(t)-u(\varphi+\omega t)\|e^{\gamma^t}$ the following estimate

$$\|y(t)-u(\varphi+\omega t)\|e^{\gamma t} \leq K(1+KN)^{i(t)}e^{KNt}\|y(0)-u(\varphi)\|, \tag{5.160}$$

where $i(t)$ denotes the number of the points $t_i(\varphi)$ in the interval $[0,t[$.

By using (5.140) we get from (5.160) that

$$\|y(t)-u(\varphi+\omega t)\| \leq Ke^{-(\gamma-KN-\frac{1}{\theta}\ln(1+KN))t}\times\|y(0)-u(\varphi)\|,$$

i.e. $\|y(t)-u(\theta+\omega t)\| \to 0$ for $t\to\infty$ and sufficiently small Lipschitz constant N.

Chapter 6

Optimum control in impulsive systems

6.1 Formulation of the problem. Auxiliary results.

In succeeding sections of this chapter, a central result of the optimum control theory, the Pontryagin maximum principle [117], is extended to problems of optimum control in systems of impulsive differential equations. Necessary conditions for optimum control are obtained for problems of optimum control in differential systems that have an impulsive effect both at fixed times and in the case when the impulsive effect occurs at the moment the phase point meets a given hypersurface in the phase space. The main results of this chapter are contained in [16, 15, 14].

Consider the problem of optimum control in a differential system subject to an impulsive effect at the moment when the phase point $x(t)$ meets a given smooth hypersurface in the phase space.

Let, in a phase space $\mathbf{M} \subseteq \mathbf{R}^n$, a smooth hypersurface S be given by an equation $s(x) = 0$. For the impulsive differential system

$$\frac{dx}{dt} = f(x, u), \qquad x \notin S,$$
$$\Delta x|_{x \in S} = g(x, w) \tag{6.1}$$

with a control, the following optimum control problem is considered: to find amongst all admissible controls $\{u(t), w_1, \dots, w_{N(T)}\}$ that transfer the system from a state x^0 into a state x^1 over a time interval $[T_0, T]$ the one that minimizes the functional

$$\mathcal{J} = \int_{T_0}^{T} f_0(x(s), u(s))ds + \sum_{i=1}^{N(T)} g_0(x(\tau_i), w_i). \tag{6.2}$$

Here τ_i, $i = \overline{1, N(T)}$ are the times at which the phase point $x(t)$ reaches the hypersurface of the trajectory discontinuity, i.e. the roots of the equation $s(x(t)) = 0$, which belong to the interval $]T_0, T[$, $N(T)$ is the number of such roots. Note that the time T is not fixed but given by the condition that the phase point $x(t)$ reaches the point x^1, i.e. $x(T) = x^1$.

The vectors $u = (u_1, \dots, u_m)$, $w_i = (w_{i_1}, \dots, w_{i_r}$ are actuating effects that take the values in the sets $U \subset \mathbf{R}^m$ and $W \subset \mathbf{R}^r$ respectively. Here U is an arbitrary set in $\mathbf{R}^m$ and W is an arbitrary convex set in $\mathbf{R}^r$. Admissible functions $u(t)$ are piecewise continuous functions defined on the interval $[T_0, T]$ and taking values in U. It is assumed that the functions $u(t)$ are continuous from the left at discontinuity points and continuous at the ends of the control interval.

For definiteness, we assume that a solution of system (6.1) is continuous from the left at points of discontinuity. Hence, at the time $t = T$, there are no impulsive effects in the system even though the phase point meets a hypersurface of discontinuity.

For the functions that define the controlled system (6.1) and functional (6.2), we make the following assumptions:

a) components of the vector $f(x,u)$ and the matrix $\frac{\partial f}{\partial x}(x,u)$ are defined and continuous on the Cartesian product $\mathbf{M} \times \bar{U}$;

b) components of the vector $g(x,w)$ and the function $g_0(x,w)$ are defined and continuously differentiable on the set $\mathbf{M} \times \bar{W}$ and have bounded second order partial derivatives;

c) for every fixed $w \in W$, the mapping $x \to x + g(x,w)$ is one-to-one;

d) the function $s(x)$ is continuously differentiable and has bounded second derivatives;

e) at the times when the phase point reaches a surface of discontinuity, for any values of the actuating effects, the following conditions hold,

$$\langle \operatorname{grad} s(x(t)),\ f(x(t),\ u(t)) \rangle|_{t=\tau_i} \neq 0,$$
$$\langle \operatorname{grad} s(\tilde{g}^{-1}(x(t),w_i), f(x(t),u(t)) \rangle|_{t=\tau_i} \neq 0, \qquad i = \overline{1, N(T)},$$

where $\tilde{g}^{-1}(x,w)$ is the mapping inverse to the mapping $x \to x + g(x,w)$.

Following [117], we will add one more coordinate x_0 to the coordinates $x_1, \ldots, x_n$ and define its dependence on time by

$$\frac{dx_0}{dt} = f_0(x,u), \qquad x \notin S,$$
$$\Delta x_0|_{x \in S} = g_0(x,w).$$

Denoting by X the vector $X = (x_0, x_1, \ldots x_n)$ we can restate the initial problem as follows. A cylindrical surface$s(X) = s(x) = 0$ and a line Π parallel to the axis X_0 and passing through the point $X' = (0, x')$ are given in a $(n+1)$-dimensional phase space. The controlled process is described by the impulsive differential system

$$\frac{dX}{dt} = F(X,u), \qquad x \notin S,$$
$$\Delta X|_{X\in S} = G(X,w), \tag{6.3}$$

where $F = (F_0, F_1, \ldots, F_n) = (f_0, f)$, $G = (G_0, G_1, \ldots, G_n) = (g_n, g)$.

Amongst all the actuating effects $\{u(t), w_1, \ldots, w_{N(T)}\}$ that map the point $X^0 = (0, x^0)$ into the line Π, we are looking for such an actuating effect, for which the point of intersection with the line Π has the least X_0 coordinate. In the sequel we will use precisely this formulation of the problem. It can be seen in [117] that a variational system plays a very important role for finding necessary conditions for the control in the form of Pontryagin maximum principle to be optimum. Introduce a similar system corresponding to a solution of system (6.3). Let $\{u(t), w_1, \ldots, w_{N(T)}$ be an admissible actuating effect that takes the point X_0 into the line Π and $X(t) = (X_0(t), X_1(t), \ldots, X_n(t))$ be the solution of system 6.3) that corresponds to this actuating effect.

For this solution $X(t)$, the system in variations can be defined to be one of the following impulsive systems,

$$\frac{dz}{dt} = \frac{\partial F(X(t), u(t))}{\partial X} z, \qquad t \neq \tau_i$$
$$\Delta z|_{t=\tau_i} = A_i z, \tag{6.4}$$

or

$$\frac{dz}{dt} = \frac{\partial F(X(t), u(t))}{\partial X} z, \qquad t \neq \tau_i$$
$$\Delta z|_{t=\tau_i} = B_i z, \tag{6.5}$$

Here τ_i are solutions of the equation $s(X(t)) = 0$, which belong to the interval $]T_0, T_1[$, the matrices A_i and B_i are given by

$$A_i = \{a^i_{\alpha\beta}\}, \; a^i_{\alpha\beta} = \frac{\partial G_\alpha(X(\tau_i), w_i)}{\partial X_\beta} +$$
$$+ \frac{1}{\langle \operatorname{grad} s(X(\tau_i)), F(X(\tau_i), u(\tau_i))\rangle} \Big\{ F(X(\tau_i), u(\tau_i)) -$$
$$- \left(E + \frac{\partial G}{\partial X}(X(\tau_i), w_i) \right) F(X(\tau_i), u(\tau_i)) \Big\}_\alpha \frac{\partial s}{\partial X_\beta}(X(\tau_i)),$$

$$E + B_i = \{b^i_{\alpha\beta}\}^{-1}, b^i_{\alpha\beta} = \frac{\partial G^{-1}_\alpha(X(\tau_i), w_i)}{\partial X_\beta} +$$

$$+ \frac{1}{\langle \operatorname{grad} s(G^{-1}(X(\tau_i), w_i)), F(X(\tau_i), u(\tau_i)) \rangle} \Big\{ F(X(\tau_i), u(\tau_i)) -$$

$$- \frac{\partial G^{-1}(X(\tau_i), w_i)}{\partial X} F(X(\tau_i), u(\tau_i)) \Big\}_\alpha \frac{\partial s(G^{-1}(X(\tau_i), w_i)}{\partial x_\beta}$$

$$(\alpha, \beta = 0, 1, \ldots, n, \ , i = 1, 2, \ldots, N(T)),$$

where $\{(\cdot)\}$ denotes the component of the vector $(\cdot)$ with the index α and $G^{-1}(X, w)$ is the mapping inverse to the mapping

$$X \to X + G(X, w).$$

Note that systems (6.4) and (6.5) are equivalent. This can be seen by proving, for example, that

$$(E + A_i)(E + B_i)^{-1} = (E + B_i)^{-1}(E + A_i) = E.$$

Suppose that the points X^0 and X' do not lie on a cylindrical surface of discontinuity and let $Y(t)$ be a solution of equations (6.3) with the initial condition $Y(T_0) = X^0 + \epsilon h + o(\epsilon)$ that corresponds to the actuating effect $\{\tilde{u}(t), w_1, \ldots w_{N(T)}\}$, where $\tilde{u}(t)$ differs from $u(t)$ only on the intervals Δ_k, which have length of order ϵ and are adjacent to the points τ_i, $i = \overline{1, N(T)}$.

Theorem 95 *For any $\eta > 0$ there exists such $\epsilon_0 > 0$ that, for all $0 < \epsilon \leq \epsilon_0$ and $T_0 \leq t \leq T$, for which $|t - \tau_i| > \eta$, the following relation holds*

$$Y(t) = X(t) + \epsilon Z(t) + O(\epsilon, t), \tag{6.6}$$

where $Z(t)$ is a solution of impulsive differential system (6.4), which satisfies the condition $Z(T_0) = h$.

Proof. Assume that the solution $Y(t)$ meets the surface of discontinuity for the first time at $\tau_1 + \Delta_1$, where $\Delta_1 \geq 0$.

If we consider the segment $[T_0, \theta_1]$, then the statement of the theorem is a fact known from the theory of ordinary differential equations. Assume that the statement of the theorem holds for an interval $[T_0, \tau_i]$ and let $Y(t)$ meet the surface of discontinuity at the moment $\tau_i + \Delta_i$, $\Delta_i \geq 0$. Then

$$Y(\tau_i) = X(\tau_i) + \epsilon Z(\tau_i) + o(\epsilon), \; Y(\tau_i + \Delta_i) = Y(\tau_i) +$$

$$+\Delta_i F(Y(\tau_i), u(\tau_i)) + o(\Delta_i) = X(\tau_i) + \epsilon Z(\tau_i) +$$
$$+\Delta_i F(X(\tau_i), u(\tau_i)) + o(\epsilon) + o(\Delta_i);$$
$$o = s(Y(\tau_i + \Delta_i)) = \epsilon\langle \text{grad}s(X(\tau_i)), Z(\tau_i)\rangle +$$
$$+\Delta_i\langle \text{grad}s(X(\tau_i)), F(X(\tau_i), u(\tau_i))\rangle + o(\epsilon)$$

and, up to the terms of order $o(\epsilon)$, we have

$$\Delta_i = -\epsilon \frac{\langle \text{grad}s(X(\tau_i)), Z(\tau_i)\rangle}{\langle \text{grad}s(X(\tau_i)), F(X(\tau_i), u(\tau_i))\rangle}, \tag{6.7}$$
$$Y(\tau_i + \Delta_i + o) = Y(\tau_i + \Delta_i) + G(Y(\tau_i + \Delta_i), w_i) =$$
$$= X(\tau_i + o) + \epsilon \left(E + \frac{\partial G(X(\tau_i w_i))}{\partial x}\right) Z(\tau_i) +$$
$$+ \left(E + \frac{\partial G(X(\tau_i), w_i)}{\partial x}\right) F(X(\tau_i), u(\tau_i))\Delta_i + o(\epsilon).$$

Because

$$X(\tau_i + \Delta_i) = X(\tau_i + o) + F(X(\tau_i + 0), u(\tau_i + 0))\Delta_i + o(\Delta_i),$$

by using (6.7), we get

$$Y(\tau_i + \Delta_i + o) = X(\tau_i + \Delta_i) + \epsilon(E + \frac{\partial G}{\partial x}(X(\tau_i), w_i)Z(\tau_i) +$$
$$+(-F(X(\tau_i), u(\tau_i)) + \left(E + \frac{\partial G}{\partial x}(X(\tau_i), w_i)\right) F(X(\tau_i), u(\tau_i))\Delta_i +$$
$$+o(\epsilon) = X(\tau_i + \Delta_i) + \epsilon(E + A_i)Z(\tau_i) + o(\epsilon) =$$
$$= X(\tau_i + \Delta_i) + \epsilon Z(\tau_i + o) + o(\epsilon).$$

By choosing ϵ sufficiently small, we can make $\Delta_i < \eta$. Hence the statement of the theorem holds for the interval $[T_0, \tau_{i+1}]$. To end the proof it remains to add that if, for some time $t = \tau_i$, $\Delta_i < 0$, then, instead of system (6.4), we use system (6.5).

□

This theorem shows that we can call variational equations the system of equations (6.4). Let us write the system adjoint to system (6.4),

$$\frac{d\psi}{dt} = -\left(\frac{\partial F(X(t), u(t))}{\partial X}\right)^T \psi, \qquad t \neq \tau_i,$$
$$\Delta\psi|_{t=\tau_i} = ((A_i^T)^{-1} - E)^{-1}\psi. \tag{6.8}$$

Recall (section 2.5) that if $z(t)$ and $\psi(t)$ are solutions of systems (6.4) and (6.5) respectively, then the scalar product $\langle\psi(t), z(t)\rangle$ is constant on the whole interval $[T_0, T]$.

The linear system (6.4) allows to assign to every vector $h = z(T_0)$ a family of vectors $h_t = z(t)$.

For each $t \in [T_0, T]$, denote by A_{t,T_0} the linear nonsingular mapping defined by this correspondence. For any solution of system (6.8) and any vector $h \in \mathbf{R}^{n+1}$, we can write

$$\langle\psi(t), A_{t,T_0}h\rangle = \text{const} \tag{6.9}$$

since the scalar product of solutions of mutually adjoint systems is constant.

Choose arbitrary points $\theta_1 \leq \theta_2 \leq \ldots \leq \theta_s$ from the interval $]T_0, T]$, at which the control function $u(t)$ is continuous, and such that $|\tau_i - \theta_j| > \eta$ for some $\eta > 0$ and all $j = \overline{1, s}$ and $i = \overline{1, N(T)}$. Choose arbitrary non-negative numbers $\delta t_1, \ldots, \delta t_s$, an arbitrary number δt, and arbitrary vectors $v_1, v_2, \ldots, v_s$ that belong to the domain V. Define the intervals which depend on ϵ, $I_\nu = \{t : \theta_\nu + \epsilon l_\nu < t \leq \theta_\nu + \epsilon(l_\nu + \delta t_\nu)\}$, where

$$l_\nu = \begin{cases} \delta t - (\delta t_\nu + \ldots + \delta t_s), & \text{if} \quad \theta_\nu = T, \\ -(\delta t_\nu + \ldots + \delta t_s), & \text{if} \quad \theta_\nu = \theta_s < T, \\ -(\delta t_\nu + \ldots + \delta t_j), & \text{if} \quad \theta_\nu = \theta_{\nu+1} = \ldots \\ & \qquad \ldots = \theta_j < \theta_{j+1}, \ j < s. \end{cases}$$

For a sufficiently small ϵ, the half-intervals $I_1, I_2, \ldots, I_s$ are mutually disjoint with the intervals $J_i = \{t : |t - \tau_i| < \eta\}$. Assuming that ϵ satisfies these conditions, we define the control function

$$u^*(t) = \begin{cases} v_\nu, & \text{if} \quad t \in I_\nu, \\ \tilde{u}(t), & \text{if} \quad t \notin I_\nu, \ \nu = \overline{1, s}, \end{cases}$$

which will be called *variational*. Here

$$\tilde{u}(t) = \begin{cases} u(\tau_i - o), & \text{if} \quad \tau_i < t \leq \tau_i + \Delta_i, \\ u(t) & \text{on the rest of} \quad [T_0, T]. \end{cases}$$

Define variations of the controls w_j. An arbitrary vector $\delta w_j \in \mathbf{R}^r$ is called *w-admissible* at the point w_j if there exists such a number $s_0 > 0$ that $w_j + s\delta w_j \in W$ for all such s that $0 \leq s \leq s_0$. All w-admissible vectors form a cone in the space R^r.

The control $w_j^* = w_j + \epsilon\delta w_j$ is called *variational.* By making ϵ small enough, we can get that $w_j^* \in W$. The actuating effect $\{u^*(t), w_1^*, \ldots, w_{N(T)}^*\}$

is called a *variational* actuating effect. If the point x^1 does not lie on the surface of discontinuity, then, for sufficiently small ϵ, the solution $Y(t)$, which starts at the point X^0 at $t = T_0$ and corresponds to the actuating effect $\{u^*(t), w_1^*, \dots, w_{N(T)}^*\}$, is defined on the entire interval $[T_0, T + \epsilon\delta t]$ where the function $u^*(t)$ is defined. The solution $Y(t)$ is called *variational.* For this solution, we have

$$Y(T + \epsilon\delta t) = X(T) + \epsilon\Delta X + o(\epsilon), \tag{6.10}$$

where

$$\Delta X = F(X(T), u(T))\delta t + \sum_{\nu=1}^{s} A_{T,\theta_\nu}(F(X(\theta_\nu), v_\nu) -$$
$$-F(X(\theta_\nu), u(\theta_\nu)))\delta t_\nu + \sum_{i=1}^{N(T)} A_{T,\tau_i+0}\left(\frac{\partial G(X(\tau_i), w_i)}{\partial w}\delta w_i\right). \tag{6.11}$$

It should be noted that these formulas are valid only for a sufficiently small ϵ.

The vector ΔX does not depend on ϵ but it depends essentially on the choice of the values θ_ν, v_ν, δt_ν, δw_j, δt. Denote by α the set of these values, $\alpha = \{\theta_\nu, v_\nu, \delta t_\nu, \delta w_j, \delta t\}$. Because some of δt_ν could be equal to zero, we can consider all θ_ν and v_ν to be the equal and taken the same number of times for any symbol α. Define a linear combination of the symbols

$$\alpha' = \{\theta_\nu, v_\nu, \delta t'_\nu, \delta w'_j, \delta t'\} \text{ and } \alpha'' = \{\theta_\nu, v_\nu, \delta t''_\nu, \delta w''_j, \delta t''\}$$

by

$$\begin{aligned}\lambda_1\alpha' + \lambda_2\alpha'' &= \{\theta_n u, v_\nu, \lambda_1\delta t'_\nu + \lambda_2\delta t''_\nu, \lambda_1\delta w'_j + \\ &\quad +\lambda_2\delta w''_j, \lambda_1\delta t' + \lambda_2\delta t''\}.\end{aligned}$$

It is clear that, for a vector $\alpha = \lambda_1\alpha' + \lambda_2\alpha''$, there corresponds the vector $\Delta X_\alpha = \lambda_1\Delta X_{\alpha'} + \lambda_2\Delta X_{\alpha''}$. Whence it follows that the set of the vectors ΔX_α that correspond to all symbols α is a convex cone with the vertex at the point $X(T)$. This cone will be denoted by $\mathcal{K}$ and called an *attainability cone.*

Lemma 46 *Let $X(t)$, $T_0 \le t \le T$, be a trajectory that corresponds to an actuating effect $\{u(t), w_1, \dots, w_{N(T)}\}$ and starts at a point X^0 and γ be a line starting at the point $X(T)$ and having a tangent ray L_T at the point*

$X(T)$. If this ray belongs to the convexity of the cone $\mathcal{K}_T$, then there exists an actuating effect $\{u^*(t), w_1^*, \ldots, w_{N(T)}^*\}$ such that the corresponding trajectory $X^*(t)$ that starts at the point X^0 intersects the line γ at a point different from $X(T)$.

The following statement readily follows from this lemma.

Lemma 47 *If an actuating effect $\{u(t), w_1, \ldots, w_{N(T)}\}$ and the corresponding trajectory $X(t)$, $T_0 \leq t \leq T$ correspond to an optimum, then the ray L_T that starts at the point $X(T)$ and directed along the negative half-axis X_0 does not lie in the interior of the attainability cone.*

Proof. Indeed, let us take the ray L_T to be the curve γ. If the ray L_T belongs to the interior of the cone $\mathcal{K}_T$, then there exists such an actuating effect $\{u^*(t), w_1^*, \ldots, w_{N(T)}^*\}$ that, for some τ', we have $X_\nu^*(\tau') = X_\nu(T)$, $\nu = \overline{1,n}$, $X_0^*(\tau') \leq X_0(T)$. But this relation is a contradiction since the actuationg effect correspond to an optimum.

□

6.2 Necessary conditions for optimum

Let us introduce the following auxiliary functions:

$$H(\psi(t), X(t), u(t)) = \langle F(X(t), u(t)), \psi(t)\rangle,$$
$$M(t) = \sup_{v\in U} H(\psi(t), X(t), v),$$
$$\mathcal{H}_i(\psi(\tau_i + 0), X(\tau_i), w) = \langle \text{grad}(x + G(X(\tau_i), w)),$$
$$\psi(\tau_i + 0)\rangle, \qquad i = \overline{1, N(T)}.$$

Theorem 96 *For an actuating effect $\{u(t), w_1, \ldots, w_{N(T)}\}$ and the corresponding trajectory $X(t)$, $T_0 \le t \le T$, to be a solution of the optimum control problem (6.1), (6.2), it is necessary that there exist such a nonzero solution of system (6.8) that*

a) the condition of the maximum of

$$H(\psi(t), X(t), u(t)) = M(t)$$

holds for all the points of the interval $[T_0, T]$ excepting possibly the point $t = T_0$;

b) at a finite time $t = T$,

$$\psi_0(T) \le 0, \qquad M(T) = 0;$$

c) the derivatives of the functions $\mathcal{H}_i(w)$, $i = \overline{1, N(T)}$, with respect to all possible directions at the points w_i relatively to the set W are nonpositive.

Proof. Suppose that a finite point x_1 does not lie on the surface of discontinuity and let $\{u(t), w_1, \ldots w_{N(T)}\}$ be an optimum actuating effect with $X(t)$, $T_0 \le t \le T$ being the corresponding optimum trajectory. Then, by Lemma 47, the ray L_T does not belong to the interior of the attainability cone $\mathcal{K}_T$, i.e. there is a hyperplane that separates them and passes through the point $X(T)$. Denote by $c = (c_0, c_1, \ldots, c_n)$ a vector normal to this hyperplane. We can assume that the ray L_T lies in the nonnegative half-space of this hypersurface and the cone $\mathcal{K}_T$ – in the nonpositive half-plane. Let $\psi(t) = (\psi_0(t), \psi_1(t), \ldots \psi_n(t))$ be a solution of equations (6.8) with the initial condition $\psi(T) = c$. Because $c \ne 0$, $\psi(t)$, $T_0 \le t \le T$, is a nontrivial solution and, moreover, $\psi_0(T) \le 0$ and

$$\langle \psi(T), \Delta X\rangle \le 0 \tag{6.12}$$

for any vector ΔX that belongs to the attainability cone $\mathcal{K}_T$. By setting $\delta t = 0$, $s = 1$, $\delta t_1 = 1$, $\delta t_2 = \ldots = \delta t_s = 0$, $\delta w_1 = \ldots = \delta w_{N(T)} = 0$ in (6.11), we get

$$\langle \psi(T), A_{T,\theta_1}(F(x(\theta_1), v_1) - F(X(\tau_1), u(\tau_1))\rangle \leq 0.$$

By Lemma 46, we have

$$\langle \psi(\tau_1), F(X(\theta_1), v_1)\rangle \leq \langle \psi(\theta_1), F(X(\theta_1), u(\theta_1))\rangle$$

for any point θ_1 from the interval $[T_0, T]$, at which both the control functon $u(t)$ and the trajectory $X(t)$ are continuous.

Let, at $t = \theta$, at least one of the functions $u(t)$ or $X(t)$ be discontinuous. Because it follows from the assumptions that the number of such points is finite, there exists a sequence of numbers less than θ, $\{\theta_j\}_{j=1}^{\infty}$, at which both functions $u(t)$ and $X(t)$ are continuous and $\lim_{j\to\infty}\theta_j = \theta$. Since $H(\psi, x, u)$ is function continuous with respect to its arguments and the functions $\psi(t)$, $X(t)$, and $u(t)$ are continuous from the left at $t = \theta$, it follows from the maximum condition, which holds at the points of the sequence $\{\theta_j\}$,

$$H(\psi(\theta_j), X(\theta_j), u(\theta_j)) = M(\theta_j),$$

that the maximum condition also holds at $t = \theta$, i.e.

$$H(\psi(\theta), X(\theta), u(\theta)) = M(\theta),$$

This proves part a) of the theorem.

By setting $\delta t_1 = \delta t_2 = \ldots = \delta t_s = 0$, $\delta w_1 = \delta w_2 = \ldots = \delta w_{N(T)} = 0$ in (6.11) and using (6.12), we get

$$\langle \psi(T), F(X(T), u(T))\rangle \Delta t \leq 0.$$

Because the maximum condition holds at $t = T$ and δt may take values with different sign, we get statement b) of the theorem.

Set $\delta t = \delta t_1 = \ldots \delta t_s = 0$, $\delta w_2 = \ldots = \delta w_{N(T)} = 0$ in (6.11) and use (6.12) to get

$$\left\langle \psi(T), a_{T,\tau_i+0}\left(\frac{\partial G(X(\theta_1 - 0), w_1)}{\partial w}\delta w_1\right)\right\rangle \leq 0.$$

By Lemma 46,

$$\left\langle \psi(\tau_j + 0), \frac{\partial G(X(\tau_1), w_1)}{\partial w}\delta w_1\right\rangle \leq 0.$$

Since

$$\left\langle \psi(\tau_1), \frac{\partial G(X(\tau_1), w_1)}{\partial w} \delta w_1 \right\rangle = \left\langle \frac{\partial \mathcal{H}_1(\psi(\tau_1 + 0), X(\tau_1), w_1)}{\partial w} \delta w_1 \right\rangle,$$

and δw_1 is an arbitrary admissible direction with respect to the set W, statement c) of the theorem is proved for $t = \tau_1$. In the same way we prove it for $t = \tau_i$, $i = 2, \ldots, N(T)$.

Let an endpoint X' lie on the surface of discontinuity, i.e. $s(X') = s(X(T)) = 0$ and an actuating effect and the corresponding trajectory $X(t)$, $T_0 \leq t \leq T$ be optimum. For a sufficiently small $\tau > 0$, the point $X^2 = X(T - \tau)$ does not lie on the surface of discontinuity. Let us consider the optimum control problem (6.1), (6.2) for an initial point x^0 and an end point x^2. The actuating effect $\{u(t), w_1, \ldots w_{N(T)}\}$ and the trajectory $X(t)$, considered on the interval $[T_0, T - \tau]$ are optimum by the optimum principle. Hence there exists a nonzero function $\psi_\tau(t)$, $T_0 \leq t \leq T - \tau$, which satisfies Theorem 95 on the interval $[T_0, T - \tau]$. Extend this solution to the interval $[T_0, T]$. Without infringing the generality, we can assume that $\|\psi(T_0)\| = 1$. Fix a sequence of positive numbers, convergent to zero, $\{\theta_j\}$. Because the unit sphere in a finite dimensional space is compact, there is a subsequence $\{\theta_\nu\}$ of the sequence $\{\theta_j\}$ such that the corresponding sequence $\{\psi_{\theta_\nu}(T_0)\}_{\nu=1}^{\infty}$ will be convergent. Let $\psi^0 = \lim_{\nu \to \infty} \psi_{\theta_\nu}(T_0)$. A solution of system (6.8) with the initial condition $\psi(T_0) = \psi^0$ will be the one, the existence of which is asserted in Theorem 6.12.

□

This theorem implies that the function $\psi_0(t)$ is constant on the whole interval $[T_0, T]$. Indeed, since $F(X(t), u(t))$ does not depend on $X_0(t)$, by analyzing the structure of the matrices A_i, $i = \overline{1, N(T)}$, one can see that $\psi_0(t)$ satisfies the impulsive differential equation

$$\frac{d\psi_0}{dt} = 0, \qquad t \neq \tau_i, \qquad \Delta\psi_0|_{t=\tau_i} = 0,$$

i.e. $\psi_0(t) = \text{const}$ for all $t \in [T_0, T]$.

Lemma 48 *If an absolutely continuous function $\psi(t)$ satisfies the equations*

$$\frac{d\psi}{dt} = -\left(\frac{\partial F(X(t), u(t))}{\partial X}\right)^T \psi$$

and the relation

$$H(\psi(t), X(t), u(t)) = M(t)$$

almost everywhere, then the function $M(t)$ is constant on the whole interval I.

It also follows from Theorem 96 that the function $M(t)$ is constant on the whole interval $[T_0, T]$.

Indeed, the preceding lemma implies that the function $M(t)$ is constant for values of t lying between the times of the impulsive effects. Because for $t = \tau_i$,

$$\begin{aligned}\lim_{t \to \tau_i - 0} H(\psi(t), X(t), u(t)) &= H(\psi(\tau_i), X(\tau_i), u(\tau_i)) = \\ &= \langle \psi(\tau_i), F(X(\tau_i), u(\tau_i)) \rangle = \langle A_i^T \psi(\tau_i + 0), f(X(\tau_i), u(\tau_i)) \rangle = \\ &= \langle \psi(\tau_i + 0), A_i F(X(\tau_i), u(\tau_i)) \rangle = \langle \psi(\tau_i + 0), F(X(\tau_i + 0), \\ u(\tau_i + 0)) \rangle &= \lim_{t \to \tau_i + 0} H(\psi(t), X(t), u(t)),\end{aligned}$$

the function $M(t)$ is continuous at $t = \tau_i$ and, consequently, it is constant on the whole interval $[T_0, T]$. Hence the conditions of part b) of Theorem 6.12 can be verified for any value of $t \in [T_0, T]$.

We will give some generalization of the foregoing main optimum control problem to problems with moving ends, fixed time, and to nonautonomous systems. A method to get such generalizations is described in many monographs on optimum control [27, 31, 117] and, without much difficulties, can be applied to the considered problem of optimum control for impulsive differential systems.

Let S_0 and S_1 be smooth manifolds in a phase space $\mathbf{M} \subset \mathbf{R}$, of dimensions r_0, r_1 which are arbitrary but less than n. Suppose that the hypersurface of discontinuity and the manifolds S_0 and S_1 do not intersect. As opposed to the problem we have been considering, assume that the initial point x_0 is not fixed but belongs to the manifold S_0. Similarly, let the endpoint x' belong to the manifold S_1.

In such a setting, the optimum control problem will be called an optimum control problem with moving ends. An analogue of Theorem 3 [117](p. 59) is the following statement.

Theorem 97 *For an actuating effect $\{u(t), w_1, \ldots, w_{N(T)}\}$ and the corresponding trajectory $X(t)$, $T_0 \le t \le T$, to be a solution of the optimum control problem (6.1) and (6.2) with moving ends, it is necessary that there exists a* *nonzero solution* $\psi(t)$ *of system (6.8), which satisfies conditions of Theorem 96 and the transversality conditions at both ends, i.e. if* P_0 *and* P_1 *are tangent planes to the manifolds* S_0 *and* S_1 *at the points* $x^0 \in S_0$ *and* $x^1 \in S_1$, *then the vector* $\tilde{\psi}(T_0) = (\psi_1(T_0), \ldots, \psi_n(T_0))$ *is orthogonal to the plane* P_0 *and the vector* $\tilde{\psi}(T) = (\psi_1(T), \ldots \psi_n(T))$ *is orthogonal to the plane* P_1.

If, in the main optimum control problem, the end time $t = T$ is fixed, then we have an optimum control problem with fixed time. A solution of this problem can be obtained from the preceding theorem if equations (6.1) are supplemented with one more equation

$$\frac{dx_{n+1}}{dt} = 1, \qquad x \notin S, \qquad \Delta x_{n+1}|_{x\in S} = 0,$$
$$x_{n+1}(T_0) = T_0.$$

This optimum control problem in an $(n+1)$-dimensional space can be considered as an optimum control problem with free time because the endpoint (x', T) can be reached only at $t = T$. Whence, by using Theorem 96, we have the following statement.

Theorem 98 *For an actuating effect* $\{u(t), w_1, \ldots, w_{N(T)}\}$ *and the corresponding trajectory* $X(t)$, $T_0 \le t \le T$ *be solutions of the optimum control problem (6.1) and (6.2) with fixed time, it is necessary that there exists a nonzero solution* $\psi(t) = (\psi_0(t), \ldots, \psi(t))$ *of system (6.8), where* $\psi_0(t)$ *is a nonpositive constant function such that*

a) the condition

$$H(\psi(t), X(t), u(t)) = M(t)$$

holds for all points of the interval $[T_0, T]$ *excepting possibly the point* $t = T_0$;

b) the derivatives of the functions $\mathcal{H}_i(w)$, $i = \overline{1, N(T)}$ *at the points* w_i *with respect to all directions relatively to the set* W *are nonpositive.*

The same method of introducing an additional coordinate x_{n+1} can also be successfully used to solve optimum control problems in a nonautonomous case, i.e. when system (6.1) is written in the form

$$\frac{dx}{dt} = f(x, u, t), \qquad s(t, x) \neq 0,$$
$$\Delta x|_{s(t,x)=0} = g(x, w, t)|_{s(t,x)=0}, \tag{6.13}$$

and functional (6.2) is

$$\mathcal{J} = \int_{T_0}^{T} f_0(x(s), u(s), s)ds + \sum_{i=1}^{N(T)} g_0(x(\tau_i), w_i). \tag{6.14}$$

6.3 Impulsive control with fixed times

In this section we will be considering a controlled differential system with an impulsive effect occurring at fixed times,

$$\frac{dx}{dt} = f(t, x, u), \qquad t \neq \tau_i,$$
$$\Delta x|_{t=\tau_i} = g_i(x, w_i). \tag{6.15}$$

The system is considered on a finite time interval $[T_0, T]$, $\{\tau_i\}$ is a finite set of points with

$$T_0 = \tau_0 \leq \tau_1 < \tau_2 < \ldots < \tau_N < \tau_{N+1} = T_1.$$

Consider the following optimum control problem for system (6.15): to find such an admissible actuating effect $\{u(t), \ldots, w_1, \ldots, w_N\}$, which transforms the system from a state x^0 into a state x^1 over a fixed time interval $[T_0, T_1]$, that minimizes the functional

$$\mathcal{J} = \int_{T_0}^{T} f_0(s, x(s), u(s))ds + \sum_{i=1}^{N} g_i^0(x(\tau_i), w_i). \tag{6.16}$$

This problem can be reduced to the optimum control problem (6.1), (6.2), but if the construction of the Pontryagin maximum principle is applied to solve this problem, then it is easy to see that the class of admissible control functions $u(t)$ can be considerably extended by taking $u(t)$ to be a measurable function bounded on $[T_0, T_1]$ and taking the values in the set $U \subset \mathbf{R}^m$.

Consider the case when the functions $f(t, x, u)$ and $f_0(t, x, u)$ do not depend on time explicitly, i.e. we consider the optimum control problem (6.15), (6.16) with an assumption that system (6.15) has the form

$$\frac{dx}{dt} = f(x, u), \qquad t \neq \tau_i,$$
$$\Delta x|_{t=\tau_i} = g_i(x, w_i), \tag{6.17}$$

functional (6.16) is

$$\mathcal{J} = \int_{T_0}^{T} f_0(x(s), u(s))ds + \sum_{i=1}^{N} g_i^0(x(\tau_i), w_i) \tag{6.18}$$

and we take the end time $t = T_1$ to be not fixed.

As in the preceding section, we introduce a new variable $x_0(t)$ that satisfies the impulsive differential equation

$$\frac{dx_0}{dt} = f_0(x, u), \qquad t \neq \tau_i,$$
$$\Delta x_0|_{t=\tau_i} = g_i^0(x, w_i)$$

thus obtaining in $(n+1)$-dimensional space the impulsive differential system

$$\frac{dX}{dt} = F(X, u), \qquad t \neq \tau_i,$$
$$\Delta X|_{t=\tau_i} = G_i(X, w_i),$$

where

$$X = (X_0, X_1, \ldots, X_n) = (x_0, x), \qquad F = (F_0, F_1, \ldots, F_n) = (f_0, f),$$
$$G_i = (G_i^0, G_i^1, \ldots, G_i^n) = (g_i^0, g_i).$$

Let $\{u(t), w_1, \ldots, w_N\}$ be an admissible control and $X(t)$ be the corresponding trajectory. Introduce the following functions

$$H(\psi, X, u) = \langle F(X, u), \psi \rangle,$$
$$M(t) = \sup_{v \in U} h(\psi(t), x(t), v),$$
$$\mathcal{H}_i(\psi, X, w) = \langle \psi, X + G_i(X, w) \rangle,$$

where the function $\psi(t) = (\psi_0(t), \psi_1(t), \ldots, \psi_n(t))$ is a solution of the impulsive differential system

$$\frac{d\psi}{dt} = -\mathrm{grad}_x H(\psi, X(t), u(t)), \qquad t \neq \tau_i,$$
$$\psi(\tau_i - 0) = \mathrm{grad}_x \mathcal{H}_i(\psi(\tau_i + 0), X(\tau_i - 0), w_i), \qquad i = \overline{1, N}. \tag{6.19}$$

Theorem 99 *If an actuating effect $\{u(t), w_1, \ldots, w_N\}$ and the corresponding trajectory $X(t)$, $T_0 \leq t \leq T_1$ are solutions of the optimum control problems (6.17) – (6.18), then there necessarily exists a nonzero solution of system (6.19) satisfying the following conditions:*

a) the maximum condition

$$H(\psi(t), X(t), u(t)) = M(t)$$

holds almost everywhere on the segment $[T_0, T_1]$;

b) at the end time $t = T_1$, *we have*

$$\psi_0(T_1) \leq 0, \qquad M(T_1) = 0;$$

c) derivatives of the functions $\mathcal{H}_i(w)$, $i = \overline{1, N}$, *at the points* w_i *with respect to all directions admissible relatively the set* W *are nonpositive.*

We do not give a proof of the theorem since it is an analogue of Theorem 8 in [117].

Returning to the optimum control problem (6.15), (6.16) let us introduce an additional variable $x_{n+1}(t)$ as a solution of the boundary value problem

$$\frac{dx_{n+1}}{dt} = 1, \qquad x_{n+1}(T_0) = T_0, \qquad x_{n+1}(T_1) = T_1.$$

The time T_1 can be considered as free because the point (x', T_1) can be reached only at $t = T_1$. Thus problem (6.15), (6.16) is reduced to problem (6.17), (6.18).

Introduce auxiliary functions

$$H(\psi, x, u, t) = \langle F(t, x, u), \psi \rangle,$$
$$M(t) = \sup_{v \in U} H(\psi(t), x(t), v, t),$$
$$\mathcal{H}_i(\psi, x, w) = \langle \psi, x + G_i(x, \psi) \rangle,$$

where $\psi(t)$ is a solution of the following impulsive differential system

$$\frac{d\psi}{dt} = -\text{grad}_x H(\psi, x(t), u(t), t), \qquad t \neq \tau_i$$
$$\psi(\tau_i - 0) = \text{grad}_x \mathcal{H}_i(\psi(\tau_i + 0), x(\tau_i), w_i), \qquad i = \overline{1, N}. \qquad (6.20)$$

By using the preceding theorem we have

Theorem 100 *If the actuating effect* $\{u(t), w_1, \ldots, w_n\}$ *and the corresponding trajectory* $X(t)$, $T_0 \leq t \leq T_1$, *are a solution of the optimum control problem (6.15), (6.16), then there necessarily exists a nonzero solution* $\psi(t) = (\psi_0(t), \ldots, \psi_n(t))$ *of system (6.20) such that*

a) $\psi_0(t)$ *is constant nonpositive function on the interval* $[T_0, T_1]$;

b) the condition

$$H(\psi(t), X(t), u(t), t) = M(t)$$

holds almost everywhere on $[T_0, T_1]$;

c) *the derivatives of the functions $\mathcal{H}_i(w)$, $i = \overline{1, N}$, at the points w_i with respect to all possible directions relatively to the set W are nonpositive.*

Suppose that system (6.15) is linear,

$$\frac{dx}{dt} = A(t)x + B(t)u, \qquad t \neq \tau_i,$$
$$\Delta x|_{t=\tau_i} = C_i x + D_i w_i, \qquad i = \overline{1, N}, \tag{6.21}$$

and functional (6.16) can be written as

$$\mathcal{J} = \int_{T_0}^{T_1} f_0(s, x(s)) + b_0(s, u(s)))ds + \sum_{i=1}^{N} (g_i^0(x), h_i^0(w_i)), \tag{6.22}$$

where $g_i^0(x)$, $h_i^0(w)$, $i = \overline{1, N}$ are convex functional and $f_0(t, x)$, $b_0(t, u)$ are convex functions when the variable t is fixed.

Because system (6.21) is linear, any solution of it, for a given actuating effect $\{u(t), w_1, \ldots, w_N\}$ is defined on the whole interval $[T_0, T_1]$.

Denote by $\tilde{\mathcal{K}}(T_1)$ a set of cones of those trajectories of system (6.21), which start for $t = T_0$ at the point x^0 and correspond to all possible actuating effects $\{u(t), w_1, \ldots, w_N\}$, $T_0 \leq t \leq T_1$, and call this set the attainability set.

Theorem 101 *If the limiting sets $U \subset \mathbf{R}^m$ and $W \subset \mathbf{R}^r$ are compact and convex, then the attainability set $\tilde{\mathcal{K}}(T_1)$ is compact and convex.*

A proof of this theorem can be found in [117].

Theorem 102 *Let the limiting sets $U \subset \mathbf{R}^m$ and $W \subset \mathbf{R}^r$ be compact and convex. For an actuating effect to take a phase point from the point x^0 to some limit point of the attainability set $\tilde{\mathcal{K}}(T_1)$, it is necessary and sufficient that there exist a nonzero solution $\tilde{\psi}(t) = (\tilde{\psi}_1(t), \ldots, \tilde{\psi}_n(t))$ of the system*

$$\frac{d\tilde{\psi}}{dt} = -A^T(t)\tilde{\psi}, \qquad t \neq \tau_i,$$
$$\Delta\tilde{\psi}|_{t=\tau_i} = -C^T(C^T + E)^{-1}\tilde{\psi}, \tag{6.23}$$

which satisfies the conditions:

a) *the maximum condition*

$$\langle\tilde{\psi}(t), B(t)u(t)\rangle = \max_{v \in V}\langle\tilde{\psi}(t), B(t)v\rangle,$$

holds almost everywhere on the interval $[T_0, T_1]$;

b) the maximum condition

$$\langle\tilde{\psi}(\tau_i+0), D_i w_i\rangle = \max_{v\in W}\langle\tilde{\psi}(\tau_i+0), D_i v\rangle.$$

holds at the points $t=\tau_i$, $i=\overline{1,N}$.

This theorem gives necessary and sufficient conditions for the endpoint of the trajectory $x(T_1)$ to lie on the boundary of the attainability set. However, in a general case the actuating effect that takes the phase point from a position x^0 into some point of the boundary of the attainability set is not unique.

Suppose that the limiting set U is a convex compact polyhedron, the matrices $A(t)$ and $B(t)$ are continuously differentiable $n-2$ and $n-1$ times respectively. Define the matrices

$$B_1(t) = B(t), \qquad B_j = -A(t)B_{j-1}(t) + \frac{dB_{j-1}(t)}{dt}, \qquad j=\overline{2,n}.$$

We will say that at time $t\in[T_0,T_1]$ the condition of generality of position holds if, for any edge ω of the polyhedron U, the vectors $B_1(t)\omega,\ldots,B_n(t)\omega$ are linearly independent in $\mathbf{R}^n$.

Theorem 103 *If the limiting set* U *is a convex polyhedron in* $\mathbf{R}^m$, W *is a strictly convex compact set in* $\mathbf{R}^r$, *at every* $t\in[T_0,T_1]$ *which is not the time when an impulsive effect occurs, the condition of generality of position holds,* $\operatorname{rank}D_i=n$, $i=\overline{1,N}$, *then the system can be transferred to any limit point of the attainability set by a single actuating effect.*

We give the following theorem on existence of an optimum control without a proof.

Theorem 104 *Let, in the optimum control problem (6.21), (6.22), the limiting sets* U *and* W *be convex and compact and the end state* $x(T_1)$ *belong to a closed set* $G\subset\mathbf{R}^r$. *If there exists at least one admissible actuating effect that takes the phase point from a position* x_0 *into an end position* $x(T_1)\in G$, *then there also exists an optimum actuating effect.*

6.4 Necessary and sufficient conditions for optimum

Consider optimum control problem (6.1), (6.2) with the assumption that system (6.1) can be written as

$$\frac{dx}{dt} = A(t)x + b(t,u), \qquad t \neq \tau_i,$$
$$\Delta x|_{t=\tau_i} = C_i x + D_i w_i, \qquad i = \overline{1,N}, \tag{6.24}$$

and functional (6.1) is

$$\mathcal{J} = \int_{T_0}^{T_1} f_0(s,x(s)) + b_0(s,u(s)))ds + \sum_{i=1}^{N}(g_i^0(x(\tau_i)), h_i^0(w_i)), \tag{6.25}$$

where $f_0(t,x)$ is a convex function for each fixed t and $g_i^0(x)$, $h_i^0(w)$, $i = \overline{1,N}$ are convex functionals.

Theorem 105 *For an actuating effect $\{u(t), w_1, \ldots, w_N\}$ and the corresponding trajectory $X(t)$, $T_0 \leq t \leq T_1$, to be a solution of the optimum control problem (6.24), (6.25) with an initial state x^0 and an end state $x(T_1)$ that belongs to a closed convex set $G \subset \mathbf{R}^r$, it is necessary and sufficient that there exist such a nonzero solution $\psi(t) = (\psi_0(t), \psi_1(t), \ldots, \psi_n(t)) = (\psi_0(t), \tilde{\psi}(t))$ of system (6.20) that $\psi_0(t)$ is constant on the interval $[T_0, T_1]$, the function $\tilde{\psi}(T_1)$ is an inner normal to the set G at the point $x(T_1)$ and*

a) the maximum condition

$$H(\psi(t), x(t), u(t), t) = M(t)$$

holds almost everywhere on the interval $[T_0, T_1]$,

b) if $t = \tau_i$, $i = \overline{1,N}$, the maximum condition

$$\mathcal{H}_i(\psi(\tau_i + 0), X(\tau_i), w_i) = \max_{v \in W} \mathcal{H}_i(\psi(\tau_i + 0), X(\tau_i), v)$$

is true.

Theorem 106 *Suppose that conditions of Theorem 103 hold. An actuating effect $\{u(t), w_1, \ldots, w_N\}$ and the corresponding trajectory $X(t)$, $T_0 \leq t \leq T_1$, are solutions of the optimum control problem (6.21) – (6.22) if and only if there exists such a nonzero solution $\psi(t) = (\psi_0(t), \tilde{\psi}(t))$ of system (6.20) that $\psi_0(t)$ is a function which is constant and nonpositive on the interval $[T_0, T_1]$ and*

a) the maximum condition

$$\psi_0 b_0(t, u(t)) + \langle \tilde{\psi}(t), B(t)u(t)\rangle = \max_{v \in U}(\psi_0 b_0(t, v) + \langle \tilde{\psi}(t), B(t)v\rangle$$

holds almost everywhere on the interval $[T_0, T_1]$,

b) if $t = \tau_i$, $i = \overline{1, N}$, *the maximum condition*

$$\psi_0 h_i^0(w_i) + \langle \tilde{\psi}(\tau_i + 0), D_i w_i\rangle \max_{v \in W}(\psi_0 h_i^0(v) + \langle \tilde{\psi}(\tau_i + 0), D_i v\rangle)$$

is true.

Theorem 107 *Let, in optimum control problem (6.21), (6.22), the end state* $x(T_1)$ *be an arbitrary vector from* $\mathbf{R}^n$. *An actuating effect* $\{u(t), w_1, \dots, w_N\}$ *and the corresponding trajectory* $X(t)$, $T_0 \le t \le T_1$ *is a solution of the optimum control problem (6.21), (6.22) if and only if there exists such a nonzero solution* $\psi(t) = (\psi_0(t), \tilde{\psi}(t))$ *of system (6.20) that* $\psi_0(t)$ *is a function which is constant and negative on the interval* $[T_0, T_1]$, $\tilde{\psi}(T_1) = 0$, *and*

a) the maximum condition

$$-b_0(t, u(t)) + \langle \tilde{\psi}(t), B(t)u(t)\rangle = \max_{v \in U}(-b_0(t, v) + \langle \tilde{\psi}(t), B(t)v\rangle$$

holds almost everywhere on the interval $[T_0, T_1]$,

b) if $t = \tau_i$, $i = \overline{1, N}$, *the maximum condition*

$$-h_i^0(w_i) + \langle \tilde{\psi}(\tau_i + 0), D_i w_i\rangle = \max_{v \in W}(-h_i^0(v) + \langle \tilde{\psi}(\tau_i + 0), D_i v\rangle)$$

is true.

Theorem 108 *An actuating effect* $\{u(t), w_1, \dots, w_N\}$ *and the corresponding trajectory* $X(t)$, $T_0 \le t \le T_1$ *is a solution of the optimum control problem (6.21) - (6.23) with an end state* $x(T_1) \in G$ *if and only if there exists such a nonzero solution* $\psi(t) = (\psi_0(t), \tilde{\psi}(t))$ *of system (6.20) that* $\psi_0(t)$ *is a function which is constant and negative on the interval* $[T - 0, T_1]$, $\tilde{\psi}_\nu(T_1) = 0$ *for* $\nu = \overline{l + 1, n}$ *and the equality*

$$u(t) = E^{-1}(t)B^T(t)\tilde{\psi}(t)$$

holds almost everywhere on the interval $[T_0, T_1]$ *and, for* $t = \tau_i$, $i = \overline{1, N}$, *the following equality holds,*

$$w_i = G_i^{-1} D_i^T \psi(\tau_i + 0).$$

Chapter 7

Asymptotic study of oscillations in impulsive systems

7.1 Formulation of the problem

Let an oscillation system be given by the equation

$$\frac{d^2x}{dt^2} + \omega^2 x = \epsilon f\left(t, x, \frac{dx}{dt}\right)$$

and subjected to an impulsive effect. The impulsive effect can occur at fixed times or when the phase point meets a subset of either the phase space or the extended phase space. It is assumed that, as a result of the impulsive effect, the velocity (impulse) is increased (decreased) by a certain value, which, generally speaking, depends on the velocity and the position of the phase point at the moment just before the impulse occurs. The impulsive effect results in no change of the coordinate x (position of the point). There are however problems, in which the position of the point also changes as a result of the impulsive effect, but we will not be dealing with them here.

We will be studying two classes of weakly nonlinear oscillating impulsive systems: the systems with an impulsive effect occurring at fixed times and when the phase point passes through a fixed position $x = x_0$. A mathematical model for the first type of impulsive systems is given by

$$\begin{aligned} \frac{d^2x}{dt^2} + \omega^2 x &= \epsilon f\left(\nu t, x, \frac{dx}{dt}\right), \qquad t \neq \tau_i, \\ \Delta \left.\frac{dx}{dt}\right|_{t=\tau_i} &= \epsilon I_i\left(x, \frac{dx}{dt}\right), \end{aligned} \tag{7.1}$$

and the second type is described by

$$\begin{aligned} \frac{d^2x}{dt^2} + \omega^2 x &= \epsilon f\left(\nu t, x, \frac{dx}{dt}\right), \qquad x \neq x_0, \\ \Delta \left.\frac{dx}{dt}\right|_{x=x_0} &= \epsilon I\left(\frac{dx}{dt}\right), \end{aligned} \tag{7.2}$$

To study oscillation processes described by equations (7.1) and (7.2), one uses sometimes the Dirac δ-function in a transformation of these equations. However, such a use of a generalized function sometimes is not well mathematically substantiated and we will not use it here. We will show that applying the δ-function formalism to a study of equations (7.1) and (7.2) often replaces the mathematical models of (7.1) and (7.2) by another one, which does not always adequately reflects the system under consideration.

It is known that to study oscillations in weakly nonlinear systems of type (7.1), it is convenient to introduce new coordinates a and φ (a and φ are correspondingly the generalized amplitude and phase of the oscillation) by

$$x = a\sin\varphi, \qquad \frac{dx}{dt} = a\omega\cos\varphi, \qquad \varphi = \omega t + \theta. \tag{7.3}$$

Let us use the δ-function to rewrite system (7.1) in the form

$$\frac{d^2x}{dt^2} + \omega^2 x = \epsilon f\left(\nu t, x, \frac{dx}{dt}\right) + \epsilon\sum_{\tau_i} I\left(x, \frac{dx}{dt}\right)\delta(t-\tau_i). \tag{7.4}$$

Now we apply change of variables (7.3) to get

$$\begin{aligned}
\frac{da}{dt} &= \frac{\epsilon}{\omega}\Big[f(\nu t, a\sin\varphi, a\omega\cos\varphi) + \\
&\quad + \sum_{\tau_i} I_i(a\sin\varphi, a\omega\cos\varphi)\delta(t-\tau_i)\Big]\cos\varphi,\\
\frac{d\varphi}{dt} &= \omega - \frac{\epsilon}{a\omega}\Big[f(\nu t, a\sin\varphi, a\omega\cos\varphi) + \\
&\quad + \sum_{\tau_i} I_i(a\sin\varphi, a\omega\cos\varphi)\delta(t-\tau_i)\Big]\sin\varphi,
\end{aligned}$$

which can be rewritten in the form

$$\begin{aligned}
\frac{da}{dt} &= \frac{\epsilon}{\omega} f(\nu t, a\sin\varphi, a\omega\cos\varphi)\cos\varphi, \\
& \qquad\qquad\qquad\qquad\qquad\qquad t \neq \tau_i,\\
\frac{d\varphi}{dt} &= \omega - \frac{\epsilon}{a\omega} f(\nu t, a\sin\varphi, a\omega\cos\varphi)\sin\varphi,\\
\Delta a|_{t=\tau_i} &= \frac{\epsilon}{\omega} I_i(a\sin\varphi, a\omega\cos\varphi)\cos\varphi,\\
\Delta\varphi|_{t=\tau_i} &= -\frac{\epsilon}{a\omega} I_i(a\sin\varphi, a\omega\cos\varphi)\sin\varphi.
\end{aligned} \tag{7.5}$$

Let now make the change of variables (7.3) in equations (7.1) but without transforming them to the form (7.4).

For $t \neq \tau_i$, we obtain the equations, which are the same as the first two equations of system (7.5).

Denote by x^+, $\dot{x}^+$, a^+, φ^+ the values of the coordinate x, derivative $\frac{dx}{dt}$, amplitude a, and the phase φ correspondingly at the time $t = \tau_i$ after an impulse has occurred. The increase of the amplitude and phase, Δa and $\Delta\varphi$, resulted from the impulsive effect, can be determined from the conditions $x^+ = x$, $\dot{x}^+ = \dot{x} + \epsilon I_i(x, \dot{x})$, i.e. from the conditions $a^+ \sin\varphi^+ = a\sin\varphi$, $a^+ \cos\varphi^+ = a\cos\varphi + \frac{\epsilon}{\omega} I_i(a\sin\varphi, a\omega\cos\varphi)$ or

$$(a + \Delta a)\sin(\varphi + \Delta\varphi) = a\sin\varphi,$$
$$(a + \Delta a)\cos(\varphi + \Delta\varphi) = \frac{\epsilon}{\omega} I_i(a\sin\varphi, a\omega\cos\varphi) + a\cos\varphi. \tag{7.6}$$

It follows from these equations that

$$(a + \Delta a)^2 = \frac{\epsilon^2}{\omega^2} I_i^2(a\sin\varphi, a\omega\cos\varphi) + \frac{2a\epsilon}{\omega} I_i(a\sin\varphi, a\omega\cos\varphi) + a^2.$$

Whence

$$a + \Delta a = a\left(1 + \epsilon\frac{2I_i(a\sin\varphi, a\omega\cos\varphi)}{a\omega} + \frac{\epsilon^2 I_i^2(a\sin\varphi, a\omega\cos\varphi)}{a^2\omega^2}\right)^{\frac{1}{2}},$$

or

$$\Delta a|_{\tau=\tau_i} = \frac{\epsilon}{\omega} I(a\sin\varphi, a\omega\cos\varphi)\cos\varphi + \frac{\epsilon^2}{a\omega^2} I^2(a\sin\varphi, a\omega\cos\varphi)\sin^2\varphi + \epsilon^3 \ldots.$$

We also find from equations (7.6) that

$$\cot(\varphi + \Delta\varphi) = \cot\varphi + \frac{\epsilon I_i(a\sin\varphi, a\omega\cos\varphi)}{a\omega\sin\varphi}$$

if $\sin\varphi$ and $\sin(\varphi + \Delta\varphi)$ are not equal to zero, or

$$\tan(\varphi + \Delta\varphi) = \frac{\sin\varphi}{\cos\varphi + \dfrac{\epsilon I(a\sin\varphi, a\omega\cos\varphi)}{a\omega}}$$

if $\cos\varphi + \frac{\epsilon}{a\omega} I(a\sin\varphi, a\omega\cos\varphi)$ and $\cos(\varphi + \Delta\varphi)$ are non-zero.

By solving one of these two equations for $\Delta\varphi$, we get

$$\Delta\varphi = -\frac{\epsilon}{a\omega} I(a\sin\varphi, a\omega\cos\varphi)\sin\varphi + \epsilon^2 \ldots$$

Hence, by applying change of variables (7.3), system (7.1) becomes

$$\begin{aligned}
\frac{da}{dt} &= \frac{\epsilon}{\omega} f(\nu t, a \sin\varphi, a\omega\cos\varphi)\cos\varphi, \\
& \qquad\qquad\qquad\qquad\qquad\qquad\qquad\qquad t \neq \tau_i, \\
\frac{d\varphi}{dt} &= \omega - \frac{\epsilon}{a\omega} f(\nu t, a\sin\varphi, a\omega\cos\varphi)\sin\varphi, \\
\Delta a|_{t=\tau_i} &= \frac{\epsilon}{\omega} I_i(a\sin\varphi, a\omega\cos\varphi)\cos\varphi + \epsilon^2 \ldots, \\
\Delta\varphi|_{t=\tau_i} &= -\frac{\epsilon}{a\omega} I_i(a\sin\varphi, a\omega\cos\varphi)\sin\varphi + \epsilon^2 \ldots,
\end{aligned}$$

which differs from system (7.5) by terms of the second order in ϵ. Clearly, this discrepancy does not influence the study of the first order approximation of system (7.1) but may be essential for constructing and studying higher order approximations.

A study of equations (7.1) and (7.2) is conducted differently depending on whether there is a resonance in the system or not. Note that there might be a resonance in system (7.2) only if the frequency of the fundamental oscillations ω is rationally commensurable with the frequency ν of the external force. Additionally, there might be a resonance in system (7.1) if ω is rationally commensurable with the number $\frac{2\pi}{T}$, where T is the period of impulsive effects. In the sequel we will consider these cases separately.

7.2 Formulas for an approximate solution of a non-resonance system

Let us construct a solution of equations (7.1) with the assumption that no resonance occurs. For simplicity, assume that the period of the function $f(\nu t, x, \frac{dx}{dt})$ coincides with the period of the impulsive effects, i.e. the times τ_i of the impulsive effects and the functions $I_i(x, \frac{dx}{dt})$ are

$$\tau_{i+p} - \tau_i = \frac{2\pi}{\nu}, \qquad I_{i+p}(x, \frac{dx}{dt}) = I_i(x, \frac{dx}{dt})$$

for all $i \in \mathbf{Z}$ and some natural p that indicates how many times the system undergoes an impulsive effect during the time $\frac{2\pi}{\nu}$. With such a simplification, the frequency of the external effect, ν, is the same for both the impulsive and continuous effects. Let ω be badly approximated by the numbers $\frac{r}{s}\nu$, where r and s are natural numbers.

By making in system (7.1) the change of variables

$$x = a\sin\varphi, \qquad \frac{dx}{dt} = a\omega\cos\varphi, \tag{7.7}$$

we get the equations

$$\begin{aligned} \frac{da}{dt} &= \frac{\epsilon}{\omega} f(\nu t, a\sin\varphi, a\omega\cos\varphi)\cos\varphi, \\ & \quad t \neq \tau_i, \\ \frac{d\omega}{dt} &= \omega - \frac{\epsilon}{a\omega} f(\nu t, a\sin\varphi, a\omega\cos\varphi)\sin\varphi, \\ \Delta a|_{t=\tau_i} &= \frac{\epsilon}{\omega} I_i(a\sin\varphi, a\omega\cos\varphi)\cos\varphi + \epsilon^2\ldots, \\ \Delta\varphi|_{t=\tau_i} &= -\frac{\epsilon}{a\omega} I_i(a\sin\varphi, a\omega\cos\varphi)\sin\varphi + \epsilon^2\ldots. \end{aligned} \tag{7.8}$$

Let us denote

$$\begin{aligned} f^{(1)}(\nu t, a, \varphi) &= f(\nu t, a\sin\varphi, a\omega\cos\varphi)\cos\varphi, \\ f^{(2)}(\nu t, a, \varphi) &= f(\nu t, a\sin\varphi, a\omega\cos\varphi)\sin\varphi, \\ I_i^{(1)}(a, \varphi) &= I_i(a\sin\varphi, a\omega\cos\varphi)\cos\varphi, \\ I_i^{(2)}(a, \varphi) &= I_i(a\sin\varphi, a\omega\cos\varphi)\sin\varphi, \end{aligned}$$

and assume that the functions $I_i^{(1)}(a,\varphi)$, $I_i^{(2)}(a,\varphi)$ are finite trigonometric polynomials in φ, and let

$$\frac{\partial I_i^{(j)}}{\partial \varphi}(a,\varphi) = \sum_{k=1}^{N}(A_{ik}^{(j)}(a)\sin k\varphi + B_{ik}^{(j)}(a)\cos k\varphi),$$
$$i=\overline{1,p}, \qquad j=1,2. \tag{7.9}$$

Denote by $Z_j(a,\varphi,t)$, $i=1,2$, the following functions

$$Z_j(a,\varphi,t) =$$
$$-\frac{1}{\pi}\sum_{i=1}^{P}\sum_{k=1}^{N}\left\{\left[A_{ik}^{(j)}(a)\sin k\varphi + B_{ik}^{(j)}(a)\cos k\varphi\right]\sum_{n=1}^{\infty}\frac{\cos n\nu(t-\tau_i)}{(k\omega)^2-n^2}+\right.$$
$$\left.+k\omega\left[B_{ik}^{(j)}(a)\sin k\varphi - A_{ik}^{(j)}(a)\cos k\varphi\right]\sum_{n=1}^{\infty}\frac{\sin n\nu(t-\tau_i)}{n[(k\omega)^2-n^2]}\right\}. \tag{7.10}$$

By a direct verification, we see that these functions, for $t\neq\tau_i$, satisfy the relation

$$\frac{\partial Z_j}{\partial\varphi}\omega + \frac{\partial Z_j}{\partial t}\nu = -\sum_{i=1}^{P}\frac{\partial I_i^{(j)}(a,\varphi)}{\partial\varphi}\left\{\frac{1}{2}-\frac{\nu(t-\tau_i)}{2\pi}\right\}, \tag{7.11}$$

where $\left\{\frac{1}{2}-\frac{t}{2\pi}\right\}$ denotes the function which is 2π-periodic and defined on the period to be $\frac{1}{\pi}\cdot\frac{\pi-t}{2}$, $0<t<2\pi$.

To get expressions for the first order approximations of solutions of equations (7.8) and to get equations for the first order approximations, change the variables in (7.8) by the formulas

$$a = b+\frac{\epsilon}{\omega}u^{(1)}(b,\theta,t), \qquad \varphi=\theta-\frac{\epsilon}{\omega b}u^{(2)}(b,\theta,t), \tag{7.12}$$

where

$$u^{(j)}(b,\theta,t) = \frac{1}{\omega}\int\!\!\int\left[f^j(\nu t,a,\varphi)+\frac{\nu}{2\pi}\sum_{i=1}^{P}I_i^{(j)}(b,\theta)-f_0^{(j)}(b)-\right.$$
$$\left.-\frac{\nu}{2\pi}\sum_{i=1}^{P}\bar{I}_i^{(j)}(b)\right]+Z_j(b,\theta,t)+\sum_{i=1}^{P}I_i^{(j)}(b,\theta)\left\{\frac{1}{2}-\frac{\nu(t-\tau_i)}{2\pi}\right\},$$
$$j=1,2.$$

Here $f_0^{(j)}(b)$ and $\bar{I}_i^{(j)}(b)$ are the mean values of the functions $f^{(j)}(\nu t, b, \theta)$ and $I_i^{(j)}(b, \theta)$ correspondingly, and the integrals mean the primitives, the mean values of which are zero. Substituting (7.12) into equations (7.8) we obtain the system

$$\begin{aligned}
\frac{db}{dt} &= \frac{\epsilon}{\omega}\left[f_0^{(1)}(b) + \frac{\nu}{2\pi}\sum_{i=1}^{P}\bar{I}_i^{(1)}(b)\right] + \epsilon^2 \ldots \\
& t \neq \tau_i \\
\frac{d\theta}{dt} &= \omega - \frac{\epsilon}{\omega b}\left[f_0^{(2)}(b) + \frac{\nu}{2\pi}\sum_{i=1}^{P}\bar{I}_i^{(2)}(b)\right] + \epsilon^2 \ldots \\
& \Delta b|_{t=\tau_i} = \epsilon^2 \ldots, \qquad \Delta\theta|_{t=\tau_i} = \epsilon^2 \ldots . \qquad (7.13)
\end{aligned}$$

If we neglect the terms of order ϵ^2 in these equations, we get equations for the first order approximations, and these equations are already without an impulsive effect,

$$\frac{db}{dt} = \frac{\epsilon}{\omega}F(b), \qquad \frac{d\theta}{dt} = \omega - \frac{\epsilon}{\omega b}\Phi(b), \qquad (7.14)$$

where

$$\begin{aligned}
F(b) &= f_0^{(1)}(b) + \frac{\nu}{2\pi}\sum_{i=1}^{P}\bar{I}_i^{(1)}(b), \\
\Phi(b) &= f_0^{(2)}(b) + \frac{\nu}{2\pi}\sum_{i=1}^{P}\bar{I}_i^{(2)}(b).
\end{aligned}$$

By using the change of variables (7.7) and (7.12), we get an expression for the first order approximation of the needed function x, which, up to terms of order ϵ^2, is given by

$$x = b\sin\theta + \frac{\epsilon}{\omega}\left(u^{(1)}(b, \theta, t)\sin\theta - u^{(2)}(b, \theta, t)\cos\theta\right), \qquad (7.15)$$

where b and θ are solutions of equations (7.14).

It should be noted that, although the functions $u^{(j)}(b, \theta, t)$, $j = 1, 2$, have first kind discontinuities at $t = \tau_j$, the function $x(t)$ is continuous.

As a simple example, consider the behavior of a linear oscillator subjected to a periodic impulsive effect which increases the moment of the oscillator by the quantity $\epsilon I(1-x^2)\dfrac{dx}{dt}$. Such an oscillator is given by the equations

$$\begin{aligned} \frac{d^2x}{dt^2}+\omega^2 x &= -\epsilon\lambda\frac{dx}{dt}, \qquad t\neq t_i, \\ \Delta\left.\frac{dx}{dt}\right|_{t=t_i} &= \epsilon I(1-x^2)\frac{dx}{dt}, \qquad \lambda>0, \qquad I>0. \end{aligned} \tag{7.16}$$

Assume that the impulsive effect occurs during the period 2π once, say at $t_i=\pi+2\pi i$. By applying change of variables (7.7) to equations (7.16), we get

$$\begin{aligned} \frac{da}{dt} = -\epsilon\lambda a\cos^2\varphi, \qquad & \frac{d\varphi}{dt} = \omega+\epsilon\lambda\cos\varphi\cdot\sin\varphi, \\ & t\neq t_i \\ \Delta a|_{t\neq t_i} = \epsilon I a(1-a^2\sin^2\varphi)&\cos^2\varphi+\epsilon^2\ldots, \\ \Delta a|_{t\neq t_i} = \epsilon I(1-a^2\sin^2\varphi)&\cos\varphi\sin\varphi+\epsilon^2\ldots. \end{aligned} \tag{7.17}$$

The corresponding averaged system will be

$$\frac{db}{dt}=\frac{\epsilon}{\omega}\left(-\frac{\lambda}{2}+\frac{I}{4\pi}-\frac{I}{16\pi}b^2\right)b, \qquad \frac{d\theta}{dt}=\omega. \tag{7.18}$$

One can see from these equations that, if $I<2\pi\lambda$, solutions of the first equations approach zero, which means that the oscillations in the system are damped.

If $I>2\pi\lambda$, then the first equation of (7.18) has the asymptotically stable stationary solution

$$b_0=2\sqrt{1-\frac{2\pi\lambda}{I}}.$$

This means that system (7.16) has a family of first order approximate solutions which are periodic and orbitally stable, $x=b_0\sin(\omega t+\theta_0)$, θ_0 is an arbitrary constant.

We will construct improved first order approximations of the corresponding stationary value of solutions of equations (7.16). To do that, we first construct the functions $Z_1(a,\varphi,t)$ and $Z_2(a,\varphi,t)$.

In the case under consideration

$$\begin{aligned} \frac{\partial I^{(1)}(a,\varphi)}{\partial\varphi} &= -Ia\omega(\sin 2\varphi+\frac{a^2}{2}\sin 4\varphi), \\ \frac{\partial I^{(2)}(a,\varphi)}{\partial\varphi} &= Ia\omega\left[\left(1-\frac{a^2}{2}\right)\cos 2\varphi+\frac{a^2}{2}\cos 4\varphi\right]. \end{aligned} \tag{7.19}$$

And so, by (7.10),

$$Z_1(a,\varphi,t) = \frac{Ia\omega}{\pi}\left[\sin 2\varphi \sum_{n=1}^{\infty} \frac{(-1)^{n-1}}{n^2-4\omega^2}\cos nt - \right.$$

$$- \cos 2\varphi \sum_{n=1}^{\infty} \frac{(-1)^{n-1}\cdot 2\omega}{n(n^2-4\omega^2)}\sin nt +$$

$$+\frac{a^2}{2}\left(\sin 4\varphi \sum_{n=1}^{\infty} \frac{(-1)^{n-1}}{n^2-16\omega^2}\cos nt - \right.$$

$$\left.\left. - \cos 4\varphi \sum_{n=1}^{\infty} \frac{(-1)^{n-1}\cdot 4\omega}{n(n^2-16\omega^2)}\sin nt\right)\right],$$

$$Z_2(a,\varphi,t) = -\frac{Ia\omega}{\pi}\left\{\left(1-\frac{a^2}{2}\right)\left[\cos 2\varphi \sum_{n=1}^{\infty} \frac{(-1)^{n-1}}{n^2-4\omega^2}\cos nt + \right.\right.$$

$$\left. + \sin 2\varphi \sum_{n=1}^{\infty} \frac{(-1)^{n-1}\cdot 2\omega}{n(n^2-4\omega^2)}\sin nt\right] +$$

$$+\frac{a^2}{2}\left[\cos 4\varphi \sum_{n=1}^{\infty} \frac{(-1)^{n-1}}{n^2-16\omega^2}\cos nt + \right.$$

$$\left.\left. + \sin 4\varphi \sum_{n=1}^{\infty} \frac{(-1)^{n-1}\cdot 4\omega}{n(n^2-16\omega^2)}\sin nt\right]\right\}.$$

Now it is easy to determine the functions $u^{(1)}(a,\varphi,t)$ and $u^{(2)}(a,\varphi,t)$. We have

$$\begin{aligned} u^{(1)}(a,\varphi,t) &= \frac{a}{4}\left(-\lambda + \frac{I}{2\pi}\right)\sin 2\varphi + \frac{Ia^3}{64\pi}\sin 4\varphi + Z_1(a,\varphi,t) + \\ &+\frac{Ia\omega}{\pi}(1-a^2\sin^2\varphi)\cos^2\varphi \sum_{n=1}^{\infty}\frac{(-1)^n \sin nt}{n}, \\ u^{(2)}(a,\varphi,t) &= \left(\lambda - \frac{I}{2\pi} + \frac{I}{4\pi}a^2\right)\frac{a}{4}\cos 2\varphi - \frac{Ia^3}{64\pi}\cos 4\varphi + Z_2(a,\varphi,t) + \\ &+\frac{Ia\omega}{\pi}(1-a^2\sin^2\varphi)\cos\varphi\sin\varphi \sum_{n=1}^{\infty}\frac{(-1)^n \sin nt}{n}. \end{aligned}$$

Hence, for the improved first order approximation, the family of stationary solutions corresponding to the equilibrium b_0 will be

$$x = b_0 \sin\theta + \frac{\epsilon}{\omega}\left\{\left(\frac{I}{2\pi} - \lambda\frac{Ib_0^2}{8\pi}\right)\frac{b_0}{4}\cos\theta + \right.$$

$$\left. + \frac{Ib_0^3}{64\pi}\cos 3\theta + Z_1(b_0,\theta,t)\sin\theta - Z_2(b_0,\theta,t)\cos\theta\right\}$$

and, using the expressions obtained for the last two terms, we will finally get

$$x = b_0\sin\theta + \frac{\epsilon}{\omega}\cdot\frac{Ib_0^3}{64\pi}\cos 3\theta +$$

$$+\frac{\epsilon I b_0}{\pi}\left\{\left(1-\frac{b_0^2}{4}\right)\left[\cos\theta\sum_{n=1}^{\infty}\frac{(-1)^{n-1}}{n^2-4\omega^2}\cos nt + \right.\right.$$

$$\left. + \sin\theta\sum_{n=1}^{\infty}\frac{(-1)^{n-1}\cdot 2\omega}{n(n^2-4\omega^2)}\sin nt\right] +$$

$$+\frac{b_0^2}{4}\left[\cos 3\theta\sum_{n=1}^{\infty}\frac{(-1)^{n-1}(n^2+8\omega^2)}{(n^2-16\omega^2)(n^2-4\omega^2)}\cos nt + \right.$$

$$\left.\left. + \sin 3\theta\sum_{n=1}^{\infty}\frac{(-1)^{n-1}\cdot 6n\omega}{(n^2-16\omega^2)(n^2-4\omega^2)}\sin nt\right]\right\}.$$

Because

$$\begin{aligned} x &= b\sin\theta + \frac{\epsilon}{\omega}\left[u^{(1)}(b,\theta,t)\sin\theta - u^{(2)}(b,\theta,t)\cos\theta\right], \\ \frac{dx}{dt} &= b\omega\cos\theta + \epsilon\left[u^{(1)}(b,\theta,t)\cos\theta + u^{(2)}(b,\theta,t)\sin\theta\right] \end{aligned}$$

up to order of ϵ^2, the one-parameter family of improved first order approximations of solutions of equations (7.16), which corresponds to the equilibrium b_0, covers the invariant set given in the extended phase space $\{x, \dot{x}, t\}$, up to order ϵ^2, by the expression $\sqrt{x^2 + \frac{1}{\omega^2}\dot{x}^2} = b_0 \frac{\epsilon}{\omega} u_1(b_0,\theta,t)$. All the solutions $x(t)$ of equations (7.16), which satisfy at a certain time $t = t_0$ the relation $\sqrt{x^2(t_0) + \frac{1}{\omega^2}\dot{x}^2(t_0)} = b_0 + \frac{\epsilon}{\omega}u_1(b_0,\theta_0,t_0)$ for some θ_0, also satisfy the relation

$$\sqrt{x^2(t) + \frac{1}{\omega^2}\dot{x}^2(t)} = b_0 + \frac{\epsilon}{\omega}u_1(b_0, \theta_0 + \omega(t-t_0), t)$$

for all $t \in \mathbf{R}$.

7.3 Substantiation of the averaging method for a non-resonance system

Now we shall prove that the averaging method can be used to study systems of type (7.8). In the preceding section we have shown that, by using change of variables (7.7) and (7.12), equations (7.7) are transformed into equations (7.13) which we will write as

$$\begin{aligned} \frac{db}{dt} &= \frac{\epsilon}{\omega}F(b) + \epsilon^2 R^{(1)}(b,\theta,t,\epsilon), \\ & t \neq \tau_i, \\ \frac{d\theta}{dt} &= \omega - \frac{\epsilon}{\omega b}\Phi(b) + \epsilon^2 R^{(2)}(b,\theta,t,\epsilon), \\ \Delta b|_{t=\tau_i} &= \epsilon^2 I_i^{(1)}(b,\theta,\epsilon), \qquad \Delta\theta|_{t=\tau_i} = \epsilon^2 I_i^{(2)}(b,\theta,\epsilon), \end{aligned} \tag{7.20}$$

where the functions $R^{(1)}(b,\theta t,\epsilon)$ and $R^{(2)}(b,\theta,t,\epsilon)$ are piecewise continuous in t with first kind discontinuities at $t = \tau_i$ and satisfy the Lipschitz conditions with respect b and θ uniformly in t and ϵ, $0 \le \epsilon < \epsilon_0$. The functions $I_i^{(1)}(b,\theta,\epsilon)$ and $I_i^{(2)}(b,\theta,\epsilon)$ satisfy the Lipschitz conditions with respect to b and θ uniformly in $i \in \mathbf{Z}$ and ϵ, $0 \le \epsilon \le \epsilon_0$.

Suppose that the equation $F(b) = 0$ has a simple positive root such that $F'(b_0) < 0$. Let $b = a(Ce^{\lambda t})$ be the general solution of the equation $\omega\frac{db}{dt} = F(b)$ in a neighborhood of the point $b = b_0$, $\lambda = \frac{1}{\omega}F'(b_0)$, C be an arbitrary constant. Denote

$$\bar{\Phi}(h) = -\int_0^h \frac{1}{\lambda\omega h}\left[\frac{\Phi(A(h))}{A(h)} - \frac{\Phi(b_0)}{b_0}\right] dh.$$

Evidently

$$\lambda h \frac{d\bar{\Phi}}{dh} = -\left(\frac{\Phi(A(h))}{\omega A(h)} - \frac{\Phi(b_0)}{\omega b_0}\right).$$

Let us make the following change of variables in (7.1),

$$b = A(h), \qquad \theta = \varphi - \bar{\Phi}(h). \tag{7.21}$$

If $t \neq \tau_i$, we get

$$\begin{aligned} \frac{dh}{dt} &= \epsilon\lambda h + \epsilon^2 P(h,\varphi,t,\epsilon), \\ \frac{d\varphi}{dt} &= \omega(\epsilon) + \epsilon^2 Q(h,\varphi,t,\epsilon), \end{aligned}$$

where

$$
\begin{aligned}
P(h,\varphi,t,\epsilon) &= \frac{1}{A'(h)}R^{(1)}(A(h),\varphi-\bar{\Phi}(h),t,\epsilon),\\
Q(h,\varphi,t,\epsilon) &= R^{(2)}(A(h),\varphi-\bar{\Phi}(h),t,\epsilon)+\\
&\quad +\frac{1}{A'(h)}\frac{d\bar{\Phi}}{dh}R^{(1)}(A(h),\varphi-\bar{\Phi}(h),t,\epsilon),\\
\omega(\epsilon) &= \omega-\frac{\epsilon}{\omega b_0}\Phi(b_0).
\end{aligned}
$$

In the case when $t=\tau_i$, we have

$$
\begin{aligned}
&A(h+\Delta h)-A(h)=\epsilon^2 J_i^{(1)}(A(h),\varphi-\Phi(h),\epsilon),\\
&\Delta\varphi-\left[\bar{\Phi}(h+\Delta h)-\bar{\Phi}(h)\right]=\epsilon^2 J_i^{(2)}(A(h),\varphi-\Phi(h),\epsilon).
\end{aligned}
$$

These equations can be solved for Δh and $\Delta\varphi$ with the solutions being of order ϵ^2. Hence change of variables (7.21) transforms (7.20) into the equations

$$
\begin{aligned}
\frac{dh}{dt} &= \epsilon\lambda h+\epsilon^2 P(h,\varphi,t,\epsilon),\\
\frac{d\varphi}{dt} &= \omega(\epsilon)+\epsilon^2 Q(h,\varphi,t,\epsilon),\\
\Delta h|_{t\neq\tau_i} &= \epsilon^2 H_i(h,\varphi,\epsilon), \qquad \Delta\varphi|_{t\neq\tau_i}+\epsilon^2 G_i(h,\varphi,\epsilon). \qquad (7.22)
\end{aligned}
$$

There exist such positive numbers ρ_0 and ϵ_0 that the functions $P(h,\varphi,t,\epsilon)$, $Q(h,\varphi,t,\epsilon)$, $H_i(h,\varphi,\epsilon)$, $G_i(h,\varphi,\epsilon)$ considered in the domain

$$
|\,h\,|\leq\rho_0, \qquad \varphi\in R, \qquad t\in R, \qquad 0<\epsilon\leq\epsilon_0, \qquad (7.23)
$$

are defined and continuous in h, φ, ϵ, piecewise continuous in t with first kind discontinuities at $t=\tau_i$, periodic in φ and νt with the period 2π, and satisfy the inequalities

$$
\begin{aligned}
&|P(h',\varphi',t,\epsilon)-P(h'',\varphi'',t,\epsilon)|+|Q(h',\varphi',t,\epsilon)-Q(h'',\varphi'',t,\epsilon)|+\\
&\quad+|H_i(h',\varphi',t,\epsilon)-H_i(h'',\varphi'',t,\epsilon)|+\\
&\quad+|G_i(h',\varphi',t,\epsilon)-G_i(h'',\varphi'',t,\epsilon)|\leq L(|h'-h''|+|\varphi'-\varphi''|), (7.24)\\
&|P(h,\varphi,t,\epsilon)|+|Q(h,\varphi,t,\epsilon)|+\\
&\quad+|H_i(h,\varphi,t,\epsilon)|+|G_i(h,\varphi,t,\epsilon)|\leq M, \qquad (7.25)
\end{aligned}
$$

where L and M are positive constants.

The assumption that the equation $F(b) = 0$ has a simple positive root implies that equations (7.14) have a stationary solution $b = b_0$, $\theta = \omega(\epsilon)t+\theta_0$, where θ_0 is an arbitrary constant. The corresponding approximate solution of equations (7.1) will be

$$\begin{aligned} x &= b_0 \sin(\omega(\epsilon)t + \theta_0) + \frac{\epsilon}{\omega}(u^{(1)}(b_0) \sin(\omega(\epsilon)t + \theta_0) - \\ &-u^{(2)}(b_0) \cos(\omega(\epsilon)t + \theta_0)). \end{aligned} \tag{7.26}$$

Theorem 109 *All solutions of impulsive differential equation (7.1), with the initial conditions being sufficiently close to the initial conditions for approximate stationary solutions (7.26), are uniformly bounded on the whole positive semiaxis $t \geq 0$ at least for sufficiently small values of the parameter ϵ.*

Lemma 49 *If the initial values h_0 are sufficiently small, then there exists such a positive number $\bar{\epsilon} \leq \epsilon_0$ that, for all $0 < \epsilon \leq \bar{\epsilon}$, the solutions $h_t(\varphi, \epsilon)$ of system (7.22) are uniformly bounded for all $t \geq t_0$.*

Proof. Fix a positive number δ_0, $\delta_0 < \rho_0$, and choose $\bar{\epsilon}$ such that

$$(1 + \bar{\epsilon}^2 L)^p \delta_0 + \frac{\bar{\epsilon} M p e^{-\epsilon\lambda T}}{T\left(|\lambda| - \bar{\epsilon}(1 + \frac{p}{T})L\right)} + \frac{\bar{\epsilon} M}{|\lambda| - \bar{\epsilon} L} \leq \rho_0.$$

By using (7.24) – (7.25), it follows from equations (7.22) that

$$\begin{aligned} e^{-\epsilon\lambda(t-t_0)}|h_t| &\leq h_0 + \epsilon^2 \int_{t_0}^{t} \left[e^{-\epsilon\lambda(\sigma-t_0)} M + L e^{-\epsilon\lambda(\sigma-t_0)}|h_\sigma|\right] d\sigma + \\ &+\epsilon^2 \sum_{t_0<\tau_i<t} \left(e^{-\epsilon\lambda(\tau_i-t_0)} M + L e^{-\epsilon\lambda(\tau_i-t_0)}|h_{\tau_i}|\right). \end{aligned}$$

Whence, by Lemma (ref. 4.2.2), we have

$$|h_t(\varphi, \epsilon)| \leq (1 + \epsilon^2 L)^P |h_0| + \frac{\epsilon M \rho e^{-\epsilon\lambda T}}{T\left[|\lambda| - \epsilon\left(1 + \frac{P}{T}\right) L\right]} + \frac{\epsilon M}{|\lambda| - \epsilon L}.$$

Consequently $|H_t(\varphi, \epsilon)| \leq \rho_0$ for all $t \geq t_0$, $0 < \epsilon \leq \bar{\epsilon}$, as long as $|h_0| < \delta$. This finishes the proof of the lemma.

□

Theorem 109 follows from Lemma 49 if changes of variables (7.7), (7.12), (7.21) are used because the initial condition h_0, expressed in terms of the variables b, has the form $b = b_0$ and so, if ϵ is sufficiently small, i.e. h_0 is close to zero, the initial conditions x_0, $\dot{x}_0$ for the initial variables are close to the corresponding initial values of approximate stationary solutions of (7.26). □

Let us show that, for small ϵ, system (7.22) has an integral set. We will look for this integral set to be in the form $\mathcal{T}(\epsilon)$: $h = u(\varphi, t, \epsilon)$, where $u(\varphi, t, \epsilon)$ is a function which is continuous in φ, piecewise continuous in t with first kind discontinuities at $t = \tau_i$, periodic with respect to φ and t with the period 2π.

Represent $\mathcal{T}(\epsilon)$ as the limit of the sequence $\mathcal{T}'(\epsilon), \ldots, \mathcal{T}^{(m)}(\epsilon), \ldots,$

$$\mathcal{T}^{(m+1)}(\epsilon) : h = u^{(m+1)}(\varphi, t, \epsilon), \qquad m = 0, 1, \ldots,$$

each of which is an integral set of the system

$$\begin{aligned}
\frac{dh}{dt} &= \epsilon\lambda h + \epsilon^2 P\left(u^{(m)}(\varphi, t, \epsilon), \varphi, t, \epsilon\right), \\
& \quad t \neq \tau_i, \\
\frac{d\varphi}{dt} &= \omega(\epsilon) + \epsilon^2 Q\left(u^{(m)}(\varphi, t, \epsilon), \varphi, t, \epsilon\right), \\
\Delta h|_{t=\tau_i} &= \epsilon^2 H_i\left(u^{(m)}(\varphi, \tau_i, \epsilon), \varphi, \epsilon\right) \\
\Delta \varphi|_{t=\tau_i} &= \epsilon^2 G_i\left(u^{(n)}(\varphi, \tau_i, \epsilon), \varphi, \epsilon\right).
\end{aligned} \tag{7.27}$$

For sufficiently small ϵ by Theorem 2 [90], the sequence of functions $u^{(m)}(\varphi, t, \epsilon)$ uniformly converges to the limit function $\lim_{m\to\infty} u^{(m)}(\varphi, t, \epsilon) = u(\varphi, t, \epsilon)$ which defines the integral set $\mathcal{T}(\epsilon)$ for equations (7.22).

We thus have proved the following theorem.

Theorem 110 *Let in the region $h \leq \rho_0$, $\varphi \in \mathbf{R}$, $t \in \mathbf{R}$, $0 < \epsilon \leq \epsilon_0$, system (7.22) satisfy inequalities (7.24) – (7.25). Then there exists such a positive number $\epsilon^0 \leq \epsilon_0$ that, for every ϵ, $0 < \epsilon \leq \epsilon^0$, system (7.22) has the integral set $\mathcal{T}(\epsilon)$: $h = u(\varphi, t, \epsilon)$, where the function $u(\varphi, t, \epsilon)$ is periodic with respect to φ and t with the period 2π, satisfies the Lipschitz condition in φ with the Lipschitz constant proportional to ϵ, piecewise continuous in ϵ, and $u(\varphi, t, \epsilon) \to 0$ as $\epsilon \to 0$. Being restricted to the integral set $\mathcal{T}(\epsilon)$, system*

(7.22) is reduced to the system

$$\begin{aligned}\frac{d\varphi}{dt} &= \omega(\epsilon)+\epsilon^2 Q(u(\varphi,t,\epsilon),\varphi,t,\epsilon), \qquad t\neq\tau_i,\\ \Delta\varphi|_{t=\tau_i} &= \epsilon^2 G_i(u(\varphi,\tau_i,\epsilon),\varphi,\epsilon).\end{aligned}$$

□

By using Theorems 109 and 110 and using change of variables (7.7) which reduces equations (7.1) to system (7.22), we can formulate the following theorem which gives a foundation for use of the averaging method to study impulsive differential equations of type (7.1).

Theorem 111 *Let, for impulsive system (7.1), the following conditions hold:*

a) the function $f(\nu t, x, \frac{dx}{dt})$ is periodic with respect to νt with the period 2π, continuously differentiable with respect to x and $\frac{dx}{dt}$ in $x^2 + \frac{1}{\omega^2}\left(\frac{dx}{dt}\right)^2 \leq \nu^2$, and the derivatives satisfy the Lipschitz condition in this region;

b) the functions $I_i\left(x, \frac{dx}{dt}\right)$, which are assumed to be finite polynomials in their variables, and the times τ_i satisfy the equalities

$$I_{i+p}\left(x,\frac{dx}{dt}\right) = I_i\left(x,\frac{dx}{dt}\right), \qquad \tau_{i+p} = \tau_i + \frac{2\pi}{\nu};$$

c) the equation

$$F(a) \equiv f_0^{(1)}(a) + \frac{\nu}{2\pi}\sum_{i=1}^{P} \bar{I}_i^{(1)}(a) = 0,$$

where

$$f_0^{(1)}(a) = \frac{1}{(4\pi)^2}\int_0^{2\pi}\int_0^{2\pi} f(\varphi, a\sin\psi, a\omega\cos\psi)\cos\psi d\varphi d\psi,$$

has such an isolated positive root $a = a_0$ that $F'(a_0) < 0$.

Then there exists such a positive number ϵ^* that, for all ϵ, $0 < \epsilon \leq \epsilon^*$,

1) *system (7.1) has the integral set* $\sqrt{x^2 + \frac{1}{\omega^2}\left(\frac{dx}{dt}\right)^2} = u^*(\varphi, t, \epsilon)$, *where the function* $u^*(\varphi, t, \epsilon)$ *satisfies the Lipschitz condition with respect to* φ, *is piecewise continuous in* φ *and* t *with first kind discontinuities at* $t = \tau_i$, *periodic with respect to* φ *and* t *with the period* 2π, *and such that* $u^*(\varphi, t, \epsilon) \to a_0$ *uniformly in* φ *and* t *as* $\epsilon \to 0$;

2) *there exists a* $\delta_0(\epsilon)$*-neighborhood of the cylinder* $x^2 + \frac{1}{\omega^2}\left(\frac{dx}{dt}\right)^2 = a_0^2$ ($\delta_0(\epsilon) \to 0$ *for* $\epsilon \to 0$) *such that any solution of system (7.1), which starts inside this neighborhood at* $t = t_0$, *and its derivative remain bounded for all* $t \geq t_0$.

7.4 Averaging in a resonance system and its substantiation

Let us study the impulsive system

$$\begin{aligned} \frac{d^2x}{dt^2} + \omega^2 x &= \epsilon f\left(\nu t, x, \frac{dx}{dt}\right), \qquad t \neq \tau_i, \\ \Delta \left.\frac{dx}{dt}\right|_{t=\tau_i} &= \epsilon I_i\left(x, \frac{dx}{dt}\right) \end{aligned} \tag{7.28}$$

in the case of a resonance. We assume that the frequency ν of the external effect is the same as the frequency of the impulsive effect, i.e.

$$\tau_{i+p} = \tau_i + \frac{2\pi}{\nu}, \qquad I_{i+p}\left(x, \frac{dx}{dt}\right) = I_i\left(x, \frac{dx}{dt}\right) \tag{7.29}$$

and, moreover, the proper oscillation frequency ω is close to one of the numbers $\frac{r}{s}\nu$, where r and s are natural numbers.

Set $\omega^2 = \frac{r^2}{s^2}\nu^2 - \epsilon\bar{\Delta}$ and, by using the formulas

$$x = a\sin\varphi, \qquad \frac{dx}{dt} = a\frac{r}{s}\nu\cos\varphi, \qquad \varphi = \frac{r}{s}\nu t + \theta, \tag{7.30}$$

write equations (7.28) in terms of the variables a and θ:

$$\begin{aligned} \frac{da}{dt} &= \frac{\epsilon s}{r\nu}\left[\bar{\Delta} a\sin\varphi + f(\nu t, a\sin\varphi, a\frac{r}{s}\nu\cos\varphi)\right]\cos\varphi, \\ & \quad t \neq \tau_i, \\ \frac{d\theta}{dt} &= -\frac{\epsilon s}{ar\nu}\left[\bar{\Delta} a\sin\varphi + f\left(\nu t, a\sin\varphi, a\frac{r}{s}\nu\cos\varphi\right)\right]\sin\varphi, \\ \Delta a|_{t=\tau_i} &= \frac{\epsilon s}{r\nu} I_i\left(a\sin\varphi, a\frac{r}{s}\nu\cos\varphi\right)\cos\varphi + \epsilon^2\ldots, \\ \Delta\theta|_{t=\tau_i} &= -\frac{\epsilon s}{ar\nu} I_i\left(a\sin\varphi, a\frac{r}{s}\nu\cos\varphi\right)\sin\varphi + \epsilon^2\ldots. \end{aligned} \tag{7.31}$$

System (7.31) is a system in standard form, 2π-periodic with respect to θ, and $2\pi s$-periodic with respect to νt. Solutions of the corresponding averaged system can be taken as first order approximations of solutions of system (7.31). The averaged system will be

$$\begin{aligned} \frac{da}{dt} &= \frac{\epsilon s}{r\nu}\left(A(a,\theta) + I^{(1)}(a,\theta)\right), \\ \frac{d\theta}{dt} &= -\frac{\epsilon s}{ar\nu}\left(\frac{\bar{\Delta} a}{2} + B(a,\theta) + I^{(2)}(a,\theta)\right), \end{aligned} \tag{7.32}$$

where

$$
\begin{aligned}
A(a,\theta) &= \frac{1}{2\pi r}\int_0^{2\pi r} f\left(\frac{s}{r}(\varphi-\theta), a\sin\varphi, a\frac{r}{s}\nu\cos\varphi\right)\cos\varphi,\\
B(a,\theta) &= \frac{1}{2\pi r}\int_0^{2\pi r} f\left(\frac{s}{r}(\varphi-\theta), a\sin\varphi, a\frac{r}{s}\nu\cos\varphi\right)\sin\varphi,\\
I^{(1)}(a,\theta) &= \frac{1}{2\pi s}\sum_{i=1}^{\rho s} I_i\left(a\sin\left(\frac{r}{s}\nu\tau_i+\theta\right), a\frac{r}{s}\nu\cos\left(\frac{r}{s}\nu\tau_i+\theta\right)\right)\times\\
&\times\cos\left(\frac{r}{s}\nu\tau_i+\theta\right),\\
I^{(2)}(a,\theta) &= \frac{1}{2\pi s}\sum_{i=1}^{\rho s} I_i\left(a\sin\left(\frac{r}{s}\nu\tau_i+\theta\right), a\frac{r}{s}\nu\cos\left(\frac{r}{s}\nu\tau_i+\theta\right)\right)\times\\
&\times\sin\left(\frac{r}{s}\nu\tau_i+\theta\right).
\end{aligned}
\tag{7.33}
$$

Improved first order approximations of a and θ are constructed according to the general method, given in [23], which is used for such constructions. So if, in equations (7.28), the function $f(\nu t, x, \frac{dx}{dt})$ does not explicitly depend on t, i.e. $f(\nu t, x, \frac{dx}{dt}) = f_0(x, \frac{dx}{dt})$, then

$$
\begin{aligned}
a(t) &= a + \frac{\epsilon s}{r\nu}\left[-\frac{\bar{\Delta}a}{4}\cos 2\varphi + \sum_{k=1}^{\infty}\frac{a_k^{(1)}\sin k\varphi - b_k^{(1)}\cos k\varphi}{k} + \right.\\
&\left. + \frac{1}{\pi}\sum_{i=1}^{\rho s} I_i^{(1)}(a,\theta)\sum_{k=1}^{\infty}\frac{\sin k\nu(t-\tau_i)}{k}\right],\\
\theta(t) &= \theta - \frac{\epsilon s}{ar\nu}\left[-\frac{\bar{\Delta}a}{4}\sin 2\varphi + \sum_{k=1}^{\infty}\frac{a_k^{(2)}\sin k\varphi - b_k^{(2)}\cos k\varphi}{k} + \right.\\
&\left. + \frac{1}{\pi}\sum_{i=1}^{\rho s} I_i^{(2)}(a,\theta)\sum_{k=1}^{\infty}\frac{\sin k\nu(t-\tau_i)}{k}\right],
\end{aligned}
\tag{7.34}
$$

where a and θ are solutions of first order approximation equations (7.32), $a_k^{(1)}$, $b_k^{(1)}$, $a_k^{(2)}$, $b_k^{(2)}$ are Fourier coefficients of the functions

$$
f_0\left(a\sin\varphi, a\frac{r}{s}\nu\cos\varphi\right)\cos\varphi \quad \text{and} \quad f_0\left(a\sin\varphi, a\frac{r}{s}\nu\cos\varphi\right)\sin\varphi
$$

correspondingly, and

$$I_i^{(1)}(a,\theta) = I_i\left(a\sin\left(\frac{r}{s}\nu\tau_i+\theta\right), a\frac{r}{s}\nu\cos\left(\frac{r}{s}\nu\tau_i+\theta\right)\right)\cos\left(\frac{r}{s}\nu\tau_i+\theta\right),$$
$$I_i^{(2)}(a,\theta) = I_i\left(a\sin\left(\frac{r}{s}\nu\tau_i+\theta\right), a\frac{r}{s}\nu\cos\left(\frac{r}{s}\nu\tau_i+\theta\right)\right)\sin\left(\frac{r}{s}\nu\tau_i+\theta\right).$$

Consider a few examples. Let a system, which is described by the Van der Pol equation, be subjected to an impulsive effect which increases the moment of the system by a constant value ϵT, $I > 0$, at the time when the impulse occurs. Such a system is given by the equations

$$\begin{aligned} \frac{d^2x}{dt^2}+\omega^2 x &= \epsilon(1-x^2)\frac{dx}{dt}, \qquad t \neq \tau_i, \\ \Delta\frac{dx}{dt}\bigg|_{t\neq\tau_i} &= \epsilon I. \end{aligned} \tag{7.35}$$

Suppose that $\tau_i = \dfrac{2\pi i}{\omega}$, i.e. the main resonance is considered. In order to get the system in standard form, let us make in system (7.35) the change of variables

$$x = a\sin(\omega t+\theta), \qquad \frac{dx}{dt} = a\omega\cos(\omega t+\theta). \tag{7.36}$$

We get

$$\begin{aligned} \frac{da}{dt} &= \frac{\epsilon}{\omega}(1-a^2\sin^2\varphi)a\cos^2\varphi, \\ &\quad \omega t \neq 2\pi i, \\ \frac{d\theta}{dt} &= -\epsilon(1-a^2\sin^2\varphi)\sin\varphi\cos\varphi, \\ \Delta a|_{t=t_i} &= \frac{\epsilon}{\omega}I\cos\theta+\epsilon^2\ldots, \quad \Delta\theta|_{t=t_i} = -\frac{\epsilon}{a\omega}\sin\theta+\epsilon^2\ldots. \end{aligned} \tag{7.37}$$

By taking the average with respect to the time explicitly contained in the system, we get the system of first order approximation

$$\begin{aligned} \frac{da}{dt} &= \frac{\epsilon a}{2}\left(1-\frac{a^2}{4}\right)+\frac{\epsilon I}{2\pi}\cos\theta, \\ \frac{d\theta}{dt} &= -\frac{\epsilon I}{2\pi a}\sin\theta. \end{aligned} \tag{7.38}$$

This system has two stationary phases $\theta_1 = 0$ and $\theta_2 = \pi$, and to each of them there correspond stationary amplitudes, given by the equations

$$a\left(1-\frac{a^2}{4}\right)+\frac{I}{\pi}=0 \quad \text{and} \quad a\left(1-\frac{a^2}{4}\right)-\frac{I}{\pi}=0. \tag{7.39}$$

It follows from the first equation that if $I > 0$, then the equation has one positive solution $a = a_0^{(1)}$. The second equation of (7.39) has two positive solutions $a = a_0^{(1)}$, $a = a_0^{(3)}$ if $0 < I < \frac{2\pi}{3\sqrt{3}}$, and has no positive solutions if $I > \frac{2\pi}{3\sqrt{3}}$.

The second equation shows that the stationary phase $\theta = \pi$ is unstable, and consequently the stationary oscillations corresponding to the stationary amplitudes $a_0^{(2)}$ and $a_0^{(3)}$, and the stationary phase $\theta = \pi$ are unstable. If we write the variational equations corresponding to the equilibrium point $(a_0^{(1)}, 0)$, then the eigenvalues of the matrices of these equations will be $\lambda_1 = 1 - \frac{3}{4}a_0^{(2)}$, $\lambda_2 = -\frac{1}{a_0^{(1)}}$. It is easy to see that $a_0^{(1)} > 2$ and, hence, λ_1 and λ_2 are negative.

Thus, if $I > 0$, system (7.38) has the unique asymptotically stable equilibrium $(a_0^{(1)}, 0)$.

The solutions of system (7.37), which correspond to this equilibrium, have the improved first order approximations:

$$\begin{aligned} a(t) &= a_0 + \frac{\epsilon a_0}{4}\left(\sin 2\omega t + \frac{a_0^2}{8}\sin 4\omega t\right) + \frac{\epsilon I}{\pi\omega}\sum_{k=1}^{\infty}\frac{\sin k\omega t}{k}, \\ \theta(t) &= \frac{\epsilon}{4}\left(1-\frac{a_0^2}{2}\right)\cos 2\omega t + \frac{\epsilon a_0^2}{32}\cos 4\omega t. \end{aligned}$$

Thus stable stationary oscillations in system (7.35), up to terms of the second order with respect to ϵ, are given by

$$\begin{aligned} x(t) &= a_0 \sin\omega t + \frac{\epsilon a_0}{4}\left[\left(1-\frac{a_0^2}{4}\right)\cos\omega t - \frac{a_0^2}{8}\cos\omega t\right] + \\ &\quad + \frac{\epsilon I}{\pi\omega}\sin\omega t \sum_{k=1}^{\infty}\frac{\sin k\omega t}{k}, \end{aligned}$$

where a_0 is a positive root of the first equation in (7.39). It should be also noted that the function $x(t)$ is continuous, 2π-periodic, and its derivative has first kind discontinuities at the points $\omega t = 2\pi i$ with the jump equal to ϵI.

Let us consider the same Van der Pol impulsive equation but with the assumption that the impulsive effect increases its kinetic energy by a constant value, i.e. the system is given as

$$\begin{aligned} \frac{d^2x}{dt^2} + \omega^2 x &= \epsilon(1-x^2)\frac{dx}{dt}, \qquad t \neq \tau_i = \frac{2\pi i}{\omega}, \\ \Delta\left(\frac{dx}{dt}\right)^2\Bigg|_{\omega t = \omega\tau_i = 2\pi i} &= 2\epsilon I. \end{aligned} \tag{7.40}$$

By solving the second equation of (7.40) for $\Delta\frac{dx}{dt}$, write the equations in the following form

$$\begin{aligned} \frac{d^2x}{dt^2} + \omega^2 x &= \epsilon(1-x^2)\frac{dx}{dt}, \qquad t \neq \tau_i, \\ \Delta x|_{t=\tau_i} &= \frac{\epsilon I}{\frac{dx}{dt}} + \epsilon^2 \ldots. \end{aligned} \tag{7.41}$$

Let us introduce the variables a and θ by applying the change of variables (7.36) to system (7.41). We have

$$\begin{aligned} &\frac{da}{dt} = \epsilon(1 - a^2\sin^2\varphi)a\cos^2\varphi, \qquad \frac{d\theta}{dt} = \epsilon(1 - a^2\sin^2\varphi)\sin\varphi\cos\varphi, \\ &\qquad\qquad t \neq \tau_i, \\ &\Delta a|_{t=\tau_i} = \frac{\epsilon I}{a\omega^2} + \epsilon^2 \ldots, \qquad \Delta\theta|_{t=\tau_i} = -\frac{\epsilon I}{a^2\omega^2}\tan\theta + \epsilon^2 \ldots. \end{aligned} \tag{7.42}$$

The averaged system corresponding to (7.42) will be

$$\frac{da}{dt} = \frac{\epsilon a}{2}\left(1 - \frac{a^2}{4}\right) + \frac{\epsilon I}{2\pi a\omega}, \qquad \frac{d\theta}{dt} = -\frac{\epsilon I}{2\pi a^2\omega}\tan\theta. \tag{7.43}$$

It is readily seen from equations (7.43) that stationary phases are $\theta = k\pi$ and the second equation implies that the stationary phases are asymptotically stable. Stationary amplitudes are determined as positive solutions of the equation

$$a^4 - 4a^2 - \frac{4I}{\pi\omega} = 0 \tag{7.44}$$

or

$$a^2 = 2 \pm 2\sqrt{1 + \frac{I}{\pi\omega}}.$$

If $I \geq 0$, then equation (7.44) has the unique positive solution

$$a = \sqrt{2 + 2\sqrt{1 + \frac{I}{\pi\omega}}}, \tag{7.45}$$

if $-\pi\omega < I < 0$, there are two such solutions – $a = \sqrt{2 \pm 2\sqrt{1 + \frac{I}{\pi\omega}}}$, and if $I < -\pi\omega$, then equation (7.44) does not have real roots. In the case when $-\pi\omega < I < 0$, the stationary amplitude $a = \sqrt{2 + 2\sqrt{1 + \frac{I}{\pi\omega}}}$ is asymptotically stable but the stationary amplitude $a = \sqrt{2 - 2\sqrt{1 + \frac{I}{\pi\omega}}}$ is unstable. In the case when $I \geq 0$, the stationary amplitude given by (7.45) is asymptotically stable.

Hence, by examining the first order approximation, one can see that if $I \geq -\pi\omega$, then system (7.40) has a family of oscillations with the amplitudes equal to $\sqrt{2 + 2\sqrt{1 + \frac{I}{\pi\omega}}}$ but all these oscillations eventually approach two asymptotically stable oscillations with the stationary phases $\theta = 0$ and $\theta = \pi$.

In other words, when considered in the extended phase space $(x, \dot{x}, t)$, system (7.40) has an asymptotically stable invariant set which is defined, up to terms of order ϵ, by the expression

$$x^2 + \frac{1}{\omega^2}\left(\frac{dx}{dt}\right)^2 = 2 + 2\sqrt{1 + \frac{I}{\pi\omega}} \tag{7.46}$$

and there exist in this set two periodic solutions which attract the other solutions. The first order approximation of these solutions is given by

$$x = \pm\sqrt{2 + 2\sqrt{1 + \frac{I}{\pi\omega}}}\sin\omega t.$$

Besides the invariant set, which is approximately given by (7.46), in the case when $-\pi\omega < I < 0$, system (7.40) also has the invariant set defined by

$$x^2 + \frac{1}{\omega^2}\left(\frac{dx}{dt}\right)^2 = 2 - 2\sqrt{1 + \frac{I}{\pi\omega}},$$

but this set is asymptotically unstable.

Periodic solutions of system (7.42), which correspond to stable equilibriums of the averaged equations (7.43), are given by the following expressions

$$a(t) = a_0 + \frac{\epsilon a_0}{4}\left(\sin 2\omega t + \frac{a_0^2}{8}\sin 4\omega t\right) + \frac{\epsilon I}{\pi a \omega^2}\sum_{k=1}^{\infty}\frac{\sin k\omega t}{k},$$

$$\theta(t) = \frac{\pi}{2}\left(1-(-1)^j\right) + \frac{\epsilon}{4}\left(1 - \frac{a_0^2}{2}\right)\cos 2\omega t + \frac{\epsilon a_0^2}{32}\cos 4\omega t, \tag{7.47}$$

where $a_0^2 = 2 + 2\sqrt{1 + \frac{I}{\pi\omega}}$, $j = 0, 1$.

Now we will give sufficient conditions for solutions of equations (7.28) to remain close, on an infinite time interval, to solutions of the corresponding averaged equations (7.32) or, more precisely, for equilibriums of the averaged system (7.32) to give rise to periodic solutions of equations (7.28).

In order to make the work of proving the corresponding statements less technical , we will consider the case when the system has the main resonance, i.e. $\omega = \nu$, and consider the question of whether system (7.28) has solutions which are 2π-periodic with respect to νt.

Assume that equations (7.28) have such a solution $x = x(t)$, for which the jumps of the derivative $\Delta\dot{x}$ at $t = \tau_i$ are given by $\Delta\dot{x}|_{t=\tau_i} = \epsilon I_i^0$, where the quantities I_i^0 are such that $I_{i+p}^0 = I_i^0$. Consequently, the periodic solution $x = x(t)$ satisfies the equations

$$\begin{aligned} \frac{d^2x}{dt^2} + \omega^2 x &= \epsilon f\left(\nu t, x, \frac{dx}{dt}\right), \qquad t \neq \tau_i, \\ \Delta \left.\frac{dx}{dt}\right|_{t=\tau_i} &= \epsilon I_i^0. \end{aligned} \tag{7.48}$$

On the other hand, let, for some I_i^0, $I_{i+p}^0 = I_i^0$, equations (7.48) have a solution $x = x(t)$, which is 2π-periodic with respect to νt. This solution is a solution of the initial equations (7.28) as long as

$$I_i^0 = I_i\left(x(\tau_i), \frac{dx}{dt}(\tau_i)\right). \tag{7.49}$$

Hence, finding periodic solutions of system (7.28) is reduced to finding solutions of system (7.48), which are 2π-periodic with respect to νt, and choosing the parameters I_i^0 that satisfy relation (7.49). Let us rewrite system

(7.48) as

$$\begin{aligned}\frac{d^2x}{dt^2}+\nu^2x &= \epsilon\left[\bar{\Delta}x+f\left(\nu t,x,\frac{dx}{dt}\right)\right], \qquad t\neq\tau_i,\\ \Delta\left.\frac{dx}{dt}\right|_{t=\tau_i} &= \epsilon I_i^0\end{aligned}$$

and make the change of variable

$$x=\epsilon\sum_{i=1}^{p}I_i^0\varphi(\nu(t-\tau_i))+z,$$

where

$$\varphi(\nu t)=\frac{1}{\pi\nu}\left(\frac{1}{2}-\sum_{k=2}^{\infty}\frac{\cos k\nu t}{k^2-1}\right).$$

We get the following ordinary differential equation in z,

$$\begin{aligned}&\frac{d^2x}{dt^2}+\nu^2z=\epsilon\Bigg[\bar{\Delta}z+\\ &+f\left(\nu t,z+\epsilon\sum_{i=1}^{\rho}I_i^0\varphi(\nu(t-\tau_i)),\frac{dz}{dt}+\epsilon\sum_{i=1}^{\rho}I_i^0\varphi'(\nu(t-\tau_i))\right)+\\ &+\bar{\Delta}\sum_{i=1}^{\rho}I_i^0\varphi(\nu t-\tau_i))\Bigg]+\frac{\epsilon\nu}{\pi}\sum_{i=1}^{\rho}I_i^0\cos(\nu t-\tau_i),\end{aligned}$$

From which, by using change of variables (7.30) with $r=s$, $x=z$, we get the system in standard form

$$\begin{aligned}&\frac{da}{dt}=\frac{\epsilon}{\nu}\Bigg[\Delta a\sin\varphi+\\ &+f\left(\nu t,a\sin\varphi+\epsilon\sum_{i=1}^{\rho}I_i^0\varphi(\nu(t-\tau_i)),a\nu\cos\varphi+\epsilon\sum_{i=1}^{\rho}I_i^0\varphi'(\nu(t-\tau_i))\right)+\\ &+\bar{\Delta}\sum_{i=1}^{\rho}I_i^0\varphi(\nu(t-\tau_i))\Bigg]\cos\varphi+\frac{\epsilon}{\pi}\sum_{i=1}^{\rho}I_i^0\cos\nu(t-\tau_i)\cos\varphi,\end{aligned}$$

$$\frac{d\theta}{dt} = -\frac{\epsilon}{a\nu}\Bigg[\Delta a \sin\varphi +$$

$$+f\left(\nu t, a\sin\varphi + \epsilon\sum_{i=1}^{p} I_i^0 \varphi(\nu t - \tau_i)), a\nu\cos\varphi + \epsilon\sum_{i=1}^{p} I_i^0 \varphi'(\nu(t-\tau_i))\right) +$$

$$+\bar{\Delta}\sum_{i=1}^{p} I_i^0 \varphi(\nu(t-\tau_i))\Bigg]\sin\varphi - \frac{\epsilon}{\pi a}\sum_{i=1}^{p} I_i^0 \cos(t-\tau_i)\sin\varphi. \tag{7.50}$$

Relations (7.49), expressed in terms of the variables a and θ become

$$I_i^0 = I_i\left(a(\tau_i)\sin(\nu\tau_i + \theta(\tau_i)) + \epsilon\sum_{j=1}^{p} I_j^0 \varphi(\nu(\tau_i - \tau_j)),\right.$$

$$\left. a(\tau_i)\nu\cos(\nu\tau_i + \theta(\tau_i)) + \epsilon\sum_{j=1}^{p} \varphi'(\nu(\tau_i - \tau_j))\right). \tag{7.51}$$

Assume that the function $f(\nu t, x, y)$ satisfy the Lipschitz condition with respect to x and y in some annular domain

$$\nu^2\alpha^2 \le \nu^2 x^2 + y^2 \le \beta\nu^2 \tag{7.52}$$

Then system (7.50) is a T-system in the strip $0 < \alpha \le a \le \beta$ ($T = \frac{2\pi}{\nu}$) and, by [131], the solution $(a(t), \theta(t))$ of (7.50), which starts at the point (a_0, θ_0) for $t = 0$, is 2π-periodic with respect to νt if (a_0, θ_0) satisfies the system

$$\frac{\epsilon}{\nu}A(a,\theta) + \frac{\epsilon}{2\pi}\sum_{i=1}^{p} I_i^0 \cos(\nu\tau_i + \theta) + \epsilon^2 \ldots = 0$$

$$-\frac{\epsilon}{a\nu}\left(\frac{\bar{\Delta}a}{2} + B(a,\theta)\right) - \frac{\epsilon}{2\pi a}\sum_{i=1}^{p} I_i^0 \sin(\nu\tau_i + \theta) + \epsilon^2 \ldots = 0, \tag{7.53}$$

where $A(a,\theta)$ and $B(a,\theta)$ are given by the first two equalities of (7.33) with $r = s = 1$.

Moreover, by [131], this periodic solution is a limit of the uniformly convergent sequence of periodic functions

$$a_{m+1}(t) =$$

$$= a_0 + \frac{\epsilon}{\nu}\int_0^t \left[F_1(\tau, a_m(\tau), \theta_m(\tau), \epsilon) - \overline{F_1(\tau, a_m(\tau), \theta_m(\tau), \epsilon)}\right] d\tau,$$

$$\theta_{m+1}(t) =$$
$$= \theta_0 + \frac{\epsilon}{\nu} \int_0^t \left[F_2(\tau, a_m(\tau), \theta_m(\tau), \epsilon) - \overline{F_1(\tau, a_m(\tau), \theta_m(\tau), \epsilon)} \right] d\tau,$$

Here $\frac{\epsilon}{\nu} F_1(t, a, \theta, \epsilon)$ and $-\frac{\epsilon}{\nu} F_2(t, a, \theta, \epsilon)$ are the right-hand sides of system (7.50), and $\bar{F}(t) = \frac{\nu}{2\pi} \int_0^{2\pi} F(t)dt$, $m = 0, 1, 2, \ldots$.

In particular, for $m = 0$, we find that, up to order ϵ^2,

$$a_1(t) = a_0 - \frac{\epsilon \bar{\Delta} a_0}{4\nu^2} (\cos 2(\nu t + \theta_0) - \cos 2\theta_0) +$$
$$+ \frac{\epsilon}{\nu^2} \int_{\theta_0}^{\nu t + \theta_0} [f(\varphi - \theta_0, a_0 \sin\varphi, a_0 \nu \cos\varphi) \cos\varphi - A(a_0, \theta_0)] d\varphi +$$
$$+ \frac{\epsilon}{4\pi\nu} \sum_{i=1}^{p} I_i^0 (\sin(2\nu t + \theta_0 - \nu\tau_i) - \sin(\theta - \nu\tau_i))$$
$$\theta(t) = \theta_0 + \frac{\epsilon \bar{\Delta}}{4\nu^2} (\sin 2(\nu t + \theta_0) - \sin\theta_0) -$$
$$- \frac{\epsilon}{a_0 \nu^2} \int_{\theta_0}^{\nu t + \theta_0} [f(\varphi - \theta_0, a_0 \sin\varphi, a_0 \nu \cos\varphi) \sin\varphi - B(a_0, \theta_0)] d\varphi +$$
$$+ \frac{\epsilon}{4\pi\nu a_0} \sum_{i=1}^{p} I_i^0 (\cos(2\nu t + \theta_0 - \nu\tau_i) - \cos(\theta - \nu\tau_i)). \qquad (7.54)$$

Assume that the functions $I_i(x, y)$, $i = \overline{1, p}$, satisfy the Lipschitz condition with respect to x, y in the domain (7.52). Then, for small ϵ, equations (7.51) can be solved,

$$I_i^0 = I_i^0(a_0, \theta_0, \epsilon), \qquad i = \overline{1, p}, \qquad (7.55)$$

and the functions $I_i^0(a_0, \theta_0, \epsilon)$ satisfy the Lipschitz conditions with respect to a_0, θ_0, ϵ in the region, to which these variables belong. It follows from (7.51) that, up to terms of order ϵ, we have

$$I_i^0(a_0, \theta_0, \epsilon) = I_i(a_0 \sin(\nu\tau_i + \theta_0), a_0 \nu \cos(\nu\tau_i + \theta_0)), \qquad i = \overline{1, p}.$$

By substituting (7.55) into (7.53), we get a system for determining a_0 and

θ_0. It can be written as

$$A(a,\theta)+\frac{\nu}{2\pi}\sum_{i=1}^{\rho}I_i(a\sin(\nu\tau_i+\theta),a\nu\cos(\nu\tau_i+\theta))\times$$
$$\times\cos(\nu\tau_i+\theta)+\epsilon^2\ldots=0,$$
$$\frac{\bar{\Delta}a}{2}+B(a,\theta)+\frac{\nu}{2\pi}\sum_{i=1}^{\rho}I_i(a\sin(\nu\tau_i+\theta),a\nu\cos(\nu\tau_i+\theta))\times$$
$$\times\sin(\nu\tau_i+\theta)+\epsilon^2\ldots=0. \qquad (7.56)$$

If ϵ is small, this system has a solution as soon as the equations

$$A(a,\theta)+\frac{\nu}{2\pi}\sum_{i=1}^{\rho}I_i(a\sin(\nu\tau_i+\theta),a\nu\cos(\nu\tau_i+\theta))\times$$
$$\times\cos(\nu\tau_i+\theta)=0,$$
$$\frac{\bar{\Delta}a}{2}+B(a,\theta)+\frac{\nu}{2\pi}\sum_{i=1}^{\rho}I_i(a\sin(\nu\tau_i+\theta),a\nu\cos(\nu\tau_i+\theta))\times$$
$$\times\sin(\nu\tau_i+\theta)=0. \qquad (7.57)$$

have solutions (a^0,θ^0), ρ-neighborhoods of which belong to the strip $\alpha < a^0 < \beta$ and the index of the mapping defined by the left-hand sides of equations (7.57) is nonzero at the point (a^0,θ^0). It follows from the form of the change of variables used in equation (7.28) that, up to terms of some order $\delta(\epsilon)$, $\delta(\epsilon)\to 0$ for $\epsilon\to 0$, the periodic solutions themselves are given by

$$\begin{aligned} x(t) &= a(t)\sin(\nu t+\theta(t))+ \\ &+\frac{\epsilon}{\pi\nu}\sum_{i=1}^{\rho}I_i(a_0\sin(\nu\tau_i+\theta_0),a_0\nu\cos(\nu\tau_i+\theta_0))\times \\ &\times\left(\frac{1}{2}-\sum_{k=2}^{\infty}\frac{\cos k\nu(t-\tau_i)}{k^2-1}\right), \end{aligned} \qquad (7.58)$$

where $a(t)$, $\theta(t)$ are the functions defined by (7.54), and a_0, θ_0 are solutions of (7.57).

We thus have proved the following theorem.

Theorem 112 *Let, in system (7.28), the functions $f(\nu t,x,y)$, $I_i(x,y)$ be continuous and satisfy the Lipschitz condition with respect to x and y in the* *domain* $a\alpha^2 \le x^2 + \frac{y^2}{\nu^2} \le \beta^2$. *Suppose that the function* $(\nu t, x, y)$ *is* 2π-*periodic with respect to* νt, $\omega \simeq \nu$, *and equalities (7.29) hold. Assume that system (7.57) has an isolated solution* $a = a_0$, $\theta = \theta_0$, *a* ρ-*neighborhood of which belongs to the strip* $\alpha < a_0 < \beta$ *and such that the index of the mapping, defined by the left-hand sides of equations (7.57), is different from zero at the point* (a_0, θ_0). *Then there exists such* ϵ_0 *that, for all* ϵ, $0 < \epsilon \le \epsilon_0$, *system (7.28) has a solution, which is* 2π-*periodic with respect to* νt *and such that, up to terms of some order* $\delta(\epsilon)$, $\delta(\epsilon) \to 0$ *if* $\epsilon \to 0$, *it is given by expression (7.58) and satisfies*

$$|a(t) - a_0| + |\theta(t) - \theta_0| < \eta(\epsilon),$$

where $\eta(\epsilon) \to 0$ *for* $\epsilon \to 0$.

7.5 Formulas for approximate solutions for an impulsive effect occuring at fixed positions

In this section we will be considering oscillation of a point, the movement of which is described by the equations

$$\frac{d^2x}{dt^2} + \omega^2 x = \epsilon f_1\left(\nu t, x, \frac{dx}{dt}\right),$$

with an impulsive effect occurring each time when the point has a position $x = x_0$. As before, we assume that the impulsive effect increases the moment by a quantity depending on the velocity at the time when the impulsive effect occurs. The system under consideration is given by

$$\begin{aligned} \frac{d^2x}{dt^2} + \omega^2 x &= \epsilon f\left(\nu t, x, \frac{dx}{dt}\right), \qquad x \neq x_0, \\ \Delta \left.\frac{dx}{dt}\right|_{x=x_0} &= \epsilon I\left(\frac{dx}{dt}\right). \end{aligned} \tag{7.59}$$

We first consider an autonomous system, i.e. when

$$f\left(\nu t, x, \frac{dx}{dt}\right) = f\left(x, \frac{dx}{dt}\right).$$

To find asymptotic solutions of this system, make the change of variables

$$x = a\sin\varphi, \qquad \frac{dx}{dt} = a\omega\cos\varphi, \qquad \varphi = \omega t + \theta \tag{7.60}$$

which transforms system (7.59) into the standard form

$$\begin{aligned} \frac{da}{dt} &= \frac{\epsilon}{\omega} f(a\sin\varphi, a\omega\cos\varphi)\cos\varphi, \\ & \quad a\sin\varphi \neq x_0, \\ \frac{d\theta}{dt} &= -\frac{\epsilon}{a\omega} f(a\sin\varphi, a\omega\cos\varphi)\sin\varphi, \\ \Delta a|_{a\sin\varphi = x_0} &= \frac{\epsilon}{\omega} I(a\omega\cos\varphi)\cos\varphi + \epsilon^2 \ldots, \\ \Delta\theta|_{a\sin\varphi = x_0} &= -\frac{\epsilon}{a\omega} I(a\omega\cos\varphi)\sin\varphi + \epsilon^2 \ldots. \end{aligned} \tag{7.61}$$

Because $a\sin\varphi = x_0$ never holds if $a < |x_0|$, system (7.61) will not be an impulsive system if $a < |x_0|$, and so we will consider equations (7.61) in the region $a \geq |x_0|$.

By solving the equation $a\sin(\omega t+\theta) = x_0$ for t, we can assume that the system undergoes an impulsive effect when the phase point meets the surface $t = t_i(a,\theta)$, where

$$t_i(a,\theta) = \frac{1}{\omega}\left(-\theta + (-1)^i \arcsin\frac{x_0}{a} + \pi i\right). \tag{7.62}$$

Hence equation (7.61) can be written as

$$\begin{aligned}
\frac{da}{dt} &= \frac{\epsilon}{\omega} f(a\sin\varphi, a\omega\cos\varphi)\cos\varphi, \\
&\quad t \neq t_i(a,\theta), \\
\frac{d\theta}{dt} &= -\frac{\epsilon}{a\omega} f(a\sin\varphi, a\omega\cos\varphi)\sin\varphi, \\
\Delta a|_{t=t_i(a,\theta)} &= (-1)^i\frac{\epsilon}{\omega a} I\left((-1)^i\omega\sqrt{a^2-x_0^2}\right)\sqrt{a^2-x_0^2} + \epsilon^2\ldots, \\
\Delta\theta|_{t=t_i(a,\theta)} &= -\frac{\epsilon x_0}{\omega a^2} I\left((-1)^i\omega\sqrt{a^2-x_0^2}\right) + \epsilon^2\ldots.
\end{aligned} \tag{7.63}$$

It should be noted that system (7.63) is periodic in θ with the period 2π and is periodic in t with the period $\frac{2\pi}{\omega}$. Indeed, the functions $t_i(a,\theta)$ are such that $t_{i+2}(a,\theta) = t_i(a,\theta) + \frac{2\pi}{\omega}$. If we denote the right-hand sides of the last two equations of (7.63) by $I_i^{(1)}(a)$ and $I_i^{(2)}(a)$ correspondingly, then it is easy to see that $I_{i+2}^{(1)}(a) = I_i^{(1)}(a)$, $I_{i+2}^{(2)}(a) = I_i^{(2)}(a)$.

By taking the average in system (7.63) with respect to the time explicitly contained in the system, we get first order approximation equations for $a(t)$ and $\theta(t)$,

$$\begin{aligned}
\frac{da}{dt} &= \frac{\epsilon\sqrt{a^2-x_0^2}}{2\pi a}[I(\omega\sqrt{a^2-x_0^2}) - I(-\omega\sqrt{a^2-x_0^2})] + \frac{\epsilon}{\omega}A(a), \\
\frac{d\theta}{dt} &= \frac{\epsilon x_0}{2\pi a^2}[I(\omega\sqrt{a^2-x_0^2}) - I(-\omega\sqrt{a^2-x_0^2})] + \frac{\epsilon}{\omega a}B(a),
\end{aligned} \tag{7.64}$$

where

$$\begin{aligned}
A(a) &= \frac{1}{2\pi}\int_0^{2\pi} f(a\sin\varphi, a\omega\cos\varphi)\cos\varphi d\varphi, \\
B(a) &= \frac{1}{2\pi}\int_0^{2\pi} f(a\sin\varphi, a\omega\cos\varphi)\sin\varphi d\varphi.
\end{aligned}$$

The improved first order approximations for $a(t)$ and $\theta(t)$

$$
\begin{aligned}
a &= \bar{a} + \frac{\epsilon\sqrt{\bar{a}^2 - x_0^2}}{\pi\omega\bar{a}} \times \\
&\times \left[I(\omega\sqrt{\bar{a}^2 - x_0^2}) \sum_{k=1}^{\infty} \frac{\sin k(\varphi - \varphi_{\bar{a}}^{(1)})}{k} - \right. \\
&\left. -I(-\omega\sqrt{\bar{a}^2 - x_0^2}) \sum_{k=1}^{\infty} \frac{\sin k(\varphi - \varphi_{\bar{a}}^{(2)})}{k} \right] + \\
&+ \frac{\epsilon}{\omega^2} \sum_{k=1}^{\infty} \frac{a_k^{(1)} \sin k\varphi - b_k^{(1)} \cos k\varphi}{k}, \\
\theta &= \bar{\theta} + \frac{\epsilon x_0}{\pi\omega\bar{a}^2} \times \\
&\times \left[I(\omega\sqrt{\bar{a}^2 - x_0^2}) \sum_{k=1}^{\infty} \frac{\sin k(\varphi - \varphi_{\bar{a}}^{(1)})}{k} - \right. \\
&\left. -I(-\omega\sqrt{\bar{a}^2 - x_0^2}) \sum_{k=1}^{\infty} \frac{\sin k(\varphi - \varphi_{\bar{a}}^{(2)})}{k} \right] - \\
&- \frac{\epsilon}{\omega^2\bar{a}} \sum_{k=1}^{\infty} \frac{a_k^{(2)} \sin k\varphi - b_k^{(2)} \cos k\varphi}{k},
\end{aligned}
$$

are given by where $\bar{a}$ and $\bar{\theta}$ are solutions of the first order approximation equations (7.64), $a_k^{(1)}$, $b_k^{(1)}$ and $a_k^{(2)}$, $b_k^{(2)}$ are the Fourier coefficients of the functions $f(a\sin\varphi, a\omega\cos\varphi)\cos\varphi$ and $f(a\sin\varphi, a\omega\cos\varphi)\sin\varphi$ correspondingly,

$$
\varphi_a^{(1)} = \arcsin\frac{x_0}{a}, \qquad \varphi_a^{(2)} = \pi - \arcsin\frac{x_0}{a}.
$$

As an example, consider small oscillations of an ordinary pendulum [11, 70]. It is known that, to compensate the attenuation of oscillations, there are impulses being sent at certain moments to the pendulum by the clock spring using a special device, known as a release hook. This kind of oscillations can be modeled as follows: when the pendulum passes by a certain angle x_0, it is effected by a weak hit impulse in the direction of its movement.

The equation describing the pendulum motion can be written as follows

$$\begin{aligned} \frac{d^2x}{dt^2}+\omega^2x &= \epsilon\lambda\frac{dx}{dt}, \\ \Delta\left.\frac{dx}{dt}\right|_{x=x_0} &= \begin{cases} \epsilon I_0, & \text{for } \dfrac{dx}{dt}\geq 0, \\ -\epsilon I_0, & \text{for } \dfrac{dx}{dt}<0, \end{cases} \end{aligned} \tag{7.65}$$

where $\omega^2=\frac{k}{J}$, $\epsilon\lambda=\frac{\lambda_1}{J}$, $\epsilon I_0=\frac{I}{J}$. Here kx is the moment of the resetting forces, $\lambda_1\frac{dx}{dt}$ is the moment of the friction force, J is the inertia moment, I – the impulse of the hitting force.

Passing in (7.65) to the amplitude and phase variables by using (7.60), we get

$$\begin{aligned} \frac{da}{dt} &= -\epsilon\lambda a\cos^2\varphi, \quad \frac{d\theta}{dt}=\epsilon\lambda\sin\varphi\cos\varphi, \quad t\neq t_i(a,\theta), \\ \Delta a|_{t=t_i(a,\theta)} &= \frac{\epsilon I_0}{\omega a}\sqrt{a^2-x_0^2}+\epsilon^2\ldots, \\ \Delta\theta|_{t=t_i(a,\theta)} &= (-1)^{i+1}\frac{\epsilon x_0 I_0}{\omega a^2}+\epsilon^2\ldots, \end{aligned} \tag{7.66}$$

where $t_i(a,\theta)$ are given by (7.62). The averaged equations, which correspond to system (7.66), will be

$$\begin{aligned} \frac{da}{dt} &= -\frac{\epsilon\lambda}{2}a+\frac{2\epsilon I_0}{2\pi a}\sqrt{a^2-x_0^2} \\ \frac{d\theta}{dt} &= 0. \end{aligned} \tag{7.67}$$

Now let us find out whether there exist stationary dynamic modes. This is related to existence of positive roots of the stationary amplitude equation

$$-\frac{\lambda a}{2}+\frac{I_0}{\pi a}\sqrt{a^2-x_0^2}=0. \tag{7.68}$$

By solving this equation, we find that

$$a_{1,2}^2=2\left(\frac{I_0}{\pi\lambda}\right)^2\pm\frac{2\pi I_0}{\pi\lambda}\sqrt{\frac{I_0^2}{(\pi\lambda)^2}-x_0^2}. \tag{7.69}$$

Whence it can be seen that, if $x_0 < \frac{I_0}{\pi\lambda}$, equation (7.68) has two positive roots. If $x_0 > \frac{I_0}{\pi\lambda}$, then there are no real roots of this equation and hence the system can not have stationary dynamic modes.

Thus, if the parameters of system (7.65) are such that $x_0 < \frac{I_0}{\pi\lambda}$, then this system has two periodic solutions with a constant amplitude. One of them, namely the one that has the amplitude equal to

$$a^0 = \sqrt{2\frac{I_0}{\pi\lambda}\left(\frac{I_0}{\pi\lambda} + \sqrt{\frac{I_0^2}{(\pi\lambda)^2} - x_0^2}\right)}, \tag{7.70}$$

is called asymptotically stable, the other – unstable.

The improved first order approximation of solutions of equations (7.66), which correspond to the stable stationary amplitude (7.70), is given by the expressions

$$\begin{aligned} a(t) = a_0 + \frac{\epsilon\sqrt{a_0^2 - x_0^2} I_0}{\pi\omega a_0}\left[\sum_{k=1}^{\infty}\frac{\sin k\left(\omega t + \theta_0 - \arcsin\frac{x_0}{a_0}\right)}{k} + \right.\\ \left. + \sum_{k=1}^{\infty}\frac{(-1)^k \sin\left(\omega t + \theta_0 + \arcsin\frac{x_0}{a_0}\right)}{k}\right] - \frac{\epsilon\lambda a_0}{4\omega}\sin 2(\omega t + \theta_0),\\ \theta(t) = \theta_0 - \frac{\epsilon x_0 I_0}{\pi\omega a_0^2}\left[\sum_{k=1}^{\infty}\frac{\sin k\left(\omega t + \theta_0 - \arcsin\frac{x_0}{a_0}\right)}{k} + \right.\\ \left. - \sum_{k=1}^{\infty}\frac{(-1)^k \sin\left(\omega t + \theta_0 + \arcsin\frac{x_0}{a_0}\right)}{k}\right] - \frac{\epsilon\lambda}{4\omega}\cos 2(\omega t + \theta_0), \end{aligned} \tag{7.71}$$

where a_0 is the stationary amplitude, given by (7.70), θ_0 is an arbitrary constant.

Thus, it follows from using the change of variables (7.60) that the family of periodic solutions of (7.65), which corresponds to the stationary solution of first equation of (7.67) is given, up to terms of order ϵ^2, by the expression

$$x(t) = a_0 \sin(\omega t + \tau) - \frac{\epsilon \lambda a_0}{4\omega} \cos(\omega t + \tau) +$$
$$+ \frac{\epsilon I_0}{\pi \omega} \left[\sin\left(\omega t + \tau - \arcsin \frac{x_0}{a_0}\right) \sum_{k=1}^{\infty} \frac{\sin k \left(\omega t + \tau - \arcsin \frac{x_0}{a_0}\right)}{k} + \right.$$
$$+ \sin\left(\omega t + \tau + \arcsin \frac{x_0}{a_0}\right) \times$$
$$\left. \times \sum_{k=1}^{\infty} \frac{(-1)^k \sin k \left(\omega t + \tau + \arcsin \frac{x_0}{a_0}\right)}{k} \right], \tag{7.72}$$

where τ is an arbitrary constant.

We see from (7.71) and (7.72) that the quantities $a(t)$ and $\theta(t)$ are discontinuous in t with first kind discontinuities at $a_0 \sin(\omega t + \tau) = x_0$, $x(t)$ is continuous but its first derivative has first kind discontinuities at $a_0 \sin(\omega t + \tau) = x_0$ with the jumps equal to I_0 or $-I_0$ depending on whether the value of $\cos(\omega t + \tau)$ is positive or negative at the time of the jump.

As in the preceding example, consider an impulsive linear oscillator but assuming that the system undergoes a change of the kinetic energy by a constant value at the time when the impulsive effect occurs. Equations for such an oscillator will be

$$\frac{d^2 x}{dt^2} + \omega^2 x = -\epsilon \lambda \frac{dx}{dt}, \qquad x \neq x_0,$$
$$\Delta \left(\frac{dx}{dt} \right)^2 \Bigg|_{x = x_0} = \epsilon 2I, \qquad I > 0. \tag{7.73}$$

If we passing to the variables a and θ in these equations by using formulas (7.60), we obtain the system

$$\frac{da}{dt} = -\epsilon \lambda a \cos^2 \varphi, \quad \frac{d\theta}{dt} = \epsilon \lambda \sin \varphi \cos \varphi, \quad t \neq t_i(a, \theta),$$
$$\Delta a|_{t = t_i(a,\theta)} = \frac{\epsilon I}{\omega^2 a}, \quad \Delta \theta|_{t = t_i(a,\theta)} = (-1)^{i+1} \frac{I x_0}{a^2 \omega^2 \sqrt{a^2 - x_0^2}}. \tag{7.74}$$

The corresponding averaged system will be

$$\frac{da}{dt} = -\epsilon \frac{\lambda a}{2} + \frac{\epsilon I}{\pi \omega a}, \quad \frac{d\theta}{dt} = 0. \tag{7.75}$$

The first equation of (7.75) has the asymptotically stable stationary solution $a_0 = \sqrt{\frac{2I}{\pi\omega\lambda}}$. This solution corresponds to the family of periodic solutions of (7.73), which, up to terms of order ϵ^2, is given by

$$x(t) = a_0 \sin(\omega t + \tau) - \frac{\epsilon\lambda a_0}{4\omega}\cos(\omega t + \tau) + \frac{\epsilon I}{\pi\omega^2\sqrt{a_0^2 - x_0^2}} \times$$

$$\times \left[\sin\left(\omega t + \tau - \arcsin\frac{x_0}{a_0}\right) \sum_{k=1}^{\infty} \frac{\sin k\left(\omega t + \tau - \arcsin\frac{x_0}{a_0}\right)}{k} + \right.$$

$$\left. + \sin\left(\omega t + \tau + \arcsin\frac{x_0}{a_0}\right) \sum_{k=1}^{\infty} \frac{(-1)^k \sin k\left(\omega t + \tau - \arcsin\frac{x_0}{a_0}\right)}{k} \right].$$

Now, let the function $f(\nu t, x, \frac{dx}{dt})$ in (7.59) depend on time explicitly.

If no resonance occurs, i.e. if $\omega \neq \frac{r}{s}\nu$, where r and s are the greatest mutually prime numbers, first order approximations of solutions of equations (7.59) are given as first order approximations of solutions of the equations

$$\begin{aligned} \frac{d^2x}{dt^2} + \omega^2 x &= \epsilon f_0\left(x, \frac{dx}{dt}\right), \qquad x \neq x_0, \\ \Delta \left.\frac{dx}{dt}\right|_{x=x_0} &= \epsilon I\left(\frac{dx}{dt}\right), \end{aligned} \tag{7.76}$$

where

$$f_0\left(x, \frac{dx}{dt}\right) = \frac{1}{2\pi}\int_0^{2\pi} f\left(\varphi, x, \frac{dx}{dt}\right) d\varphi.$$

System (7.76) is autonomous, and so the first order approximations of its solutions can be determined by using the preceding scheme.

Now consider a resonance system, i.e. when ω is rationally depended on the frequency of the external effect or is close to this frequency, $\omega \simeq \frac{r}{s}\nu$.

Introduce in (7.59) the difference between the proper and the external frequencies,

$$\epsilon\bar{\Delta} = \omega^2 - \left(\frac{r}{s}\nu\right)^2, \tag{7.77}$$

and write this equation as

$$\begin{aligned} \frac{d^2x}{dt^2} + \left(\frac{r}{s}\nu\right)^2 x &= \epsilon f\left(\nu t, x, \frac{dx}{dt}\right) - \epsilon\bar{\Delta}x, \quad x \neq x_0, \\ \Delta \left.\frac{dx}{dt}\right|_{x=x_0} &= \epsilon I\left(\frac{dx}{dt}\right). \end{aligned} \tag{7.78}$$

Use the change of variables

$$x = a\sin\varphi, \quad \frac{dx}{dt} = \frac{r}{s}\nu a\cos\varphi, \quad \varphi = \frac{r}{s}\nu t + \theta \tag{7.79}$$

to write (7.78) in standard form,

$$\begin{aligned} \frac{da}{dt} &= \frac{\epsilon s}{r\nu}\left[f\left(\nu t, a\sin\varphi, a\frac{r}{s}\nu\cos\varphi\right)\cos\varphi - \bar{\Delta}a\sin\varphi\cos\varphi\right], \\ & \quad t \neq t_i(a,\theta), \\ \frac{d\theta}{dt} &= \frac{\epsilon s}{r\nu a}\left[f\left(\nu t, a\sin\varphi, a\frac{r}{s}\nu\cos\varphi\right)\sin\varphi - \bar{\Delta}a\sin^2\varphi\right], \\ \Delta a|_{t=t_i(a,\theta)} &= (-1)^i\frac{\epsilon s}{\nu r a}I\left(-1)^i\frac{r\nu}{s}\sqrt{a^2-x_0^2}\right)\sqrt{a^2-x_0^2} + \epsilon^2 \ldots, \\ \Delta\theta|_{t=t_i(a,\theta)} &= (-1)^i\frac{\epsilon x_0 s}{\nu r a^2}I\left(-1)^i\frac{r\nu}{s}\sqrt{a^2-x_0^2}\right) + \epsilon^2 \ldots, \end{aligned} \tag{7.80}$$

where

$$t_i(a,\theta) = \frac{s}{r\nu}\left(-\theta + (-1)^i\arcsin\frac{x_0}{a} + i\pi\right). \tag{7.81}$$

By taking in (7.80) the average with respect to the explicitly contained time and neglecting the terms which contain the second degree of ϵ, we get the equations for the first order approximations

$$\begin{aligned} \frac{da}{dt} &= \epsilon\frac{\sqrt{a^2-x_0^2}}{2\pi a}\times \\ &\times \left[I\left(\frac{r\nu}{s}\sqrt{a^2-x_0^2}\right) - I\left(-\frac{r\nu}{s}\sqrt{a^2-x_0^2}\right)\right] + \frac{\epsilon s}{r\nu}A(a,\theta), \\ \frac{d\theta}{dt} &= \omega - \frac{r}{s}\nu - \frac{\epsilon x_0}{2\pi a^2}\times \\ &\times \left[I\left(\frac{r\nu}{s}\sqrt{a^2-x_0^2}\right) + I\left(-\frac{r\nu}{s}\sqrt{a^2-x_0^2}\right)\right] - \frac{\epsilon s}{r\nu a}B(a,\theta) \end{aligned} \tag{7.82}$$

where

$$A(a,\theta) = \\ = \frac{1}{2\pi s}\int_0^{2\pi s} f\left(\varphi, a\sin\left(\frac{r}{s}\varphi+\theta\right), a\frac{r\nu}{s}\cos\left(\frac{r}{s}\varphi+\theta\right)\right)\cos\left(\frac{r}{s}\varphi+\theta\right)d\varphi,$$

$$B(a,\theta) = \\ = \frac{1}{2\pi s}\int_0^{2\pi s} f\left(\varphi, a\sin\left(\frac{r}{s}\varphi+\theta\right), a\frac{r\nu}{s}\cos\left(\frac{r}{s}\varphi+\theta\right)\right)\sin\left(\frac{r}{s}\varphi+\theta\right)d\varphi.$$

The improved first order approximations of a and θ are

$$\begin{aligned}
a &= \bar{a} + \frac{\epsilon\sqrt{\bar{a}^2 - x_0^2}s}{\pi\bar{a}r\nu}\left[I\left(\frac{r\nu}{s}\sqrt{\bar{a}^2-x_0^2}\right)\sum_{k=1}^{\infty}\frac{\sin k\left(\varphi - \arcsin\frac{x_0}{\bar{a}}\right)}{k} - \right.\\
&- \left. I\left(-\frac{r\nu}{s}\sqrt{\bar{a}^2-x_0^2}\right)\sum_{k=1}^{\infty}\frac{(-1)^k\sin k\left(\varphi + \arcsin\frac{x_0}{\bar{a}}\right)}{k}\right] - \\
&-\frac{\epsilon s^2\bar{\Delta}\bar{a}}{4r^2\nu^2}\cos 2\varphi + \epsilon u_1\left(\nu t, \bar{a}, \frac{r}{s}\nu t + \bar{\theta}\right),\\
\theta &= \bar{\theta} + \frac{\epsilon x_0 s}{\pi\bar{a}^2 r\nu}\left[I\left(\frac{r\nu}{s}\sqrt{\bar{a}^2-x_0^2}\right)\sum_{k=1}^{\infty}\frac{\sin k\left(\varphi - \arcsin\frac{x_0}{\bar{a}}\right)}{k} + \right.\\
&+ \left. I\left(-\frac{r\nu}{s}\sqrt{\bar{a}^2-x_0^2}\right)\sum_{k=1}^{\infty}\frac{(-1)^k\sin k\left(\varphi + \arcsin\frac{x_0}{\bar{a}}\right)}{k}\right] + \\
&+\frac{\epsilon^2 s^2\bar{\Delta}}{4r^2\nu^2}\sin 2\varphi + \epsilon u_2\left(\nu t, \bar{a}, \frac{r}{s}\nu t + \bar{\theta}\right),
\end{aligned} \tag{7.83}$$

where $\bar{a}$ and $\bar{\theta}$ are solutions of the averaged system (7.82), $u_1(\nu t, a, \varphi)$, $u_2(\nu t, a, \varphi)$ are corrections due to the term $f\left(\nu t, x, \frac{dx}{dt}\right)$, determined by the usual procedure [23].

As an example of the system that leads to equations (7.78), let us consider a clock, controlled by a sinusoidal electric current, with an autonomous releasing mechanism [32],

$$\begin{aligned}
\frac{d^2x}{dt^2} + \frac{g}{l}x &= -\frac{\gamma}{ml^2}\frac{dx}{dt} + \frac{A}{ml^2}\sin\nu t, \quad x \neq 0,\\
\Delta\left(\frac{dx}{dt}\right)^2\Bigg|_{x=0} &= \frac{2I_0}{m},
\end{aligned} \tag{7.84}$$

where m is the mass of the pendulum, l – its length, γ – the friction coefficient, ν – the frequency of an external force, I_0 is the change of the kinetic energy at the time of the impulse.

By introducing the notations $\frac{g}{l} = \omega^2$, $\frac{\gamma}{ml^2} = \epsilon\lambda$, $\frac{A}{ml^2} = \epsilon A_0$, $\frac{I_0}{m} = \epsilon I$, we can rewrite the equation in the needed form

$$\begin{aligned}
\frac{d^2x}{dt^2} + \omega^2 x &= \epsilon\left(-\lambda\frac{dx}{dt} + A_0 \sin\nu t\right), \quad x \neq 0, \\
\Delta \left.\frac{dx}{dt}\right|_{x=0} &= \frac{\epsilon I}{\frac{dx}{dt}} + \epsilon^2 \ldots . \qquad (7.85)
\end{aligned}$$

Let us study equations (7.85) in the case when the main resonance occurs, i.e. when $\omega \simeq \nu$.

Introduce in (7.85) the variables a and θ by

$$x = a\sin(\nu t + \theta), \qquad \frac{dx}{dt} = a\nu\cos(\nu t + \theta).$$

We get

$$\begin{aligned}
\frac{da}{dt} &= \frac{\epsilon}{\nu}[-\lambda a\nu\cos(\nu t + \theta) - \bar{\Delta} a \sin(\nu t + \theta) + A_0 \sin\nu t]\cos(\nu t + \theta), \\
\frac{da}{dt} &= \frac{\epsilon}{\nu}[-\lambda a\nu\cos(\nu t + \theta) - \bar{\Delta} a \sin(\nu t + \theta) + A_0 \sin\nu t]\cos(\nu t + \theta), \\
\Delta a|_{\nu t+\theta=i\pi} &= \frac{\epsilon I}{\nu^2 a}, \quad \Delta\theta|_{\nu t+\theta=i\pi} = 0, \qquad (7.86)
\end{aligned}$$

where $\epsilon\bar{\Delta} = \omega^2 - \nu^2$.

To get a system for first order approximation, take the average in equations (7.86) with respect to the explicitly contained time:

$$\begin{aligned}
\frac{da}{dt} &= \epsilon\left(-\frac{\lambda a}{2} - \frac{A_0}{2\nu}\sin\theta + \frac{I}{\pi\nu a}\right), \\
\frac{d\theta}{dt} &= \epsilon\xi - \frac{\epsilon A_0}{2\nu a}\cos\theta, \quad \epsilon\xi = \omega - \gamma. \qquad (7.87)
\end{aligned}$$

Let us find out when there exist stationary oscillations in system (7.85). To do that, set the right-hand sides of equations (7.87) to zero and eliminate θ from the obtained system. We get the equation for stationary amplitudes,

$$\left(\frac{\lambda^2}{4} + \xi^2\right)a^4 - \left(\frac{\lambda I}{\pi\nu} + \frac{A_0^2}{4\nu^2}\right)a^2 + \frac{I^2}{\pi^2\nu^2} = 0. \qquad (7.88)$$

If the difference between the frequencies ξ is sufficiently small, namely if

$$\xi^2 \leq \frac{\pi A_0^2}{8\nu I} + \frac{\pi^2 A_0^4}{64 I^2 \nu^2}, \tag{7.89}$$

this equation has two positive roots, and so the system given by (7.86) may have two stationary modes, determined by the corresponding stationary amplitudes and phases,

$$\begin{cases} a_1^2 = \dfrac{4\lambda I\nu + A_0^2\pi - \sqrt{8\pi\lambda\nu I A_0^2 + \pi^2 A_0^4 - 64\nu^2\xi^2 I^2}}{2\pi\nu^2(\lambda^2 + 4\xi^2)}, \\ \tan\theta_1 = \dfrac{I}{\pi\nu\xi a_1^2} - \dfrac{\lambda}{2\xi}, \quad \dfrac{\cos\theta_1}{\xi} > 0, \end{cases} \tag{7.90}$$

$$\begin{cases} a_0^2 = \dfrac{4\lambda I\nu + A_0^2\pi - \sqrt{8\pi\lambda\nu I A_0^2 + \pi^2 A_0^4 - 64\nu^2\xi^2 I^2}}{2\pi\nu^2(\lambda^2 + 4\xi^2)}, \\ \tan\theta_2 = \dfrac{I}{\pi\nu\xi a_1^2} - \dfrac{\lambda}{2\xi}, \quad \dfrac{\cos\theta_2}{\xi} > 0, \end{cases} \tag{7.91}$$

The first mode is not stationary, the second – is.

Thus we see that, in the case of the resonance, the clock has a stationary synchronous mode of oscillations, the amplitude of which does not depend on the initial conditions. This property, discovered by using an approximate solution, agrees well with the property of a real clock [11].

The periodic solution that describes stable oscillations is given, up to terms of order ϵ^2, by the expression

$$\begin{aligned} x(t) &= a_0 \sin(\nu t + \theta_0) - \frac{\epsilon\lambda a_0}{4\nu}\cos(\nu t + \theta_0) - \frac{\epsilon\xi a_0}{2\nu}\sin(\nu t + \theta_0) + \\ &+ \frac{\epsilon a_0}{4\nu^2 a_0}\sin(\nu t + \theta_0) + \frac{\epsilon I}{\pi\nu^2 a}\sin(\nu t + \theta_0)\sum_{k=1}^{\infty}\frac{\sin 2k(\nu t + \theta_0)}{k}, \end{aligned}$$

where a_0 and θ_0 give a stable stationary position of the averaged system (7.87).

7.6 Substantiation of the averaging method for systems with impulses occurring at fixed positions

To make grounds for applications of asymptotical methods to study weakly nonlinear oscillating impulsive systems considered in the preceding section, we give theorems on proximity of exact solutions of the considered equations to solutions of exact equations. These theorems are proved in [136, 62].

Theorem 113 *Let the functions $f(\nu t, x, y)$ and $I(y)$ that determine system (7.59) be defined, continuous, 2π-periodic with respect to $\frac{r}{s}\nu t$, and satisfy Lipschitz conditions in x and y for t, x, and y from the region $t \in \mathbf{R}$, $\alpha^2 \leq x^2 + \frac{y^2}{\omega^2} \leq \beta^2$, where α and β are some positive constants which satisfy the inequality $\beta > \alpha > x_0$. Suppose that system (7.82) has an isolated stationary solution $a = a^0$, $\theta = \theta^0$ which, together with a ρ-neighborhood, belongs to the strip $\alpha < a^0 < \beta$ and such that the index of the mapping given by the right-hand sides of the equations (7.82) is nonzero at the point (a^0, θ^0). Then there exists such $\epsilon_0 > 0$ that, for all ϵ, $0 < \epsilon < \epsilon_0$, system (7.59) has a solution $x = a(t)\sin\left(\frac{r}{s}\nu t + \theta(t)\right)$, which is 2π-periodic with respect to $\frac{r}{s}\nu t$ and such that*

$$|a(t) - a^0| + |\theta(t) - \theta^0| < \eta(\epsilon),$$

where $\eta(\epsilon)$ is a constant approaching to zero as $\epsilon \to 0$.

Let us find out what an isolated equilibrium of the averaged system, which corresponds to equations (7.59), is related to if the system is non-resonance. We also assume that the system undergoes an impulsive effect if the phase point meets the line given by the equation $x = \epsilon x_0$, i.e. the considered system is

$$\begin{aligned} \frac{d^2x}{dt^2} + \omega^2 x &= \epsilon f\left(t, x, \frac{dx}{dt}\right), \quad x \neq \epsilon x_0, \\ \Delta \left.\frac{dx}{dt}\right|_{x=\epsilon x_0} &= \epsilon I\left(\frac{dx}{dt}\right), \end{aligned} \tag{7.92}$$

where the function $f(t, x, y)$ is a polynomial in $\sin t$, $\cos t$, x, y, and ω is an irrational number badly approximated by rationals.

Theorem 114 *Let, in impulsive system (7.92), the function $f(t,x,y)$ be a trigonometric polynomial, the coefficients of which are polynomials in x and y. Let the function $I(y)$ be continuously differentiable and its derivative satisfy the Lipschitz conditions in some interval $\alpha \leq y \leq \beta$. Assume that the equation*

$$F(a) \equiv F_1(a) + \frac{\omega}{2\pi}[I(a\omega) - I(-a\omega)] = 0,$$

where

$$F_1(a) = \frac{1}{4\pi^2}\int_0^{2\pi}\int_0^{2\pi} f(t, a\sin\psi, a\omega\cos\psi)\cos\psi dt d\psi,$$

has an isolated positive solution $a = a_0$, for which $F'(a_0) < 0$. Then there exists such a positive ϵ^* that, for all $0 < \epsilon < \epsilon^*$,

1) system (7.92) has the integral set

$$\sqrt{(x - \epsilon x_0)^2 + \frac{1}{\omega}^2\left(\frac{dx}{dt}\right)^2} = u^*(\varphi, t, \epsilon),$$

where the function $u^*(\varphi,t,\epsilon)$ satisfies the Lipschitz conditions with respect to t, is piecewise continuous in φ with first kind discontinuities at $\varphi = k\pi$, 2π-periodic in φ and t, and $u^*(\varphi,t,\epsilon) \to a_0$ as $\epsilon \to 0$;

2) there exists a $\delta_0(\epsilon)$-neighborhood of the cylinder $\omega^2x^2 + \left(\frac{dx}{dt}\right)^2 = \omega^2a_0^2$ ($\delta_0(\epsilon) \to 0$ as $\epsilon \to 0$) such that all the solutions that start in this neighborhood and their first derivatives remain bounded for all $t \geq 0$.

As an example consider non-resonance oscillations of a linear oscillator, which is subject to a sinusoidal external force and an impulsive effect occurring when the oscillator passes by its lower equilibrium with a positive velocity and increasing the kinetic energy of the oscillator by a constant value.

The motion of such an oscillator is described by the system

$$\begin{aligned}
\frac{d^2x}{dt^2} + \omega^2 x &= -\lambda\frac{dx}{dt} + U\sin t, \quad x \neq 0, \\
\Delta\left(\frac{dx}{dt}\right)^2\Big|_{x=0} &= \begin{cases} 2I, & \text{if} \quad \dfrac{dx}{dt} > 0, \\ 0, & \text{if} \quad \dfrac{dx}{dt} \leq 0. \end{cases}
\end{aligned}$$

Let us study the motion in the case when there is no resonance. Considering λ, U, and I to be small positive quantities, set $\lambda = \epsilon\lambda_0$, $U = \epsilon U_0$, $I = \epsilon I_0$ and rewrite these equations as

$$\begin{aligned} \frac{d^2x}{dt^2} + \omega^2 x &= \epsilon\left[-\lambda_0\frac{dx}{dt} + U_0 \sin t\right], \quad x \neq 0, \\ \Delta \left.\frac{dx}{dt}\right|_{x=0} &= \begin{cases} \dfrac{\epsilon I_0}{\frac{dx}{dt}} + \epsilon^2 \ldots, & \dfrac{dx}{dt} > 0 \\ 0, & \dfrac{dx}{dt} \leq 0. \end{cases} \end{aligned}$$

Whence we get the equations

$$\begin{aligned} \frac{da}{dt} &= \frac{\epsilon}{\omega}(-\lambda_0 a \cos\psi + U_0 \sin t)\cos\psi, \\ &\quad \psi \neq 2k\pi \\ \frac{d\psi}{dt} &= \omega - \frac{\epsilon}{\omega a}(-\lambda_0 a\omega \cos\psi + U_0 \sin t)\sin\psi, \\ \Delta a|_{\psi=2k\pi} &= \frac{\epsilon I_0}{\omega a} + \epsilon^2 \ldots. \end{aligned}$$

The equation for stationary amplitudes of this system will be

$$-\frac{\lambda_0 a}{2} + \frac{I_0}{2\pi a\omega} = 0.$$

This equation has the unique positive solution $a_0 = \sqrt{\dfrac{I_0}{\pi\lambda_0\omega}}$, to which there corresponds an integral set of the initial equation and it lies in a neighborhood of the cylinder

$$x^2 + \frac{1}{\omega^2}\left(\frac{dx}{dt}\right)^2 = \frac{I_0}{\pi\lambda_0\omega}.$$

Up to terms of order ϵ^2, this set is given by the equation

$$\sqrt{x^2 + \frac{1}{\omega^2}\left(\frac{dx}{dt}\right)^2} = a_0 + \frac{\epsilon}{\omega}U(a_0, \varphi, t) + \frac{\epsilon I_0}{\omega a_0}\left\{\frac{1}{2} - \frac{\varphi}{2\pi}\right\},$$

where

$$U(a_0, \varphi, t) = -\frac{\lambda_0 a_0}{4}\sin 2\varphi - \frac{U_0}{2}\left[\frac{\cos(t+\varphi)}{1+\omega} + \frac{\cos(t-\varphi)}{1-\omega}\right].$$

The solutions of the initial equations that lie on this set are given, up to terms of order ϵ^2, by the expression

$$x = a_0 \sin(\omega t + \theta) - \frac{\epsilon \lambda_0 a_0}{4\omega} \cos(\omega t + \theta) - \frac{\epsilon U_0}{1 - \omega^2} \sin t + \\ + \frac{\epsilon I_0}{a_0 \omega^2} \left\{ \frac{1}{2} - \frac{\omega t + \theta}{2\pi} \right\} \sin(\omega t + \theta),$$

which, by expanding the function $\left\{ \frac{1}{2} - \frac{\varphi}{2\pi} \right\}$ in the Fourier series, can be finally written as

$$x = a_0 \sin(\omega t + \theta) - \frac{\epsilon \lambda_0 a_0}{4\omega} \cos(\omega t + \theta) - \frac{\epsilon U_0}{1 - \omega^2} \sin t + \\ + \frac{\epsilon I_0}{\pi a_0 \omega^2} \sin(\omega t + \theta) \sum_{k=1}^{\infty} \frac{\sin k(\omega t + \theta)}{k}.$$

7.7 General averaging scheme for impulsive systems

We will be considering an impulsive differential system in standard form. Assume that an impulsive effect occurs when the phase point meets a certain sequence of hypersurfaces in the phase space.

Such a system can be written as

$$\begin{aligned} \frac{dx}{dt} &= \epsilon X(t,x), \quad t \neq \tau_j(x) \\ \Delta x|_{t=\tau_j(x)} &= \epsilon I_j(x), \end{aligned}$$
$$x = (x_1, \ldots, x_n), \qquad X = (X_1, \ldots, X_n), \qquad I_j = (I_j^1, \ldots, I_j^1), \tag{7.93}$$

where $\tau_j(x)$ are scalar functions, $j = 0, \pm 1, \ldots$.

For the sake of simplicity, assume that the system is periodic with respect to t with the period 2π. For an impulsive system, this means that the function $X(t,x)$ is periodic with respect to t with the period 2π and there exists such a natural number p that

$$I_{j+p}(x) = I_j(x), \qquad \tau_{j+p}(x) = \tau_j(x) + 2\pi \tag{7.94}$$

for all $j = 0, \pm 1, \ldots$.

Denote by $X_0(x)$ and $I^0(x)$ the following expressions

$$X_0(x) = \frac{1}{2\pi} \int_0^{2\pi} X(t,x)dt, \qquad I^0(x) = \frac{1}{2\pi} \sum_{i=1}^{p} I_i(x). \tag{7.95}$$

The system

$$\frac{d\xi}{dt} = \epsilon \left[X_0(\xi) + I_0(\xi) \right] \tag{7.96}$$

will be called averaged with respect to system (7.93).

Note that, if system (7.93) is not periodic, $X_0(x)$ and $I^0(x)$ in the corresponding averaged system are defined as follows

$$\begin{aligned} X_0(x) &= \lim_{T\to\infty} \frac{1}{T} \int_t^{t+T} X(\tau,x)d\tau, \\ I^0(x) &= \lim_{T\to\infty} \frac{1}{T} \sum_{t<\tau_i(x)<t+T} I_i(x), \end{aligned}$$

if, of course, these limits exist.

Now we will construct approximations of solutions of system (7.93) and substantiate the averaging method.

We take a solution of the averaged system (7.96),

$$x = \xi(t), \tag{7.97}$$

to be a first degree approximation of solutions of system (7.93).

There is a reason for such a choice of the solution. Let us make, in system (7.93), the following change of variables

$$x = \xi + \epsilon \tilde{X}(t, \xi) + \epsilon \sum_{k=1}^{p} I_k(\xi) \left\{ \frac{1}{2} - \frac{t - t_k^{(1)}(\xi)}{2\pi} \right\}, \tag{7.98}$$

where $\tilde{X}(t,\xi)$ denotes the first integral of the function $X(t,\xi)$ such that its average with respect to T is equal to zero, $\left\{ \frac{1}{2} - \frac{t - \tau_j^{(1)}(\xi)}{2\pi} \right\}$ is a periodic function, given on its period by

$$\left\{ \frac{1}{2} - \frac{t - \tau_j^{(1)}(\xi)}{2\pi} \right\} = \frac{1}{2} - \frac{t - \tau_j^{(1)}(\xi)}{2\pi},$$
$$\tau_j^{(1)}(\xi) < t \leq \tau_j^{(1)} + 2\pi, \tag{7.99}$$

and $\tau_j^{(1)}(\xi)$, $j = 1, 2, \ldots, p$, are solutions of the equations

$$\tau_j^{(1)}(\xi) = \tau_j(\xi + \epsilon \tilde{X}(\tau_j^{(1)}(\xi), \xi) \quad + \quad \epsilon \sum_{k=1}^{p} I_k(\xi) \left(\frac{1}{2} - \frac{\tau_j^{(1)}(\xi) - \tau_k^{(1)}(\xi)}{2\pi} \right)$$
$$(j = 1, 2, \ldots, p) \tag{7.100}$$

and belong to the interval $[0, 2\pi]$.

If the functions $\tau_j(\xi)$, $j = 1, 2, \ldots, p$, are smooth, and ϵ is sufficiently small, then system (7.100) can be solved for $\tau_j^{(1)}(\xi)$ and its solutions can be written as

$$\tau_j^{(1)}(\xi) = \tau_j(\xi) +$$
$$+ \epsilon \left\langle \frac{\partial \tau_j(\xi)}{\partial \xi}, \tilde{X}(\tau_j(\xi), \xi) + \sum_{k=1}^{p} I_k(\xi) \left(\frac{1}{2} - \frac{\tau_j(\xi) - \tau_k(\xi)}{2\pi} \right) \right\rangle + \epsilon^2 \ldots,$$
$$(j = 1, 2, \ldots, p) \tag{7.101}$$

By substituting expression (7.98) into (7.93) and solving for $\frac{d\xi}{dt}$ if $t \neq \tau_j^{(1)}(\xi)$, we have

$$\frac{d\xi}{dt} = \epsilon[X_0(\xi) + I^0(\xi)] + \epsilon^2 \dots . \tag{7.102}$$

By using the jump condition at $t = \tau_j^{(1)}(\xi)$, we get the equation for finding $\Delta\xi$,

$$\Delta\xi + \epsilon[\tilde{X}(\tau_j^{(1)}(\xi), \xi + \Delta\xi) - \tilde{X}(\tau_j^{(1)}(\xi), \xi)] + \\ + \epsilon \sum_{k=1}^{p} [I_k(\xi + \Delta\xi) - I_k(\xi)] \left(\frac{1}{2} - \frac{\tau_j^{(1)}(\xi) - \tau_k^{(1)}(\xi)}{2\pi} \right) = \epsilon^2 \dots ,$$

where the dots denote terms that are independent of $\Delta\xi$. If the value of ϵ is small, then this equation can be solved for $\Delta\xi$, and its solution is proportional to ϵ^2.

Hence, by making in system (7.93) the change of variables (7.98), we are led to the equations

$$\begin{aligned} \frac{d\xi}{dt} &= \epsilon[X_0(\xi) + I^0(\xi)] + \epsilon^2 \dots , \quad t \neq \tau_j^{(1)}(\xi) \\ \Delta\xi|_{t=\tau_j^{(1)}(\xi)} &= \epsilon^2 . \end{aligned} \tag{7.103}$$

Neglecting the terms of the second order in ϵ, we get from (7.103) equations of first order approximations, which are equations (7.96).

We can take (7.98) to be an improved first order approximation of solutions of equations (7.93) with ξ being a solution of (7.96). However, if we recall that exact solutions of equations (7.96) are needed only to make the change of variables (7.98) rigorous, we can use approximate solutions in the improved first degree approximation and, without increasing the order of error, replace the quantities $\tau_j^{(1)}(\xi)$ by $\tau_j(\xi)$.

Finally, the improved first degree approximation of solutions of system (7.93) becomes

$$x = \xi + \epsilon\tilde{X}(t, \xi) + \epsilon \sum_{j=1}^{p} I_k(\xi) \left\{ \frac{1}{2} - \frac{t - \tau_k(\xi)}{2\pi} \right\} , \tag{7.104}$$

where $\xi = \xi(t)$ is a solution of the first degree approximation equation (7.104).

Relation (7.104) can also be written in the following form

$$x = \xi + \epsilon\tilde{X}(t,\xi) + \frac{\epsilon}{\pi}\sum_{j=1}^{p}\sum_{k=1}^{\infty}\frac{\sin k(t-\tau_j(\xi))}{k}. \tag{7.105}$$

To get a second degree approximation, we need to find a change of variables that is similar to (7.98), which would transform the variable x to the variable ξ which satisfies the equations

$$\begin{aligned} \frac{d\xi}{dt} &= \epsilon[X_0(\xi) + I^0(\xi)] + \epsilon^2 P(\xi) + \epsilon^3 \ldots, \quad t \neq \tau_j^{(2)}(\xi,\epsilon) \\ \Delta\xi\big|_{t=\tau_j^{(2)}(\xi,\epsilon)} &= \epsilon^3. \end{aligned} \tag{7.106}$$

To find this change of variables, let us find the functions $F(t,\xi)$, $J_j(\xi)$, $j = 1,2,\ldots,p$, such that the expression

$$\begin{aligned} x &= \xi + \epsilon\tilde{X}(t,\xi) + \epsilon^2 F(t,\xi) + \\ &\quad + \sum_{j=1}^{p}(\epsilon I_j(\xi) + \epsilon^2 J_j(\xi))\left\{\frac{1}{2} - \frac{t-\tau_j^{(2)}(\xi)}{2\pi}\right\}, \end{aligned} \tag{7.107}$$

where ξ satisfies the equation

$$\frac{d\xi}{dt} = \epsilon[X_0(\xi) + I^0(\xi)] + \epsilon^2 P(\xi), \tag{7.108}$$

would make equations (7.93) hold up to terms of order ϵ^2.

Note that $\tau_j^{(2)}(\xi,\epsilon)$ are solutions of the equations

$$\begin{aligned} \tau_j^{(2)}(\xi,\epsilon) &= \tau_j\Big(\xi + \epsilon\tilde{X}(\tau_j^{(2)}(\xi,\epsilon),\xi) + \epsilon^2 F(\tau_j^{(2)}(\xi,\epsilon),\xi) + \\ &\quad + \sum_{k=1}^{p}(\epsilon I_k(\xi) + \epsilon^2 J_k(\xi))\Big(\frac{1}{2} - \frac{\tau_j^{(2)}(\xi,\epsilon) - \tau_k^{(2)}(\xi,\epsilon)}{2\pi}\Big)\Big) \\ &\qquad (j = 1,2,\ldots,p). \end{aligned} \tag{7.109}$$

Here, as in the construction of first order approximations, we need exact solutions of system (7.109) only to make the calculations rigorous. In the expression that give second order approximations, we replace them by the approximate solutions that satisfy (7.109) up to the order of ϵ^2.

After some calculations that are similar to [23], we get the functions $P(\xi)$, $F(t,\xi)$, and $J_j(\xi)$, $j = 1,2,\ldots,p$, given by the expressions

$$P(\xi) = \frac{\partial X(t,\xi)}{\partial \xi}\left(\tilde{X}(t,\xi) + \sum_{k=1}^{p} I_k(\xi)\left\{\frac{1}{2} - \frac{t-\tau_k(\xi)}{2\pi}\right\}\right) + \\ + \frac{1}{2\pi}\sum_{k=1}^{p} J_k(\xi), \tag{7.110}$$

$$J_j(\xi) = \frac{\partial I_j(\xi)}{\partial \xi}\left[\tilde{X}(\tau_j(\xi),\xi) + \sum_{k=1}^{p} I_k(\xi)\left(\frac{1}{2} - \frac{\tau_j(\xi)-\tau_k(\xi)}{2\pi}\right)\right] \tag{7.111}$$

where $F(t,\xi)$ is the first integral with the zero mean of the function

$$\frac{\partial X(t,\xi)}{\partial \xi}\left[\tilde{X}(t,\xi) + \sum_{k=1}^{p} I_k(\xi)\left\{\frac{1}{2} - \frac{t-\tau_k(\xi)}{2\pi}\right\}\right] - \frac{\partial X(t,\xi)}{\partial \xi}\times \\ \times\left[\tilde{X}(t,\xi) - \sum_{k=1}^{p} I_k(\xi)\left\{\frac{1}{2} - \frac{t-\tau_k(\xi)}{2\pi}\right\}\right](X_0(\xi) + I^0(\xi)). \tag{7.112}$$

Substitute now the values in (7.108) – (7.112) into (7.107) and consider the obtained expression as a formula of the change of variables that transforms the variable x, which is defined by the exact equations (7.93), into a new variable ξ. After some calculations, we see that this variable, for $t \neq \tau_j^{(2)}(\xi,\epsilon)$, satisfies the equation

$$\frac{d\xi}{dt} = \epsilon[X_0(\xi) + I^0(\xi)] + \epsilon^2 P(\xi) + \epsilon^3 \ldots . \tag{7.113}$$

If $t = \tau_j^{(2)}(\xi,\epsilon)$, then it has a first kind discontinuity with the jump determined from the equation

$$\begin{aligned}
\Delta\xi &= \epsilon\big[\tilde{X}(\tau_j^{(2)}(\xi,\epsilon),\xi+\Delta\xi) - \tilde{X}(\tau_j^{(2)}(\xi,\epsilon),\xi)\big] + \\
&+\epsilon^2\big[F(\tau_j^{(2)}(\xi,\epsilon),\xi+\Delta\xi) - F(\tau_j^{(2)}(\xi,\epsilon),\xi)\big] + \\
&+\sum_{\substack{k=1\\k\neq j}}^{p}\Big[\epsilon(I_k(\xi+\Delta\xi)-I_k(\xi)) + \epsilon^2(J_k(\xi+\Delta\xi)-J_k(\xi))\times \\
&\times\left(\frac{1}{2} - \frac{\tau_j^{(2)}(\xi,\epsilon)-\tau_k^{(2)}(\xi,\epsilon)}{2\pi}\right)\Big] + \epsilon I_j(\xi) + \epsilon^2 J_j(\xi) = \\
&= \epsilon I_j\Bigg(\xi + \epsilon\tilde{X}(\tau_j^{(2)}(\xi,\epsilon),\xi) + \epsilon^2 F(\tau_j^{(2)}(\xi),\xi) + \\
&+\sum_{k=1}^{p}(\epsilon I_k(\xi)+\epsilon^2 J_k(\xi))\left(\frac{1}{2} - \frac{\tau_j^{(2)}(\xi,\epsilon)-\tau_k^{(2)}(\xi,\epsilon)}{2\pi}\right)\Bigg). \quad (7.114)
\end{aligned}$$

For small values of the parameter ϵ, equation (7.114) can be solved with respect to $\Delta\xi$ and its solution is a value of order ϵ^2. Thus, by making in system (7.93) the change of variables (7.107), we get the following equations

$$\begin{aligned}
\frac{d\xi}{dt} &= \epsilon[X_0(\xi) + I^0(\xi)] + \epsilon^2 P(\xi) + \epsilon^3\ldots, \quad t \neq \tau_j^{(2)}(\xi,\epsilon) \\
\Delta\big|_{t=\tau_j^{(2)}(\xi,\epsilon)} &= \epsilon^3\ldots. \quad (7.115)
\end{aligned}$$

Thus, if ξ satisfies (7.115), the right-hand sides of which differs from the right-hand side of equations (7.108) by values of order ϵ^3, then expression (7.107) gives an exact solution of equations (7.93). So, we can take

$$x = \xi + \epsilon\tilde{X}(t,\xi) + \epsilon\sum_{k=1}^{p} I_k(\xi)\left\{\frac{1}{2} - \frac{t-\tau_k(\xi)}{2\pi}\right\} \quad (7.116)$$

to be a second degree approximation of solutions of equations (7.93) with $\xi = \xi(t)$ being a solution of (7.108). An improved second degree approximation of solutions of system (7.93) will be

$$\begin{aligned}
x &= \xi + \epsilon\tilde{X}(t,\xi) + \epsilon^2 F(t,\xi) + \\
&+\sum_{k=1}^{p}[\epsilon I_k(\xi) + \epsilon^2 J_k(\xi)]\left\{\frac{1}{2} - \frac{t-\bar{\tau}_k^{(1)}(\xi)}{2\pi}\right\}, \quad (7.117)
\end{aligned}$$

where $\xi = \xi(t)$ is a solution of equations (7.108), $J_k(\xi)$ are defined by (7.111), $F(t,\xi)$ is the first integral of the function

$$\frac{\partial X}{\partial \xi}\left[\tilde{X}(t,\xi) + \sum_{k=1}^{p} I_k(\xi)\left\{\frac{1}{2} - \frac{t-\tau_k(\xi)}{2\pi}\right\}\right] -$$

$$- \left[\frac{\partial \tilde{X}}{\partial \xi} - \sum_{k=1}^{p} \partial I_k \partial \xi \left\{\frac{1}{2} - \frac{t-\tau_k(\xi)}{2\pi}\right\}\right](X_0(\xi) + I^0(\xi)),$$

with the mean value with respect to t equal to zero, and $\bar{\tau}_j^{(1)}(\xi)$, $j = 1, 2, \ldots, p$, are values satisfying (7.109) up to terms of order ϵ^2. The functions $\bar{\tau}_j^{(1)}(\xi)$ can easily be found if we write them as

$$\bar{\tau}_j^{(1)}(\xi) = \tau_j(\xi) + \epsilon\tau_j^0(\xi) \tag{7.118}$$

and substitute into equations (7.109) requiring that the equations hold up to the order of ϵ^2. Having done this, we find that

$$\tau_j(\xi) = \left\langle \frac{\partial \tau_j}{\partial \xi}, \tilde{X}(\tau_j(\xi),\xi) + \sum_{k=1}^{p} I_k(\xi)\left(\frac{1}{2} - \frac{\tau_j(\xi)-\tau_k(\xi)}{2\pi}\right)\right\rangle. \tag{7.119}$$

Everything that has been done can be immediately generalized to the equations

$$\begin{aligned} \frac{dx}{dt} &= \epsilon X(t,x) + \epsilon^2 Y(t,x), \qquad t \neq \tau_j(x), \\ \Delta x|_{t=\tau_j(x)} &= \epsilon I_j(x) + \epsilon^2 J_j(x), \end{aligned} \tag{7.120}$$

which contains terms of the second degree in ϵ.

In this case, equations for the second order approximation will be

$$\frac{\epsilon}{dt}[X_0(\xi) + I_0(\xi)] + \epsilon^2[Y_0(\xi) + J_0(\xi) + P(\xi)], \tag{7.121}$$

where

$$Y_0(\xi) = \frac{1}{2\pi}\int_0^{2\pi} Y(t,\xi)dt, \qquad J^0(\xi) = \frac{1}{2\pi}\sum_{k=1}^{p} 2\pi J_k(\xi),$$

and $P(\xi)$ is defined by (7.110).

Second order approximations of solutions of system (7.118) is given by expression (7.116), where ξ is a solution of (7.121).

For the improved second order approximation, we find

$$
\begin{aligned}
x &= \xi + \epsilon\tilde{X}(t,\xi) + \epsilon^2[F(t,\xi) - \tilde{Y}(t,\xi)] + \\
&+ \sum_{k=1}^{p}[\epsilon I_k(\xi) + \epsilon^2 J_k(\xi) + \epsilon^2 J_k(\xi)]\left\{\frac{1}{2} - \frac{t - \tilde{\tau}_j^{(1)}(\xi)}{2\pi}\right\}. \qquad (7.122)
\end{aligned}
$$

Now we briefly discuss how to obtain higher order approximations.
Let the equations of an impulsive system be given in the standard form

$$
\begin{aligned}
\frac{dx}{dt} &= \epsilon X(t,x) + \epsilon^2 X_1(t,x) + \ldots + \epsilon^m X_{m-1}(t,x), \\
& \quad t \neq \tau_j(x) \\
\Delta x|_{t=\tau_j(x)} &= \epsilon I_j(x) + \epsilon^2 I_j^{(1)}(x) + \ldots + \epsilon^m I_j^{m-1}(x), \qquad (7.123)
\end{aligned}
$$

where $X_k(t,x)$ and $I_j^{(k)}(x)$ are functions of the same type as $X(t,x)$ and $I_j(x)$ respectively.

To get an mth approximation, consider the expression

$$
\begin{aligned}
x &= \xi + \epsilon F_1(t,\xi) + \epsilon^2 F_2(t,\xi) + \ldots + \epsilon^m F_m(t,\xi) + \\
&+ \sum_{k=1}^{p}(\epsilon J_k^{(1)}(\xi) + \epsilon^2 J_k^{(2)}(\xi) + \ldots + \epsilon^m J_k^{(m)}(\xi)) \times \\
&\times \left\{\frac{1}{2} - \frac{t - \tau_k^{(m)}(\xi,\epsilon)}{2\pi}\right\}, \qquad (7.124)
\end{aligned}
$$

where the functions $F_k(t,\xi)$ are periodic in t with the period 2π and have the zero mean, $\xi = \xi(t)$ are solutions of the equation

$$
\frac{d\xi}{dt} = \epsilon P_1(\xi) + \epsilon^2 P_2(\xi) + \ldots + \epsilon^m P_m(\xi), \qquad (7.125)
$$

and $\tau_j^{(m)}(\xi,\epsilon)$, $j = 1,2,\ldots,p$ are solutions of the equations

$$
\begin{aligned}
\tau_j^{(m)}(\xi\epsilon) &= \tau_j(\xi) + \sum_{\nu=1}^{m}\epsilon^\nu F_\nu(\tau_j^{(m)}(\xi,\epsilon),\epsilon) + \\
&+ \sum_{k=1}^{p}\sum_{\nu=1}^{m}\epsilon^\nu J_k^{(\nu)}(\xi)\left(\frac{1}{2} - \frac{\tau_j^{(m)}(\xi,\epsilon) - \tau_k^{(m)}(\xi,\epsilon)}{2\pi}\right), \\
&\qquad (j = 1,2,\ldots,\rho) \qquad (7.126)
\end{aligned}
$$

and belong to the interval $[0, 2\pi]$.

Represent an approximate solution of (7.126) as a series in the parameter ϵ,

$$\tau_j^{(m)}(\xi,\epsilon) = \tau_j(\xi) + \epsilon\tau_j^{(1)}(\xi) + \epsilon^2\tau_j^{(2)}(\xi) + \ldots + \epsilon^m\tau_j^{(m)}(\xi) + \ldots. \quad (7.127)$$

Substituting expressions (7.124) and (7.127) into equations (7.123) and (7.126) and equating the coefficients at equal powers of ϵ up to degree m, we choose the functions $F_1(t,\xi)$, ..., $F_m(t,\xi)$, $P_1(\xi)$, ..., $P_m(\xi)$, $J_j^{(1)}(\xi)$, ..., $J_j^{(m)}(\xi), \tau_j^{(1)}(\xi), \ldots, \tau_j^{(m-1)}(\xi)$, $\psi = 1, 2, \ldots, p$, in such a way that expression (7.124) satisfies equations (7.123) and expression (7.124) satisfies (7.126) up to terms of order ϵ^{m+1}. As a result we get

$$\begin{aligned}
J_j^{(1)}(\xi) &= J_j(\xi), \\
J_j^{(2)}(\xi) &= I_j^{(1)}(\xi) + \\
&\quad + \frac{\partial I_j(\xi)}{\partial \xi}\left[X(\tau_j(\xi),\xi) + \sum_{k=1}^{p} I_k(\xi)\left(\frac{1}{2} - \frac{\tau_j(\xi) - \tau_k(\xi)}{2\pi}\right)\right], \\
&\quad (j = 1, 2, \ldots, \rho), \\
F_1(t,\xi) &= \tilde{X}(t,\xi), \\
F_2(t,\xi) &= \tilde{X}_1(t,\xi) + \mathcal{F}(t,\xi),
\end{aligned} \quad (7.128)$$

where $\mathcal{F}(t,\epsilon)$ is the first integral of the function

$$\begin{aligned}
&\frac{\partial X}{\partial \xi}\cdot\left[\tilde{X}(t,\xi) + \sum_{k=1}^{p} I_k(\xi)\left\{\frac{1}{2} - \frac{t - \tau_k(\xi)}{2\pi}\right\}\right] - \\
&- \left[\frac{\partial \tilde{X}}{\partial \xi} - \sum_{k=1}^{p}\frac{\partial I_k}{\partial \xi}\left\{\frac{1}{2} - \frac{t - \tau_k(\xi)}{2\pi}\right\}\right](X_0(\xi) + I^0(\xi)), \\
&P_1(\xi) = X_0(\xi) + I^0(\xi), \\
&P_2(\xi) = X_0^{(1)}(\xi) + I_0^{(1)}(\xi) + P(\xi),
\end{aligned} \quad (7.129)$$

and the mean value of which is zero. Here

$$\begin{aligned}
X_0^{(1)}(\xi) &= \frac{1}{2\pi}\int_0^{2\pi} X_1(t,\xi)dt, \qquad I_0^{(1)}(\xi) = \frac{1}{2\pi}\sum_{k=1}^{p} I_k^{(1)}(\xi), \\
P(\xi) &= \overline{\frac{\partial X}{\partial \xi}\left(\tilde{X}(t,\xi) + \sum_{k=1}^{p} I_k(\xi)\left\{\frac{1}{2} - \frac{t-\tau_k(\xi)}{2\pi}\right\}\right)} + \frac{1}{2\pi}\sum_{k=1}^{p} J_k^{(2)}(\xi), \\
\tau_j^{(1)}(\xi) &= \left\langle \frac{\partial \tau_j(\xi)}{\partial \xi}, \tilde{X}(\tau_j(\xi),\xi) + \sum_{k=1}^{p} I_k(\xi)\left(\frac{1}{2} - \frac{\tau_j(\xi)-\tau_k(\xi)}{2\pi}\right)\right\rangle, \\
&\qquad (j = 1,2,\ldots,p). \qquad (7.130)
\end{aligned}$$

If now, having determined the functions $F_1(t,\xi)$, ..., $F_m(t,\xi)$, $P_1(\xi)$, ..., $P_m(\xi)$, $J_j^{(1)}(\xi), \ldots, J_j^{(m)}(\xi)$, $\tau_j^{(1)}(\xi), \ldots, \tau_j^{(m)}(\xi)$, we consider expression (7.124) as a certain formula for a change of variables, which transforms the variable x into a new variable ξ, for this variable, we get the equations

$$\begin{aligned}
\frac{d\xi}{dt} &= \epsilon P_1(\xi) + \epsilon^2 P_2(\xi) + \ldots + \epsilon^m P_m(\xi) + \epsilon^{m+1} R(t,\xi,\epsilon), \quad t \neq \tau_j^{(m)}(\xi,\epsilon) \\
&\Delta\xi|_{t=\tau_j^{(m)}(\xi,\epsilon)} = \epsilon^{m+1} \ldots . \qquad (7.131)
\end{aligned}$$

Hence, the expression

$$\begin{aligned}
x &= \xi + \epsilon F_1(t,\xi) + \ldots + \epsilon^m F_m(t,\xi) + \\
&\sum_{k=1}^{p}[\epsilon J_k^{(1)}(\xi) + \ldots + \epsilon^m J_k^{(m)}(\xi)]\left\{\frac{1}{2} - \frac{t-\bar{\tau}_k^{m-1}(\xi,\epsilon)}{2\pi}\right\} \qquad (7.132)
\end{aligned}$$

can be taken as the mth degree approximation of solutions of system (7.123) with ξ being determined from the mth approximation equation (7.125).

7.8 On correspondence between exact and approximate solutions over an infinite time interval

Before substantiating the general scheme of the averaging method given in the preceding section, consider the value of the difference between an exact solution of system (7.93) and its approximation which is a solution of the averaged equations (7.96). It has been shown in [134] that, with rather general restrictions imposed on the functions that define equations (7.93), for any $\eta > 0$ there exists such $\epsilon_0 > 0$ that, for all $0 < \epsilon < \epsilon_0$, system (7.93) has a solution $x(t, x_0)$, $x(0, x_0) = x_0$, which is defined for $t \in [0, \frac{L}{\epsilon}]$ and such that $\|x(t, x_0) - \xi(\epsilon t, x_0)\| < \epsilon$ for $t \in [0, \frac{L}{\epsilon}]$, where $\xi(\epsilon t, x_0)$, $\xi(0, x_0) = x_0$, is a solution of the averaged equations (7.96), and $L > 0$. A qualitative comparison between solutions of system (7.93) and solutions of the averaged system (7.96) has also been studied in [134] with the assumption made that the averaged equations have an asymptotically stable equilibrium. Sufficient conditions for existence of a family of bounded solutions of equations (7.93) in a neighborhood of the isolated equilibrium of the averaged system are also given there.

In this section, we will formulate theorems that extend the results obtained by A.M. Samoilenko in [134] concerning the substantiation of the averaging method applied to impulsive systems in standard form.

First, let us consider a differential system subjected to an impulsive effect, which occurs at fixed times,

$$\begin{aligned} \frac{dx}{dt} &= \epsilon X(t, x), \qquad t \neq \tau_i, \\ \Delta x|_{t=\tau_i} &= \epsilon I_i(x). \end{aligned} \tag{7.133}$$

The following assumptions concerning the functions $X(t, x)$, $I_i(x)$ and the times of the impulsive effects $\{\tau_i\}$ are made:

a) $X(t, x)$, $I_i(x)$ are functions, continuous with respect to their arguments and having the second order continuous derivatives with respect to x in the region

$$t \in \mathbf{R}, \qquad x = \mathrm{col}(x_1, \ldots, x_n) \in D \subset \mathbf{R}^n; \tag{7.134}$$

b) the times τ_i, at which the impulsive effect occurs, are indexed by integers such that $\tau_i \to -\infty$ if $i \to -\infty$ and $\tau_i \to +\infty$ if $i \to +\infty$, and there are such numbers L and d that, on any time interval of length L, there are at most d points of the sequence $\{\tau_i\}$;

c) the following finite limits

$$\begin{aligned} \lim_{T\to\infty} \frac{1}{T}\int_t^{t+T} X(\tau, x)d\tau &= X_0(x), \\ \lim_{T\to\infty} \frac{1}{T} \sum_{t<\tau_i<t+T} I_i(x) &= I^0(x) \end{aligned} \tag{7.135}$$

exist and are uniform with respect to $t \in \mathbf{R}$ and $x \in D$.

Together with system (7.133), we will be considering the corresponding averaged system

$$\frac{dx}{dt} = \epsilon[X_0(x) + I^0(x)] \equiv \epsilon F(x) \tag{7.136}$$

and assume that this system has an isolated equilibrium $x = x_0$, which belongs together with its ρ-neighborhood to the region D, and the real parts of the eigenvalues of the matrix $A = \dfrac{\partial F}{\partial x}(x_0)$ are nonzero.

Theorem 115 *Let impulsive differential system (7.133) satisfy conditions a) – c). Suppose that the corresponding averaged system (7.136) has an equilibrium $x = x_0$ and the real part of all the roots of the characteristic equation* $\det|\lambda E - F'_x(x_0)| = 0$ *are nonzero at the equilibrium point.*

Then there exist such positive constants ϵ', σ_1, $D(\epsilon)$, $D(\epsilon) \le \sigma_1 < \rho_1$, $D(\epsilon) \to 0$ if $\epsilon \to 0$, that, for any positive ϵ, $0 < \epsilon < \epsilon'$, the following statements hold:

1) System (7.133) has a unique solution $x = x^*(t)$ defined for all $t \in \mathbf{R}$ and satisfying the inequality $\|x^*(t) - x^0\| < \sigma$;

2) If system (7.133) is periodic (almost periodic) with respect to t, then the solution $x^*(t)$ is also periodic (almost periodic);

3) Let $x(t)$ be any solution of (7.133), which satisfies, for some $t = t_0$, the inequality $\|x(t_0) - x^0\| < \sigma$. If the real parts of all the roots of the characteristic equation $\det|\lambda E - F'_x(x_0)| = 0$ *are negative, then*

$$\|x(t) - x^*(t)\| \le ce^{-\gamma\epsilon(t-t_0)}, \qquad t \ge t_0, \tag{7.137}$$

where c and γ are positive constants. If the real parts of all the roots of the characteristic equation are positive, then there exists such $t' > t_0$ that $\|x(t') - x^0\| > \sigma_1$. And, finally, if r roots have negative real parts and the remaining $n-r$ roots have positive real parts, then a σ_0-neighborhood of the point x^0 contains such r-dimensional set $\mathcal{M}_{t_0}$ that $x(t^0) \in \mathcal{M}_{t_0}$ implies that (7.137) holds, and, if $x(t^0) \notin \mathcal{M}_{t_0}$, then there exists $t' > t_0$ with $\|x(t') - x^0\| > \sigma_1$.

This theorem is an extension of the known Bogolyubov's theorem [23] on substantiation of the averaging method used in impulsive systems on an infinite time interval. Its proof is given in [137].

Now consider a system of differential equations, which is subject to an impulsive effect occurring when the integral curve intersects certain sets given in the extended phase space,

$$\frac{dx}{dt} = \epsilon X(t,x), \qquad t \neq \tau_i(x),$$
$$\Delta x|_{t=\tau_i(x)} = \epsilon I_i(x). \tag{7.138}$$

We impose the following restrictions on the functions defining system (7.138):

a) $X(t,x)$, $I_i(x)$, $\tau_i(x)$ are functions, continuous with respect to their variables and satisfying the Lipschitz condition with respect to x in the region $t \geq 0$, $x = (x_1, \ldots, x_n) \in D \subset \mathbf{R}$;

b) for $t \geq 0$, $x \in D$, the limits

$$\begin{aligned} \lim_{T\to\infty} \frac{1}{T}\int_t^{t+T} X(t,x)dt &= X_0(x), \\ \lim_{T\to\infty} \frac{1}{T} \sum_{T<\tau_i(x)<t+T} I_i(x) &= I_0(x), \\ \lim_{T\to\infty} \frac{i(t,t+T,x)}{T} &= d, \end{aligned} \tag{7.139}$$

exist and are uniform with respect to t and x. Here $i(t,t+T,x)$ is the number of the hypersurfaces $t = \tau_i(x)$ which intersect the axis t within the interval $[t, t+T]$.

Together with system (7.138), we will consider the corresponding averaged system

$$\frac{dx}{dt} = \epsilon[X_0(x) + I_0(x)]. \tag{7.140}$$

In [134], a substantiation of the averaging method is given for systems (7.138), the value for the difference between an exact solution of system (7.138) and its approximation, which is a solution of the averaged system (7.140), is obtained, and the behavior of solutions of (7.138) in a neighborhood of an asymptotically stable equilibrium of the corresponding averaged system is studied. Let us consider the behavior of solutions of system (7.138) in a neighborhood of a trajectory of an orbitally asymptotically stable periodic solution of the averaged system (7.140).

Suppose that system (7.140) has an orbitally asymptotically stable solution $x = x^*(\nu t)$, which is 2π-periodic with respect to νt. Let C be the trajectory of this solution, i.e. a closed curve defined in the space $\mathbf{R}^n$ by the vector function $x^*(\nu t)$,

$$C = \{x \in D : x = x^*(\varphi) = x^*(\varphi + 2\varphi),\ 0 \leq \varphi \leq 2\pi\}. \tag{7.141}$$

We assume that the curve C and its ρ-neighborhood belong to the region D. A ρ-neighborhood of the curve C is the set $x \in D$, such that $d(x, C) \equiv \inf_{y \in C} \|x - y\| < \rho$. Let us prove the following statement.

Theorem 116 *Let system (7.138) satisfy conditions a) and b). If the averaged system (7.140) has an orbitally asymptotically stable periodic solution $x = x^*(\nu t)$, which, together with its ρ-neighborhood, belongs to the region D, and*

$$\tau_i(x) \geq \tau_i(x + I_i(x)) \tag{7.142}$$

for all $i = 1, 2, \ldots$ and all x from the ρ_0-neighborhood of the curve C, then there exists such ρ-neighborhood D_ρ of this curve ($\rho \leq \rho_0$) and such $\epsilon^0 > 0$ that, for all ϵ, $0 < \epsilon \leq \epsilon^0$, and all $x \in D_\rho$, the solution $x_t(x)$, $x_0(x) = x$, of system (7.138) is uniformly bounded for $t \in [0, \infty[$.

Proof. It should be noted at once that, because of inequality (7.142), there are no beatings of solutions of system (7.138) at the surfaces $t = \tau_i(x)$ at least for solutions from some ρ^0-neighborhood of the curve C. Let $\bar{x}(t, x)$, $\bar{x}(0, x) = x$, be a solution of the averaged system (7.140) for $\epsilon = 1$. Because the solution $x = x^*(\nu t)$ is orbitally asymptotically stable and solutions of the averaged system depend on initial conditions continuously, there exists such a positive number ρ' that

$$d(\bar{x}(t,x),C) \leq \rho_0, \quad \text{if} \quad d(x,C) \leq \rho', \quad t \geq 0. \tag{7.143}$$

This inequality shows that, if

$$x \in T_{\rho'} = \{x \in D : d(x,C) \leq \rho'\}, \tag{7.144}$$

the solutions $\bar{x}(\epsilon t, x)$ satisfy condition c) of Theorem 1 [134] for all $t \geq 0$. This theorem implies that there exists such $\epsilon^0 = \epsilon^0(L,\rho)$ that, for all $0 < \epsilon \leq \epsilon^0$, all $t \in \left[0, \frac{L}{\epsilon}\right[$, and some ρ ($\rho \leq \rho^0$, $\rho \leq \rho'$), we have

$$\|x_{t+\tau}(\tau,x) - \bar{x}(\epsilon t, x)\| \leq \frac{\rho}{2}, \tag{7.145}$$

where $x_{t+\tau}(\tau,x)$ is a solution of equations (7.138) that passes through the point $x \in T_\rho$.

Let us choose ρ such that T_ρ belong to the region where the trajectory C is asymptotically stable and let L be such that

$$d(\bar{x}_t(x \in T_\rho), C) \leq \frac{\rho}{2}, \qquad t \geq L. \tag{7.146}$$

By using inequalities (7.143) – (7.145), we get the following estimates

$$\begin{aligned} d(x_{t+\tau}(x \in T_\rho, \tau), C) \;&\leq\; \|x_{t+\tau}(x \in T_\rho, \tau) - \bar{x}(\epsilon t, x \in T_\rho)\| + \\ &+\; d(\bar{x}(\epsilon t, x \in T_\rho), C) \leq \frac{\rho}{2} + \rho_0, \qquad t \in \left[0, \frac{L}{\epsilon}\right[, \\ &\quad d\left(x_{\frac{L}{\epsilon}}(x \in T_\rho, 0), C\right) \leq \rho, \end{aligned} \tag{7.147}$$

The latter estimate means that

$$x_{\frac{L}{\epsilon}}(x \in T_\rho, 0) \in T_\rho. \tag{7.148}$$

Since $x_{t+\tau}(x,0) = x_{t+\tau}(x_\tau(x,0),\tau)$, by using (7.148), we get

$$x_{t+\frac{L}{\epsilon}}(x \in T_\rho, 0) = x_{t+\frac{L}{\epsilon}}\left(x_{t+\frac{L}{\epsilon}}(x \in T_\rho, 0), \frac{L}{\epsilon}\right) = x_{t+\frac{L}{\epsilon}}\left(x' \in T_\rho, \frac{L}{\epsilon}\right).$$

Whence, using (7.147), it follows that

$$\begin{aligned} d\left(x_{t+\frac{L}{\epsilon}}(x \in T_\rho, 0), C\right) &\leq \frac{\rho}{2} + \rho_0, \qquad t \in \left[0, \frac{L}{\epsilon}\right[, \\ d\left(x_{\frac{2L}{\epsilon}}(x \in T_\rho, 0), C\right) &\leq \rho. \end{aligned}$$

The latter inequality shows that $x_{\frac{2L}{\epsilon}}(x \in T_\rho, 0) \in T_\rho$ and that

$$\begin{aligned} d\left(x_{t+k\frac{L}{\epsilon}}(x \in T_\rho, 0), C\right) &\leq \frac{\rho}{2} + \rho_0, \\ d\left(x_{(k+1)\frac{L}{\epsilon}}(x \in T_\rho, 0), C\right) &\leq \rho, \end{aligned} \tag{7.149}$$

if $t \in \left[0, \frac{L}{\epsilon}\right[$, $k = 0, 1, 2, \ldots$. Inequalities (7.149) mean that

$$d\left(x_t(x \in T_\rho), C\right) \leq \frac{\rho}{2} + \rho_0. \tag{7.150}$$

for all $t \in [0, \infty[$.

□

Theorem 116 shows that there exist solutions of system (7.138), which are bounded for $t \geq 0$. Let us now find conditions which would imply that this system has solutions that are bounded for all $t \in \mathbf{R}$.

Suppose that system (7.138) satisfies conditions a) and b) for $t \geq 0$ as well as for $t < 0$.

Set

$$\begin{aligned} X^0(x) &= \lim_{T \to \infty} \frac{1}{T} \int_{t-T}^{t} X(t, x)dt, \\ I^0(x) &= \lim_{T \to \infty} \frac{1}{T} \sum_{t-T<\tau_i(x)<t} I_i(x), \qquad t < 0. \end{aligned}$$

Assume that system (7.140), averaged for $t \geq 0$, has an orbitally asymptotically stable solution $x = x^*(\nu t)$ and, for x from a ρ_0-neighborhood of the trajectory C, inequality (7.142) holds. Let, in the given by Theorem 116 ρ-neighborhood $D\delta$ of the curve C, the system

$$\frac{dx_1}{dt} = \epsilon[X^0(x_1) + I^0(x_1)], \tag{7.151}$$

averaged for $t \leq 0$, have a periodic solution $x_1 = x_1^*(\nu, t)$, C_1 be the trajectory of this solution, and suppose that

$$\tau_i(x) \geq \tau_i(x + I_i(x)) \tag{7.152}$$

for all $i = -1, -2, \dots$ and all x from some ρ_0'-neighborhood of the curve C_1.

If the stated above conditions hold, then we have the following theorem.

Theorem 117 *If the periodic solution $x_1 = x_1^*(\nu, t)$ of equations (7.151) is orbitally asymptotically unstable (orbitally asymptotically stable for $t < 0$), then there exist such $\epsilon^0 > 0$ and such a domain D_{ρ_1} containing C and C_1 that, for all $0 < \epsilon \leq \epsilon^0$, all the solutions $x_t(x)$, for which $x \in D_{\rho_1}$, are uniformly bounded for $t \in \mathbf{R}$.*

If the periodic solution $x_1 = x_1^*(\nu t)$ is orbitally asymptotically stable, then there exists such $\epsilon^0 > 0$ that, for all ϵ, $0 < \epsilon \leq \epsilon^0$, system (7.138) has an integral set.

Proof. Let a solution $x_1 = x_1^*(\nu, t)$ be orbitally asymptotically unstable. By applying Theorem 7.134 to the intervals $t \geq 0$ and $t \leq 0$, we see that there exist the corresponding ρ- and $\bar{\rho}$-neighborhoods of the curves C and C_1 such that

$$\begin{aligned} \|x_t(x \in D_\rho)\| &\leq m_1, \qquad t \in [0, \infty[, \\ \|x_t(x \in T_{\bar{\rho}})\| &\leq m_2, \qquad t \in]-\infty, 0]. \end{aligned}$$

Because $C \subset D_\rho$, the set $D_\rho \cup T_{\bar{\rho}} = D_{\rho_1}$ is not empty and, consequently, $\|x_t(x \in D_{\rho_1})\| \leq m = \max(m_1, m_2)$, $t \in \mathbf{R}$.

Let now the solution $x_1 = x_1^*(\nu, t)$ be orbitally asymptotically stable. By applying Theorem 7.134 to system (7.138), we see that, for $t \in \left[0, \frac{L}{\epsilon}\right]$, the solution $x_t(x \in T'_{\rho_1}, \tau)$ exists and satisfies the estimates

$$\begin{aligned} d\left(x_{t+\tau}(x \in T_{\rho'_1}, \tau), C\right) &\leq \frac{\rho'_1}{2} + \rho_1^0, \qquad t \in \left[0, \frac{L}{\epsilon}\right[, \qquad \tau \geq -\frac{L}{\epsilon}, \\ d\left(x_0\left(x \in T_{\rho'_1}, -\frac{L}{\epsilon}\right), C\right) &\leq \rho'_1, \end{aligned} \tag{7.153}$$

where $T_{\rho'_1} = \{x \in D : d(x, C_1) \leq \rho'_1$.

Inequalities (7.153) lead to the estimates

$$\begin{aligned} d\left(x_{t-k\frac{L}{\epsilon}}\left(x \in T_{\rho'_1}, -k\frac{L}{\epsilon}\right), C_1\right) &\leq \frac{\rho'_1}{2} + \rho_1^0, \\ d\left(x_0\left(x \in T_{\rho'_1}, -k\frac{L}{\epsilon}\right), C_1\right) &\leq \rho'_1, \end{aligned} \tag{7.154}$$

for $t \in \left(-k\frac{L}{\epsilon}, 0\right]$, $k = 1, 2, \ldots$.

Take now a closed smooth curve γ_0 in $T_{\rho'_1}$:

$$\gamma_0 = \left\{x \in T_{\rho'_1} : x = \tilde{x}(\theta) = \tilde{x}(\theta + 2\pi)\right\}$$

By using inequalities (7.154), we have $d\left(x_0(x \in \gamma_0, -\frac{kL}{\epsilon}), C_1\right) \leq \rho'_1$, $k = 0$,-1, 2, ..., i.e. all the curves $\gamma_k = \left\{x \in T_{\rho'_1} : x = x_0\left(x \in \gamma_0, -\frac{kL}{\epsilon}\right) = \tilde{x}_k(\theta)\right\}$, obtained from γ_0 by a translation of γ_0 on $k\frac{L}{\epsilon}$ along the trajectories of system (7.138), belong to $T_{\rho'_1}$. Without loss of generality, we can assume that, for any $k = 1, 2, \ldots$, the times $t_k = -\frac{kL}{\epsilon}$ lie inside the interval $[\overline{\tau}_i, \underline{\tau}_{i+1}]$ for some $i = i(k)$, where $\overline{\tau}_i = \max_{x \in T_{\rho'_1}} \tau_i(x)$, $\underline{\tau}_{i+1} = \min_{x \in T_{\rho'_1}} \tau_{i+1}(x)$, and so the curves γ_k are continuous and satisfy the Lipschitz condition with the constant $l - l(\epsilon)$. Because solutions of (7.138) are uniformly continuous with respect to initial conditions on the intervals $[\overline{\tau}_i, \underline{\tau}_{i+1}]$ and the functions $X(t, x)$ and $I_i(x)$ satisfy the Lipschitz condition in x with a constant proportional to ϵ, the functions $\tilde{x}_k(\theta)$ satisfy the inequality

$$\|\tilde{x}_{k+1}(\theta) - \tilde{x}_k(\theta)\| \leq \lambda(\epsilon)\|\tilde{x}_k(\theta) - \tilde{x}_{k-1}(\theta)\|$$

for all $k = 1, 2, \ldots$, $\lambda(\epsilon) \to 0$, $\epsilon \to 0$.

Whence it follows that there exists ϵ^0 such that, for every fixed ϵ, $0 < \epsilon \leq \epsilon^0$, a sequence of the closed curves γ_k converges to a limit closed curve $\lim_{k\to\infty} \gamma_k = \gamma_\epsilon$.

Let us show that the solutions of equations (7.138) that start at a point $x \in \gamma_\epsilon$ when $t = 0$, i.e. $x_t(x \in \gamma_\epsilon, 0)$, exist for all $t \in \mathbf{R}$ and that they are bounded. First we show that these solutions are defined for all $t \in]-\infty, 0]$ and are bounded. Indeed, the sequence of solutions $x_t\left(x \in \gamma_0, -\frac{kL}{\epsilon}\right)$, defined for $-\frac{kL}{\epsilon} \leq t \leq 0$, is uniformly bounded for $k = 1, 2, \ldots$ because it coincides with the sequence $x_{\bar{t}-\frac{kL}{\epsilon}}\left(x \in \gamma_k, -k\frac{L}{\epsilon}\right)$ for $t \in \left[-k\frac{L}{\epsilon}, 0\right]$.

Assume that, for some $t = t^0 < 0$,

$$d(x_t(x \in \gamma_\epsilon), C_1) > \frac{\rho'_1}{2} + \rho_1^0. \tag{7.155}$$

Because $x_t(x \in \gamma_\epsilon)$ is a piecewise continuous function of t on any interval and inequality (7.155) holds for $t = t^0$, it follows that this inequality also holds on some interval $[t_0, t_0']$, on which the solution $x_t(x \in M_\epsilon)$ is continuous.

Because the points of discontinuity of the solution $x_t(x \in \gamma_\epsilon)$ depend continuously on x if the time interval is finite, one can choose such $\tau_1 \in [t_0, t_0']$ that

$$x_{\tau_1}(x \in \gamma_k) \to x_{\tau_1}(x \in \gamma_\epsilon) \qquad \text{for} \qquad k \to \infty.$$

Relation (7.155) leads to the inequality

$$d\left(x_{\tau_1}(x \in \gamma_k), C_1\right) > \frac{\rho'_1}{2} + \rho_1^0 \tag{7.156}$$

for all sufficiently large k. But (7.156) contradicts to (7.154) because

$$x_{\tau_1}\left(x \in \gamma_0, -k\frac{L}{\epsilon}\right) = x_{\bar{\tau}-k\frac{L}{\epsilon}}\left(x \in \gamma_k, -k\frac{L}{\epsilon}\right).$$

This contradiction shows that $x_t(x \in \gamma_\epsilon)$ is bounded for $t \in]-\infty, 0]$. But, since $x \in \gamma_\epsilon \subset D_\rho$, by Theorem 116, $x_t(x \in \gamma_\epsilon)$ is also bounded for $t \geq 0$ and, consequently, it is bounded for all $t \in \mathbf{R}$.

□

Supplement A

Periodic and almost periodic impulsive systems

(by S.I. Trofimchuk)

A.1 Impulsive systems with generally distributed impulses at fixed times

In this chapter, we will be considering impulsive systems with impulses that occur at fixed times, which are more general than the systems considered in preceding sections. Before going to general statements, we consider two examples.

As the first example, consider a discontinuous dynamical system that describes a vertical motion of a resilient ball which centrally hits a stationary horizontal platform. We suppose that the mass of the ball is much smaller than the mass of the platform. We also take the collision coefficient to be $k \in (0,1)$. If x and x' are the coordinates (the altitude with respect to the platform and the velocity of the ball respectively), g is the gravity acceleration, then a mathematical model for the system will be

$$x'' = -q,$$
$$x'(t_i+0) = -kx' \quad \text{if} \quad x(t_i+0) = 0 \quad \text{and} \quad x'(t_i+0) \leq 0,$$
$$x(0) = H, \qquad x'(0) = 0,$$
$$x \geq 0.$$

The phase plane and one of trajectories of this discontinuous dynamical system is given in the figure below.

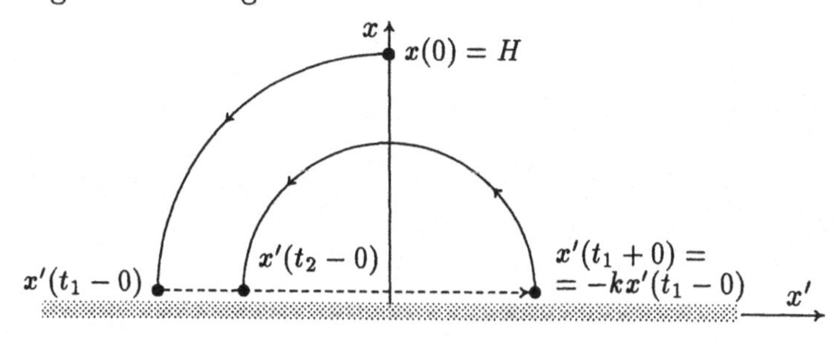

Let t_i be the moments at which the ball hits the platform the ith time. It is easy to calculate that

$$\Delta_1 = t_1 = \left(\frac{2H}{q}\right)^{1/2}, \quad \Delta_2 = t_2 - t_1 = 2kt_1, \quad \Delta_{n+1} = t_{n+1} - t_n = kt_n$$

for $n > 1$. Moreover, because the series $\sum_{i\geq 1} \Delta_i$ converges to the number $\tau = (2H/g)^{1/2}(1+k)/(1-k)$ and $x'(t_{i+1}-0) = k^i(2gH)^{1/2} \to 0$, the ball will stay motionless on the platform after the finite time τ. Hence, the equation of the considered trajectory is an impulsive system with the impulse times t_i accumulating at the point τ:

$$x'' = -g, \qquad \Delta x'|_{t_i} = -(k+1)k^i(2gH)^{\frac{1}{2}}.$$

The second example is given by a description of a sequence of free harmonic oscillators, each of which undergoes a single impulsive effect at a given point τ_i:

$$\frac{d^2x}{dt^2} + \omega_i^2 x_i = 0, \qquad \Delta \left.\frac{dx}{dt}\right|_{t_i} = h_i \neq 0, \qquad x = (x_1, x_2, \dots).$$

Let m_i be the mass of the ith oscillator, x_i – its coordinate. Assume that the total energy of the free system (without impulsive effects) is finite. (Hence we can assume that the phase space of the system is isomorphic to l_2 and is a set of sequences $(x, dx/dt)$, for which the sum $\sum m_i[(dx_i/dt)^2 + \omega_i^2 x_i^2]$ is finite). Suppose that $\sum m_i h_i^2 < \infty$. Then the change of the total energy H in the impulsive system is described by

$$\frac{dH}{dt} = 0, \qquad \Delta H|_{t_i} = H_i \neq 0,$$

where the countable set $T = \{t_i\}$ is arbitrary. So, we see that it is necessary to study generalized impulsive systems

$$\frac{dx}{dt} = f(t,x), \quad t \neq t_i, \quad t \in [a,b], \quad x(a) = x_0, \tag{A.1}$$

$$\Delta x|_{t_i} = h_i(x), \quad i \in N, \quad x \in \Omega \tag{A.2}$$

when the set $T = \{t_i : t_i \neq t_j \text{ if } i \neq j\}$ may have finite limit points. Excluding unnecessary times of impulsive effects, we will assume that $h_i \overset{\text{def}}{=} \sup_{x\in\Omega} |h_i(x)| > 0$. Let us consider a simple system of this type with $T = \{t_n = 1 - 1/n : n \geq 2\}$,

$$\frac{dx}{dt} = 0, \quad t \in [0,2]/T, \quad x \in \mathbf{R},$$
$$\Delta x|_{t_i} = c_i, \quad c_i \in \mathbf{R} \setminus \{0\}, \quad i \leq 2, \tag{A.3}$$

in a greater detail.

A solution of the Cauchy problem $x(0) = x_0$ can be determined for (A.3) straight away on any segment $[0, 1-\epsilon]$, $0 < \epsilon < 1$, as a solution of an ordinary impulsive system with a finite number of impulses: if we set $c_1 = 0$, then

$x(t) = x_0 + \sum_{i=1}^{n} c_i$ for $1 - 1/n \leq t < 1 - 1/(n+1)$. To extend the solution $x(t)$ to $[0, 1]$, it is clearly necessary that the finite limit $\lim_{t \to 1-0} x(t)$ exist, i.e. the series $\sum_{i=1}^{+\infty} = c_\infty$ converge. If $c_\infty \in \mathbf{R}$, then, by setting $x(t) = c_\infty$ for $t \in [1, 2]$, we obtain a function which satisfies the initial condition and (A.3) everywhere but the point $t = 1$ (where it is not clear whether the derivative $x'(1)$ equal to 0 or not. Because $x'(1) = \lim_{\epsilon \to 0}(1/\epsilon) \sum_{t_i : 1-t_i \leq \epsilon} c_i$, the derivative $x'(1)$ will be equal to zero if and only if

$$\sum_{1-t_i \leq \epsilon} c_i = o(\epsilon) \quad \text{for} \quad \epsilon \to 0.) \tag{A.4}$$

For the Cauchy problem to be correctly defined, it is necessary that, for small translations of the times of impulsive effects, its solution be changed insignificantly. Applied to system (A.3), this means that, if terms of the series $\sum_{i=1}^{+\infty} c_i$ are rearranged and some of them, which have sufficiently large index, are omitted, the sum is little changed, i.e. the series is absolutely convergent.

To make the presentation even more simple and to avoid having conditions of type (A.4), we will require that a solution $x(t)$ satisfy equation (A.1) not for all $t \neq t_i$ but almost all t. Correspondingly, system (A.1 - A.2) will be a Caratheodory type system and its solution $x(t)$ will be defined as a left continuous function which is given as a sum of an absolutely continuous function and a jump function satisfying equation (A.1) almost everywhere and equation (A.2) for all i. Naturally, $x(t)$ may have discontinuities only on the set T. If, from the beginning, the assumption suggested by the first example is made,

$$\mu(b) \stackrel{\text{def}}{=} b + \sum_{a < t_i \leq b} h_i < +\infty, \qquad \text{where} \quad \sup_{\Omega} |h_i(x)| = h_i$$

(and we will assume that), then we can give an equivalent definition for a Caratheodory solution of the impulsive system.

Definition 3 *A function $x(t)$, which is μ-absolutely continuous will be called a (Caratheodory) solution of (A.1 - A.2) if $x(t)$ satisfies relations (A.1 - A.2) μ-almost everywhere.*

Among other conditions that we need to be imposed on system (A.1 - A.2), we give the following:

a) the functions $h_i(x)$ are continuous and the function $f(t, x)$ is continuous in x for almost all $t \in [a, b]$ and is measurable in t for all x;

b) $|f(t,x)| \le m(t)$, where $m(t)$ is a locally integrable function.

Let $x(t)$ be a solution of problem (A.1 – A.2), $x(0) = x_0$. The function $x(t)$ is differentiable almost everywhere and, at the point t_i, where it is left continuous, it has a finite jump ξ_i such that the series $\sum_{a<t_i\le\sigma} \xi_i$ is absolutely convergent. Consider the left continuous jump function $s(t) = \sum_{a<t_i<t} \xi_i$. Clearly, $s'(t) = 0$ almost everywhere. Because $s(t_i+0) - s(t_i) = \sum_{a<t_j\le t_i} \xi_i - \sum_{a<t_j<t_i} \xi_i = \xi_i$, the function $l(t) = x(t) - s(t)$ is continuous (indeed, $l(t_i+0) - l(t_i) = x(t_i+0) - s(t_i+0) - (x(t_i) - s(t_i)) = 0$). By the condition, the function $l(t)$ is absolutely continuous and its derivative $l'(t)$ is equal almost everywhere to $f(t, x(t)) \overset{\text{def}}{=} q(t)$. It is clear that the function $r(t) - l(t) - \int_a^t q(t)dt$ is absolutely continuous on $[a, \sigma)$ and $r'(t) = 0$ almost everywhere. Hence $r(t) \equiv r(a) = l(a)$. Because $\xi_i = h_i(x(t_i))$, the function $x(t)$ satisfies the integral equation

$$x(t) = x_0 + \int_a^t f(u, x(u))du + \sum_{a<t_i<t} h_i(x(t_i)). \tag{A.5}$$

It is evident that any solution of (A.5) will be a solution of problem (A.1 – A.2). Hence, for an impulsive system with the assumptions made, we get an integral equation which is equivalent to the system.

If the point a is not a limit point of the set T, then a local existence and uniqueness of a solution of problem (A.1 – A.2) is equivalent to local existence and uniqueness of a solution of problem (A.1). Now suppose that a is a limit point of the set T, $t \in [a,b]$, and, for the sake of simplicity, $a \notin T$.

Theorem 118 *A Caratheodory solution of (A.1 – A.2) exists.*

Proof. Consider the integral equation

$$x_m(t) = x_0 + \int_a^t f(u, x_m(u))du + \sum_{\substack{a<t_i<t \\ i\le m}} h_i(x_m(t_i)). \tag{A.6}$$

From the preceding considerations, it follows that it is equivalent to the impulsive system

$$\frac{dx}{dt} = f(t,x), \quad t \ne t_i, \quad i = \overline{1,m}, \quad \Delta x|_{t_i} = h_i(x), \quad i = \overline{1,m},$$

with a finite number of impulsive effects. Under the assumptions made, the initial value problem $x(0) = x_0$ for this system has, at least, one piecewise continuous solution $x_m(t)$. For a sufficiently small value of $\mu = t - a > 0$, the value of $x_m(t)$, for all m, lies in the region $\Omega : |x_m(t) - x_0| \leq M$ for $t \in [a, a+\mu)$. Moreover, the variation of $x_m(t)$ on $[a, a+\mu)$ is bounded by the quantity $\int_a^\sigma m(u)du + \sum_{a<t_i\leq\sigma} \sup|h_i(x)|$ which does not depend on m.

So, all the conditions of the Helly theorem [19] are satisfied and there exists a subsequence $x_{m_k}(t)$ which converges at every point of the interval $[a, a+\mu)$ to a function $x^*(t)$ (which is integrable by the Lebesgue theorem). Moreover, it follows from the conditions of the theorem that

$$\sum_{\substack{a<t_i<t \\ i\leq m_k}} h_i(x_{m_k}(t_i)) \to \sum_{a<t_i<t} h_i(x^*(t_i)).$$

for all $t \in [a, a+\mu)$. By passing in (A.6) to the limit as $m_k \to +\infty$ (this can be done by Lebesgue theorem), we finish the proof.

□

One could suppose that the initial value problem for an impulsive system will have a unique solution if the functions $f(t,x)$ and $h_i(x)$ satisfy the Lipschitz condition with some Lipschitz constant. The following example shows that such a statement would be false.

Example. Let the monotone decreasing sequences of positive numbers $\{t_n\}_{n=1}^{+\infty}$ and $\{c_n\}_{n=0}^{+\infty}$ be such that $\lim_{n\to\infty} c_n = \lim_{n\to\infty} t_n = 0$ and $|(c_{n-1}/c_n)| \leq P$. Consider the impulsive system

$$\begin{cases} \dfrac{dx}{dt} = 0, & t \neq t_i, \\ \Delta x|_{t_i} = h(x_i), & i \in \mathbf{N}, \end{cases} \quad \text{for}$$

$$h_n(x) = \begin{cases} ((c_{n-1} - c_n)/c_n)x, & x \geq c_n, \\ c_{c-1} - c_n, & x \leq c_n, \end{cases}$$

defined in the first quadrant. Because $|h_n(x)| \leq c_{n-1} - c_n$, we have that $\sum \sup|h_i(x)| \leq c_0$ and hence all the conditions are satisfied. Moreover, the function $h_n(x)$ is Lipschitzian with the Lipschitz constant $l_n = c_{n-1}/c_n - 1 \leq P + 1$. At the same time, the Cauchy problem $x(0) = 0$ has at least two different solutions $x(t) \equiv 0$ and $x(t) = c_{n-1}$, $t_n < t \leq t_{n-1}$. Let us show that, under the assumptions made, the series $\sum_i l_i$ diverges. Clearly, $c_n = c_0/\prod_{i=1}^n(l_i+1)$. If $\sum_i l_i < +\infty$, then $\prod_{i=1}^n(l_i+1) \to d \in \mathbf{R}\setminus\{0\}$ and so $c_n \to c_0/d \neq 0$, which contradicts to the assumption.

It turns out that, for this example, convergence of the series $\sum l_i$ will be a necessary and sufficient condition for the initial value problem to have a unique solution.

Lemma 50 *Let $u(t)$ have the bounded variation on $[a,b]$ and a nonnegative integrable function $\beta(s) : [a,b] \to \mathbf{R}$ be such that*

$$u(t) \leq \alpha + \int_a^t \beta(s)u(s)ds + \sum_{a \leq t_i < t} \gamma_i u(t_i), \tag{A.7}$$

where the constants α, γ_i are nonnegative, $\sum_{a \leq t_i \leq b} \gamma_i < +\infty$. Then

$$u(t) \leq \alpha \prod_{\alpha \leq t_i < t} (\gamma_i + 1) \exp(\int_a^t \beta(s)ds), \quad t \in [a,b].$$

Proof. The series $\sum_{i=0}^{+\infty} \gamma_i u(t_i)$ converges and $c_N = \sum_{i=N}^{+\infty} \gamma_i u(t_i) \to 0$ as $N \to +\infty$. From (A.7), we have the estimate

$$u(t) \leq (\alpha + c_N) + \int_a^t \beta(s)u(s)ds + \sum_{\substack{a \leq t_i < t \\ i < N}} \gamma_i u(t_i) \stackrel{\text{def}}{=} v(t). \tag{A.8}$$

On the other hand, by (A.8), the function $v(t)$ satisfies a similar inequality,

$$v(t) \leq (\alpha + c_N) + \int_a^t \beta(s)v(s)ds + \sum_{\substack{a \leq t_i < t \\ i < N}} \gamma_i v(t_i).$$

Because the function $v(t)$ is nonnegative, piecewise continuous, by Lemma 1, it follows that the estimate

$$v(t) \leq (\alpha + c_N) \prod_{\substack{a \leq t_i < t \\ i < N}} (\gamma_i + 1)\exp(\int_a^t \beta(s)ds), \quad t \in [0,T] \tag{A.9}$$

holds. It follows from (A.8) and (A.9) that

$$u(t) \leq (\alpha + c_N) \prod_{\substack{a \leq t_i < t \\ i < N}} (\gamma_i + 1)\exp(\int_a^t \beta(s)ds), \quad t \in [0,T].$$

By passing to the limit in N, we end the proof of the lemma.

□

Theorem 119 *If the following conditions hold*

a) $|f(t,x) - f(t,y)| \leq L(t)|x-y| \qquad \forall x, y \in \Omega,\ t \in [a,\sigma);$
b) $|h_i(x) - h_i(y)| \leq l_i|x-y|;$
c) $\int_a^\sigma L(s)ds + \sum_{a \leq t_i \leq \sigma} l_i < +\infty,$

then a solution of Cauchy problem (A.1 - A.2) is unique.

Proof. Suppose that there exist two distinct solutions $x(t)$ and $y(t)$ in a small neighborhood of the point a, $[a, a+\mu)$. They both satisfy equation (A.5), so conditions of Theorem 119 insure that the inequality

$$|x(t) - y(t)| \leq \int_a^t L(s)|x(s) - y(s)|ds + \sum_{a<t_i<t} l_i|x(t_i) - y(t_i)|.$$

holds. By applying the preceding lemma, we get $x(t) \equiv y(t)$.

□

We end this section with the following remark. In practice, one can give the elements $f(t,x)$, $h_i(x)$, t_i which define system (A.1 - A.2), only with a certain degree of precision. Moreover, the index i may run over a finite subset N. Thus, instead of integration of the exact system (A.1 - A.2), it is necessary to find a solution $x^{(n)}(t)$ of the approximate system

$$\begin{aligned} \frac{dx}{dt} &= f^{(n)}(t,x), \quad t \neq t_i^{(n)}, \\ \Delta x|_{t_i^{(n)}} &= h_i^{(n)}(x), \quad i \in \mathcal{A}(n) \subset N, \\ x(a) &= x_0^{(n)}, \quad t \in [a,\sigma). \end{aligned} \tag{A.10}$$

Correctness of the Cauchy problem for system (A.1 - A.2) means, firstly, that it has a unique solution and, secondly, that the mapping, which assigns to system (A.1 - A.2) its solution, is continuous for a suitable choice of topologies in the space of the right-hand sides of (A.1 - A.2) and the space of solutions. It turns out that, under certain conditions, this problem is correct. Thus we define convergence in the space of right-hand sides of (A.1 - A.2) as follows: sequence (A.10) converges to system (A.1 - A.2) if, for $n \to +\infty$,

a$_1$) $x_0^n \to x_0$; $t_i^{(n)} \to t_i$ uniformly with respect to $i \in \mathcal{A}(n)$;

a$_2$) $\sum_{a<t\leq\sigma} \sup_\Omega |h_i(x) - h_i^{(n)}(x)| \to 0$, where $h_i^{(n)}(x) \equiv 0$ for $i \notin \mathcal{A}(n)$;

a$_3$) for all x, $\int_a^t f^{(n)}(t,x)dt \to \int_a^t f(t,x)dt$ uniformly with respect to $t \in [a,\sigma)$.

We define the distance between two solutions of impulsive systems on $[a,\sigma]$, $\rho(x(t),y(t))$, to be the Hausdorff distance between the closures of their graphs in $[a,\sigma] \times \Omega$.

Theorem 120 (see [162]) *Let, for every system (A.10), conditions of Theorem 119 be satisfied uniformly with respect to $n \in \mathbf{N}$ and conditions a_1) – a_3) hold. If problem (A.1 – A.2) has a solution $x(t)$ on $[a,\sigma]$, then, for large n, a solution $x_n(t)$ of problem (A.10) also exists on $[a,\sigma]$ and $x_n(t) \to x(t)$ in the metric ρ.*

A.2 Periodic impulsive systems with impulses located on a surface

In this section we will consider T-periodic impulsive evolutionary systems. This case is significantly simpler than the case of an almost periodic system. In particular, here a study of the system can be reduced to a study of the Poincare map. We will be considering systems with unfixed impulsive effects and assume that the number of hypersurfaces where the "shocks" are located is finite on the period. We will be looking for a piecewise continuous solution. Admitting the phase space to be infinite dimensional does not require new ideas. We will construct a piecewise continuous solution step by step by using different solutions of the system and making them agree on the hypersurfaces of "shocks", given by the discrete part of the impulsive system. Here, considering a more simple case of ordinary differential equations in a Banach space, we mention only some results which give a sufficiently complete account on the methods used to study certain classes of such impulsive systems.

So, assume that a certain process is described by the T-periodic differential equation

$$\frac{dx}{dt} = f(t,x), \qquad x \in U \subset X \tag{A.11}$$

in a Banach space X, where the vector valued function $f(x,t)$ and the Frechet derivative $f'_x(t,x) \in \mathcal{L}(X)$ are continuous with respect to their variables. Let a smooth hypersurface Γ be given in $W = [0,T] \times \mathcal{U}$, have codimension 1, and be defined by the scalar equation $\varphi(t,x) = 0$ (hence $\varphi'_t(t,x)$ and $\varphi'_x(t,x)$ are not both zero at points of Γ). We will assume that, after the integral curve (A.11) meets Γ at a certain point (τ,ξ), its phase coordinate instantaneously changes into $H(\tau,\xi)$ and then the process is again determined by (A.11) until the point meets Γ once more. (We assume that there is no problem concerning extension of solutions). Consequently, the second equation of the impulsive system can be written as

$$x^+ = H(t,x^-), \quad o = \varphi(t,x^-). \tag{A.12}$$

Let us study the question on preservation and stability of T-periodic solutions of such a system (with respect to small perturbations of the right-hand sides of (A.11), (A.12), denoted in the sequel by ϵ).

Poincare mapping. So consider the perturbed system

$$\frac{dx}{dt} = f(t,x,\epsilon), \qquad x \in U \subset X, \tag{A.13}$$

$$x^+ = H(t, x^-, \epsilon) \qquad \text{on} \quad \Gamma = \{\varphi(t, x, \epsilon) = 0\}, \tag{A.14}$$

where $H_t(t, x, \epsilon)$, $H_x(t, x, \epsilon)$ exist and are continuous on a neighborhood of Γ. Suppose that the unperturbed system (A.11) has a T-periodic solution $x = x(t)$, $x(0) = \vartheta$. To make it more simple, assume that the solution undergoes one impulsive effect at time $\mu \in (0, T)$ and at a point $\xi \in U$. Assume that the vector field $(1, f(t, x, \epsilon))$ is tranversal at the point $(\mu, \xi) \in \Gamma$ to the tangent plane $\varphi'_t(\mu, \epsilon)\Delta t + \varphi'_x(\mu, \xi)\Delta x = 0$, i.e.

$$\Phi(\mu, \xi) \overset{\text{def}}{=} \varphi'_t(\mu, \xi) + \varphi'_x(\mu, \xi) f(\mu, \xi) \neq 0. \tag{A.15}$$

Then, in some neighborhood $\mathcal{U}$ of the point ϑ, there is a mapping $\mu(x, \epsilon)$: $\mathcal{U} \to \mathbf{R}$, $\mu(\vartheta, 0) = \mu$, which, for the initial value problem $x(0, 0, \epsilon, x) = x$ for (A.13), (A.14), assigns the moment of the first intersection of the integral curve and Γ. Clearly, it can be found from the equation

$$\varphi(\mu, x(\mu, o, \epsilon, x), \epsilon) = 0,$$

which, by (A.15) and the implicit function theorem, can be solved for a sufficiently small diameter of $\mathcal{U}$ and small ϵ (if $\varphi'_t(, x, \epsilon)$ and $\varphi'_x(t, x, \epsilon)$ are continuous with respect to their variables). There

$$\begin{aligned} \mu'_x(x, \epsilon) &= -\Phi^{-1}(\mu(x, \epsilon), x(\mu, 0, \epsilon, x), \epsilon) \times \\ &\times \varphi'_x(\mu(x, \epsilon), x(\mu, 0, \epsilon, x), \epsilon) x'_x(\mu, 0, \epsilon, x); \\ \mu'_x(\vartheta, 0) &= -\Phi^{-1}(\mu, \vartheta, 0)\varphi'_x(\mu, \vartheta, 0) x'_x(\mu, 0, 0, \vartheta). \end{aligned}$$

Suppose that smooth perturbations of $f(t, x)$, $H(t, x)$, and $\varphi(t, x)$ are such that the integral curve of (A.13) that passes through the point

$$(\mu(x, \epsilon), H(\mu(x, \epsilon), x(\mu, \epsilon), 0, \epsilon, x))$$

will not intersect Γ on $(\mu(x, \epsilon), T)$. Then there is a Poincare mapping defined for small ϵ on a subdomain $\mathcal{U}' \subset \mathcal{U}$,

$$S(x, \epsilon) = x(T, \mu(x, \epsilon), \epsilon, H(\mu(x, \epsilon), x(\mu(x, \epsilon), \epsilon, x), \epsilon)) : \mathcal{U}' \to \mathcal{U} \tag{A.16}$$

and it is clear that its fixed points define periodic solutions of (A.13 – A.14). Taking into account the equality $S(\vartheta, 0) = \vartheta$ and differential properties of S, we can use the implicit function theorem. To do that, we need to calculate

$S'_x(\vartheta,0)$ by directly using (A.16):

$$S'_x(\vartheta,0) = -x'_\mu(T,\mu,0,H(\mu,\xi))\Phi^{-1}(\mu,\vartheta,0)\varphi'_x(\mu,\vartheta,0)x'_x(\mu,0,0,\vartheta)+ \\ +x'_H(T,\mu,0,H(\mu,\xi))\times \\ \times\{-H'_\mu(\mu,\xi)\Phi^{-1}(\mu,\vartheta,0)\varphi'_x(\mu,\vartheta,0)x'_x(\mu,0,0,\vartheta)+H'_\xi(\mu,\xi)\times \\ \times[-f(\mu,\xi)\Phi^{-1}(\mu,\vartheta,0)\varphi'_x(\mu,\vartheta,0)x'_x(\mu,0,0,\vartheta)+x'_x(\mu,0,0,\vartheta)]\}.$$

Because

$$-x'_\mu(t,\mu,0,H(\mu,\xi)) = x'_H(T,\mu,0,H(\mu,\xi))f(\mu,H(\mu,\xi)),$$

we see that

$$S'_x(\vartheta,0) = x'_H(T,\mu,0,H(\mu,\xi))\{[f(\mu,H(\mu,\xi))-H'_\mu(\mu,\xi)- \\ -H'_\xi(\mu,\xi)f(\mu,\xi)]\Phi^{-1}(\mu,\vartheta,0)\varphi'_x(\mu,\vartheta,0)+H'_\xi(\mu,\xi)]\}x'_x(\mu,0,0,\vartheta).$$

Now the implicit function theorem immediately implies the following theorem.

Theorem 121 *Assume that $S'_x(\vartheta,0)-E$ is invertible. Then, for sufficiently small ϵ, system (A.3 - A.4) has a T-periodic solution $x(t,\epsilon)$, which approaches $x(t)$ in the Hausdorff metric as $\epsilon\to 0$. For $x(t,\epsilon)$ to be exponentially stable, it is sufficient that the spectrum of $S'_x(\vartheta,0)$ lie inside the unit disc.*

Remark. It is clear that $S'_x(\vartheta,0)$ coincides with the monodromy matrix for the linear T-periodic impulsive system (the variational system!)

$$\frac{dz}{dt} = f'_x(t,x(t))z(t) \\ z(nT+\mu+) = \{\Phi^{-1}(\mu,\vartheta,0)[f(\mu,H(\mu,\xi))-H'_\mu(\mu,\xi)- \\ -H'_\xi(\mu,\xi)f(\mu,\xi)]\varphi'_x(\mu,\vartheta,0)+H'_\xi(\mu,\xi)]\}z(nt+\mu-).$$

Also note that the case when the impulsive effect occurs during the period many times can be considered in the same way.

Example. (A model of Rusakov - Kharkevich, see [67]). A ball moves along a line with the motion described by $Px = x''+ax'+bx-f(t)=0$ and centrally collides (the collision coefficient is σ) with a massive constraint the movement of which is given by $x=\varphi(t)$ (see picture).

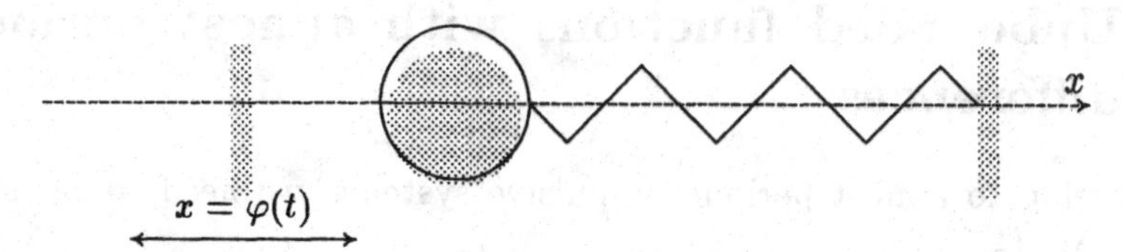

Assume that $\varphi(t)$ and $f(t)$ are T-periodic and there exists a T-periodic mode of this system, with one collision during the period occurring at a pont (ξ_1, ξ_2) $(z = (x, x'))$, and such that there are no beatings in the system. In this case it is easy to calculate the characteristic equation for $S'_x(\vartheta, 0)$:

$$\lambda^2 - \mathrm{sp}D\lambda + \sigma^2 e^{-aT} = 0, \qquad \text{where}$$

$$D = \begin{pmatrix} -\sigma & 0 \\ -(1+\sigma)\dfrac{P\varphi(\mu)}{\vartheta_2 - \varphi'(\mu)} & -\sigma \end{pmatrix} \exp\left[\begin{pmatrix} 0 & 1 \\ -b & -a \end{pmatrix} T\right].$$

By Schur criterion, for the periodic motion to be asymptotically stable, it is necessary and sufficient that the condition

$$|\mathrm{sp}D| < 1 + \sigma^2 e^{-aT}, \quad \sigma^2 < e^{aT}.$$

holds. It should be noted that, in a number of important cases, it is possible to get explicit equations for a T-periodic mode in the Rusakov – Kharkevich model. The question of existence of such modes for this model in the general case remains unexplored.

A.3 Unbounded functions with almost periodic differences

Before passing to almost periodic impulsive systems, we need to get some auxiliary results.

In this section we will consider scalar valued functions $F(t)$, continuous on $\mathbf{R}$, such that, for any fixed $y \in \mathbf{R}$, the difference $\psi_y(t) = F(t+y) - F(t)$ is Bohr almost periodic.

It is known that if $\psi_y(t) \in AP(\mathbf{R})$ for every $y \in \mathbf{R}$ and $F(t)$ is bounded, then $F(t)$ is also an almost periodic function (a theorem of Loomis [78]). S. Bochner has also proved [21] that $F(t) \in AP(\mathbf{R})$ if $F(t)$ is bounded and uniformly continuous on $\mathbf{R}$, and $\psi_z(t) \in AP(\mathbf{R})$ for some $z \neq 0$. We assume here, in general, that $F(t)$ is unbounded.

Theorem 122 *If, for any $N > 0$, the functions $\psi_y(t)$ are Bohr almost periodic uniformly with respect to $|y| \leq N$, then the function $F(t)$ is uniformly continuous and there exists a sequence $f_n(t) \in AP(\mathbf{R})$, $n \in \mathbf{N}$, such that the limit*

$$\lim_{n\to\infty} \int_a^t f_n(t)dt = F(t) - F(a) \tag{A.17}$$

is uniform with respect to $t, a \in \mathbf{R}$. Moreover, the finite limit

$$\lim_{T\to+\infty} T^{-1}(F(T+a) - F(a)) = \langle F \rangle \tag{A.18}$$

exists and is uniform with respect to a.

Conversely, if for $f_n(t) \in AP(\mathbf{R})$, condition (A.17) holds for some a uniformly with respect to t, then the family $\{\psi_y(t), y \in K\}$ is uniformly almost periodic if K is a compact set.

The functions $\psi_y(t)$ are almost periodic uniformly with respect to y if and only if $F(t) = mt + f(t)$, $m \in \mathbf{R}$, $f(t) \in AP(\mathbf{R})$.

Theorem 122 allows to describe integral almost periodic functions (N-integral almost periodic functions) introduced by D. Wexler in [55] (see also Definition 5) and, in particular, to describe the structure of numeric sequences $\{t_n\}_{n\in\mathbf{Z}}$ such that the totality of sequences $\{t_n^j\} = \{t_{n+j} - t_n\}_{n\in\mathbf{Z}}$, $j \in \mathbf{Z}$ are equipotentially almost periodic.

Definition 4 *A continuous function $F(t) : \mathbf{R} \to \mathbf{R}$ is called Δ-almost periodic ($N - \Delta$-almost periodic) if, for every $\epsilon > 0$ (and every $N > 0$), one* *can find* $l = l(\epsilon) > 0$ *(*$l = l(\epsilon, N) > 0$*) such that there exists such* ω *in any interval of length* l *that*

$$|(F(t+\omega) - F(t)) - (F(t'+\omega) - F(t'))| < \epsilon \qquad \forall\, t, t' \in \mathbf{R}$$

(for $t, t' \in \mathbf{R}$ *satisfying* $|t - t'| < N$ *correspondingly).*

It is clear that a Δ-almost periodic function is also $N - \Delta$-almost periodic, and, when $f(t)$ belongs to $AP(\mathbf{R})$, the function $mt + f(t)$ will be Δ-almost periodic. The introduced notions are generalizations of (N-) integral almost periodic functions.

Definition 5 *A locally integrable function* $f(t) : \mathbf{R} \to \mathbf{R}$ *is integral almost periodic (*N*-almost periodic if, for every* $\epsilon > 0$ *(and every* $N > 0$*) one can find* $l = l(\epsilon) > 0$ *(*$l = l(\epsilon, N) > 0$*) such that there exists such* ω *in any interval of length* l *that*

$$\left| \int_t^{t+\omega} f(u)du - \int_{t'}^{t'+\omega} f(u)du \right| < \epsilon \qquad \forall\, t, t' \in \mathbf{R}$$

(for $t, t' \in \mathbf{R}$ *satisfying* $|t - t'| < N$ *correspondingly).*

Evidently, if $f(t)$ is integral almost periodic (N-integral almost periodic), then the function $F(t) = \int_a^t f(u)du$ is Δ-almost periodic ($N - \Delta$-almost periodic). By using the identity

$$\psi_y(t+\omega) - \psi_y(t) = (F(t+y+\omega) - F(t+y)) - (F(t+\omega) - F(t)),$$

we see that $F(t)$ is Δ-almost periodic ($N - \Delta$-almost periodic) if and only if the family of functions $\psi_y(t)$ is almost periodic uniformly with respect to $y \in \mathbf{R}$ (with y, respectively, running over a set compact in $\mathbf{R}$).

Let us show that $N - \Delta$-almost periodic functions are uniformly continuous. Fix $N = 1$ and an arbitrary $\epsilon > 0$. Let $l = l(\epsilon, 1) > 1$. Then, in any interval of length l, one can find such ω that, for every $t, t' : |t - t'| \leq 1$, we have

$$|(F(t+\omega) - F(t)) - (F(t'+\omega) - F(t'))| < \epsilon/2. \tag{A.19}$$

The function $F(t)$ is uniformly continuous on $[-1, 1]$ and, by using $\epsilon > 0$, find $\delta(\epsilon) < 1$ such that

$$|F(t_1) - F(t_2)| < \epsilon/2 \tag{A.20}$$

for all t_1, t_2 with $|t_1 - t_2| < \delta$. Let now $t^{(1)} < t^{(2)}$ with $|t^{(1)} - t^{(2)}| < \delta$. Find in $[t^{(1)}, t^{(1)} + l]$ an almost period ω such that (A.19) holds. If we set $t = t^{(1)} - \omega \in [-l, 0]$, $t' = t^{(2)} - \omega \in [-l, 1]$, then, by (A.19, A.20),

$$|F(t+\omega) - F(t'+\omega)| - |F(t) - F(t')| < \epsilon/2$$

and $|F(t^{(1)}) - F(t^{(2)})| < \epsilon$.

Now it is not difficult to prove the first part of the theorem. Indeed, let $f_n(t) = 2^n(F(t+2^{-n}) - F(t))$. Then $f_n(t) \in AP(\mathbf{R})$,

$$\int_a^t f_n(u)du = 2^n \int_0^{2^{-n}} F(t+u)du - 2^n \int_0^{2^{-n}} F(a+u)du \overset{n\to\infty}{\rightrightarrows} F(t) - F(a)$$

uniformly with respect to $t, a \in \mathbf{R}$ because $F(t)$ is uniformly continuous.

Let m be such that

$$\left|\int_a^{t+a} f_m(u)du - (F(t+a) - F(a))\right| < 1.$$

for all $t, a \in \mathbf{R}$. By dividing both sides of this inequality by $|t|$ and using that the finite limit $\lim_{t\to\infty} t^{-1} \int_a^{t+a} f_m(u)du$ exists and is uniform with respect to a, we see that the limit (A.18) also exists and is uniform with respect to a.

Conversely, let (A.17) hold and be uniform with respect to t for some $a \in \mathbf{R}$. Because $f_n(t) \in AP(\mathbf{R})$, $n \in \mathbf{N}$, by [55] (p. 297), the almost periodic function $f_n(t)$ is N-integral almost periodic. But then $F_n(t) = \int_a^t f_n(u)du$ are $N - \Delta$-almost periodic functions. Because $N - \Delta$-almost periodic functions form a linear space, which is closed with respect to the limit uniform in $t \in \mathbf{R}$, we see that the first part of the theorem holds.

As its consequence note that, for an arbitrary N-integrable almost periodic function $f(t)$, the finite mean value

$$\langle f \rangle = \lim_{t\to\infty} t^{-1} \int_0^{t+a} f(u)du$$

exists and is uniform with respect to a.

The second part of the theorem will be proved if we could show that the function $f(t) = F(t) - \langle F \rangle t$ is Bohr almost periodic, i.e. if for an arbitrary $\epsilon > 0$ there exists a dense set of numbers ω such that, for all $t \in \mathbf{R}$,

$$|f(t+\omega) - f(t)| = |F(t+\omega) - F(t) - \langle F \rangle \omega < \epsilon.$$

Taking into account Definition 4 and that the function $F(t)$ is Δ-almost periodic, it will be sufficient to show that, for an arbitrary $\omega \in \mathbf{R}$, one can find $t' \in \mathbf{R}$ such that

$$F(t' + \omega) - F(t') = \langle F \rangle \omega.$$

If this would not be so, then, for some $\omega \in \mathbf{R}$, we would have one of the two cases:

a) $F(t + \omega) - F(t) = \psi_\omega(t) > \langle F \rangle \omega$ for all t,

b) $\psi_\omega(t) < \langle F \rangle \omega$ for all t.

Suppose that condition a) holds. Then the mean value $\langle \psi_\omega \rangle$ of the Bohr almost periodic function $\psi_\omega(t)$ must be greater than $\langle F \rangle \omega$. But, by using (A.18), we have

$$\begin{aligned}
\langle \psi_\omega \rangle &= \lim_{n\to\infty} \frac{1}{n\omega} \int_0^{n\omega} (F(u+\omega) - F(u))du = \\
&= \lim_{n\to\infty} \frac{1}{n\omega} \left[\sum_{k=0}^{n-1} \left(\int_{k\omega}^{(k+1)\omega} F(u+\omega)du - \int_{k\omega}^{(k+1)\omega} F(u)du \right) \right] = \\
&= \lim_{n\to\infty} \frac{1}{n\omega} \left[\int_{n\omega}^{(n+1)\omega} F(u)du - \int_0^{\omega} F(u)du \right] = \\
&= \lim_{n\to\infty} \int_0^{\omega} \frac{F(u+n\omega) - F(u)}{n\omega} du = \langle F \rangle \omega.
\end{aligned}$$

This contradiction finishes the proof of the theorem.

□

Corollary 5 *Let a number sequence $\{t_n\}_{n\in\mathbf{Z}}$ be such that the totality of sequences $\{t_n^j\} = \{t_{n+j} - t_n\}_{n\in\mathbf{Z}}$, $j \in \mathbf{Z}$, be equipotentially almost periodic. Then $t_n = an + c_n$, where $\{c_n\}_{n\in\mathbf{Z}}$ is an almost periodic sequence, a is a real number.*

Proof. If conditions of the corollary hold, the function $\psi(t)$, which is equal to $t_n^1 = t_{n+1} - t_n$ for $t \in [n, n+1)$, is integral almost periodic. Indeed, because the sequences $\{t_n^j\}$, $j \in \mathbf{Z}$ are equipotentially almost periodic, for each $\eta > 0$ there exists $l(\eta) > 0$ such that one can find $h \in \mathbf{Z}$ in any interval of length $l(\eta)$ so that

$$\left|\sum_{n=k}^{k+h} t_n^1 - \sum_{n=k'}^{k'+h} t_n^1\right| < \eta \qquad \forall k,\ k' \in Z. \tag{A.21}$$

For $k' = k+1$, it follows from (A.21) that

$$|t_k^1 - t_{k+h+1}^1| < \eta. \tag{A.22}$$

Because $\sum_{n=k}^{k+h} t_n^1 = \int_k^{k+h+1} \psi(u)du$, then for all $t, t' \in \mathbf{R}$, it follows from (A.21, A.22) that

$$\left|\int_t^{t+h+1} \psi(u)du - \int_{t'}^{t'+h+1} \psi(u)du\right| \leq \left|\int_{[t]}^{[t]+h+1} \psi(u)du - \right.$$
$$\left. - \int_{[t']}^{[t']+h+1} \psi(u)du)\right| + \left|\int_{[t]+h+1}^{t+h+1} \psi(u)du - \int_{[t]}^{t} \psi(u)du\right| +$$
$$+ \left|\int_{[t']+h+1}^{t'+h+1} \psi(u)du - \int_{[t']}^{t'} \psi(u)du\right| < \eta + \{t\}|t_{[t]}^1 - t_{[t]+h+1}^1| +$$
$$+\{t'\}|t_{[t']}^1 - t_{[t']+h+1}^1| < 3\eta,$$

where $[t]$ and $\{t\}$ are the integer and fractional parts of t. Consequently, $\psi(t)$ is an integral almost periodic function and, according to the theorem, $\int_0^t \psi(u)du = at + c(t)$, where $a \in \mathbf{R}$, $c(t) \in AP(\mathbf{R})$. But then, $t_n - t_0 = \int_0^n \psi(u)du = an + c(n)$ and, because $\{c(n)\}$ is an almost periodic sequence, the corollary is proved.

□

Remark. Let $f(t) > 0$ be such a Bohr almost periodic function that the function $r(t) = \int_0^t f(u)du - \langle f\rangle t$ be unbounded (for example, $f(t) = 5 + \sum_{n=1}^{+\infty} \frac{\cos(tn^{-2})}{n^2}$). Then, by setting $t_n = \int_0^n f(u)du$, we get such a strictly increasing sequence of numbers that the totality of almost periodic sequences $\{t_{n+j} - t_n\}_{n\in\mathbf{Z}}$, $j \in \mathbf{Z}$, will not be equipotentially almost periodic. Indeed, if this is not so, we would have $\int_0^n f(u)du = \langle f\rangle n + q(n)$ for some almost periodic sequence $\{q(n)\}$. But since $r(t) = \int_{[t]}^t (f(u) - \langle f\rangle)du + q([t])$, we get that $|r(t)| \leq (\|f(u)\| + |\langle f\rangle|) + \sup_n |q(n)| < \infty$.

There is a natural question: how necessary the requirement in the statement of Theorem 122 that the family of functions $\psi_y(t)$ be uniformly almost periodic is, and would it not be sufficient that this family be pointwise almost periodic. We complete the study by answering this question, assuming only that, in addition, the function $F(t)$ is continuous. This is equivalent to boundedness of $F(t)$ on some interval, since any additive function Γ (i.e. $\Gamma(a+b) = \Gamma(a)+\Gamma(b)$, $\forall a, b \in \mathbf{R}$) which is bounded on an interval is linear, and because of

Proposition 2 *([30]). If $\psi_a(t) \in C(\mathbf{R})$ for all $a \in \mathbf{R}$, then there exists such an additive function Γ that $F(t) - \Gamma(t) \in C(\mathbf{R})$.*

Theorem 123 *If, for all $a \in \mathbf{R}$, $\psi_a(t)$ is uniformly continuous and bounded on $\mathbf{R}$, then, if the function $F(t)$ is continuous, it is uniformly continuous.*

Proof. Let X be the Banach space of all functions $g(t) : \mathbf{R} \to \mathbf{R}$, bounded, uniformly continuous on $\mathbf{R}$, with the norm $|g|_X = \sup_{t\in\mathbf{R}} |g(t)|$. Define a new function $\Psi(a) : \mathbf{R} \to X$ as $\Psi(a)(t) = \psi_a(t)$. Then

$$\Psi(a+\Delta)(t) - \Psi(a) = \Psi(t+a). \tag{A.23}$$

For a fixed Δ, the function $M_\Delta(a) : \mathbf{R} \to X$, $M_\Delta(a) = \Psi(\Delta)(t+a)$ will be bounded and uniformly continuous on $\mathbf{R}$ because

$$\sup_{a\in\mathbf{R}} |M_\Delta(a)|_X = \sup_{a\in\mathbf{R}} |\psi_\Delta(a)| \quad \text{and}$$
$$\sup_{a\in\mathbf{R}} |M_\Delta(a+\mu) - M_\Delta(a)|_X = \sup_{a\in\mathbf{R}} |\psi_\Delta(a+\mu) - \psi_\Delta(a)|.$$

The function $F(t)$ is uniformly continuous if and only if the function $\Psi(a)$ is continuous at some point $a_0 \in \mathbf{R}$. Indeed, in this case, $\Psi(a)$ is also continuous at 0 because

$$|\Psi(a) - \Psi(a_0)|_X = |\Psi(a - a_0)|_X.$$

Let us first prove that the function $\nu(a) = |\Psi(a)|_X$ is continuous at some point $b \in \mathbf{R}$. Consider the sequence of functions $r_n(a) = \sup_{|t|\le n} |\psi_a(t)|$. It is clear that $r_n(a) \in C(\mathbf{R})$ and $\lim_{n\to\infty} r_n(a) = \nu(a)$ pointwise. Consequently the function $\nu(a)$ belongs to the first Baire class (moreover, it is lower semicontinuous) and thus it has a set of points of continuity, which is dense everywhere in $\mathbf{R}$ (see [17]): if b is one of the points, then the function $\nu(a)$ is bounded, $0 \le \nu(a) \le L < +\infty$, for all a in some interval Σ, $b \in \Sigma$.

Now let us prove that $\Psi(a)$ is weakly continuous on $[0,1]$. Let $m \in X^*$, $\varphi(a) = m\Psi(a)$. Then $\varphi(a+\Delta) - \varphi(a) = mM_\Delta(a)$, the function $mM_\Delta(a)$ is continuous with respect to a for any fixed Δ. Moreover,

$$\sup_{a\in\Sigma} |\varphi(a)| \le \sup_{a\in\Sigma}(|m||\Psi(a)|_X) = |m| \sup_{a\in\Sigma} \nu(a) \le |m|L < +\infty$$

and hence $\varphi(a)$ is continuous by the previous proposition.

We use the following result of [44]: any weakly continuous function that takes the values in a Banach space is a function of the first Baire class with respect to the strong convergence. From this it follows that $\Psi(a) : [0,1] \to X$ is continuous at a certain point.

□

Corollary 6 *Let a function $F(t)$ be continuous and, for every $a \in \mathbf{R}$, the difference $\psi_a(t)$ is Bohr almost periodic. Then, for any set B, bounded on $\mathbf{R}$, the family of functions $\{\psi_a(t), a \in B\}$ is uniformly almost periodic.*

An example of the function $F(t) = t^2$ shows that it is essential to require in the statement of Theorem 123 that $\psi_a(t)$ be bounded.

A.4 Spaces of almost periodic functions on the line

Almost periodicity of an impulsive system can be understood in different ways: we can, for example, require that each of its parts be almost periodic. In this section, we continue to study countable almost periodic sets on the line, where the impulsive effects are concentrated.

Let T be a countable set of real numbers, which includes arbitrarily large positive and negative numbers, and such that, for any $m > 0$, the set $\{t \in T : |t| \leq m\}$ is finite. The totality of such sets will be denoted by $\mathfrak{U}$. It follows from the definition that a number $t \in T$ could be included in T only a finite number of times, and this number will be called the *multiplicity* of t. The quantity

$$\rho(T_1, T_2) = \inf_{\varphi} \sup_{t \in T_1} |\varphi(t) - t|,$$

where the infimum is taken over all the bijections $\varphi : T_1 \to T_2$, $T_1, T_2 \in \mathfrak{U}$, defines a distance on $\mathfrak{U}$ (which will be proved later).

Let $\mathfrak{B}$ be the set of all increasing, unbounded from below and above sequences of numbers $\{t_i\}$, which do not have finite limit points, and such that a number $t_0 \geq 0$ is either $t_0 = 0$, $t_1 > 0$ or $t_0 > 0$ and $t_i \notin [0, t_0)$ for all i. We introduce in $\mathfrak{B}$ a distance d (which may equal to ∞), given for any of its elements $\{t_i\}$ and $\{t_i'\}$ by

$$d(\{t_i\}, \{t_i'\}) = \inf_{\varphi} \sup_{i} |\varphi(t_i) - t_i|,$$

where the infimum is taken over all the bijections $\varphi : \{t_i\} \to \{t_i'\}$ that preserve the order (any such bijection has the form $\varphi(t_i) = t'_{i+m}$).

To see that d is a distance, we use the following chain of inequalities showing that the triangle axiom holds (the other axioms are obviously true):

$$\begin{aligned} d(\{t_i\}, \{t_i'\}) &= \inf_{m} \sup_{i} |t'_{i+m} - t_i| \leq \inf_{m} \sup_{i} \{|t'_{i+m} - t''_{i+k}| + \\ &+ |t_i - t''_{i+k}|\} \leq \inf_{m} \sup_{i} \{|t'_{i+m} - t''_{i+k}|\} + \sup_{i} |t_i - t''_{i+k}| = \\ &= d(\{t_i''\}, \{t_i'\}) + \sup_{i} |t''_{i+k} - t_i|, \end{aligned}$$

but then, because k is arbitrary,

$$\begin{aligned} d(\{t_i''\}, \{t_i'\}) &\leq d(\{t_i''\}, \{t_i'\}) + \inf_{k} \sup_{i} |t''_{i+k} - t_i| = \\ &= d(\{t_i''\}, \{t_i'\}) + d(\{t_i''\}, \{t_i\}). \end{aligned}$$

Moreover, $(\mathfrak{B}, d)$ is a complete metric space. Let us show that. Let $T^{(n)}$, $n \geq 0$ be a Cauchy sequence, $T^{(n)} = \{t_j^{(n)}\}_{j \in \mathbf{Z}}$. Consider the set A_j, $j \geq 0$, of limit points of the sequence $\{t_j^{(n)}\}_{n \geq 0}$. It is clear that either $A_k \neq \{0\}$ for $k = 0$ or $A_k \neq \{0\}$, $A_{k-1} = \ldots = A_0 = \{0\}$ for some $k > 0$. Then there will exist such $\tau > 0$ that either $A_k = \{\tau\}$ or $A_k = \{0; \tau\}$. In the first case, $t_k^{(n)} \to \tau$, $t_{k-1}^{(n)}; \ldots t_0^{(n)} \to 0$ as $n \to \infty$ and hence

$$\sup_i |t_i^{(n)} - t_i^{(m)}| \to 0 \quad \text{for} \quad n, m \to \infty. \tag{A.24}$$

In the second case, relation (A.24) remains valid for all sequences $\{t_j^{(n)}\}_{j \in \mathbf{Z}}$, some of which may require a shift in j by -1.

In the same way as one proves completeness of l_∞, relation (A.24) implies that there exists a vector $\{t_i\} \in \mathfrak{B}$, for which we have

$$\sup_i |t_i^{(n)} - t_i| \to 0 \qquad \text{for} \qquad n \to \infty.$$

This means that $\mathfrak{B}$ is complete.

Lemma 51 *The spaces $(\mathfrak{U}, \rho)$ and $(\mathfrak{B}, d)$ are isometric. The isomorphism $\lambda : \mathfrak{U} \to \mathfrak{B}$ is defined by indexing elements of $T \in \mathfrak{U}$ in the increasing order (counting multiplicities) and by a choosing suitable element t_0 from T.*

Proof. It is clear that λ is one-to-one and $\rho(T_1, T_2) \leq d(\lambda(T_1), \lambda(T_2))$. Let now the quantity $d(T_1, T_2)$ be finite and prove that then $\rho_{12} = \rho(T_1, T_2) = d(\lambda(T_1), \lambda(T_2))$. By the definitions given above, for every $\epsilon > 0$, there exists a bijection $\varphi : \lambda(T_1) \to \lambda(T_2)$ such that $\rho_{12} \leq \sup_i |\varphi(t_i) - t_i'| \leq \rho_{12} + \epsilon$. By using φ, let us build such a bijection $\psi : \lambda(T_1) \to \lambda(T_2)$, which preserves the order (i.e. $\psi(t_i) \leq \psi(t_{i+1}) \Leftrightarrow t_i \leq t_{i+1}\ \forall i$), that

$$\rho_{12} \leq \sup_i |\psi(t_i) - t_i'| \leq \rho_{12} + \epsilon \tag{A.25}$$

(by the definition of ρ_{12}, the existence of the lower bound in (A.25) is clear).

If we set $\varphi(t_i) = t'_{\varphi^*(i)}$, then giving the bijection $\varphi : \lambda(T_1) \to \lambda(T_2)$ is equivalent to giving the bijection $\varphi^* : \mathbf{Z} \to \mathbf{Z}$. Let $P_0 = \{i > 0 : \varphi^*(i) < \varphi^*(0)\}$, $M_0 = \{k < 0 : \varphi^*(k) > \varphi^*(0)\}$. The set P_0 is finite because, for $i \in P_0$,

$$\begin{aligned}
&\varphi(t_0) = t'_{\varphi^*(0)} \geq t'_{\varphi^*(1)} = \varphi(t_i), \\
&0 \leq \varphi(t_0) - \varphi(t_i) \leq \varphi(t_0) - \varphi(t_i) - t_0 + t_i = \\
&\qquad = (\varphi(t_0) - t_0) - (\varphi(t_i) - t_i) < 2(\rho_{12} + \epsilon),
\end{aligned}$$

and, if P_0 is infinite, $\lambda(T_2) \notin \mathfrak{B}$. Similarly, if $k \in M_0$, then $\varphi(t_k) \geq \varphi(t_0)$ and M_0 is finite. We will assume that

$$L = M_0 \cup P_0 \cup \{0\} = \{k_n < k_{n-1} < \ldots < k_1 < 0 < i_1 < \ldots < i_m\}.$$

Replace the restriction $\varphi^* : L \to \varphi^*(L)$ by the mapping $\bar{\varphi}^* : L \to \bar{\varphi}^*(L)$ which preserves the order ($\bar{\varphi}^*(k_n)$ is the smallest number from $\varphi^*(L)$, etc). If we set $\bar{\varphi}^* : \mathbf{Z} \to \mathbf{Z}$ with $\bar{\varphi}^*|_L = \bar{\varphi}^*$, $\bar{\varphi}^*|_{\mathbf{Z}\setminus L} = \varphi$, and $\bar{\varphi} : \lambda(T_1) \to \lambda(T_2)$, where $\bar{\varphi}(t_i) = t'_{\bar{\varphi}^*(i)}$, then $\bar{\varphi}$, $\bar{\varphi}^*$ are bijective mappings, and

$$|\bar{\varphi}(t_i) - t_i| < \rho_{12} + \epsilon \qquad \forall i. \tag{A.26}$$

It is sufficient to prove this relation only for $p \in L$. Consider the case $m < n$ (other cases: $m = n$ and $m > n$ are treated similarly). If $m < n$, then $\bar{\varphi}^*(P_0 \cup \{0\}) \subset \varphi^*(M_0)$ and so, for $\bar{\varphi}(t_{i_k}) \geq t_{i_k}$, we have the following

$$\begin{aligned} 0 \leq \bar{\varphi}(t_{i_k}) - t_{i_k} &= t'_{\bar{\varphi}^*(i_k)} - t_{i_k} = t'_{\varphi^*(k_r)} - t_{i_k} = \\ &= \varphi(t_{k_r}) - t_{i_k} \leq \varphi(t_{k_r}) - t_{k_r} < \rho_{12} + \epsilon; \end{aligned}$$

If $\bar{\varphi}(t_{i_k}) < t_{i_k}$, then

$$\begin{aligned} 0 < t_{i_k} - \bar{\varphi}(t_{i_k}) &= t_{i_k} - t'_{\bar{\varphi}^*(i_k)} = t_{i_k} - t'_{\varphi^*(k_p)} \leq \\ &\leq t_{i_k} - t'_{\varphi^*(i_k)} = t_{i_k} - \varphi(t_{i_k}) < \rho_{12} + \epsilon. \end{aligned}$$

Further, $\bar{\varphi}^*(\{k_n, \ldots, k_{n-m+1}\}) = \varphi^*(P_0)$, and so, if $\bar{\varphi}(t_\delta) \geq t_\delta$ and $\delta \in \{k_n, \ldots, k_{n-m+1}\}$, we have

$$0 \leq \bar{\varphi}(t_\delta) - t_\delta = \varphi(t_{i_q}) - t_\delta \leq \varphi(t_\delta) - t_\delta < \rho_{12} + \epsilon;$$

if $\bar{\varphi}(t_\delta) < t_\delta$, then

$$\begin{aligned} 0 < t_\delta - \bar{\varphi}(t_\delta) &= t_\delta - t'_{\bar{\varphi}^*(\delta)} = t_\delta - \varphi(t_{i_d}) \leq \\ &\leq t_{i_d} - \varphi(t_{i_d}) < \rho_{12} + \epsilon. \end{aligned}$$

And finally, because $\bar{\varphi}^*(\{k_{n-m}, \ldots, k_1\}) \subset \varphi^*(M_0 \cup \{0\})$, we see that, if $\bar{\varphi}(t_{k_\delta}) \leq t_{k_\delta}$, the following estimate holds:

$$0 \leq t_{k_\delta} - \bar{\varphi}(t_{k_\delta}) = t_{k_\delta} - t'_{\varphi^*(k_\alpha)} \leq t_0 - \varphi(t_0) < \rho_{12} + \epsilon.$$

Moreover, if $\bar{\varphi}^*(k_\delta) \subset \varphi^*(M_0 \cup \{0\})$, then there exists $\mu \geq \delta$ such that $\varphi^*(k_\mu) \geq \bar{\varphi}^*(k_\delta)$ because if not,

$$\begin{aligned} \varphi^*(k_\mu) < \bar{\varphi}^*(k_\delta),\ \bar{\varphi}^*(\{k_n, \ldots, k_\delta\}) = \{\varphi^*(s) : s \in L, \\ \varphi^*(s) \leq \bar{\varphi}^*(k_\delta)\} \supset \varphi^*(P_0 \cup \{0\} \cup \{k_n, \ldots, k_\delta\}) \end{aligned}$$

for all $\mu \geq \delta$. Hence, for $\bar{\varphi}(t_{k_\delta}) > t_{k_\delta}$, we have

$$0 < \bar{\varphi}(t_{k_\delta}) - t_{k_\delta} = t'_{\bar{\varphi}^*(k_\delta)} - t_{k_\delta} \leq t'_{\varphi^*(k_\mu)} - t_{k_\mu} =$$
$$= \varphi(t_{k_\mu}) - t_{k_\mu} < \rho_{12} + \epsilon,$$

whence (A.26) follows.

The aforesaid implies the relation

$$\bar{\varphi}^*(N) = \{m \in Z : m > j_0\}, \tag{A.27}$$

which will be important in the sequel.

Let now $j_0 = \bar{\varphi}^*(0)$, set $\psi(t_s) = t'_{j_0+s}$, and prove (A.25) for such ψ. Let, for example, $s > 0$ and $j_0 + s = \bar{\varphi}^*(\alpha)$. Then, if $\psi(t_s) \geq t_s$, we have:

a1) for $\alpha \leq s$,

$$0 \leq \psi(t_s) - t_s = \bar{\varphi}(t_\alpha) - t_s = t'_{\bar{\varphi}^*(\alpha)} - t_s \leq \bar{\varphi}(t_\alpha) - t_\alpha < \rho_{12} + \epsilon,$$

a2) for $\alpha > s$, there exists a positive $\beta \leq s$ such that $\bar{\varphi}^*(\beta) \geq \bar{\varphi}^*(\alpha)$ because if not, by using (A.27), we will have $\bar{\varphi}^*(\beta) < \bar{\varphi}^*(\alpha)$ for all $0 \leq \beta \leq s$ and $\{j_0, j_0 + 1, \ldots, j_0 + s\} \supset \bar{\varphi}^*(\{0, 1, \ldots, s, \alpha\})$. Hence,

$$0 \leq \psi(t_s) - t_s = \bar{\varphi}(t_\alpha) - t_s \leq \bar{\varphi}(t_\beta) - t_\beta < \rho_{12} + \epsilon.$$

If $\psi(t_s) < t_s$, then

b1) for $\alpha \geq s$,

$$0 \leq t_s - \psi(t_s) = t_s - \bar{\varphi}(t_\alpha) \leq t_\alpha - \bar{\varphi}(t_\alpha) < \rho_{12} + \epsilon,$$

b2) for $\alpha < s$ there is $\beta \geq s$ such that $\bar{\varphi}^*(\beta) \leq \bar{\varphi}^*(\alpha)$ for otherwise, by using (A.27) we see that $\bar{\varphi}^*(\beta) > \bar{\varphi}^*(\alpha)$ for all $\beta \geq s$ and $\{j_0, j_0 + 1, \ldots, j_0 + s\} \subset \bar{\varphi}^*(\{0, 1, \ldots, s - 1\})$. So

$$0 < t_s - \psi(t_s) = t_s - \bar{\varphi}(t_\alpha) \leq t_\beta - \bar{\varphi}(t_\beta) < \rho_{12} + \epsilon.$$

A similar argument, used for $s < 0$, finishes the proof of the estimate (A.25) and Lemma 51.

□

Corollary 7 *The space* $(\mathfrak{U}, \rho)$ *is complete.*

If $T \in \mathfrak{U}$, then, for $\tau \in \mathbf{R}$, the set $T + \tau$, the elements of which ar precisely elements of T increased by τ, also belongs to $\mathfrak{U}$ and $\rho(T, T+\tau) \leq \tau$, $\rho(T_1 + \tau, T_2 + \tau) = \rho(T_1, T_2)$ for all $T_1, T_2 \in \mathfrak{U}$, $\tau \in \mathbf{R}$.

The mapping $\theta_s : \mathfrak{U} \times \mathbf{R} \to \mathfrak{U}$, given by the formula $\theta_s(T) = T + s$, is continuous in its arguments and defines an isometrical dynamical system on $(\mathfrak{U}, \rho)$:

$$\rho(\theta_s(T_1), \theta_s(T_2)) = \rho(T_1, T_2) \qquad \forall T_1, T_2 \in \mathfrak{U},\ s \in R.$$

Definition 6 *1. A set $T \in \mathfrak{U}$ is* Bohr almost periodic *if for every $\epsilon > 0$ there exists such $l(\epsilon) > 0$ that one can find in any interval of length $l(\epsilon)$ such a number τ that $\rho(T, T + \tau) < \epsilon$.*

2. A set $T \in \mathfrak{U}$ is Bochner almost periodic *if in any sequence $\{h_i\}$ there is a subsequence $\{h_{i_k}\}$ such that $T + h_{i_k} \to T' \in \mathfrak{U}$ for $k \to +\infty$.*

3. A set T is strongly almost periodic *if its elements could be indexed, $T = \{t_i\}_{i=-\infty}^{+\infty}$, in such a way that the totality of the sequences $\{t_i^j\} = \{t_{i+j} - t_i\}$ is almost periodic in i equipotentially with respect to $j \in \mathbf{Z}$. Such an indexing of elements of T will be called an* almost periodic representation *of the almost periodic set T.*

Example. The set of integers $\mathbf{Z} \subset \mathbf{R}$ is strongly almost periodic with the natural representation $\mathbf{Z} = \{n\}$ because the sequences $\{t_i^j\} = \{j\}$ are periodic with the period 1. Any strongly almost periodic set has infinitely many almost periodic representations, for example, $\mathbf{Z} = \{n+(-1)^{n+1}2\}$ but, as we will show later, there always exists such a representation $\{t_i\}$ (unique up to a translation of indices) such that $t_i \geq t_{i-1}$ for all i.

Because $(\mathfrak{U}, \rho)$ is complete, the criterion of Bokhner holds [75]:

Corollary 8 *The Bochner almost periodicity and the Bohr almost periodicity are equivalent. If $T_1, \ldots, T_m$ are sets which are Bohr almost periodic, then the set $T_1 \sqcup \ldots \sqcup T_m$ as also almost periodic, where $A \sqcup B$ denotes the free union of A and B.*

Proof. It is sufficient to consider the case $m = 2$. First note that, if $P, Q, L, S \in \mathfrak{U}$, $h \in \mathbf{R}$, then

a) $P \sqcup Q \in \mathfrak{U}$;

b) $(P \sqcup Q) + h = (P + h) \sqcup (Q + h)$;

c) $\rho(P \sqcup Q, L \sqcup S) \leq \max\{\rho(P, L); \rho(Q, S)\}$.

Because T_1, T_2 are almost periodic sets, for any sequence $\{h_i\}$ of reals there is a subsequence $\{h_{i_k}\}$ such that $T_j + h_{i_k} \to T'_j \in \mathfrak{U}$ for $k \to +\infty$, $j = 1, 2$.

Let $T_{12} = T_1 \sqcup T_2$, $T'_{12} = T'_1 \sqcup T'_2$. Then $\rho(T_{12} + h_{i_k}, T'_{12}) = \rho((T_1 + h_{i_k}) \sqcup (T_2 + h_{i_k}), T'_1 \sqcup T'_2) \leq \max\{\rho(T_1 + h_{i_k}, T'_1), \rho(T_2 + h_{i_k}, T'_2)\} \to 0$, and so the set T_{12} is almost periodic.

□

Corollary 5 can be restated in new terms as: a set T is strongly almost periodic with an almost periodic representation $\{t_i\}$ if and only if $t_i = ia + c_i$, where $\{c_i\}$ is an almost periodic sequence, $a \neq 0$.

Denote by $i(\alpha, \beta)$ the number of elements of the set T which lie in (α, β). Clearly the limit

$$\lim_{q \to \infty} q^{-1} i(\alpha, \alpha + q) = 1/a$$

exists and is uniform with respect to α. The quantity a will be called a *growth index* of T.

Theorem 124 *The Bohr (Bochner) almost periodicity is equivalent to its strong almost periodicity. Moreover, an almost periodic representation of an almost periodic set can be obtained by indexing its elements in an increasing order (counting multiplicities).*

We will first need a generalization of one result [55].

Lemma 52 *For a set T to be strongly almost periodic with an almost periodic representation $\{t_i\}$, it is necessary and sufficient that, for every $\eta > 0$, the set Ω_η of all the numbers ω such that*

$$|t_n^{h_\omega} - \omega| < \eta \tag{A.28}$$

for some $h_\omega \in \mathbf{Z}$ and all $n \in \mathbf{Z}$ be relatively dense in $\mathbf{R}$.

The lemma was first proved in [55] for $t_{n+1} > t_n$. The following proof is a modification of the argument given in [55] for the general case.

Proof. *Necessity.* The necessity of density of Ω_η is obvious because of Corollary 5 (if h_ω is a η-period for the almost periodic sequence $\{c_i\}$ of the corollary, then (A.28) holds with $\omega = h_\omega a$).

Sufficiency. Let $H_\eta = \{h_\omega : \omega \in \Omega_\eta\}$. Order the integers in H_η in a strictly increasing sequence $\{h_i\}_{i=-\infty}^{+\infty}$. Clearly if we set $h_0 = 0$, then $h_i = -h_i$ for all i. Let us prove that $\{h_i\}_{i=-\infty}^{+\infty}$ is relatively dense. To show that it will be sufficient to prove that $\lim_{t \to \pm\infty} h_i = \pm\infty$ and that the sequence $\{h_{i+1} - h_i\}$ is bounded from above. But $\lim_{t \to \pm\infty} h_i = \pm\infty$

for, otherwise, as it follows from $|h_i| \leq m_0$ because of the symmetry of H_η, $-c \leq t_0^{h_\omega} \leq c$ for all $\omega \in H_\eta$ and, by (A.28), $|\omega| \leq c + \eta$, which is a contradiction because Ω_η is relatively dense.

Let $\Omega_i = \{\omega : h_\omega = h_i\}$ and show that, for arbitrary $j \geq i$,

$$\omega_j - \omega_i \geq -2\eta \tag{A.29}$$

for $\omega_k \in \Omega_k$. Because

$$t^{h_j}_{-h_i+s} - t^{h_i}_{-h_i+s} = t_{h_j-h_i+s} - t_s, \tag{A.30}$$

and since, for (j,i) there always exists such $\bar{s} = s(i,j)$ that $t_{h_j-h_i+\bar{s}} \geq t_{\bar{s}}$ (if not, $t_{h+s} < t_s$ for all $s \in \mathbf{Z}$ with $h = h_j - h_i \geq j - i \geq 0$, and $\lim_{n\to\infty} t_{nh+s} < +\infty$), we see that $t^{h}_{-h_i+\bar{s}} - t^{h_i}_{-h_i+\bar{s}} \geq 0$. Taking into account the relation

$$-\eta < t^{h_k}_{r} - \omega_k < \eta, \tag{A.31}$$

which holds for all k and r, we get

$$\omega_j - \omega_i \geq (t^{h_j}_{-h_i+\bar{s}} - \eta) - (t^{h_i}_{-h_i+\bar{s}} + \eta) \geq -2\eta.$$

Now, if $\overline{\lim}_{i\to\infty}(h_{i+1} - h_i) = +\infty$, then, by using (A.30) for $s = 0$, $j = i+1$, we can assert that for any L there is d such that

$$t^{h_d+1}_{-h_d} - t^{h_d}_{-h_d} \geq L + 6\eta.$$

But then, by (A.31), $\omega_{i+1} - \omega_i \geq L + 4\eta$, which is a contradiction since $\Omega_\eta = \cup_{i\in\mathbf{Z}}\Omega_i$ is relatively dense and the interval $(\omega_d + 2\eta, \omega_{d+1} - 2\eta)$ of length L does not contain ω_k for any $k \in \mathbf{Z}$ because of (A.29).

So, the sequence $\{h_{i+1} - h_i\}$ is bounded and the set $H_\eta = \{h_i\}$ is relatively dense. The integer $h_i \in H_\eta$ is a 2η-almost period, common to all the sequences $\{t^j_h\}$, $j \in \mathbf{Z}$, because $t^j_{n+h_i} - t^j_n = t^{h_i}_{n+j} - t^{h_i}_n$ for all j and n. The proof of Lemma 52 is completed.

□

Now we finish the proof of Theorem 124. Let T be a set which is Bochner (Bohr) almost periodic. Then $\lambda(T)$ is its almost periodic representation. Indeed, for any $\epsilon > 0$ there exists such $T(\epsilon) > 0$ that any interval of length $T(\epsilon)$ contains such τ that $\rho(T, T+\tau) < \epsilon$, But then we also have

$$|-t_{n+d} + t_n - \tau| = d(\lambda(T), \lambda(T+\tau)) = \rho(T, T+\tau) < \epsilon.$$

Here $\lambda(T) = \{t_n\}$, $\lambda(T+\tau) = \{t_{n-d}+\tau\}$. By applying Lemma 52, we see that T is a strongly almost periodic set with an almost periodic representation $\lambda(T)$.

If the set T is strongly almost periodic, the conditions of Lemma 52 are fulfilled and for every $\eta > 0$ there exists a relatively dense set Ω_η such that, for $\omega \in \Omega_\eta$,

$$\rho(T, T+\omega) \leq \sup_n |\varphi(t_n+\omega) - t_n - \omega| < \eta,$$
$$\varphi(t_n+\omega) = t_{n+h_\omega},$$

and hence T is a Bohr almost periodic set. Theorem 124 is completely proved.

□

Corollary 9 *If $T_1, \ldots, T_n$ are sets with the growth index equal to $a_1, \ldots, a_n$ respectively, then $T = T_1 \sqcup \ldots \sqcup T_n$ is also almost periodic with the growth index equal to a, where $1/a = 1/a_1 + \ldots + 1/a_n$.*

A.5 Spaces of piecewise continuous almost periodic functions

In this section we study properties of piecewise continuous almost periodic functions. This class includes almost periodic solutions of impulsive systems. Spaces of such functions (which are also S-almost periodic functions[1] under additional conditions that are not restrictive) have the topology stronger than the topology of the space of Stepanov almost periodic functions. This fact explains to a large extent our interest to these spaces.

1. First we consider a general construction. Let $T \in \mathfrak{U}$. Arrange the numbers from T into a strictly increasing sequence $\{t_i\}$. The set $s(T) = \{t_i\}$ will be called a *support* of T. Thus there defined a mapping $s : \mathfrak{U} \to \mathfrak{U}$ and $s(\mathfrak{U})$ is invariant with respect to the translation $\theta_s(T) = T + s$.

It is possible to introduce a generalized Hausdorff metric in $s(\mathfrak{U})$, χ, by

$$\chi(P, Q) = \inf\{a : F_a(P) \supset Q,\ F_a(Q) \supset P\} \subset R \cup \{+\infty\}.$$

Here and in the sequel, $F_\alpha(P)$ denotes a closed α-neighborhood of the set P.

Let $\mathfrak{S}$ be such a subset of $\mathfrak{U}$ that $\mathfrak{S} \supset s(\mathfrak{U})$ and $\theta_s(\mathfrak{S}) = \mathfrak{S}$ for all s, and let δ be a metric on $\mathfrak{S}$ having the hollowing properties:

a1) $\delta(\theta_s(T), \theta_s(Q)) = \delta(T, Q)$ for all s;

a2) $\chi(T, Q) \leq \delta(T, Q)$;

a3) $\delta(\theta_s(Q), Q) \leq |s|$ for all s, Q.

Moreover, we will need a commutative binary operation in $\mathfrak{S}$, the sum of sets denoted by $\overset{*}{\sqcup}$, which possess the following properties:

b1) $(T \overset{*}{\sqcup} P) + a = (T + a) \overset{*}{\sqcup} (P + a)$;

b2) $s(T \overset{*}{\sqcup} P) = s(T) \cup s(P)$;

[1]Recall that a locally Bochner integrable function $f(t) : \mathbf{R} \to X$ with the values in a Banach space X is called Stepanov almost periodic (S-almost periodic) if for every $\epsilon > 0$ there exists a relatively dense set of ϵ-almost periods $\tau \in \Omega_\epsilon$, such that

$$\sup_{t \in \mathbf{R}} \int_t^{t+1} f(u + \tau) - f(u)\, du < \epsilon.$$

b3) $\delta(T_1 \overset{*}{\sqcup} T_2, P_1 \overset{*}{\sqcup} P_2) \le \max(\delta(T_1, P_1), \delta(T_2, P_2))$.

The translation mapping θ_s defines a continuous flow in $(\mathfrak{S}, \delta)$ because, by a1) and a3), $\delta(\theta_s(T), \theta_r(Q)) \le |s - r| + \delta(T, Q)$.

The spaces $(\mathfrak{U}, \rho, \sqcup, \theta_s)$, studied in Section A.4, and $(s(\mathfrak{U}), \chi, \cup, \theta_s)$, where $\cup$ is the union of sets, could serve as examples of the spaces $(\mathfrak{S}, \delta, \overset{*}{\sqcup}, \theta_s)$.

Similar to the definitions given in Section A.4, we can introduce Bohr almost periodic sets and Bochner almost periodic sets in $(\mathfrak{S}, \delta, \overset{*}{\sqcup}, \theta_s)$. Note that these notions may not coincide because $(\mathfrak{S}, \delta)$ is not complete, but Bochner δ-almost periodicity always implies Bohr δ-almost periodicity.

2. Fix a space $(\mathfrak{S}, \delta, \overset{*}{\sqcup}, \theta_s)$. In this and the subsequent sections we will assume that $x(t) : \mathbf{R} \to \mathbf{R}$ is a piecewise continuous function having, on a set $T \in \mathfrak{S}$, jump discontinuities.

Definition 7 *The pair $X = (x(t), T)$ is called Bohr δ-almost periodic function if*

c1) for every $\epsilon > 0$ there exists a relatively dense set of almost periods, Ω_ϵ:

$$|x(t + \tau) - x(t)| < \epsilon \qquad \forall t \in \mathbf{R} \setminus F_\epsilon(s(T)),\ \tau \in \Omega_\epsilon; \tag{A.32}$$

c2) T is a Bohr δ-almost periodic set;

c3) for all $a > 0$, the function $x(t) : \mathbf{R} \setminus F_\alpha(s(T)) \to \mathbf{R}$ is uniformly continuous.

We denote the set of all Bohr δ-almost periodic functions by $AP(\mathfrak{S}, \delta)$.

Let us prove that, for a pair X, the set Ω_ϵ can be chosen in such a way that any $\tau \in \Omega_\epsilon$ will at the same time be an ϵ-almost period of the Bohr δ-almost periodic set T.

Lemma 53 *Let T be a Bohr δ-almost periodic set. Then for all $\eta > 0$ there exists $L(\eta) > 0$ such that, for every positive $\gamma < \eta$ and every interval J of length $L(\eta)$, one can find an integer m such that $m\gamma \in J$ and*

$$\delta(T, T + m\gamma) < \eta. \tag{A.33}$$

Proof. Indeed, it follows from the definition of a Bohr δ-almost periodic set that for all $\eta > 0$ there exists $L_1(\eta/2) > 0$ such that any interval (a, b) of length $L_1(\eta/2)$ contains ω satisfying $\delta(T, T + \omega) < \eta/2$. Let $L(\eta) = L_1(\eta/2) +$

η. In any interval (α, β) of length $L(\eta)$, choose a subinterval $(\alpha+\eta/2, \beta+\eta/2)$ which has length $L_1(\eta/2)$ and contains ω. Then $\Delta = [\omega - \eta/2, \omega + \eta/2] \subset (\alpha, \beta)$. Use γ from $(0, \eta)$ to find m, for which $m\gamma \in \Delta \subset (\alpha, \beta)$. Having done this, we get

$$\delta(T, T + m\gamma) \leq \delta(T, T + \omega) + |\omega - m\gamma| < \eta.$$

□

Lemma 54 *Let* $X = (x(t), T) \in AP(\mathfrak{S}, \delta)$. *Then for all* $\eta > 0$ *there exist* $L(\eta), \delta(\eta) > 0$ *such that for every* $\gamma \in (0, \delta(\eta))$ *and every interval* J *of length* $L(\eta)$ *one can find an integer* m *such that* $m\gamma \in J$ *and*

$$|x(t + m\gamma) - x(t)| < \eta \qquad \forall t \in \mathbf{R} \setminus F_\eta(s(T)). \tag{A.34}$$

Proof. By c1), for every $\eta > 0$ there exists $L_1(\eta/2) > 0$ such that any interval of length $L_1(\eta/2)$ contains τ, for which

$$|x(t + \tau) - x(t)| < \eta/2 \qquad \forall t \in \mathbf{R} \setminus F_{\eta/2}(s(T)). \tag{A.35}$$

The function $x(t)$ is uniformly continuous on $\mathbf{R} \setminus F_{\eta/2}(s(T))$. Let $\delta(\eta) < \eta/2$ be such a positive number that $t', t'' \in \mathbf{R} \setminus F_{\eta/2}(s(T))$ and $|t' - t''| < \delta(\eta)$ (in this case t' and t'' necessarily belong to the same interval from $\mathbf{R} \setminus F_{\eta/2}(s(T))$). This would imply that

$$|x(t') - x(t'')| < \eta/2.$$

Let $L(\eta) = L_1(\eta/2) + \eta$ and (α, β) be an arbitrary interval of length $L(\eta)$. The subinterval $(\alpha + \eta/2, \beta + \eta/2)$ of length $L_1(\eta/2)$ contains such τ that (A.35) holds. For $\gamma \in (0, \delta(\eta))$, it is always possible to find an integer m such that $m\gamma \in [\tau - \eta/2, \tau + \eta/2] \subset (\alpha, \beta)$. Besides, if $t \in \mathbf{R} \setminus F_\eta(s(T))$, then, for $t' = t + m\gamma - \tau$, we have $|t' - t| = |m\gamma - \tau| < \delta/2 < \eta/4$. Hence $t' \in \mathbf{R} \setminus F_{\eta/2}(s(T))$ and t, t' lie in the same interval from $\mathbf{R} \setminus F_{\eta/2}(s(T))$. Consequently, for $t \in \mathbf{R} \setminus F_\eta(s(T))$,

$$|x(t + m\gamma) - x(t)| \leq |x(t + m\gamma - \tau) - x(t)| + \\ + |x(t + m\gamma - \tau) - x(t + m\gamma)| < \eta.$$

□

Lemma 55 *Let the pair $X = (x(t), T)$ be a δ-almost periodic function. Then for all $\eta > 0$ there exist $L(\eta), \delta(\eta) > 0$ such that, for every $\gamma \in (0, \delta(\eta))$ and every interval J of length $L(\eta)$, there is an integer m such that $m\gamma \in J$ and both inequalities (A.33) and (A.34) hold.*

Proof. We follow [55]. Take $\eta/8 > 0$ and use Lemmas 53 and 54 to choose $L_1, \delta, L_2 > 0$. Let $L = \max\{L_1, L_2\}$. For every positive $\gamma < \delta$, any interval of length L contains points $m\gamma$, $m'\gamma$ for some integers m, m' such that

$$\begin{aligned} &\delta(T, T + m\gamma) < \eta/8, \\ &|x(t + m'\gamma) - x(t)| < \eta/8 \qquad \forall t \in \mathbf{R} \setminus F_{\eta/8}(s(T)). \end{aligned}$$

Because $|m'\gamma - m\gamma| < L$, the differences $m' - m$ assume only a finite number of values n_i. For each n_i there exists a pair (m_i, m_i') which satisfies (A.35). Fix this pair. Set $\lambda = \max|m_i'\gamma|$, $l = L + 2\lambda$. Let $J = (\alpha, \alpha + l)$ be an arbitrary interval. In the subinterval $J' = (\alpha + \lambda, \alpha + \lambda + L)$, by the preceding argument, there will exist numbers $m\gamma$, $m'\gamma$ which satisfy (A.35) and $m\gamma - m'\gamma = n_i\gamma = (m_i - m_i')\gamma$. Let $q = m - m_i$. Clearly, $q\gamma \in J$. Moreover,

$$\begin{aligned} \delta(T, T + q\gamma) &= \delta(T, T + (m - m_i)\gamma) \leq \\ &\leq \delta(T, T + m\gamma) + \delta(T + m\gamma, T + m\gamma - m_i\gamma) = \\ &= \delta(T, T + m\gamma) + \delta(T, T + m_i\gamma) < \eta/4. \end{aligned}$$

Let us now show that, for $t \in \mathbf{R}\backslash F_\eta(s(T))$, $t + q\gamma \in \mathbf{R}\backslash F_\eta(s(T))$. Indeed, if this does not hold, then

$$|t + q\gamma - s| < \eta/2$$

for some $s \in T$. Because $\delta(T, T + q\gamma) < \eta/4$, using property a2) of the metric δ, we can find such a number $r \in T$ for this s that $|(r + q\gamma) - s| < \eta/2$. Finally,

$$|t - r| \leq |t + q\gamma - s| + |(s - (r + q\gamma))| < \eta \quad \text{and} \quad t \notin \mathbf{R}\backslash F_\eta(s(T)).$$

So, if $t \in \mathbf{R}\backslash F_\eta(s(T))$, then

$$\begin{aligned} |x(t + q\gamma) - x(t)| \leq\ &|x(t + m'\gamma - m_i'\gamma) - x(t + m'\gamma)| + \\ &+ |x(t + m'\gamma) - x(t)| < \eta. \end{aligned}$$

□

Corollary 10 *Conditions c1), c2), c3) are equivalent to the condition*

c) *for every $\epsilon > 0$ there exists a set Ω_ϵ, relatively dense in $\mathbf{R}$, of such numbers τ that $\delta(T, T+\tau) < \epsilon$ and*

$$|x(t+\tau)-x(t)| < \epsilon \qquad \forall t \in \mathbf{R} \backslash (F_\epsilon(s(T-\tau)) \cup F_\tau(s(T-\tau))). \quad \text{(A.36)}$$

Indeed, suppose c) holds. Then the set T is δ-almost periodic and, for ϵ, one can choose a relatively dense set $\Omega_{\epsilon/4}$ such that

$$|x(t+\tau)-x(t)| < \epsilon/4 \qquad \forall t \in M = \mathbf{R} \setminus (F_{\epsilon/4}(s(T)) \cup F_{\epsilon/4}(s(T-\tau))),$$
$$\delta(T, T+\tau) < \epsilon/4 \qquad \forall \tau \in \Omega_{\epsilon/4}.$$

But then $\chi(T, T+\tau) \leq \epsilon/4$ and so $F_{\epsilon/2}(s(T)) \subset F_\epsilon(s(T))$. Consequently, $F_{\epsilon/4}(s(T)) \cup F_{\epsilon/4}(s(T-\tau)) \subset F_\epsilon(s(T))$ and, if $t \in \mathbf{R} \backslash F_\epsilon(s(T))$, then $t \in M$. As a result we get (for $\tau \in \Omega_{\epsilon/4}$),

$$|x(t+\tau)-x(t)| < \epsilon/4 < \epsilon \qquad \forall t \in \mathbf{R} \setminus F_\epsilon(s(T)).$$

Let us check that c3) holds. Fix $a > 0$ and $\epsilon \in (0, a)$. Let $L > 0$ be such that an $\epsilon/4$-almost period τ be contained in an interval of length L. Then

$$\left.\begin{aligned} |x(t+\tau)-x(t)| < \epsilon/4 \qquad \forall t \in \mathbf{R} \setminus F_{\epsilon/4}(s(T)), \\ \delta(T, T+\tau) < \epsilon/4. \end{aligned}\right\} \quad \text{(A.37)}$$

Consider a function $x(t)$, which is piecewise continuous on $[-\epsilon, L]$, and for ϵ, find such a positive $\delta(\epsilon) < \epsilon/2$ that, for $t', t'' \in [0, L]$, $|t'-t''| < \delta(\epsilon)$, $(t', t'') \cap T = \emptyset$, we will have $|x(t')-x(t'')| < \epsilon/2$. Let now the numbers $t_1 \leq t_2 \in \mathbf{R} \backslash F_a(s(T))$ be such that $|t_1 - t_2| < \delta(\epsilon)$ (in this case, t_1 and t_2 lie in the same interval from $\mathbf{R} \backslash F_a(s(T))$).

If $t_i \in \mathbf{R} \backslash F_a(s(T)) \subset \mathbf{R} \backslash F_\epsilon(s(T))$, then, by the inequality $\chi(T, T+\tau) \leq \delta(T, T+\tau) < \epsilon/4$, we will certainly have that $t_i - \tau \in \mathbf{R} \backslash F_{\epsilon/4}(s(T))$ and hence

$$|x(t_1)-x(t_2)| \leq |x(t_1)-x(t_1-\tau)| +$$
$$+|x(t_1-\tau)-x(t_2-\tau)| + |x(t_2-\tau)-x(t_2)| < \epsilon$$

$((t_1-\tau, t_2-\tau) \cap T = \emptyset,\ \tau \in [t_2 - L(\epsilon/4), t_2],\ \epsilon < 2L(\epsilon/4))$.

Conversely, let conditions c1) – c3) hold. By Lemma 55, the set Ω_ϵ can be considered consisting of ϵ-almost periods, which are common to T and $x(t)$ and (A.32) implies (A.36).

□

Definition 8 *If* $X = (x(t), T)$, $Y = (y(t), P)$, *set*

$$\alpha X = (\alpha x(t), T), \quad X + Y = (x(t) + y(t), T \overset{*}{\sqcup} P),$$
$$XY = (x(t)y(t), T \overset{*}{\sqcup} P).$$

If $x(t) \neq 0$ *for all* $t \in \mathbf{R}$, *then*

$$Y/X = (y(t)/x(t), T \overset{*}{\sqcup} P).$$

Lemma 56 (Weak boundedness of a function $X \in AP(\mathfrak{S}, \delta)$). *If* $X = (x(t), T) \in AP(\mathfrak{S}, \delta)$, *then for all* $\epsilon > 0$ *there exists* $M(\epsilon) > 0$ *such that* $|x(t)| \leq M(\epsilon)$ *for all* $t \in \mathbf{R} \backslash F_\epsilon(s(T))$.

Proof. Let $t \in \mathbf{R} \backslash F_\epsilon(s(T))$. By Lemma 55, for every $\epsilon > 0$ there exists $L(\epsilon/4) > 0$ such that any interval $(t - L(\epsilon/4), t)$ will contain τ such that (A.37) holds. The function $x(t)$ is bounded on $[0, L(\epsilon/4)]$: $|x(t)| < M(\epsilon/4)$. Let us show that

$$s = t - \tau \in (0, L(\epsilon/4)) \cap \mathbf{R} \setminus F_{\epsilon/2}(s(T)).$$

If not, $s \in F_{\epsilon/2}(s(T))$ and so, $|s - r| \leq \epsilon/2$ for some $r \in T$, and, by (A.37), there exists $p \in T$ for this r such that $|r + \tau - p| < \epsilon/2$ and, consequently,

$$|t - p| = |t - \tau - r + r + \tau - p| < \epsilon \qquad \forall t \in \mathbf{R} \setminus F_\epsilon(s(T)).$$

As a result, we get that $|x(t) - x(s)| = |x(t - \tau) - x(t)| < \epsilon/4$, whence

$$|x(t)| \leq |x(t) - x(s)| + |x(s)| < \epsilon/4 + m(\epsilon/4) = M(\epsilon).$$

□

The following lemma follows from Lemma 55.

Lemma 57 (On uniform almost periodicity of a finite number of functions from $AP(\mathfrak{S}, \delta)$). *Let* $X_i = (x_i(t), T_i) \in AP(\mathfrak{S}, \delta)$. *Then for every* $\epsilon > 0$, $i = \overline{1, m}$, *there exists a set* Ω_ϵ, *which is relatively dense in* $\mathbf{R}$, *consisting of* ϵ-*almost periods* τ *such that* $\delta(T_i, T_i + \tau) < \epsilon$ *for* $i \leq m$ *and*

$$|x_i(t + \tau) - x_i(t)| < \epsilon \qquad \forall t \in \mathbf{R} \setminus F_\epsilon(s(T_i)).$$

Theorem 125 *The space $AP(\mathfrak{S},\delta)$ is an* **R**-*algebra*[2]. *If $X = (x(t), T) \in AP(\mathfrak{S},\delta)$, $x(t) \neq 0$ for all t, and for all $a > 0$ there exists $\mu > 0$ such that $|x(t)| \geq \mu$ for all $t \in \mathbf{R}\backslash F_a(s(T))$, then $X^{-1} = (1/x(t), T) \in AP(\mathfrak{S},\delta)$.*

Proof.

A (addition). By Lemma 57, for every $\epsilon > 0$ there exists an almost periodic set of almost periods τ such that

$$\begin{array}{ll} \delta(T, T+\tau) < \epsilon/2, & \delta(P, P+\tau) < \epsilon/2, \\ |x(t+\tau) - x(t)| < \epsilon/2 & \forall t \in \mathbf{R} \setminus F_{\epsilon/2}(s(T)); \\ |y(t+\tau) - y(t)| < \epsilon/2 & \forall t \in \mathbf{R} \setminus F_{\epsilon/2}(s(P)). \end{array}$$

But then $\delta((T \overset{*}{\sqcup} P)+\tau, T \overset{*}{\sqcup} P) = \delta((T+\tau) \overset{*}{\sqcup} (P+\tau), T \overset{*}{\sqcup} P) \leq \max(\delta(T, T+\tau), \delta(P, P+\tau)) < \epsilon$ and

$$\begin{array}{c} |x(t+\tau) + y(t+\tau) - (x(t) + y(t))| < \epsilon \\ \forall t \in (\mathbf{R} \setminus F_{\epsilon/2}(s(T) \cup s(P))) \supset \mathbf{R} \setminus F_\epsilon(s(T \overset{*}{\sqcup} P)). \end{array}$$

B (multiplication). Let us fix $\epsilon > 0$. By Lemma 56, the function $|x(t)|$ (correspondingly, $|y(t)|$) is bounded by $M(\epsilon/2)$ on the set $\mathbf{R}\backslash F_{\epsilon/2}(s(T))$ (correspondingly, $\mathbf{R}\backslash F_{\epsilon/2}(s(P))$). Apply Lemma 57 to $\eta = \min\{\epsilon/4, \epsilon M^{-1}/2\}$ to see that there exists an almost periodic set Ω_η such that, for $\tau \in \Omega_\eta$,

$$\begin{array}{ll} \delta(T, T+\tau) < \eta, & \delta(P, P+\tau) < \eta; \\ |x(t+\tau) - x(t)| < \eta & \forall t \in \mathbf{R} \setminus F_\eta(s(T)) \supset \mathbf{R} \setminus F_\epsilon(s(T)), \\ |y(t+\tau) - y(t)| < \eta & \forall t \in \mathbf{R} \setminus F_\eta(s(P)) \supset \mathbf{R} \setminus F_\epsilon(s(P)). \end{array}$$

Further, if $t \in \mathbf{R}\backslash F_\epsilon(s(T))$, then $t+\tau \in \mathbf{R}\backslash F_{\epsilon/2}(s(T))$ (for, otherwise, $|t+\tau-p| \leq \epsilon/2$ for some $p \in T$, $|p-(r+\tau)| < 2\eta \leq \epsilon/2$ for some $r \in T$, and, hence, $|t-r| = |t+\tau-p+p-(r+\tau)| \leq \epsilon$). But then if $\tau \in \Omega_\eta$,

$$\begin{array}{l} |x(t+\tau)y(t+\tau) - x(t)y(t)| \leq \\ \leq |x(t+\tau)|\,|y(t+\tau) - y(t)| + |y(t)|\,|x(t+\tau) - x(t)| \leq \epsilon \\ \forall t \in (\mathbf{R} \setminus F_\epsilon(s(T))) \cap (\mathbf{R} \setminus F_\epsilon(s(P))) = \mathbf{R} \setminus F_\epsilon(s(T \overset{*}{\sqcup} P)). \end{array}$$

[2]To have distributivity in $AP(\mathfrak{S},\delta)$, we will identify δ-almost periodic functions $A = (a(t), T)$ and $B = (a(t), P)$ if $s(T) = s(P)$.

C (existence of X^{-1}). Fix $\epsilon > 0$. By the assumption, $|x(t)| \geq \mu(\epsilon/2)$ on $\mathbf{R} \backslash F_{\epsilon/2}(s(T))$. Let $\eta = \min\{\epsilon/4, \epsilon\mu^2\}$. Because X is δ-almost periodic, there exists relatively dense set Ω_η such that for all $\tau \in \Omega_\eta$,

$$\delta(T, T+\tau) < \eta, \quad |x(t+\tau) - x(t)| < \eta \qquad \forall t \in \mathbf{R} \setminus F_\eta(s(T)).$$

If $t \in \mathbf{R} \backslash F_\epsilon(s(T))$, then $t + \tau \in \mathbf{R} \backslash F_{\epsilon/2}(s(T))$ (see item **B**) and hence

$$|x^{-1}(t+\tau) - x^{-1}(t)| = |x(t+\tau) - x(t)| \left(|x(t)x(t+\tau)|\right)^{-1} <$$
$$< \eta\mu^{-2}(\epsilon/2) \leq \epsilon \qquad \forall t \in \mathbf{R} \setminus F_\epsilon(s(T))$$

for $\tau \in \Omega_\eta$.

3. Let the properties of $x(t)$ and T be the same as in part **2**. Consider the space $\mathfrak{M}$ of all pairs $X = (x(t), T)$. It is clear that $\mathfrak{M}$ is invariant with respect to the translations $\theta_s(X) \stackrel{\text{def}}{=} (x(t+s), T - s) \in \mathfrak{M}$.

Equip $\mathfrak{M}$ with the uniform topology and study its subspace $AP(\mathfrak{S}, \delta)$. Let a be a positive number, $U_\infty \stackrel{\text{def}}{=} \mathfrak{M} \times \mathfrak{M}$,

$$U_a \stackrel{\text{def}}{=} \{(X,Y) = [(x(t,T); (y(t), P)] \in \mathfrak{M} \times \mathfrak{M} :$$
$$\delta(T, P) < a, \quad |x(t) - y(t)| < a \qquad \forall t \in \mathbf{R} \setminus F_a(s(T \sqcup P))\}.$$

The following properties are evident:

p1) for all X, a, we have $(X, X) \in U_a$;

p2) if $(X,Y) \in U_a$, then $(Y,X) \in U_a$ (i.e. $U_a^{-1} = U_a$);

p3) if $(X,T) \in U_{a/2}$, $(Y,Z) \in U_{a/2}$, then $(X,Z) \in U_a$;

p4) $U_a \cap U_b = U_{\min(a,b)}$.

Hence [63], the family U_a form a base of some uniformity on $\mathfrak{M}$. The space $\mathfrak{M}$ is endowed with the uniform topology. By Aleksandrov-Uryson theorem [63], the uniform space $(\mathfrak{M}, \mathfrak{U})$ is metrizable for it is a Hausdorff space with a countable base of uniformity. A sequence $(x_n(t), T_n)$ from this space converges to $(x(t), T) \in \mathfrak{M}$ if and only if, for any $\epsilon > 0$ there exists n_0 such that, for all $n \geq n_0$, the following inequalities hold:

$$\delta(T, T_n) < \epsilon, \quad |x(t) - x_n(t)| < \epsilon \qquad \forall t \in \mathbf{R} \setminus F_\epsilon(s(T_n \sqcup T)).$$

Because the relations $(\theta_s(X), \theta_s(Y)) \in U_a$ and $(X, Y) \in U_a$ are equivalent, we can assume that the metric d, which generate the uniformity $\mathfrak{U}$, is invariant, $d(\theta_s s(X), \theta_s(Y)) = d(X, Y)$ for all s, X, Y.

The triple $(\mathfrak{M}, \theta_s, \mathbf{R})$ is a group of continuous transformations of $\mathfrak{M}$ if $\mathbf{R}$ is endowed with the discrete topology. The definition of Bohr δ-almost periodicity can be restated in new terms as follows (using Corollary of Lemma 55):

Definition 9 *A pair $X = (x(t), T) \in \mathfrak{M}$ is called Bohr δ-almost periodic if for every element V of the uniformity $\mathfrak{U}$ there exists a set, relatively dense in $\mathbf{R}$, of such numbers τ that $(\theta_\tau(X), X) \in V$.*

We make a few more definitions.

Definition 10 *A pair $X = (x(t), T)$ will be called a Bochner δ-almost periodic function if any sequence $\theta_{s_n}(X)$ is precompact in $(\mathfrak{M}, \mathfrak{U})$. The subspace of $(\mathfrak{M}, \mathfrak{U})$ consisting of such functions will be denoted by $AP_h(\mathfrak{S}, \delta)$*[3].

It is clear that the property of a function X to be Bochner almost periodic allows to compensate the incompleteness of $(\mathfrak{M}, \mathfrak{U})$ and to prove the following theorem by using standard methods.

Theorem 126 *A function X belongs to the set $AP_h(\mathfrak{S}, \delta)$ if and only if the set $H(X) = [\theta(\mathbf{R})X]$ is compact in $(\mathfrak{M}, \mathfrak{U})$. Also, for all $Y \in H(X)$, we have $H(X) = H(Y) \subset AP_h(\mathfrak{S}, \delta) \subset AP(\mathfrak{S}, \delta)$.*

Definition 11 *By $APH(\mathbf{R})[(\mathfrak{N}, \mathfrak{U})]$, we denote the subspace of $AP(\mathfrak{U}, \rho)$-$[(\mathfrak{M}, \mathfrak{U})]$ consisting of such functions that*

$$\forall \epsilon > 0 \ \exists \delta(\epsilon) > 0 : |x(t') - x(t'')| < \epsilon \ \forall t', t'' \in \mathbf{R} : \\ |t' - t''| < \delta(\epsilon), [t', t''] \cap \{t_i\} = \emptyset. \quad (A.38)$$

It turns out that, for an important class of functions from $AP(\mathfrak{U}, \rho)$, the notions of Bochner and Bohr almost periodicity coincide if we consider only piecewise equipotentially continuous functions.

Theorem 127 $APH(\mathbf{R}) = AP(\mathfrak{U}, \rho) \cap \mathfrak{N} = AP_h(\mathfrak{U}, \rho) \cup \mathfrak{N}$.

[3] $AP_h(\mathfrak{S}, \delta) \neq AP(\mathfrak{S}, \delta)$. To see that, consider the pair $X = (0, T)$ from $AP(\mathfrak{S}, \delta)$, where $T \in (s(\mathfrak{U}), \chi)$ is given by $T = \mathbf{Z} \cup_{k=0}^{+\infty} \{2^k n + 2^{-k-1} \sin^2 n : n \in \mathbf{Z}\}$.

Proof. It will be sufficient to show that $AP(\mathfrak{U},\rho) \subseteq AP_h(\mathfrak{U},\rho) \cap \mathfrak{N}$. Let $A = (a(t), L) \in APH(\mathbf{R})$, $\Pi(A) = \{\theta_s(A), s \in \mathbf{R}\}$, $\omega_A(s) = \inf\{q : (\theta_s(A), A) \in U_q\}$, $\sigma_A(\epsilon) = \sup |a(t') - a(t)|$, where the supremum is taken for all $t < t'$ such that $|t' - t| \le \epsilon$, $[t, t'] \cap L = \emptyset$. By (A.38), $\lim_{\epsilon \to +0} \sigma_A(\epsilon) = 0$ and, because $\omega_A(s) \le \max\{|s|, \sigma_A(|s|)\}$, we see that $\lim_{\epsilon \to 0} \omega_A(\epsilon) = 0$.

The mapping $\theta_s : \Pi(A) \times \mathbf{R} \to \Pi(A)$ is uniformly continuous. Indeed, fix an arbitrary $\epsilon > 0$ and chose $\delta > 0$ such that $|s| + \sigma_A(|s|) < \epsilon/2$ for $|s| \le \delta$. Then, for any $|t - r| < \delta$, $X, Y \in \Pi(A) : (X, Y) \in U_{\epsilon/2}$, we have

$$(\theta_{t-r}(A),\, A) \in U_{\epsilon/2} \Leftrightarrow (\theta_t(X),\, \theta_r(X)) \in U_{\epsilon/2}$$
$$(X,\, Y) \in U_{\epsilon/2} \Leftrightarrow (\theta_r(Y),\, \theta_r(X)) \in U_{\epsilon/2},$$

and thus $(\theta_t(X), \theta_r(Y)) \in U_\epsilon$.

Let us show that $H(A) = [\theta(\mathbf{R})A] \subset \mathfrak{N}$. If $B = (b(t), K) \in H(A)$, then there exists a number sequence $\{s_n\}$ such that

$$\theta(s_n)(A) = (a(t + s_n),\, L - s_n) \to B \quad \text{for} \quad n \to \infty.$$

Let $[t, t'] \cap K = \emptyset$ with $t < t'$. There exists $\kappa > 0$ with $[t - \kappa, t' + \kappa] \cap K = \emptyset$. Use this κ to find n_0 in such a way that $(\theta(s_n)(A), B) \in U_{\kappa/4}$ for all $n \ge n_0$. But then $[t - \kappa/4, t' + \kappa/4] \cap (L - s_n) = \emptyset$ and hence $|a(t' + s_n) - a(t + s_n)| \le \sigma_A(|t - t'|)$ for all $n \ge n_0$. Since the functions $a(t + s_n)$ converge to $b(t)$ uniformly on $[t - \kappa/4, t' + \kappa/4]$, we get $|b(t') - b(t)| \le \sigma_A(|t - t'|)$ and $B \in \mathfrak{N}$. Moreover, we have shown that $\sigma_B(\epsilon) \le \sigma_A(\epsilon)$ for $\epsilon \ge 0$, where $\sigma_B(\epsilon) = \sup |b(t') - b(t)|$ with the supremum taken for all $t < t'$ such that $|t' - t| \le \epsilon$, $[t, t'] \cap K = \emptyset$.

The mapping $\theta_s : H(A) \times \mathbf{R} \to H(A)$ is also uniformly continuous. (Indeed, if $\omega_B(s) = \inf\{q : (\theta_s(B), B) \in U_q\}$ for $B \in H(A)$, then $\omega_B(s) \le \max\{|s|, \sigma_B(|s|)\} \le \max\{|s|, \sigma_A(|s|\}$. For $\epsilon > 0$ find $\delta \in (0, \epsilon/2)$ such that $\sigma_A(|s|) < \epsilon/2$ for $|s| \le \delta$. Then, for any $|t - r| < \delta$, $X, Y \in H(A)$, $(X, Y) \in U_{\epsilon/2}$, we have $(\theta_t(X), \theta_r(X)) \in U_{\epsilon/2}$ and $(\theta_r(Y), \theta_r(X)) \in U_{\epsilon/2}$, and hence $(\theta_t(X), \theta_r(Y)) \in U_\epsilon$.

Let us show that the space $(H(A), \mathfrak{U})$ is complete. If $Y_n = (y_n(t), T_n)$ is a Cauchy sequence, then, because the space $(\mathfrak{U}, \rho)$ is complete, $\rho_n = \rho(T_n, T) \to 0$ for some $T \in \mathfrak{U}$. We can assume that ρ_n is monotonically decreasing. Then, for $n \ge m$, a sequence of continuous functions $y_n(t)$, which uniformly converges to $y(t) \in C(G_m)$, is defined on the set

$$G_m = \mathbf{R} \setminus F_{2\rho_m}(s(T)) \subset \mathbf{R} \setminus (F_{\rho_m}(s(T_m)) \cup F_{\rho_m}(s(T))).$$

Moreover, for every t_1, t_2 belonging to the same connected component, we have

$$|y_n(t_1) - y_n(t_2)| \le \sigma_A(|t_1 - t_2|) \Longrightarrow |y(t_1) - y(t_2)| \le \sigma_A(|t_1 - t_2|).$$

Because $G_m \subset G_{m+1}$ and $\cup_m G_m = \mathbf{R}\backslash T$, we can consider $y(t)$ to be defined on $\mathbf{R}\backslash T$ and $y_n(t) \to y(t)$ on every G_m, and that $(A.38)$ holds for $y(t)$. Define $y(t)$ at points of $s(T)$ so it be left continuous. Then $Y = (y(t), T) \in \mathfrak{N}$, $Y_n \to Y$, whence $(H(A), \mathfrak{U})$ is complete. Theorem 127 now follows from the Bochner theorem [75].

□

4. We give the most important properties of the space $APH(\mathbf{R}) \subset AP(\mathfrak{U}, \rho)$.

By Lemma 56, the function $X \in APH(\mathbf{R})$ is weakly bounded. There exist unbounded functions in $APH(\mathbf{R})$[4]. The space $APH(\mathbf{R})$ is closed with respect to addition, multiplication by a scalar, but is not closed with respect to multiplication[5]. If $A, B \in APH(\mathbf{R})$ and are bounded, then $AB \in APH(\mathbf{R})$.

---

[4]Let $T = \{n\} \cup \{n + c_n\}_{n \in \mathbf{Z}}$, where $\{c_n\}$ is such an almost periodic sequence that $0 < c_n < 1/4$ and $\inf_n c_n = 0$, and the function $a(t)$ is

$$a(t) = \begin{cases} (t-n) + c_n^{-1} & \text{for} \quad t \in (n, n + c_n]; \\ 0, & \qquad t \in (n + c_n, n + I]. \end{cases}$$

It is clear that $A = (a(t), T) \in \mathfrak{N}$, the function $a(t)$ is unbounded, and T is a ρ-almost periodic set. We prove c1) for A. Fix $\epsilon \in (0, 0.25)$. The sequence $\{q_n = \max(c_n, \epsilon)\}$ is almost periodic and, because $q_n \ge \epsilon > 0$, the sequence $\{q_n^{-1}\}$ will also be almost periodic. Let Ω_ϵ be the set of $\epsilon/2$ almost periods common to $\{q_n^{-1}\}$ and $\{c_n\}$. Let us show that if $p \in \Omega_\epsilon$, the following inequality holds

$$|a(t+p) - a(t)| < \epsilon/2 \qquad \forall t \in \mathbf{R} \setminus F_\epsilon(T). \tag{A.39}$$

Indeed, if $t \in \mathbf{R}\backslash F_\epsilon(s(T))$, then: 1) either $t \in (m + c_m + \epsilon, m + 1 - \epsilon]$ for some m and then, by using the estimate

$$|c_{m+p} - c_m| < \epsilon/2,$$

we have $t + p \in (m + p + c_m + \epsilon, m + p + 1 - \epsilon] \subset (m + p, m + p + c_{m+p}]$, whence (A.39) follows; 2) or $t \in (m + \epsilon, m + c_m - \epsilon]$ and then $c_m > 2\epsilon$. In the latter case, we also have $t + p \in (m + p + \epsilon, m + p + c_m - \epsilon] \subset (m + p, m + p + c_{m+p}]$ and $c_{m+p} \ge c_m - |c_m - c_{m+p}| > \epsilon$, whence $q_m^{-1} = c_m^{-1}$, $q_{m+p}^{-1} = c_{m+p}^{-1}$. As the result,

$$|a(t+p) - a(t)| = |c_m^{-1} - c_{m+p}^{-1}| = |q_m^{-1} - q_{m+p}^{-1}| < \epsilon/2.$$

Consequently $A \in APH(\mathbf{R})$.

[5]Take the function A, constructed in the previous footnote. Then $A^2 \notin \mathfrak{N}$ because $a^2(n + c_n) - a^2(n) = 2 + c_n^2 > 2$.

If $A \in APH(\mathbf{R})$ and $|a(t)| \geq \mu > 0$, then $A^{-1} \in APH(\mathbf{R})$. Not all the functions form $APH(\mathbf{R})$ have the mean value (in the usual sense)[6].

Lemma 58 *Let $B = (b(t), T) \in AP(\mathfrak{S}, \delta)$,*

$$\lambda_\epsilon(t) = mes(F_\epsilon(s(T)) \cap [t, t+1)).$$

Suppose that $|b(t)| \leq m$ for all $t \in \mathbf{R}$, $\lambda_\epsilon(t) \to 0$ for $\epsilon \to 0$ uniformly with respect to $t \in \mathbf{R}$. Then $b(t)$ is an S-almost periodic Stepanov function[7].

Proof. Let τ be an ϵ-almost period of B, $\epsilon < 1/2$. Then

$$|b(t+\tau) - b(t)| < \epsilon \qquad \forall t \in \mathbf{R} \setminus F_\epsilon(s(T))$$

and, if $\Phi' = F_\epsilon(s(T)) \cap [t, t+1]$, $\Phi'' = [t, t+1] \backslash \Phi'$ then

$$\int_t^{t+1} |b(u+\tau) - b(u)| du = \int_{\Phi'} |b(u+\tau) - b(u)| du +$$
$$+ \int_{\Phi''} |b(u+\tau) - b(u)| du \leq 2m\lambda_\epsilon(t) + \epsilon.$$

□

5. In the sequel however, it will be sufficient to use a less formal definition of a piecewise continuous almost periodic function and consider Wexler almost periodic functions (W-functions), named after D. Wexler who was the fist to consider this class of functions [55], which are piecewise continuous solutions of an impulsive system $x : \mathbf{R} \to M$ with the values in a complete metric space (M, ρ) and with the jumps on T satisfying a modified inequality (A.36)[8]: for every $\epsilon > 0$ there exists a relatively dense set of the numbers such that

$$\rho(x(t+\tau), x(t)) < \epsilon \qquad \forall t \in \mathbf{R} : |t - t_k| > \epsilon, \quad |t + \tau - t_p| > \epsilon.$$

[6] Let $T = \{n\} \cup \{n + 1/2\cos^2 n\}$ and the function $f(t) : \mathbf{R} \to \mathbf{R}$ be such that

$$f(t) = \begin{cases} 2|\cos n|^{-4} & \text{for} \quad t \in (n, n + 1/2\cos^2 n]; \\ 0, & \quad t \in (n + 1/2\cos^2 n, n+1]. \end{cases}$$

Then $(f(t), T) \in APH(\mathbf{R})$ but $\frac{1}{n+1}\int_0^{n+1} f(u)du = \frac{1}{n+1}\sum_{k=0}^n \cos^{-2} k \to \infty$.

[7] In particular, all bounded functions from $APH(\mathbf{R})$ are S-almost periodic.

[8] Note that if T is strongly almost periodic, then the inequality $|t + \tau - t_k| > \epsilon$ does not change the region of the estimate and if $t_i^1 \geq \delta > 0$, then this inequality only slightly decreases the region of the estimate so, in both cases, the sense of the definition will not change if this inequality is omitted.

We will call such solutions strongly W-almost periodic if X is a normed space with the topology induced by the norm. The function $x(t)$ will also be called scalar W-almost periodic if $\langle x(t), f\rangle$ is W-almost periodic for all $f \in X^*$.

Note that the notion of a W-almost periodic function permits us to leave out a description of the set $\{t_i\}$ but it does make sense if the points $\{t_i\}$ are not very densely distributed, for example, if

$$\zeta \stackrel{\text{def}}{=} \sup_{\alpha \in \mathbf{R}} \operatorname{Card}(\{t_i\} \cap (\alpha, \alpha+1)) < +\infty.$$

A next important step in studying almost periodic systems is that, although, following [55], one proves for the most part that they have W-almost periodic solutions, we will be looking for S-almost periodic solutions $x(t)$ of the impulsive systems, which, in infinite dimensional spaces, are also S-bounded, scalar S-almost periodic (a scalar S-almost periodic function is the function, for which $\langle x(t), f\rangle$ is S-almost periodic for all $f \in X^*$). The advantages of this approach are clear. Instead of an incomplete metrizable uniform space, we work with a complete normed Stepanov space. The relation between strongly[9] S-almost periodic and W-almost periodic solutions of impulsive systems turns out to be very close. In the remaining part of this section, we will be investigating this relation as well as considering Levitan N-almost periodic functions.

Lemma 59 *For a function $x(t) : \mathbf{R} \to X$, which is uniformly continuous on the set $\{t : |t - t_j| \geq \epsilon \ \forall j\} \subset \mathbf{R}$ for all $\epsilon > 0$, to be S-almost periodic, it is necessary and, for $\zeta < +\infty$ and $\sup_{t \in \mathbf{R}} |x(t)| < +\infty$, sufficient that it be W-almost periodic.*

Proof. Assume that $x(t)$ is almost periodic. Fix $\epsilon \in (0,1)$ and let $\kappa < \epsilon/2$ be such that, if $|t - s| < \kappa$, then $|x(t) - x(s)| < \epsilon/8$ for all t and s which belong to the set $\{t : |t - t_j| \geq \epsilon/2 \ \forall j\}$. Let $\epsilon/2 \geq \mu_n \to 0$ and Ω_n be the set, which is relatively dense in $\mathbf{R}$, of almost periods μ_n of the S-almost periodic function $x(t)$. Let us show that if n is sufficiently large, the set Ω_n consists of ϵ-almost periods of the W-almost periodic function $x(t)$. Indeed, if this is not the case, then there exists a sequence $z_n : |z_n - t_j| \geq \epsilon, |z'_n - t_p| \geq \epsilon$ for all n, j, p such that $|x(z'_n) - x(z_n)| \geq \epsilon$, where $z'_n = z_n + \tau_n$, $\tau_n \in \Omega_n$. The sequences $X_1 = \{x(t + z'_n)\}$ and $X_2 = \{x(t + z_n)\}$, $t \in [0, \epsilon/2]$ are jointly

[9] scalar almost periodic solutions will be considered in the last section

equipotentially continuous and $|x(t+z'_n)-x(t+z_n)| \geq \epsilon/2-\epsilon/8-\epsilon/8=\epsilon/4$ for $t \in [0,\kappa]$. But then

$$|x(u+z_n)-x(u+z'_n)|_S \geq \epsilon\kappa/4,$$

which contradicts the assumption.

The second (converse) part of Lemma 59 follows from Lemma 58.

□

We will see in the sequel how important the notion of the Levitan almost periodicity is in studying almost periodic impulsive systems. Recall that a continuous function $x(t) : \mathbf{R} \to X$, which takes values in a Banach space X, is called Levitan N-almost periodic if for all $\epsilon > 0$, $N > 0$ there exists a relatively dense set of almost periods, $\Omega_{\epsilon,N}$, such that

$$|x(t+\tau)-x(t)| < \epsilon \qquad \forall t \in \mathbf{R} : |t| \leq N, \ \tau \in \Omega_{\epsilon,N},$$

and if for all $\delta > 0$, $N > 0$ there exists $\epsilon > 0$ with $\Omega_{\epsilon,N} \pm \Omega_{\epsilon,N} \subset \Omega_{\delta,N}$.

The following result is given without a proof (see [75]).

Theorem 128 *Let $(M, h(t)m)$ be a minimal compact isometric flow, $K = \{h(t)m_0, t \in \mathbf{R}\}$ - a subspace of M with the induced topology ("irrational winding"), X - a Banach space, and a function $\Gamma : K \to X$ be continuous. Then the function $\gamma(t) = \Gamma(h(t)m_0) : \mathbf{R} \to X$ is Levitan N-almost periodic.*

Because we are studying discontinuous solutions, we will generalize somewhat the notion of N-almost periodicity (just as Stepanov S-almost periodic functions generalize abstract continuous Bohr functions).

Definition 12 *A locally integrable function $x(t) : \mathbf{R} \to X$ will be called S-N-almost periodic if the continuous function*

$$X(t) : \mathbf{R} \to L([0,1], X), \qquad X(t) = x(t+s), \ s \in (0,1)$$

is Levitan N-almost periodic in the usual sense.

Similarly to Lemma 59, one can prove the following (see [151]).

Lemma 60 *For a function $x(t) : \mathbf{R} \to X$, which is uniformly continuous on the set $\{t : |t-t_j| \geq \epsilon \ \forall j\} \subset \mathbf{R}$ for all $\epsilon > 0$, to be S-N-almost periodic, it is necessary and, if $\zeta < +\infty$ and $\sup_{t\in\mathbf{R}} |x(t)| < +\infty$, sufficient that the following conditions hold:*

1) *for all $\epsilon > 0$ and $N > 0$ there exists a relatively dense set $\mathcal{W}_{\epsilon,N}$ such that, for all $\tau \in \mathcal{W}_{\epsilon,N}$,*

$$|x(t+\tau) - x(t)| < \epsilon \qquad \forall t \in \mathbf{R} : |t| \leq N, \, |t - t_k| > \epsilon;$$

2) *for all $\delta > 0$ and $N > 0$ there exists $\epsilon > 0$ such that $\mathcal{W}_{\epsilon,N} \pm \mathcal{W}_{\epsilon,N} \subset \mathcal{W}_{\delta,N}$.*

A.6 Almost periodic measures on the line

Before discussing the problem of almost periodic solutions of the system

$$\begin{aligned} \frac{dx}{dt} &= f(t,x) \\ \Delta x|_{t_i} &= h_i(x), \qquad t \in \mathbf{Z}, \end{aligned} \tag{I}$$

we formulate the main requirements imposed on the system. One can suppose that the weakest requirement that needs to be imposed on system (I) to get acceptable results is to assume that the family of distributions $f(t,x) + \sum h_i(x)\delta(t-t_i)$ is almost periodic with respect to t. Here we make a more strict assumption that the function $f(t,x)$ and the measure $h(x) = \sum h_i(x)\delta(t-t_i)$ are almost periodic uniformly on compact sets.

A study of the first equation of (I)

Our goal is to weaken the condition of Bohr almost periodicity in t almost periodic in x, which is usually imposed on $f(t,x)$. Consider a finite dimensional case: $x \in \Omega$, Ω is a domain in $\mathbf{R}^n$. In the remaining part of this paragraph, we assume that

$$(\mathcal{A}) \quad \begin{cases} \mathcal{A}.1. & \text{the function } f(t,x) \text{ is continuous in } x \text{ for almost all } t \in R \\ & \text{and is measurable with respect to } t \text{ for every } x \in \Omega; \\ \mathcal{A}.2. & \text{for every compact } K \subset \Omega \text{ there exists S-almost periodic} \\ & \text{function } \mu_{K,f} \text{ such that } sup_{x\in K}|f(t,x)| \le \mu_{K,f}(t); \\ \mathcal{A}.3. & \text{the function } f(t,x) \text{ is S-almost periodic uniformly} \\ & \text{with respect to } x \in K. \end{cases}$$

Fix a compact set $K \subset \Omega$. Let Q_1 consist of restrictions of all the functions which satisfy conditions $(\mathcal{A})$ to the set $\mathbf{R} \times K$ and let $\rho(f,g) = \sup_{x\in K}|f(\cdot,x) - g(\cdot,x)|_s$ for $f, g \in Q_1$.

Lemma 61 *The set*

$$H(f) = \left[\{\tau_s f = f(t+s,x),\, s \in R\}\right] \subset Q_1$$

is compact in the metric space (Q_1,ρ) and $(H(f),\tau_s)$ is a minimal isometric flow.

Proof. Let us denote by Π the set of functions defined on the compact set $Z = [0,1] \times K$ and satisfying the Caratheodory conditions on Z with positive majorants $m(t)$, which belong to a compact subset of $B \subset L[0,1]$ (thus Π depends on the choice of B). Let us define a distance in Π by

$$d(g,f) = \sup_{x\in K} |f(u,x) - g(u,x)|_{L[0,1]}.$$

The space (Π, d) is complete. Indeed, if a sequence f_n in Π is Cauchy, then

$$\exists f(t,x) \in L[0,1] \qquad \forall x : f_n(t,x) \overset{K}{\rightarrow} f(t,x) \quad \text{in} \quad L[0,1],$$

consequently the sequence also converges with respect to the measure. Hence, by Riesz theorem (see, for example [19]), there exists a subsequence f_{n_k} that converges in $\mathbf{R}^n$ to f uniformly with respect to $x \in K$ for almost all $t \in [0,1]$. Consequently, the function $f(t,x)$ is continuous with respect to x for almost all t. Moreover, $|f_{n_k}(t,x)| \leq m_{n_k}(t)$ and, because B is compact, by Riesz theorem, we can assume that m_{n_k} converges pointwise on $[0,1]$ to a certain function $m(t)$ from B.

So, as a result, we get that $|f(t,x)| \leq m(t)$ and Π is complete.

Let now $f \in Q_1$ and B be the closure in $L[0,1]$ of the set $\{\mu_{k,f}(t+s) | s \in \mathbf{R}\}$. Let us show that the function $F(s) = f(t+s,x) : \mathbf{R} \to \Pi$ is continuous. To do that, we first prove that the mapping $f(t,x) : K \to S^1(\mathbf{R})$ is continuous. Let $t \in [-A, A+1]$, $x_n \to x_0$, $x_n \in K$. Conditions $(\mathcal{A}.1)$ – $(\mathcal{A}.2)$ and Lebesgue theorem on bounded convergence imply that $f(t,x_n) \to f(t,x_0)$ in $L[-A,A+1]$. Fix $\epsilon > 0$. Because f is almost periodic, there exists such a segment $[-A,A]$ that, for any t, one can find $\epsilon/3$-almost period of the function, τ, such that $t - \tau \in [-A,A]$. Consider the following inequalities, which hold for all t and x_n,

$$\int_t^{t+1} |f(u,x_0) - f(u,x_n)|du \leq \int_0^1 |f(t+u,x_0) - f(t-\tau+u,x_0)|du +$$
$$+ \int_0^1 |f(t-\tau+u,x_n) - f(t+u,x_n)|du + \int_0^1 |f(t-\tau+u,x_0) -$$
$$-f(t-\tau+u,x_n)|du \leq \epsilon/3 + \epsilon/3 + \int_{-A}^{A+1} |f(u,x_0) - f(u,x_n)|du,$$

Because the last term is less then $\epsilon/3$ for large n, the function $f(t,x) : K \to S^1(\mathbf{R})$ is continuous. Further,

$$\int_s^{s+1} |f(u+\delta,y) - f(u,y)|du \;\leq\; 2|f(u,y) - f(u,x)|_s +$$
$$+ \int_s^{s+1} |f(u+\delta,x) - f(u,x)|du$$

and, because the almost periodic function $f : \mathbf{R} \to L[0,1]$ is uniformly continuous, for all $\epsilon > 0$ and $x \in K$ there exists $\delta_1 > 0$ such that the last term will be less than $\epsilon/3$ if $|\delta| < \delta_1$. So there exist $\delta(x,\epsilon) > 0$ and such a neighborhood of the point x, $O(x,\epsilon)$, that for all $y \in O(x,\epsilon)$, $|\delta| < \delta(x,\epsilon)$, we will have that $|f(u+\delta,y) - f(u,y)|_S < \epsilon$. But then,

$$|F(s+\delta) - F(s)|_\Pi = \sup_{x\in K} \int_s^{s+1} |f(u+\delta,x) - f(u,x)|du < \epsilon$$
$$\forall |\delta| < \min\{\delta(x_i,\epsilon),\, i = \overline{1,q}\},$$

where $\{O(x_i,\epsilon), i = \overline{1,q}\} \subset \{O(x,\epsilon), x \in K\}$ is a finite subcover of K.

So, the function $F(s)$ is almost periodic and hence, by Bochner theorem [75], the set of function $\{F^h = F(s+h), h \in \mathbf{R}\}$ is precompact in the space $C(\mathbf{R},\Pi)$ equipped with the metric $m(G,F) = \sup_{s\in\mathbf{R}} d(G(s),F(s))$. Now notice that if $F^{h_n} \to G \in C(\mathbf{R},\Pi),$, $G(s) = g(u,x,s)$, the equality $F^{h_n}(s) = f(u+s+h_n,x)$ implies that there is a function $g \in Q_1$ such that $g(u,x,s) = g(u+s,x)$.

Because $m(G,F) = \rho(g,f)$, the lemma is proved.

□

Definition of almost periodicity of the discrete measure $h(a) = \sum h_i(a)\delta(t-t_i)$, viewed as an operator with the values in some linear topological space, and conditions for this measure to be almost periodic uniform in $a \in A$.

We will suppose that $h_i(a) \in C(A)$ for all i and A is a metric compact space. Let $(W, |\cdot|_n)$ be a Banach space. We will also equip W with the topology τ, generated by a countable set of seminorms $|\cdot|_i$ which separate points, and assume that the norm topology majorizes τ. It is known that τ is metrizable with the metric

$$\rho(\nu_1,\nu_2) = \sum_{i\geq 1} 2^{-i} \min\{1, |\nu_1 - \nu_2|_i\}.$$

In addition, we assume that the balls

$$F_q(0) = \{x : |x|_n \leq q\}$$

are complete in this metric.

Let $\mathcal{D}^m_{L_1}$ be the space of $m-1$ times continuously differentiable functions $\varphi : \mathbf{R} \to \mathbf{R}$, the $(m-1)^{\text{th}}$ derivative of which is absolutely continuous, and $D^j\varphi(t) \in L_1(\mathbf{R})$, $0 \leq j \leq m$.

The space $\mathcal{D}^m_{L_1}$ with the norm $|\varphi|_m = \max_{0\leq j\leq m} \int_{-\infty}^{+\infty} |D^j\varphi(t)|dt$ is a Banach space (see, for example, [55]).

We denote by $\mathcal{L}_m(W) = \mathcal{L}(\mathcal{D}^m_{L_1}, W)$ the set of all continuous linear operators $h : \mathcal{D}^m_{L_1} \to W$ (for W equipped with the norm topology). It is clear that $\mathcal{L}_m \subset \mathcal{L}_k$ for $m < k$. In $\mathcal{L}_m(W)$, consider the operator norm $|\cdot|_{\mathcal{L}_m}$ and the metric

$$\sigma(h_1, h_2) = \sup_{|\varphi|_m=1} \rho(h_1\varphi, h_2\varphi).^{10}$$

This metric is invariant with respect to the translation $(\tau_s h).\varphi = h.\tau_{-s}\varphi$ and the subspace $(\mathcal{L}^\alpha_m(W) = \{h : |h|_{\mathcal{L}_m} \leq \alpha\}, \sigma)$ is complete[11] and $\tau_s\mathcal{L}^\alpha_m(W) = \mathcal{L}^\alpha_m(W)$ for all s.

Definition 13 *We call a distribution $g \in \mathcal{L}^\alpha_m(W)$ m-almost periodic if the function $\tau_s g : \mathbf{R} \to \mathcal{L}^\alpha_m(W)$ is Bohr almost periodic.*

It is clear that the subspace $AP^\alpha_m \subset \mathcal{L}^\alpha_m(W)$ of all m-almost periodic distributions is closed and the hull $H(h)$ (the closure of $\{\tau_s h, s \in \mathbf{R}\}$ in $(\mathcal{L}^\alpha_m(W), \sigma)$) belongs to AP^α_m.

Definition 14 *A family of distributions $h(a) \in \mathcal{L}^\alpha_m(W)$, $a \in A$, will be called almost periodic uniformly with respect to a if the family of functions $\tau_s h(a) : \mathbf{R} \to \mathcal{L}^\alpha_m(W)$ is almost periodic uniformly with respect to a.*

In the sequel, a distribution $h(a)$ is assumed to be almost periodic uniformly with respect to a. If $h = \sum_i h_i\delta(t-t_i)$, we set $h.\varphi = h\varphi = \sum_i h_i\varphi(t_i)$.

Lemma 62 *Let*

$$\sup_{t\in R}\sup_{a\in A} \sum_{t_i\in[t,t+1]} |h_i(a)|_W \leq C \tag{A.40}$$

[10] Because τ is majorized by the norm topology of W, i.e. $\forall i\ \exists C_i > 0$: $|x|_i \leq C_i|x|_W$, the topology defined in $\mathcal{L}_m(W)$ by σ is majorized by the norm topology of $\mathcal{L}_m(W)$.

[11] Let $\sup_{|\varphi|_m=1} \rho(h_n\varphi, h_m\varphi) \to 0$. Because the subsequence $h_n.\varphi$ is bounded, $|h_n.\varphi|_W \leq \alpha|\varphi|_m = \alpha$, and, by the assumption, the space $(U_\alpha(0), \rho)$ is complete, we see that $h_n\varphi \to v \in U_\alpha(0)$. Define $h\varphi = v$, $\varphi \in S^1$, and extend h linearly to $\mathcal{D}^m_{L_1}$. Because

$$|h.\varphi|_w = |\nu|_w \leq \alpha \qquad \forall\varphi \in S^1,$$

we see that $h \in \mathcal{L}^\alpha_m(W)$.

for some positive C. Then $h(a) \in \mathcal{L}_1^{2C}(W) \subset \mathcal{L}_2^{2C}(W)$, and the distribution $h(a)$ will be uniformly 2-almost periodic if and only if the vector valued function

$$h(t,a) = h(a) * \varphi = \sum_i h_i(a)\varphi(t - t_i) : R \to W$$

is almost periodic uniformly with respect to each of the seminorms $|\cdot|_i$ for an arbitrary function $\varphi(t)$ from $\mathcal{D}^2_{L_1}$ [12].

Proof. *1)* If $\varphi(t) \in \mathcal{D}^2_{L_1}$, $j = 0, 1$, $k \in \mathbf{R}$, $t \in A_k = [k, k+1]$, the following inequality holds:

$$|D^j\varphi(t)| \leq \int_k^{k+1} (|D^j\varphi(t)| + |D^{j+1}\varphi(t)|)dt. \tag{A.41}$$

Thus

$$|h.\varphi|_W = |\sum_i h_i\varphi(t_i)|_w \leq \sum_k \sum_{t_i \in A_k} |h_i|_W \max_{A_k} |\varphi(s)| \leq 2C|\varphi|_1.$$

2) Let $\psi(t) \in \mathcal{D}^m_{L_1}$, $m \geq 2$. If $j = 1$, by using (A.41), one can prove similarly to *1)* that the function $h(t,a)$ is continuous uniformly on $\mathbf{R}$ with respect to the norm (we set $t, s \in [m, m+1]$, $B_k = [m-k-1, m-k]$, $C_k = [m-k-1, m-k+1]$):

$$\begin{aligned}
|h(t,a) - h(s,a)|_\omega &\leq \sum_i |h_i(a)|_\omega |\varphi(t - t_i) - \varphi(s - t_i)| \leq \\
&\leq \sum_i |h_i(a)|_\omega |\varphi'(\theta_i - t_i)|\, |t - s| \leq \\
&\leq |t - s| \sum_k \sum_{t_i \in A_k} |h_i(a)|_\omega \max_{C_k} |\varphi'(u)| \leq \\
&\leq |t - s| C \sum_k (\max_{B_k} |\varphi'(u)| + \max_{B_{k+1}} |\varphi'(u)|) \leq 4C|\varphi|_2\, |t - s|.
\end{aligned}$$

[12] i.e. $h(t)$ is continuous with respect to the seminorm $|\cdot|_p$, and for an arbitrary $\epsilon > 0$ and $p \in \mathbf{N}$ there exists a relatively dense subset Ω^p_ϵ that, for all $\tau \in \Omega^p_\epsilon$,

$$|h(t+\tau, a) - h(t,a)|_p < \epsilon \qquad \forall t \in R,\ a \in A.$$

For $\epsilon \in (0, |\psi|_m)$ and fixed p, find ω, $2^{-p}\epsilon|\psi|_m^{-1}$-almost period of the function $\tau_s h(a)$. Because

$$\begin{aligned}
2^{-p}\epsilon|\psi|_m^{-1} &> \sigma(\tau_{s+\omega}h(a), \tau_s h(a)) = \sigma(\tau_\omega h(a), h(a)) = \\
&= \sup_{|\varphi|_m=1} \rho((\tau_\omega h(a)).\varphi, h(a).\varphi) = \\
&= \sup_{|\varphi|_m=1} \sum_{i\geq 1} 2^{-i}\min\{1, |(\tau_\omega h(a)).\varphi - h(a).\varphi|_i\} \geq \\
&\geq 2^{-p} \sup_{|\varphi|_m=1} |(\tau_\omega h(a)).\varphi - h(a).\varphi(s)|_p = \\
&= 2^{-p} \sup_{|\varphi|_m=1} |(\tau_\omega h(a)).\varphi(z-s) - h(a).\varphi(z-s)|_p \geq \\
&\geq 2^{-p}|\psi|_m^{-1} \sup_{z\in R} |(\tau_\omega h(a)).\psi(z-s) - h(a).\psi(z-s)|_p = \\
&= 2^{-p}|\psi|_m^{-1} \sup_{z\in R} |h(z-\omega, a) - h(z,a)|_p,
\end{aligned}$$

ω will be ϵ-almost period of the function $h(t,a)$ with respect to the seminorm $|\cdot|_p$.

3) Now let $\{e_j\}$ be a δ-like sequence of positive functions from $\mathcal{D}^2_{L_1}$ with the support in $[-1/j, 1/j]$. Any continuous function $a(t)$, which if uniformly bounded with respect to the norm $|\cdot|_w$, is an element of $\mathcal{L}_1(W)$ which acts by $a.\varphi = \int_{-\infty}^{+\infty} a(t)\varphi(t)dt$, and so the uniformly almost periodic functions $h_j(t,a) = h(a) * e_j$ belong to $\mathcal{L}_1(W)$ for all j. Moreover, it is easy to see that they will be uniformly 2-almost periodic. Denote the interval $[t_i - 1/j, t_i + 1/j]$ by $s_{i,j}$. Then

$$\begin{aligned}
&|h_j(t,a) - h(a)|_{\mathcal{L}_2} = \\
&= \sup_{|\varphi|_2=1} \left| \sum_i h_i(a)\varphi(t_i) - \int_{-\infty}^{+\infty} \sum_i h_i(a)e_j(u-t_i)\varphi(u)du \right|_\omega = \\
&= \sup_{|\varphi|_2=1} \left| \sum_i \int_{-\infty}^{+\infty} (\varphi(t_i) - \varphi(u))h_i(a)e_j(u-t_i)\varphi(u)du \right|_\omega \leq \\
&\leq \sup_{|\varphi|_2=1} \sum_i |h_i(a)|_\omega \left| \int_{-\infty}^{+\infty} |\varphi(t_i) - \varphi(u)|e_j(u-t_i)\varphi(u)du \right|_\omega \leq \\
&\leq \sup_{|\varphi|_2=1} \sum_i |h_i(a)|_\omega \int_{s_{i,j}} j^{-1} \max_{s_{i,j}} |\varphi'(\theta)| e_j(u-t_i)du = \\
&= \sup_{|\varphi|_2=1} j^{-1} \sum_i |h_i(a)|_\omega \max_{s_{i,j}} |\varphi'(\theta)| \leq
\end{aligned}$$

$$\leq \sup_{|\varphi|_2=1} j^{-1} \sum_p \sum_{t_i \in A_p} |h_i(a)|_\omega \max_{[p-1,p+2]} |\varphi'(\theta)| \leq$$

$$\leq \sup_{|\varphi|_2=1} Cj^{-1} \sum_p \left\{ \max_{A_{p-1}} |\varphi'(\theta)| + \max_{A_p} |\varphi'(\theta)| + \max_{A_{p+1}} |\varphi'(\theta)| \right\} \leq \frac{6C}{j}.$$

Whence it follows that the sequence of operators $h_j(t,a)$ from $\mathcal{L}_2$ is bounded by the number $1 + 2C$ uniformly with respect to $j \geq 2C$ and a. Thus $h(a) \in AP_2^{2C+1}$.

□

Consider the space $F(K, \mathcal{L}_2^\alpha)$ of all uniformly almost periodic distributions $g(a) : K \to \mathcal{L}_2^\alpha$ endowed with the metric $\sup_{a \in K} \sigma(g(a), h(a))$, with respect to which the space is complete, and the set $[\{\tau_s h(a), s \in \mathbf{R}\}] = H(h) \subset F(K, \mathcal{L}_2^\alpha)$.

Lemma 63 *Assume that (A.40) holds and $h(a)$ is uniformly almost periodic. The function $\tau_s h$ is continuous on $H(h) \times \mathbf{R}$ and hence $(H(h), \tau_s h)$ is a compact minimal isometric flow.*

Proof. First, let us make an estimate for $\lambda = |h(a) - \tau_{\Delta s} h(a)|_{\mathcal{L}_2}$:

$$\lambda = \sup_{|\varphi|_2=1} |h(a).\varphi - \tau_{\Delta s} h(a).\varphi| = \sup_{|\varphi|_2=1} |h(a).(\varphi - \tau_{-\Delta s}\varphi)| \leq$$

$$\leq |h(a)|_{\mathcal{L}_1} \sup_{|\varphi|_2=1} |\varphi - \tau_{-\Delta s}\varphi|_1 \leq$$

$$\leq 2C \sup_{|\varphi|_2=1} \max_{j=0,1} \int_{-\infty}^{+\infty} |\varphi^j(u - \Delta s) - \varphi^j(u)| du.$$

Because

$$\int_{-\infty}^{+\infty} |\varphi(u - \Delta s) - \varphi(u)| du \leq$$

$$\leq \int_{-\infty}^{+\infty} \left| \int_{u-\Delta s}^{u} \varphi'(\theta) d\theta \right| du \leq$$

$$\leq \int_{-\infty}^{+\infty} \int_{u-\Delta s}^{u} |\varphi'(\theta)| d\theta du = \int_{-\infty}^{+\infty} d\theta \int_{u}^{u+\Delta s} |\varphi'(\theta)| du =$$

$$= \Delta s \int_{-\infty}^{+\infty} |\varphi'(\theta)| d\theta \leq \Delta s |\varphi|_1 \leq \Delta s |\varphi|_2,$$

we see that

$$\int_{-\infty}^{+\infty} |\varphi'(u - \Delta s) - \varphi'(u)| du \leq \Delta s |\varphi|_2.$$

So $\lambda \le \Delta s 2C$ and Lemma 62 follows from

$$\begin{aligned}\sigma(\tau_s h(a), \tau_{s_1} h_1(a)) &\le \sigma(\tau_{s_1} h(a), \tau_{s_1} h_1(a)) + \sigma(\tau_s h(a), \tau_{s_1} h_1(a)) = \\ &= \sigma(h(a), h_1(a)) + \sigma(\tau_{\Delta s} h(a), h(a)),\end{aligned}$$

where $\Delta s = s - s_1$.

□

Examples

System (I) is uniquely determined by a vector $F = (d, B, h)$. To point out the dependence of (I) on F, we will write $(\mathrm{I})_F$. There is a continuous isometric flow τ_s defined on $H(d) \times H(B) \times H(h)$, and $H(F) = [\{\tau_s F, s \in \mathbf{R}\}] \subset H(d) \times H(B) \times H(h)$ is a compact minimal set. It was assumed in the preceding considerations that, for $i \neq j$, $t_i \neq t_j$. But when studying a limit of a sequence of impulsive systems (even if the times of impulsive effects are strictly increasing in each of the system), we can obtain in the limit system that $t_i = t_j$ for some $i \neq j$ because, when passing to the limit, strict inequalities are not preserved. In this case, assuming that T does not have limit points, we will do the following: let the support of the set $T = \{t_i\}$, $s(T) = \{\tau_i\}$, be a strictly increasing sequence of all the real numbers that belong to T. For all $\tau_q \in s(T)$ there exists a finite set of indexes, $I(q) = \{i_1 < \ldots < i_r\}$, such that $t_{i_1} = \ldots = t_{i_r} = \tau_q$ and, if $i \notin I(q)$, then $t_i \neq \tau_q$. Let $H_i(x) = x + h_i(x)$ and $\bar{H}_q = H_{i_r} \cdot \ldots \cdot H_{i_1}$, $\bar{h}_q = \bar{H}_q - x$. Then (I) is a formal notation for an ordinary impulsive system

$$\frac{dx}{dt} = f(t, x), \quad \Delta x|_{\tau_q} = \bar{h}_q(x), \quad i \in \mathbf{Z}.$$

As it can be seen, the order of indexing the points of T is important because, in the general case, the mappings H_i do not commute.

Let us note that, even with these reservations, the meaning of the notation $(\mathrm{I})_G$ is not always defined if $G \in H(F) \backslash F$. We will discuss in the sequel when the notation $(\mathrm{I})_G$ is correct.[13]

[13] In this case, the structure of $G \in H(F)$ can be made more detailed. Instead of G, one can consider the vector $G_1 = (g(t, x), \{g_i(x)\}, \{p_i\})$. We make correspond the impulsive system to G_1.

Proposition 3 *Suppose*

$$(V)\quad \begin{cases} (V1) & t_n = an + c_n, \quad \text{where } a \neq 0 \text{ and } \{c_n\} \text{ is an almost periodic sequence;} \\ (V2) & \text{the sequences } \{h_i(a)\} \text{ are Bohr almost periodic uniformly with respect to each } |\cdot|_i \\ & \text{and } \sup_i \sup_{a\in K} |h_i(a)|_w \leq C. \end{cases}$$

Then the family $h(a) = \sum_i h_i(a)\delta(t - t_i) \in AP_2(W)$ *is almost periodic uniformly with respect to* $a \in K$.

Proof. This proposition follows from Lemma 62.

□

Lemma 64 *Assume that (V) holds. If* $g(a) \in H(h(a))$, *then we have that* $g(a) = \sum_i g_i(a)\delta(t - p_i)$ *for some* $\{g_i(a)\}$, $\{p_i\}$ *which have property (V). The convergence* $\tau_{\mu_n} h(a) \to g(a)$ *in* $H(h(a))$ *implies that there exists such a sequence* $\{\alpha(n), n \geq 0\}$ *that*

$$\lim_{n\to\infty} h_{i+\alpha(n)}(a) \stackrel{\rho}{=} g_i(a), \quad \lim_{n\to\infty} (t_{i+\alpha(n)} - \mu_i) = p_i$$

and the convergence is uniform with respect to the integers i *and* $a \in K$.

Proof. Let $\tau_{\mu_n} h(a) \to g(a)$ in $H(h(a)$. By Theorem 124, there exist a sequence $\{p_i\}$ satisfying property $(V1)$ and a sequence $\{\alpha(n)\}_{n\geq 0}$ such that $\lim_{n\to\infty} \sup_i |t_{i+\alpha(n)} - \mu_n - p_i| = 0$. From the sequence $\{h_{i+\alpha(n)}(a)\} \in l_\infty(W)$, choose a subsequence $\{h_{i+\alpha(n_k)}(a)\}$ that converges to $\{q_i(a)\}$ uniformly in $l_\infty(W)$ (the sequence $\{q_i(a)\}$ is also uniformly almost periodic). A direct verification shows that $\tau_{\nu_k} h(a) \to q(a) = \sum_i q_i(a)\delta(t - p_i)$ in $H(h(a))$, where $\nu_k = \mu_{n_k}$. As a result, $q(a) = g(a)$ and the lemma is proved.

□

Remark. If in Lemma 64, we assume that

$$t_{i+1} - t_i \geq 0 \qquad \forall i \in \mathbf{Z}, \tag{V3}$$

then $p_{i+1} - p_i \geq 0$ for all $i \in \mathbf{Z}$.

We adopt the notation $T(B, g) = supp\, g \cap (-B, B)$.

Suppose that, for a uniformly almost periodic measure h, (A.40) holds and also

$$t_{i+1} - t_i \geq \delta > 0 \qquad \forall i \in \mathbf{Z}. \tag{A.42}$$

It can be shown that in this case, it is not necessary to require that $(V1)$ hold.

Lemma 65 *If $g(a) \in H(h(A))$, then $g(a) = \sum g_i(a)\delta(t-s_i)$ with $s_{i+1} - s_i \geq \delta$. Let $h^n(a) \to g(a)$ in $H(h(a))$, $A > 0$, and $T(A, g(a)) = \{p_1, \ldots, p_N\}$. Then the set $T(A, h^k(a))$ can be indexed in such a way, $\{p_1^k, \ldots, p_N^k, \ldots, p_M^k\}$, that $\lim_{k\to\infty} p_i^k = p_i$, and the convergence $h_i^k(a) \xrightarrow{\rho} g_i(a)$ for all i, $1 \leq i \leq N$ and $h_j^k(a) \xrightarrow{\rho} 0$ as $k \to \infty$, $j > N$, is uniform in $a \in K$ (if the sequence $h_j^k(a)$ is finite for some j, we continue it with 0).*

Proof. Let $\{s_1, s_2\} \in supp\, g(a)$ and $0 < s_2 - s_1 < \delta$. Then we can find such functions $\varphi_i(t)$ and $\epsilon > 0$ that $supp\, \varphi_i(t) \subset T_i = [s_i - \epsilon, s_i + \epsilon]$, $s_2 - s_1 + 2\epsilon < \delta$, $T_1 \cap T_2 = \emptyset$, and $g(a).\varphi_i \neq 0$, $i = 1, 2$. Because $g(a) \in H(h(a))$, we have that, for some sequence μ_n, $\tau_{\mu_n} h(a) = h^n(a) \to g(a)$ in $(\mathcal{L}_2^{2C}(W), \sigma)$ uniformly with respect to $a \in K$. But then also $h^n(a).\varphi_i \neq 0$ for all $n \geq n_0$, $i = 1, 2$. Consequently $T_i \cap supp\, h^n(a) \neq \emptyset$ for all $n \geq n_0$, which is a contradiction to (A.42). Further, for all $\epsilon \in (0, \delta/2)$, there exists $k_0 \in \mathbf{N}$ such that for all $n \geq k_0$ and $p_i \in T(A, g(a))$ there exists in the neighborhood $O_\epsilon(p_i)$ a unique point p_i^n from $T(A, h^n(a))$ because otherwise, for ψ having support in $O_\epsilon(p_i)$, $\psi(s) \equiv 1$ in $O_{\epsilon/2}(p_i)$, $|\psi|_2| = 1$, we would have

$$0 \neq \rho(g(a).\psi, 0) \leq \rho(g(a).\psi, h^n(a).\psi) \leq \sigma(g(a), h^n(a)) \to 0.$$

Finally we have $p_i^n \to p_i$ for $i \leq N$ and thus $h_i^n(a) = h^n(a)\psi \xrightarrow{\rho} g(a).\psi = g_i(a)$. Moreover, the assumption $\lim_{k\to\infty} h_{N+1}^k(a) \neq 0$ contradicts $p_{N+1} \notin T(A, g(a))$.

□

It should be noted that the classes of almost periodic distributions which were considered are not closed under addition.

Examples (continued).

Let X be a reflexive separable Banach space, $\{f_i\}$ – an everywhere dense subset of X^*. The scheme described above can be applied to the distribution

$$h = \sum_j h_j \delta(t - t_j), \qquad h_j \in W$$

if we take the pair $(W, |\cdot|_i)$ to be one of the following:

a) $W = X$, $|h_j|_i = |\langle h_j, f_i\rangle|$;

b) $W = \mathcal{L}(X^*)$, $|h_j|_i = |h_j f_i|_{X^*}$;

c) W is the space $\mathcal{L}(X)$ equipped with the uniform topology;

d) W is the space X with the norm topology.

Definition 15 *A distribution will be called weakly (w-), strongly (s-), or uniformly (u-) almost periodic if the space W is chosen to be a) or b), or c), d) correspondingly.*

Lemma 62 implies the following

Corollary 11 *Suppose that (A.40) holds. For the distribution h to be weakly (strongly) 2-almost periodic it is necessary and sufficient that the functions $a(t) = \sum_i h_i f \varphi(t - t_i)$ be (strongly) almost periodic for arbitrary functions $\varphi(t) \in \mathcal{D}^2_{L_1}$ with compact support and for all $f \in X^*$. The distribution h is uniformly 2-almost periodic if and only if, for all $\varphi(t) \in \mathcal{D}^2_{L_1}$ with compact support, the function $a_1(t) = \sum_i h_i \varphi(t - t_i)$ is almost periodic with respect to the norm of the space W.*

A.7 Almost periodic solutions of impulsive ordinary differential equations

In this section we will find out when the impulsive system (I), defined in the domain $\mathcal{F} = \mathbf{R} \times \Omega \subset \mathbf{R}^{n+1}$, has almost periodic or Levitan almost periodic solutions in this domain[14]. We assume that condition $(\mathcal{A})$ given in Section A.6 and (A.40) hold ($|x|_W$ is one of norms on $\mathbf{R}^n$), and the measure

$$h(x) = \sum_{i \in Z} h_i(x)\delta(t - t_i)$$

is assumed to be almost periodic uniformly on compact sets.

Naturally, these conditions are not necessary for an impulsive system to have S-almost periodic solution. Let us consider the S-almost periodic function, shown in the figure below.

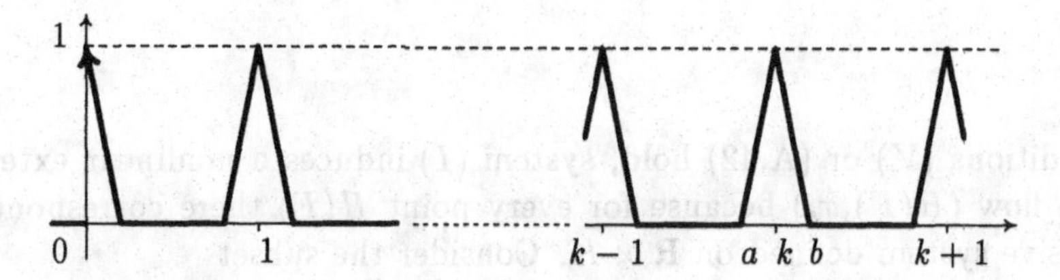

[14]Consider an example of an impulsive system that has an unbounded S-almost periodic solution $x(t)$. Let $\{t_i\} = \{i\} \cup \{i + c_i^2\}$, where $c_n > 0$ is such an almost periodic sequence that $c_n < 0.25$ and $\inf_n c_n = 0$. We set $x(t) = c_n^{-1}$ on the intervals $(n, n + c_n^2]$ and $x(t) \equiv 0$ on their complements. Then, by the construction, $x(t)$ is a solution of the impulsive system

$$\frac{dx}{dt} = 0, \quad \Delta x|_n = c^{-1}, \quad \Delta x|_{n+c_n^2} = -c_n^{-1}.$$

If an integer n is ϵ-almost period of the sequence c_k, then

$$\sup_{t \in R} \int_t^{t+1} |x(u+n) - x(u)| du = \sup_k \big\{ |c_k^2 - c_{k+n}^2| \min(c_k^{-1}, c_{k+n}^{-1}) +$$
$$+ \min(c_k^2, c_{k+n}^2 |c_k^{-1} - c_{k+n}^{-1}| \big\} \leq$$
$$\leq \sup_k \big\{ 2\epsilon \max(c_k, c_{k+n}) \min(c_k^{-1}, c_{k+n}^{-1}) +$$
$$+ \min(c_k^2, c_{k+n}^2) |c_k - c_{k+n}| (c_k c_{k+n})^{-1} \big\} \leq 2\epsilon + \epsilon = 3\epsilon.$$

Because $x(t)$ is S-almost periodic, the measure h, which defines the right-hand side of the system, is 2-almost periodic but condition (A.40) from Section A.6 does not hold. Moreover, for the sequence $T = \{t_i\}$, indexed in the increasing order, condition $(V1)$ holds and $(x(t), T) \in APH(\mathbf{R})$.

The function $x(t)$ has the saw-tooth shape. Each tooth is located over an integer on the x-axis and has the height equal to 1. There is one jump discontinuity at 0 and the length of the segments ak and kb equal to $c_k = 0.25\sin^2 k$ and 0.25 respectively. If n is ϵ-almost period of $\{c_k\}$, then n is also $\epsilon/2$-almost period of the function $x(t)$:

$$\sup_{t\in R}\int_t^{t+1}|x(u+n)-x(u)|du = 0,5\sup_k|c_k - c_{k+n}| \leq \epsilon/2$$

It remains only to notice that the generalized derivative of the S-almost periodic function $\big(x(t)\big)$ is the 2-almost periodic function $\big(h(t) = f(t)+\delta(t)\big)$ and hence the impulsive system

$$\frac{dx}{dt} = f(t); \qquad \Delta x|_0 = 1$$

has S-almost periodic solution.

If conditions (V) or (A.42) hold, system (I) induces a nonlinear extension of the flow $(H(F), \tau_s)$ because for every point $H(F)$ there corresponds an impulsive system defined on $\mathbf{R}\times K$. Consider the subset

$$X = \{(F,\bar{x}) \in X \Leftrightarrow \bar{x}(t) = x(t)\}$$

of $H(F)\times L[0,1]$, where $\bar{x}(t)$ denotes the restriction of the solution $x(t) : \mathbf{R}\to\mathbf{R}$ of system $(I)_G$ to $[0,1]$.

Lemma 66 *The set X is compact. If we assume that, for any two bounded solutions of system $(I)_G$, $G \in H(F)$, $x(t)$, $y(t)$, the equality $x(q) = y(q)$ at some point $q \in \mathbf{R}$ implies that $x(t) = y(t)$ for all $t \geq q$, then there is a continuous semiflow*

$$S^\mu : S^\mu(F, x(t) : [0,1] \to K) = (\tau_\mu F, x(t) : [\mu,\mu+1] \to K), \qquad \mu \geq 0$$

defined on X.

It should be noted that S^t is an extension of $(H(F), \tau_t)$ defined by the projection $pr : X \to H(F)$.

Proof. Let $X \neq \emptyset$, $(F_n, x_n) \in X$. Because $H(F)$ is compact, we can assume that $F_n \to F$. Fix an arbitrary positive A such that $\{A, -A\} \notin$

supp h. By using properties of the measure $h_n(x)$, we get $x_{n_k}(t) \to x(t)$ in $L[-A, A]$, where $x(t)$ is a solution of $(I)_F$. Because A can be arbitrarily large, the function $x(t)$ can be extended on $\mathbf{R}$ and hence X is compact.

Let now $\mu_n \to \mu$, $F_n \to F$ in $H(F)$, $\bar{x}_n \to \bar{x}$ in $L[0,1]$. Then, by continuity of the flow τ_s, we have that $\tau_{\mu_n} F_n \to \tau_\mu F$. If $\bar{x}_n \to \bar{x}$ in $L[0,1]$, then also $x_n \to x$ in $L[0, A]$ for all $A > 0$.

□

For the measures (systems) of the considered type, we have the following theorem.

Theorem 129 (Zhikov) *Let all solutions of system $(I)_G$, $\{x(t,\alpha,G), t \in \mathbf{R}, \alpha \in A\}$, defined in the domain $K \times \mathbf{R}$, be mutually separated:*

$$\inf_{t \in R} \int_t^{t+1} |x(u,\alpha_1,G) - x(u,\alpha_2,G)| du > 0 \qquad \forall \alpha_1 \neq \alpha_2 \ \forall G \in H(F)$$

and the set $M = \{x(s,\alpha,G), \alpha \in A\} \subset \mathbf{R}^n$ be zero dimensional for some $G \in H(F)$ and $s \in \mathbf{R}$. Then bounded solutions of each system $(I)_G$, $G \in H(F)$, will be S-almost periodic (W-almost periodic).

Proof. Conditions of the theorem imply that the fiber $X_1 = pr^{-1}(\tau_s G) \subset L[0,1]$ is zero dimensional because the mapping $j : X_1 \to M$, $j(x(s,\alpha)) = x(s+t,\alpha)|_{t \in [0,1]} \in L[0,1]$, is a homeomorphism. By using Lemma 66 and Zhikov theorem (see the subsequent Bibliographical notes), we see that this theorem holds true.

□

Theorem 129 implies the theorems of Favard and Amerio for these special classes of almost periodic impulsive systems: in the first class, the set of indices A consists of a single element, in the second class – of a finite number of elements. Moreover, by Lemma 66, Theorem 138 in Bibliographical notes takes the following particular form.

Corollary 12 *Let $x(t)$ be a bounded uniformly Lyapunov stable almost periodic solution of system (I). Let, moreover, it be asymptotically stable. Then this solution is S-almost periodic.*

Theorem 130 (Levitan) *Let almost periodic system (I) have only one bounded solution in the domain $K \times \mathbf{R}$. Then this solution is S-N-almost periodic.*

A proof of this theorem follows directly from the abstract Levitan theorem (see Bibliographical notes).

Example.

For $(t, x) \in \mathbf{R}^2$, consider the following system

$$\frac{dx}{dt} = 0, \quad \Delta x|_{t_i = i} = -x/2 + 1, \quad \Delta x|_{\tau_i = i + c_i} = -x/2, \quad c_i = \cos(i)/4.$$

It is easy to check that this almost periodic linear impulsive system is exponentially dichotomic and has a unique bounded solution $x(t)$, $0 < x(t) < 2$. Moreover, the set $T = \{t_i, \tau_i\}$ is strongly almost periodic. But this system does not have S-almost periodic solution. If such a solution $z(t)$ existed, the properties of the system would imply that it must be bounded and hence it would coincide with $x(t)$. The continuous function $\zeta(t) = 2\int_0^{0.5} z(t+u)du$ is Bohr almost periodic and thus the sequence $\zeta_n = \zeta(n + 0.25) = z(n + 0.25+)$ is also almost periodic. But then, for $\epsilon \in (0, 0.1]$, one can find such an everywhere dense set $\Sigma_\epsilon \subset \mathbf{Z}\backslash\{0\}$ that, for all $p \in \Sigma_\epsilon$, $|\zeta_{n+p} - \zeta_n| < \epsilon$ for all $n \in \mathbf{Z}$. Let us fix $p \in \Sigma_\epsilon$ and choose $m \in \mathbf{Z}$ such that $\cos m > 0$ and $\cos(m + p) < 0$. We will have

$$\begin{aligned}
\zeta_m &= z(m + 0.25+) = \\
&= 0.5\, z(m + c_m) = 0.5\, z(m+) = 0.5\,(z(m)\, 0.5 + 1) = \\
&= 0.5(z(m + c_{m-1} - 1+)0.5 + 1) = \zeta_{m-1}/4 + 0.5; \\
\zeta_{m+p} &= z(m + p+) = 0.5\, z(m + p) + 1 = \\
&= 0.5\,(z(m + p + c_{m+p})0.5) + 1 = \\
&= 0.5\,(z(m + p - 3/4)0.5) + 1 = 0.25\, \zeta_{m+p-1} + 1.
\end{aligned}$$

As a result, we have obtain a contradiction because

$$\begin{aligned}
1/2 &= (\zeta_{m+p} - \zeta_m) - 1/4(\zeta_{m+p-1} - \zeta_{m-1}) \le \\
&\le |\zeta_{m+p} - \zeta_m| + 1/4|\zeta_{m+p-1} - \zeta_{m-1}| \le 0.125.
\end{aligned}$$

The subsequent Theorem 131 shows that the bounded solution in this example will be Levitan S-N-almost periodic.

In this paragraph, we admit the support T of the measure μ to have a nonempty limit set. Suppose also that

$$\sup_{t \in R} \sum_{t_i \in [t, t+1]} \sup_{x \in K} |h_i(x)| = \nu < \infty.$$

As we saw, the structure of the measure $h(x)$ is of crucial importance regarding the properties of solutions of system (I). Because of this, we make the following

Definition 16 *An almost periodic measure $h(x)$ is called simple if, for every $A > 0$ and $\tau_{s_k} h(x) \to h(x)$ in $H(h)$, there exists k_0 in* **N** *such that, for all $k \geq k_0$, the set $T(A, \tau_{s_k} h(x))$ can be indexed in such a way, $\{p_1^k, \dots, p_N^k, \dots, p_{m(k)}^k\}$, that we have $p_i^k \to p_i$ for all $i = \overline{1, N}$ and also* $\lim_{k\to\infty} \sum_{j>N}^{m(k)} \sup_{x\in K} |h_j^k(x)| = 0$, *where* $T(A, h) = \{p_1 < \dots < p_N\}$, $\tau_{s_k} h(x) = \sum_{i\in \mathbf{Z}} h_i^k(x)\delta(t - p_i^k)$, $h(x) = \sum_{i\in\mathbf{Z}} h_i(x)\delta(t - p_i)$.

It follows from this definition that $\lim_{k\to\infty} h_j^k(x) \overset{K}{=} h_j(x)$ for all $j \geq N$.

Examples

1. If (A.42) or $(V1)$ folds (and $t_i < t_{i+1}$ for all i), then $h(x)$ is simple.

2. The almost periodic measure $\sum_{i\neq 0}(\delta(t-i) + \delta(t - \pi i)) + 2\delta(t)$ is not simple.

3. Let $h = \sum_{k=0}^{+\infty} h_k(x) \sum_{n\in\mathbf{N}} \delta(t - 2^k n - 2k + 2^{-k-1})$, $0 < h_k(x) < 2^{-k}$ for all x. This measure is simple although the function $Card\{t_i \in [t, t+1]\}$ is unbounded on **R**.

We will be considering discrete, independent of x measures, the structure of which is more complicated:

$$h = \sum h_i \delta(t - t_i), \quad h_i \in R^n, \tag{A.43}$$

or the measures satisfying condition (S) given by

$$(S) \quad \left\{ \begin{array}{l} \text{Suppose that } h = \sum_{k\geq 1} h^k(x), \text{ where } h^k(x) \text{ are simple almost} \\ \text{periodic measures, } h^k(x) = \sum_{i\in Z} h_i^k(x)\delta(t - t_i^k). \text{ Let } t_i^k \neq t_s^j \\ \text{for all } k \neq j \text{ and } sup_{t\in R} \sum_{\substack{t_i^k \in [t,t+1] \\ k\geq d}} sup_{x\in K} |h_i^k(x)| = \mu(d) \to 0 \\ \text{for } d \to \infty. \end{array} \right.$$

Because $|h - \sum_{k\geq 1}^d h^k|_2 \leq \mu(d)$, $h(x)$ is a uniformly almost periodic measure.

Theorem 131 *Suppose (S) or (A.43) holds. Then, if system (I) has a unique bounded solution, then this solution will be Levitan S-N-almost periodic.*

Proof. Let $B \subset (\mathcal{L}_2)^N$ be the set of all such sequences that $\sum_k \frac{|h^k|_2}{k^2} < +\infty$. Because (S) implies that $|h^k|_2 \leq \mu(0)$, there is a vector $\eta \in B$ corresponding to the distribution $h = \sum_{k \geq 1} h^k(x)$. In the same way as before, we can introduce, for η and the hull $\bar{H}(\eta)$, the notion of almost periodicity (uniform in x from K). If condition (S) holds, the vector η will be almost periodic. We will also keep the notations F and $H(F)$ for the pair $(f(t,x),\eta)$ and the set $H(f(t,x),\eta)$ respectively.

Lemma 67 *Let conditions (S) (or (A.43)) and $\tau_{s_k}F = F^k \to F$ in $H(F)$ hold. Then, for any $\alpha < \beta$, $\alpha, \beta \notin T \cup_k (T - s_k)$, one can choose from the sequence of solutions $x_k(t) : A = (\alpha,\beta) \to K$ of system $(I)_{F^k}$ a subsequence converging on A to some solution of system $(I)_F$, $x(t) : A \to K$.*

Proof. (In the proof for subsequences, we will keep the notations of initial sequences, for example, we can assume $x_k(\alpha+) \to x$).

1) Let us denote $f(t+s_k,x) = f_k(t,x)$. Then, for all $t, \alpha \in A$,

$$\lim_k \int_a^t f_k(t,x_k(t))dt = \int_a^t f(t,x(t))dt. \tag{A.44}$$

Indeed, the sequences $\sup_A |x_k(t)|$ and $Var_\alpha^\beta x_k(t)$ are bounded by $diam\, K$ and $([\beta-\alpha]+1)(\nu + |\mu_{K,f}|_{S^1})$ correspondingly. Hence, by the second Helly theorem, there exists a subsequence that converges on A pointwise to a function $y(t)$ with bounded variation (not necessarily left continuous). Because $f_k(u,x) \overset{K}{\to} f(u,x)$ on A almost everywhere and condition $(\mathcal{A})$ from Section A.6 holds, we deduce that $f_k(u,x_k(u)) \to f(u,x(u))$ on A almost everywhere, $|f_k(u,x_k(u))| \leq \mu_{K,f}(u+s_k)$. Formula (A.44) holds because of Vitali theorem on passing to the limit under the integral sign [19].

2) Assume S. Let $\{p_1^n,\ldots,p_{N(n)}^n\} = supp\, h^n \cap (\alpha,\beta)$. Because $\tau_{s_k}\eta \to \eta$, we see that $\tau_{s_k}h^n \to h^n$ in $H(h^n)$. Thus, because h^n is simple, the set of points $(supp\, h^n - s_k) \cap A$ can be indexed in such a way, $\{p_1^{n,k},\ldots,p_{N(n)}^{n,k},\ldots,$-$p_{m(n,k)}^{n,k}\}$, that $p_i^{n,k} \to p_i^n$ for all $i \leq N(n)$ and $\lim_{k\to\infty} \sum_{j>N}^{m(n,k)} \sup_{x\in K} |h_j^{n,k}(x)|$-$= 0$. Moreover, passing to a subsequence for every n, $i \leq N(n)$, we can assume that $x_k(p_i^{n,k}) \to \rho_i^n$. Fix p_i^τ from A and a converging to 0 sequence ϵ_n such that $p_i^\tau - \epsilon_n$ are the points, at which $y(t)$ is continuous. Because $x_k(t)$ is a solution of the equation

$$x_k(t) = x_k + \int_\alpha^t f_k(t,x_k(t))dt + \sum_{\alpha < p_i^{n,k} < t} h_i^{n,k}(x_k(p_i^{n,k})), \tag{A.45}$$

we have that

$$|x_k(p_i^{r,k}) - x_k(p_i^r - \epsilon_n)| \leq \delta(|p_i^{r,k} - p_i^r| + \epsilon_n) + \sum \sup_{x\in K} |h_j^{m,k}(x)|,$$

where the summation is taken over all $p_j^{m,k} \in [p_i^r - \epsilon_n, p_i^{r,k})$ and the function $\delta(\epsilon)$, defined by $\mu_{K,f}(t)$ decreases monotonically to zero as $x \downarrow 0$. Since h^n is simple, the last inequality implies for $k \to \infty$ that

$$|\rho_i^r - y(p_i^r - \epsilon_n)| \leq \delta(\epsilon_n) + \sum \sup_{x\in K} |h_j^m(x)|,$$

with the summation taken over all p_j^m from $[p_i^r - \epsilon_n, p_i^r)$. Making n approach ∞, we find that $\rho_i^r = y(p_i^r-)$. As a result, using the first part of the proof, we pass to the limit in k at points where $y(t)$ is continuous and get from (A.45) that

$$y(t) = x + \int_\alpha^t f(t, y(t))dt + \sum_{\alpha<p_i^r<t} h_i^r(y(p_i^r-)).$$

This proves the lemma in this case.

3) Suppose that (A.43) holds. On $I = [\alpha, \beta]$ consider the functions

$$\xi_k(t) = \sum_{\alpha<t_i-s_k<t} h_i, \quad \xi_k(\alpha) = 0; \qquad \xi(t) = \sum_{\alpha<t_i<t} h_i, \quad \xi(\alpha) = 0.$$

By Helly theorems, $\xi_k(t) \to \xi^*(t)$ on I pointwise (by changing the values of $\xi^*(t)$ at a countable number of points, we assume that this function is left continuous) and

$$\lim_k \int_I \varphi(t)d\xi_k(t) = \int_i \varphi(t)d\xi^*(t) \qquad \forall\varphi(t) \in C(I).$$

But for all $\psi(t) \in \mathcal{D}_{L_1}^2$, $supp\,\psi \in A$, we have $\langle\tau_{s_k}h.\psi\rangle = \int_I \psi(t)d\xi_k$ and so $\lim_k \int_I \psi(t)d\xi_k = \int_I \varphi(t)d\xi$ (since $\tau_{s_k}h \to h$). Setting $\zeta(t) = \xi(t) + \xi^*(\alpha+)$, $\zeta(\alpha) = 0$ for $t \in (\alpha, \beta]$, we have

$$\int_I \psi(t)d\zeta = \int_I \psi(t)d\xi^* \qquad \forall\psi(t) \in \mathcal{D}_{L_1}^2 : supp\,\psi \in A.$$

Whence it follows that $\zeta(t) = \xi^*(t)$ on I for, if not, then $\int_I \lambda(t)d\zeta(t) \neq \int_I \lambda(t)d\xi^*(t)$ for some function $\lambda(s) \in C(I)$. Consider the sequence $\lambda_n \in$

$\mathcal{D}^2_{L_1}$: $supp\lambda_n \in [\alpha + 1/n, \beta - 1/n]$, $|\lambda_n(t)| \leq M$ $\forall t \in I$, $|\lambda(t) - \lambda_n(t)| \leq 1/n$ $\forall t \in I^n = [\alpha + 2/n, \beta - 2/n]$. We have

$$|\int_I \lambda(t)d\zeta - \int_I \lambda(t)d\xi^*| = |\int_I (\lambda(t) - \lambda_n(t))d(\zeta - \xi^*)| \leq$$
$$\leq |\int_{I^n} (\lambda(t) - \lambda_n(t))d(\zeta - \xi^*)| + |\int_{I \setminus I^n} (\lambda(t) - \lambda_n(t))d(\zeta(t) - \xi^*)| \leq$$
$$\leq \mathrm{Var}_\alpha^\beta(\zeta(t) - \xi^*(t))/n + 2M\mathrm{Var}(I^n, \zeta(t) - \xi^*(t)) \to 0$$

where the last term goes to zero because the function $\zeta(t) - \xi^*(t)$ is continuous at the points α and β.

By using the integral equation

$$x_k(t) = x_k + \int_\alpha^t f_k(t, x_k(t))dt + \xi_k(t),$$

we get that

$$y(t) = x + \xi^*(\alpha+) + \int_\alpha^t f(t, y(t))dt + \xi(t)$$

for almost all t from I. Denoting by $x(t)$ the function which is on the right-hand side of the preceding identity, we end the proof of Lemma 67.

□

Remark. Examining parts 1) and 2) of the preceding proof and using Lemmas 64 and 65, it is easy to see that Lemma 67 still holds if we replace its first assumption by the following: let conditions (V) or (A.42) from Section A.6 hold, $G^k \in H(F)$ and $G^k \to F$ in $H(F)$.

Now we will continue the proof of the theorem. Let $x(t)$ be a unique bounded solution of (I). On the trajectory $\Sigma = \{\tau_s F, s \in \mathbf{R}\} \subset H(F)$, define the function $r(\tau_s F) = x(t + s)|_{t \in [0,1]} \in L[0,1]$. By Theorem 128, to prove the theorem it is sufficient to show that $r : \Sigma \to L[0,1]$ is continuous if Σ is regarded as a subspace of $H(F)$.

Suppose that r is not continuous. Then, for some numbers s, s_n, and ϵ, $\tau_{s_n} F \to \tau_s F$ in $H(F)$. But

$$\int_0^1 |x(u + s_n) - x(u + s)|du \geq \epsilon \qquad \forall n. \tag{A.46}$$

Fix an interval (α, β), $\alpha \notin T$, containing $[0,1]$. Lemma 67 implies that there exists a subsequence of

$$x_n(t) = \{x(t + s_n), t \in (\alpha, \beta)\}$$

that converges in $L[0,1]$ to a certain solution of system $(I)_{\tau_s F}$, $y(t)$. Because the numbers α and β are arbitrary, $y(t)$ is a bounded solution of $(I)_{\tau_s F}$ and, hence, $y(t) = x(t+s)$, which is a contradiction to (A.46).

□

If h does not depend on x, system (I) can be studied to a greater extend because, in this case, the operators of the impulsive effect, $H_i = x + c_i$, are commutative.

Theorem 132 *Consider the linear impulsive system*

$$\frac{dx}{dt} = A(t)x + g(t), \tag{A.47}$$

$$\Delta x|_{t_i} = h_i, \quad i \in \mathbf{Z}, \quad t \in \mathbf{R}, \quad x \in \mathbf{R}^n, \tag{A.48}$$

where $A(t)$, $g(t)$ are locally integrable functions, $A(t)$ is S-almost periodic function, the measure $h = g(t) + \sum_i h_i \delta(t - t_i)$ is 2-almost periodic, and the function $\sum_{t_i \in [t,t+1]} |h_i| + \int_t^{t+1} |g(u)|du$ is bounded on $\mathbf{R}$.

System (A.47), (A.48) has a bounded S-almost periodic solution $x(t)$ (which is also W-almost periodic if $\zeta < +\infty$ and the function $\int_0^t g(u)du$ is uniformly continuous on $\mathbf{R}$) if one of the following conditions is satisfied:

1) equation (A.47) is exponentially dichotomic (then $x(t)$ will be this unique solution);

2) system (A.47), (A.48) has a solution, bounded for $t \geq 0$, and $\|\mathcal{U}(t,\tau)\|$-$\leq C$ for all $t \geq \tau$, $\mathcal{U}(t,\tau)$ is the Cauchy function for system (A.47).

Proof. The conditions of the theorem imply that the Cauchy problem $x(s+) = x_0$ for (A.47), (A.48), i.e. the integral equation

$$x(t) = x_0 + \int_s^t A(u)x(u)du + \sum_{s<t_i<t} h_i + \int_s^t g(u)du,$$

has a unique solution. For system (A.47), (A.48) there correspond a vector $F(A(t), h)$ and its hull $H(F)$ consisting of the pairs $G = (B(t), g)$, where $B(t)$ is S-almost periodic function and g is 2-almost periodic measure, and if $F_n \to G$, then $A_n \to B$ in S^1 and $h_n \to g$ in $\mathcal{L}_2$. There is a continuous minimal isometric dynamical system of translations τ_s acting on $H(F)$. Let us describe the structure of g. Because τ_s is minimal, $h_n = \tau_{s_n} h \to g$ in $\mathcal{L}_2$ for some subsequence $\{s_n\}$. Since the structure of h_n is si,ilar to h, $h_n.\varphi = \int \varphi(t) dh_n(t)$, where $h_n(t) = \int_0^t g_n(u)du + \sum_{0<t_i<t} h_i^n$ if $t > 0$ and $h_n(t) = \int_0^t g_n(u)du - \sum_{t \leq t_i \leq 0}$ if $t < 0$, $h_n(0+) = 0$.

Let $t \in \mathcal{K} = [-k, k]$. The functions $h_n(t)$ and their variations are bounded on $\mathcal{K}$ uniformly with respect to n. Consequently, by using the second Helly theorem, we can choose a subsequence $h_{n_j}(t)$, which converges pointwise on $\mathcal{K}$ to some function with bounded variation, $q(t)$. But then, for every continuous function with compact support in $\mathcal{K}$, $h_{n_j}.\varphi \to q.\varphi$ and hence $q = g$ on $(-k, k)$ (in the sense of distribution theory). Because k is arbitrary, there is a function $q(t)$, corresponding to g, having locally bounded variation (the function is unique if $q(0+) = 0$) such that $q = g$ on $\mathbf{R}$ and $q.\varphi = \int \varphi dq$. So, for every vector $G = (B(t), g)$ there corresponds a unique family of integral equations

$$y(t) = y_0 + \int_0^t B(u)y(u)du + q(t), \quad q(0+) = 0. \tag{A.49}$$

Each of these equations has a unique solution. If equation (A.47) is exponentially dichotomic, system (A.47), (A.48) has a unique solution $z(t) \in U_r(0)$ bounded on $\mathbf{R}$. Consider the product $H(F) \times L(0, 1)$ and its subset

$$L = \{(G = (B(t), q(t)); \bar{x}) \in L \Leftrightarrow \bar{x}(t) = x(t)\},$$

where $\bar{x}(t)$ denotes the restriction of the solution $x(t) : \mathbf{R} \to U_r(0)$ of system (A.49) to $[0, 1]$.

We define the semigroup S^t on L as in Lemma 66. Then all the statements of Lemma 66 are true. Let us prove, for example, the most important – the compactness of L (assuming that $L \neq \emptyset$). So let $x_n(t)$ be a sequence of solutions of impulsive system (A.47), (A.48), norm bounded by a number r, $F_n = (A_n, h_n)$. We can assume at once that $A_n \to B$ in S^1 and $h_n \to g$ in $\mathcal{L}_2$ (or $h_n(t) \to q(t)$ pointwise; for subsequences, we will keep the notations used for the initial sequences). So

$$x_n(t) = x_{0n} + \int_0^t A_n(u)x_n(u)du + h_n(t).$$

The sequence $|x_n(t)|$ is uniformly bounded by r on any interval $[-k, k]$, so it follows that the sequence of variations of $x_n(t)$ is uniformly bounded and $x_n(t) \to y(t)$ pointwise. Vitali theorem on passing to the limit under the integral sign implies that $y(t)$ will be a solution of equation (A.48) (we can assume that $q(0+) = 0$ for otherwise, we can set $q_1(t) = q(t) - g(0+)$, $y_1 = y_0 - g(0+)$). Evidently $|y(t)|$ is bounded by r and, since k is arbitrary, it can be extended to $\mathbf{R}$.

So the statements of Lemma 66 hold for S^t and Theorem 132 follows from the first and second Favard theorem in abstract form (see Bibliographical notes)[15].

---

[15]Indeed, if equation (A.47) is exponentially dichotomic, equation (A.49) has exactly one bounded solution. If condition 2) of Theorem 132 holds, the difference $\Delta(t)$ of any two solutions of (A.49) will be a solution of (A.47) and, hence, it will be separated from 0 for all G.

A.8 Linear abstract impulsive systems and their almost periodic solutions

1. In this section we study the linear impulsive system

$$\frac{dx}{dt} = A(t)x + f(t), \tag{A.50}$$

$$\Delta x|_{t_i} = B_i x + h_i, \quad i \in \mathbf{Z}, \quad t \in \mathbf{R}, \tag{A.51}$$

where $x \in X$, $(X, |\cdot|)$ is a reflexive separable Banach space, $B_i \in \mathcal{L}(X)$, $A(t)$ is a family of closed operators with a domain $D(A(t)) \equiv D$ for all t, D is a Banach space imbedded densely and continuously in X, and the increasing sequence $\{t_i\}$ does not have finite limit points. We will be concerned with almost periodic solutions of (A.50), (A.51). Here we will restrict our attention to statements similar to the Favard minimax theorem. Taking into account the excellently developed abstract Favard theory [166] (see also Bibliographical notes), we replace impulsive system (A.50), (A.51) by some continuous semiflow. The connections that appear there will be one of the prime points for our study (thus, in particular, we do not touch the problem of N-almost periodic solutions because it can be studied using the same scheme as in Section A.7).

2. Main assumptions (P).

$P0$. System (A.50), (A.51) has at least one solution $z(t) : \mathbf{R}_+ \to U_r(0)$, bounded for $t \geq 0$.

$P1$. Suppose that for a certain positive C,

$$\sup_{t \in R} \sum_{t_i \in [t.t+1]} (|h_i| + \|B_i\|_{\mathcal{L}(X)}) \leq C, \tag{A.52}$$

and the distributions $h = \sum_i h_i \delta(t - t_i)$ and $B^* = \sum_i B_i^* \delta(t - t_i)$ are, correspondingly, weakly and strongly 2-almost periodic (see Section A.6).

$P2$. The functions $A(t) : \mathbf{R} \to Z_1$ and $f(t) : \mathbf{R} \to Z_2$ are almost periodic taking values in some complete metric spaces Z_1 and Z_2, for example, $Z_1 = \mathcal{L}(D, X)$, $Z_2 = X$.

$P3$. Let $d = (A(t), f(t))$ and denote equation (A.50) by $(A.50)_d$. Consider the closure of the trajectory $\{(A(t+s), f(t+s)), s \in \mathbf{R}\}$ in $C(\mathbf{R}, Z_1) \times C(\mathbf{R}, Z_2)$ with respect to the metric of uniform convergence. For every element e from this closure $H(d)$, there corresponds an equation of type (A.50) and there is a minimal dynamical system of translations $\tau_s(e)$ defined on

$H(d)$. Suppose that for $t \geq 0$, the abstract Cauchy problem $x(0) = x$ for this equation has a unique strongly continuous solution $S_e(t)x$, defined for $t \geq 0$. Hence there is the function $S_e(t)x$ defined on $H(d) \times X$, which is assumed to be continuous in e, x for fixed t if X is endowed with the weak topology (concerning this condition, see also Bibliographical notes).

3. Solution of impulsive systems which are scalar Stepanov and Wexler almost periodic

The relationship between scalar S-almost periodic and W-almost periodic solutions is clarified by the following lemma.

Lemma 68 *Let $x(t)$ be a solution of an impulsive system, which is uniformly bounded with respect to t, condition P3 hold, and $\zeta < +\infty$. Then the function $x(t)$ will be scalar S-almost periodic if and only if it will be scalar W-almost periodic.*

Proof. Suppose that the function $\langle x(s), f\rangle$ is S-almost periodic. Fix $\epsilon \in (0,1)$, $f \in X^*$. Let $\epsilon/2 \geq \mu_n \to 0$ and Ω_n be a set, relatively dense in $\mathbf{R}$, of μ_n-almost periods that are common to $\langle x(s), f\rangle$ and $\tau_s F$ such that $\tau_{\mu_n} F \to F$. Let us show that for sufficiently large n, the set Ω_n consists of ϵ-almost periods of the W-almost periodic function $\langle x(t), f\rangle$. Indeed if this were not so, there would be points z_n, $|z_n - t_j| \geq \epsilon$ and $|z'_n - t_p| \geq \epsilon$ for all n, j, p, such that $|\langle x(z'_n), f\rangle - \langle x(z_n), f\rangle| \geq \epsilon$, where $z'_n = z_n + \tau_n$, $\tau_n \in \Omega_n$. But then the sequences

$$X_1 = \{\langle x(t+z_n), f\rangle\} \quad \text{and} \quad X_2 = \{\langle x(t+z'_n), f\rangle\},$$

$t \in [0, \epsilon/2]$, consist of continuous functions. Moreover, X_i are precompact, as implied by $(P3)$[16]. Hence, without loss of generality, we can assume that

[16] Let, for all $t \in I = [0, A]$, $S(t)$ be a continuous transformation of the compact metric space (K, ρ) and $S(t)$ be a local semigroup,

$$S(t)S(s) = S(t+s) \qquad \forall t, s, t+s \in I; \quad S(0) = E.$$

Suppose that the function $S(t)x : I \to K$ is continuous for every fixed x. Then $S(t)x$ is continuous with respect to the totality of the variables.

Proof.[75] Denote by $Z = K \times K$ and consider the Banach space $C(Z)$ of continuous functions equipped with the *sup*-norm. There is a semigroup $\varphi(t)$ of "translations" acting on $C(Z)$: $\varphi(t)A(z) = A(S(t)z)$. By the Lebesgue theorem on bounded convergence, it is clear that $w - \lim_{t\to 0} \varphi(t)A = A$ for all $A \in C(Z)$, and so applying the known theorem of theory of semigroups [179], we get that $s - \lim_{t\to 0} \varphi(t)A = A$ for all $A \in C(Z)$. (To prove this theorem it is sufficient to use only semigroup properties of $S(t)$). Set $A = \rho$. Then for $t \to 0$, $\rho(S(t)x, S(t)y) \to \rho(x, y)$ uniformly with respect to x, y. Because $\rho(S(t_n)x_n, x) \leq$

$x(z_n) \xrightarrow{w} A$, $x(z'_n) \xrightarrow{w} B$, $A \neq B$, and $w - \lim_{n\to\infty} x(t+z_n) = x(t, A)$, $w - \lim_{n\to\infty} x(t+z'_n) = x(t, B)$ uniformly with respect to t. But this contradicts the equality

$$\lim_{n\to\infty} |\langle x(u+z_n), f\rangle - \langle x(u+z'_n), f\rangle|_S = 0.$$

One can prove the other part of the lemma similarly to Lemma 58 because $\lambda_\epsilon(t) \leq 2\epsilon\zeta \to 0$ if $\epsilon \to 0$.

□

4. Main results.

Throughout this paragraph we will assume that the conditions (P), $(V1 - V3)$ (or instead of $(V1 - V3)$, condition (A.42)) hold. But then, system (A.50), (A.51) induces a linear extension of the flow $(H(F), \tau_s)$ because, for every point of $H(F)$, by Lemmas 64 and 65, there is a corresponding linear impulsive system. In the product $H(F) \times L_1([0,1]; X)$, where $L_1([0,1]; X)$ is the separable space of functions which are Lebesgue-Bochner integrable, let us consider the subset

$$K = \{(F, \bar{x}) \in K \Leftrightarrow \bar{x}(t) = x(t)\},$$

where $\bar{x}(t)$ denotes the restriction of the solution $x(t) : \mathbf{R} \to U_r(0)$ of system $(A.50, A.51)_F$ to $[0,1]$. (Recall that by $(P0)$, $U_r(0)$ contains the values $\{z(t) : t \geq 0\}$).

Lemma 69 *The set K is nonempty and weakly compact. There is a continuous semiflow S^μ, $\mu \geq 0$, defined on K:*

$$S^\mu(F, x(t) : [0,1] \to U_r(0)) = (\tau_\mu F, x(t) : [\mu, \mu+1] \to U_r(0)). \tag{A.53}$$

Proof. Let $K \neq \emptyset$, $(F_n, x_n) \in K$. Because $H(F)$ is compact, without loss of generality, we can assume that $F_n \to G = (e, C, g)$. Fix an arbitrary $A > 0$ such that $\{A; -A\} \notin supp\, g = \{p_i\}$. Suppose, for example, that (A.42) holds. By Lemma 65 and $(P3)$, for sufficiently large m, we can consider $x_n(t) \xrightarrow{w} x(t)$ uniformly on $[-A+1/m, p_1 - 1/m)$, $x_n(p_1^n-) \xrightarrow{w} x(p_1-)$ (continuity with respect to all the variables follows from $(P3)$), where $x(t)$ is a solution of

$|\rho(S(t_n)x_n, S(t_n)x) - \rho(x_n, x)| + \rho(x, x_n) + \rho(S(t_n)x, x)$, the needed statement is proved. In particular, for all $\epsilon > 0$ there exists $\delta > 0$ such that

$$\rho(S(t)x, x) < \epsilon \qquad \forall x \in X, \quad 0 \leq t \leq \delta.$$

$(A.50)_e$ and $T(A, h^k) = \{p_1^k, \ldots, p_N^k, \ldots, p_M^k\}$ so that $\lim_{k\to\infty} p_i^k = p_i$. But because $(B_i^n)^* \xrightarrow{s} C_i^*$, $h_i^n \xrightarrow{w} g_i$ for all i, we have

$$(B_1^n + E)x_n(p_1^n) + h_1^n \xrightarrow{\omega} (C_1 + E)x(p_1) + g_1.$$

But then we also see that $x_n(t) \xrightarrow{w} x(t)$ uniformly on $(p_1 + 1/m, p_2 - 1/m)$ (m is sufficiently large). After a finite number of such steps, we see that $\langle x_n(t), f\rangle \to \langle x(t), f\rangle$ pointwise on $\mathcal{A} = (-A, A)$ (with an exception of the points of support of g). So for all $f \in X^*$, $\langle x_n(t), f\rangle \to \langle x(t), f\rangle$ in $L_1(\mathcal{A}, \mathbf{R})$. It is known that $(L_1(\mathcal{A}, X))^* = L_\infty(\mathcal{A}, X_\omega^*)$ (see [28]). Let us show that

$$\langle \xi, x_n\rangle = \int_{\mathcal{A}} \langle x_n(t), \xi(t)\rangle \to \int_{\mathcal{A}} \langle x(t), \xi(t)\rangle dt$$

for all $\xi(t) \in L_\infty(\mathcal{A}, X_\varrho^*)$. Indeed, the sequence of the integrands converges pointwise to $\langle x(t), \xi(t)\rangle$ and

$$|\langle x_n(t), \xi(t)\rangle| \le |\xi|_\infty |x_n(t)| \le r|\xi|_\infty.$$

Note that, by construction, $x(t)$ is a solution of the impulsive system and $x(t) : \mathcal{A} \to U$ because a norm is lower semicontinuous in the weak topology. Since A can be arbitrarily large, $x(t)$ can be extended to $\mathbf{R}$, and so K is weakly compact. Further, one can use the preceding argument for $F_n = \tau_{A_n} F$ and $A_n \to +\infty$ to see that $(P0)$ implies that $K \ne \emptyset$.

Because K is weakly compact, it is sufficient to show that $S^\mu k$ is separately continuous but 1) $S^\mu k$ is strongly continuous with respect to μ because the Lebesgue-Bochner integral is continuous in mean and 2) because $S^\mu k = (\tau_\mu F, x(t) : [\mu, \mu+1] \to U)$, $\mu \ge 0$, it easily follows from the weak convergence $x_n(t) \to x(t)$ in $L_1([0,1], X)$ (by almost repeating the argument at the beginning of the proof) that $x_n(t) \to x(t)$ weakly in $L_1([0,\alpha], X)$ for every positive α[17].

□

Thus the results of the Favard theory (see Bibliographical notes) can be applied to S^t.

[17] to be more precise, we first show that $x_{n_j}(t) \to x(t)$ weakly in $L_1([0,\alpha], X)$ for some subsequence and then use compactness of K and uniqueness of the limit set $(x(t))$ for $\{x_n(t)\}$

Theorem 133 *Let any nonzero bounded solution $x = x(s)$ of each homogeneous impulsive system G_{hom} be semi-separated from zero*[18]*:*

$$\varliminf_{t\to-\infty} \int_t^{t+\theta} |x(s)ds > 0, \quad \theta > 0 \qquad \forall G_{hom} \tag{A.54}$$

Then there exists a scalar S-almost periodic (W-almost periodic) solution of (A.50), (A.51).

Theorem 133 is a consequence of the previous lemma and Theorem 139 from Bibliographical notes. In the sequel, we will somewhat strengthen Zhikov lemma on compactness [75]. Let a semiflow $S^t(e,x) = (h(t)e, S_e(t)x)$ be an extension (not necessarily linear) of the compact minimal dynamical system $(H, h(t))$. We make the following assumption.

(C) The function $S_e(t)x$ is continuous with respect to e for any fixed t, x and is continuous with respect to $t \geq 0$ for any fixed e, x. Moreover, suppose that $S_e(t)x$ is equipotentially continuous with respect to x for x belonging to a bounded subset of X, uniformly with respect to $e \in H$, $t \geq 0$, i.e. the function $\mu_r(a) = \sup_B |S_e(t)x - S_e(t)y|$, where $B = \{t \geq 0, e \in H\, x, y \in U_r(0) : |x-y| \leq a\}$, is finite and monotonically decreases as a approaches to 0.

Lemma 70 *Let condition (C) hold and the function $u(e) : H \to X$, which is weakly continuous on H, define an invariant set for a semiflow S, i.e. $S_e(t)u(e) = u(h(t)e)$. Then $u(e)$ is strongly continuous.*

Proof. Because $u(e)$ is weakly continuous on compact sets, by Gelfand theorem [44], the set of points where it is strongly continuous is not empty (it is of the second category in H). Moreover, $u(e)$ is weakly compact and so $\sup_H |u(e)| = r < \infty$. Suppose that $u(e)$ has discontinuities on H. Denote by $\omega(u(e))$ the oscillation of the function $u(e)$ at the point e and consider the set $D(t) \subset H$, which consists of precisely the points e such that $\omega(u(h(t)e)) \geq \delta > 0$, where δ is so small that $D(0) \neq \emptyset$. We have $D(t) = h(t)D(0) \neq \emptyset$. Because $\omega(u(h(t)e))$ is upper semicontinuous with respect to e and $u(h(t)e)$ is weakly continuous with respect to e, the set $D(t)$ is closed and nowhere dense (see [17]). Use δ to choose $\epsilon < \delta$ so that $\mu_r(\epsilon+) < \delta/2$. The set $D' = \{e : \omega(u(e)) \geq \epsilon\} \supset D(0)$ is also closed and nowhere dense. We will

[18]the uniqueness condition for an extension to the right of a solution of an impulsive system may not hold

also show that $D(t) \subset D'$ for all $t \geq 0$. Indeed, if $m \in D(t) \setminus D'$, then $\omega(u(m)) < \epsilon$ and thus

$$\begin{aligned}
\delta &\leq \omega(u(h(t)m)) = \omega(S_m(t)u(m)) = \overline{\lim_{\substack{a\to m\\ c\to m}}} |S_a(t)u(a) - S_c(t)u(c)| \leq \\
&\leq \overline{\lim_{a\to m}} |S_a(t)u(a) - S_m(t)u(m)| + \overline{\lim_{c\to m}} |S_c(t)u(c) - S_m(t)u(m)| = \\
&= 2\,\overline{\lim_{a\to m}} |S_a(t)u(a) - S_m(t)u(m)| \leq \\
&\leq 2(\overline{\lim_{a\to m}} |S_a(t)u(m) - S_m(t)u(m)| + \\
&+ \overline{\lim_{a\to m}} |S_a(t)u(a) - S_a(t)u(m)|) = 2\,\overline{\lim_{a\to m}} |S_a(t)u(a) - S_a(t)u(m)| \leq \\
&\leq 2\,\overline{\lim_{a\to m}} \mu_r(|u(a) - u(m)|) \leq 2\mu_r(\omega(u(m))+) \leq 2\mu_r(\epsilon+) < \delta.
\end{aligned}$$

This gives a contradiction. So the closed nowhere dense D' contains the half-trajectories $h(t)e$ for all $e \in D(0)$ which is a contradiction since $h(t)$ is minimal. So the assumption that $u(e)$ is discontinuous does not hold.

□

Corollary 13 *Let condition (P0) hold for (A.50), i.e.* $B_i = 0$, $h_i = 0$ *and also condition (P3) is fulfilled. Let* $S_e(t)x$ *be strongly continuous for any fixed* t. *Let also, for the homogeneous equation (A.50), its evolution operator be uniformly norm bounded in the half-plane* $t \geq \tau$ *by a number* l. *Then (A.50) has not less than one almost periodic solution.*

Proof. Because all assumptions of Favard theorem are fulfilled and this theorem implies that an equation from $H(d)$ has a weakly almost periodic solution, there exists a weakly continuous function $u(e) : H(d) \to X$ that induces these solutions and it is sufficient to prove that it is strongly continuous. Denote by $L_e(t)$ the difference $S_e(t)x - S_e(t)0$. Because $L_e(t)x$ is a solution of the homogeneous equation (A.50), by $(P3)$, it is a weakly continuous linear operator for any t, e and, hence, it is continuous. By the conditions of the lemma, $S_e(t)0$ is continuous with respect to e for all t, so we see that $L_e(t)$ is strongly continuous with respect to e for all t and $\|L_e(t)\| \leq k$ for all $e, t \geq 0$. But then condition (C) holds with $\mu_r(a) = ka$, $r = \sup_{H(d)} |u(e)|$ and so $u(e)$ is strongly continuous.

□

Corollary 14 *Let condition (C) hold and* $(H, h(t))$ *be a minimal isometric flow. Then the trajectory* $(h(t)e, x(t))$ *of the semigroup* $S(t)$, *which is Birkhoff weakly recurrent, will be compact.*

Proof. Let A be the weak closure of the set $\{(h(t)e, x(t)), t \in \mathbf{R}\}$ in $H \times X$. Then A is a minimal weakly compact set of $S(t)$. Let us define the function $\mathcal{U}(a) = \mathcal{U}((d, z)) = z$ on A and the semigroup $S(a, x) = (S(t)a, S_d(t)x)$ on the product $A \times X$. All conditions of Lemma 70 will hold for S. It is clear that $\mathcal{U}(a)$ is weakly continuous and $\mathcal{U}(S(t)a) = \mathcal{U}((h(t)d, S_d(t)z)) = S_d(t)z = S_d(t)\mathcal{U}(a)$, and so the set $\{x(s), s \in \mathbf{R}\}$ is precompact in X by Lemma 70.

□

Remark. Considerations similar to the ones given above allow to weaken the sufficient conditions for discrete analogues of Lemma 70, Theorem 2.4 from [99] to hold. There one can give up condition (d_1) and weaken condition (d_2) to require only the sequence $\{T_n\}$ to be strongly almost periodic.

Now we will need to generalize somewhat the construction of the flow S^t given in Lemma 66. We assume additionally that the following condition $(P4)$ holds.

$P4$. The function $S_e(t)$ defined in $(P3)$ is strongly continuous with respect to the totality of the variables and the inequality

$$|S_e(t)x| \leq \lambda_1(t)|x| + r_1(t)$$

holds uniformly in e with the functions $\lambda_1(t)$, $r_1(t)$ being nondecreasing and positive.

It should be noted that condition $(P4)$ is fulfilled if $f(t)$ is an almost periodic function and 1) $A(t) = A + B(t)$, where A is a generator of a strongly continuous semigroup and $B(t) : \mathbf{R} \to \mathcal{L}(X)$ is uniformly almost periodic or 2) the function $A(t) : \mathbf{R} \to \mathcal{L}(D, X)$ is almost periodic, the Cauchy problem for (A.50) is uniformly well posed on any interval, and the evolution operator of (A.50) satisfies the exponential growth condition with respect to the norms in $\mathcal{L}(X)$ and $\mathcal{L}(D)$ (for more details see [99]).

Denote by $\Lambda_F(t)x$ a solution of the initial value problem $x(0+) = x$ for impulsive system $(A.50, A.51)_F$, $F = (e, c, h)$. It is clear that the function $\Lambda_F(t)x$ is piecewise continuous with respect to $t \geq 0$ for all F, x. Let $supp\, h \cap (0, t] = \{p^{(j)}, j = \overline{1, q}\}$. It follows from

$$\Lambda_F(t)x = \prod_{i=1}^{q+1} (H_i \circ S_{h(p^{(i-1)})e})x, \quad \text{where} \quad p^{(0)} = 0, \quad H_{q+1}x = x \tag{A.55}$$

that the function is continuous with respect to x for all F, t if condition $(P4)$ holds. By using (A.55) and condition $(P4)$, we also find that the inequality

$$|\Lambda_F(t)x| \leq \lambda(t)|x| + r(t), \tag{A.56}$$

where $\lambda(t)$, $r(t)$ are nondecreasing positive functions, holds uniformly with respect to F.

Lemma 71 *Assume that the distributions B and h are, respectively, strongly and uniformly almost periodic, and conditions (P1) - (P4) hold. Then $\Lambda_F(t)x$ is continuous with respect to F for all x, $t \neq t_i$ at the points $H(F)$, at which $0 \notin supp\, h = T$.*

Proof. Assume, for example, the (V) holds and let $F_n(e_n, c_n, h_n) \to F$. If $T \cap (0,t] = \emptyset$, then for large n, $\Lambda_{F_n}(t)x = S_{e_n}(t)x \to S_e(t)x = \Lambda_F(t)x$. In the case when $T \cap (0,t] = p^{(1)} = \ldots = p^{(r)}$, counting multiplicities, then for large n, $(0,t] \cap supp\, h_n = \{p_n^{(j)}, j = \overline{1,r}\}$ and $p_n^{(j)} \to p^{(1)} \in (0,t)$ for all j and now we can apply (A.55) to see that

$$\Lambda_{F_n}(t)x \to \Lambda_F(t)x^{19}.$$

Let now $T \cap (0,t] = \{p^{(1)} = \ldots = p^{(r)}; q^{(1)} = \ldots = q^{(s)}\}$ and $p^{(1)} < z < q^{(1)}$. Then $\Lambda_{F_n}(t)x = \Lambda_{\tau(z)F_n}(t-z)\Lambda_{F_n}(z)x$ and, because $\Lambda_{F_n}(z)x \to \Lambda_F(z)x$ by the second step of the proof and $(z,t) \cap T = \{q^{(1)} = \ldots = q^{(s)}\}$, by repeating the second step from the beginning, we find that $\Lambda_{F_n}(t)x \to \Lambda_F(t)x$. At this point, by using induction, we finish the proof of the lemma.

□

In the product $W = H(F) \times L_1([0,1], X)$, consider the mapping defined by $\Pi(a)(F,x) = (\tau_a(F), \Pi_F(a)x)$, where $y(s) = \Pi_F(a)x = \Lambda_{\tau_s(F)}(t)x(s)$ and $x(s)$ is an arbitrary function from the equivalence class of x. Clearly, if $x(s)$ will run over the functions x, $y(s)$ will run over a certain set of functions y which differ on a set of measure zero. Let us prove that $y \in L_1([0,1], X)$. To do that, consider the function $f(s,x) = \Lambda_{\tau_s(F)}(t)x$ for an arbitrary t. By (A.56), $|f(s,x)| \leq \lambda(t)|x| + r(t)$. Moreover $f(s,x)$ is piecewise continuous with respect to s for all x and is continuous with respect to x for all s^{20}. But then, as it is known, $y(s) = f(s, x(s)) \in L_1([0,1], X)$.

It is easy to see that $\Pi(a)$ defines a semigroup of transformations of W.

Lemma 72 *If conditions of Lemma 71 hold, the function $\Pi_F(t)x$ is strongly continuous with respect to $t \geq 0$ for any F, x and is strongly continuous with respect to F for any fixed x, $t \geq 0$.*

---

[19] If $S_{e_n}(t_n)x \to S_e(t)x$ for all x as $e_n \to e$, $t_n \to t$, we have that $S_{e_n}(t_n)x_n \to S_e(t)x$ as soon as $x_n \to x$. To see that, it is sufficient to note that $S_e(t)x = L_e(t)x + S_e(t)0$, where $L_e(t)$ is strongly continuous with respect to the totality of the variables.

[20] to see α), we can use Lemma 65 from Section A.7 for $s_n \to s$ and $\tau(s_n)F \to \tau(s)F$, $s, s+t \notin T$; β) is obvious.

Proof. Let $t_n \to t \geq 0$. Then

$$f_n(s,x) = \Lambda_{T_s(F)}(t_n)x = S_{h(t^*+s)e}(t_n - t^*)\Lambda_{T_s(F)}(t^*)x \to f(s,x)$$
$$\forall s \notin T - t,$$

where t^* is such that $[t^* + s, t) \cap T = \emptyset$. Let $F_n \to F$. Then $f_n(s,x) = \Lambda_{T_s(F_n)}(t)x \to f(s,x)$ by Lemma 71 as soon as $s \notin T \cup T - t$. Lemma 72 now follows from the Lebesgue theorem for abstract functions.

□

Theorem 134 *Let conditions of Lemma 72 and (P0) hold and let there exist such $l > 0$ that, for any solution of the homogeneous impulsive system that corresponds to (A.50), (A.51), the following inequality holds:*

$$|x(t)| \leq l|x(t_0)| \qquad \forall t \geq t_0.$$

Then system (A.50), (A.51) has no less than one S-almost periodic solution.

Proof. The function $\Pi_F(t)x$ satisfies the continuity conditions of Lemma 70 and there is an analogue of decomposition of $S_e(t)$ into the linear component $P_F(t)x = \mathfrak{L}_{T_s(F)}(t)x(s)$ and the component $\Lambda_{T_s(F)}(t)0$, which does not depend on x. Here $\mathfrak{L}_{T_s(F)}(t)x(s)$ denotes a solution of the initial value problem for the homogeneous impulsive system (A.50, A.51)$_F$ with $x_0 = x(s)$. Moreover

$$|P_F(t)x|_{L_1} = \int_0^1 |\mathfrak{L}_{T_s(F)}(t)x(s)|ds \leq l \int_0^1 |x(s)|ds = |x(s)|_{L_1}.$$

Consequently, the weakly almost periodic solution will be strongly almost periodic.

□

Theorem 135 *Let conditions of Theorem 133 be fulfilled and $(D, \ \cdot\ )$ be a Banach space compactly imbedded in X and*

$$S_e(t)x \leq C_1(t)|x| + C_2(t) \qquad \forall t > 0$$

uniformly with respect to e. Then there exists S-almost periodic solution of system (A.50), (A.51).

Proof. By Theorem 133, there exists a weakly S-almost periodic solution $z(t)$ of system $(A.50, A.51)_F$. Let us show that the conditions of the theorem imply that it will also be S-almost periodic. Consider Π, the weak closure of the trajectory $S^{\mu}(F, z(t) : [0,1] \to U)$ in K, and show that it is strongly compact. If $F_n = (e_n, C_n, g_n) \to G = (e, C, g)$, $x_n(t) \to x(t)$ weakly in $L_1([0,1]; X)$, then by Lemma 69, for all $t \notin supp\, g$, $x_n(t) \to x(t)$ weakly in X. Fix $s \notin supp\, g$. We have $x_n(s) \in D$ for large n. Let $\epsilon > 0$ be so small that $(s, s+\epsilon) \cap supp\, g = \emptyset$. Then $x_n(s+\epsilon) = S_{e_n}(\epsilon)x_n(s)$ and $x_n(s+\epsilon) \leq C_1(\epsilon)|x_n(s)| + C_2(\epsilon) \leq rC_1(\epsilon) + C_2(\epsilon)$. But then $x_n(s+\epsilon) \to x(s+\epsilon)$ strongly in X and, hence, $x_n(t) \to x(t)$ strongly almost everywhere in $[0,1]$ (and thus, in $L_1([0,1]; X)$).

□

5. Remark. Instead of (A.50), let us consider a homogeneous evolution system **e**, i.e. a family of such linear operators $\{\mathcal{U}(t,s), t \geq s\}$, bounded and strongly continuous with respect to the totality of the variables, that

$$\mathcal{U}(t,t) = E \quad \forall t; \qquad \mathcal{U}(t,s)\mathcal{U}(s,\tau) = \mathcal{U}(t,\tau) \quad \forall t \geq s \geq \tau.$$

For the family of the adjoint operators $\mathbf{e}^* = \{\mathcal{U}^*(t,s), s \leq t\}$, similar relations hold:

$$\mathcal{U}^*(t,t) = E^*; \quad \mathcal{U}^*(s,t)\mathcal{U}^*(t,s) = \mathcal{U}^*(t,\tau), \quad \text{if} \quad t \geq s \geq \tau.$$

If the function $\mathcal{U}^*(t,s)$ is strongly continuous with respect to the totality of variables for $t \geq s$, then $\mathbf{e}^*$ is a strongly continuous inverse evolution system. Let also the operator valued functions $\mathcal{U}(t+h, \tau+h)$ and $\mathcal{U}^*(t+h, \tau+h)$, defined on compact sets, be strongly almost periodic with respect to h uniformly in $t-\tau$ from compact sets. Then the following statement holds.

There exists such a linear extension of the compact isometric flow $\varphi_t(\varphi)$, $(\varphi_t(\varphi), \Omega_0^t(\varphi)x)$, that the function $\Omega_0^t(\varphi)$ is strongly continuous with respect to $t \geq 0$ for all φ and is strongly continuous with respect to φ if t belongs to a compact subset of $\mathbf{R}_+$. We also have that for some $\varphi_0 \in K$,

$$\Omega_0^{t-\tau}(\varphi_{h+\tau}(\varphi_0)) = \mathcal{U}(t+h, \tau+h).$$

The function $(\Omega_a^b(\varphi))^* : K \to \mathcal{L}(X^*)$ has the same continuity properties as $\Omega_a^b(\varphi)$ and so condition $(P3)$ holds if we set $S_{\varphi}(t)x = \Omega_0^t(\varphi)x$, $H(d) = K$[21].

[21] Consider a sequence of positive numbers $C = \{c_i\}$ and endow $\mathcal{L}(X)$ with the topology

Further, let all the conditions given in this section be satisfied, $B_i \equiv 0$ for all i, $\|\mathcal{U}(t+s,\tau+s)\| \le a$ for all s, $t \ge \tau$ (then also $\|\Omega_0^t(\varphi)\| \le a$ for all $t \ge 0$, φ). Then it is clear that the statements of Theorems 133, 134 (if condition ($P4$) is satisfied), and Theorem 135 hold and hence the questions posed in Remarks 3.4 and 3.7 of [99] and [100], page 163, can be answered in the affirmative even if in the conditions of Theorem 3.6 in [99], $f(t)$ is assumed to be weakly almost periodic (instead of strongly almost periodic) and the requirements of condition C_3 and continuity of φ are omitted.

---

of pointwise convergence on a countable set $\{x^i\}$ everywhere dense in X. This topology is given by a metric d, with respect to which any (norm) bounded ball in $\mathcal{L}(X)$ is a complete metric space. Let the space $\Pi(C)$ consist of such mappings $\nu(t,\tau) : \Delta \to \mathcal{L}(X)$, strongly continuous on Δ, that for all k,

$$\sup_{\Delta(k)} \|\nu(t,\tau)\| \le c_k \qquad (\Delta(k) \overset{\text{def}}{=} \{0 \le t-\tau \le k\}).$$

Endow $\Pi(C)$ with the metric

$$\rho(\nu',\nu) = \sum_{i \ge 1} 2^{-i} \min\{1, \sup_{\Delta(k)} d(\nu(t,\tau),\nu'(t,\tau))\},$$

which makes it a complete metric space.

Fix x, $|u| \le k$, and $\epsilon = 1$. There exist a segment I and a 1-almost period $\omega = \omega_t$ for all $t \in \mathbf{R}$ such that $t+\omega_t \in I$ and

$$|\mathcal{U}(t+u,t)x - \mathcal{U}(t+\omega+u,t+\omega)x| < 1.$$

Hence $|\mathcal{U}(t+u,t)x| \le |\mathcal{U}(t+\omega+u,t+\omega)x| + 1 \le C_{x,k}$, whence, in virtue of the uniform boundedness principle, $\sup_{\Delta(k)} \|\mathcal{U}(t,\tau)\| \le a_k$ for some constant a_k, $k \in \mathbf{N}$.

So $\mathcal{U}(t+s,\tau+s) \in \Pi(\{a_k\}) = \Pi(A)$ for all s. Let the mapping $\pi(s) : \mathbf{R} \to \Pi(A)$ be given by $\pi(s) = \mathcal{U}(t+s,\tau+s)$. The function $\pi(s)$ is almost periodic (the proof easily follows from the form of ρ and d). The reasonings, standard in the theory of almost periodic functions, show that one can construct such a compact minimal isometric flow $\varphi_t(\varphi)$ with the phase space K that the continuous function $\pi(\varphi) : K \to \Pi(A)$ will be such that $\pi(\varphi_s(\varphi_0)) = \mathcal{U}(t+s,\tau+s)$ for some $\varphi_0 \in K$.

If we set $\pi(\varphi)|_{t=b,\tau=a} = \Omega_a^b(\varphi) \in \mathcal{L}(X)$, then $\Omega_a^b(\varphi)$ is strongly continuous with respect to φ uniformly on $\Delta(k)$ and, if φ is fixed, it is strongly continuous on $\Delta(k)$. Let us prove that $\Omega_0^{t+s}(\varphi) = \Omega_0^s(\varphi_t(\varphi))\Omega_0^t(\varphi)$. Indeed, because the trajectory $\varphi_t(\varphi_0)$ is dense, for an arbitrary φ there is a sequence $\varphi_{t_n}(\varphi_0) \to \varphi$ and since

$$\begin{aligned}\Omega_0^{t+s}(\varphi_{t_n}(\varphi_0)) &= \pi(\varphi_{t_n}(\varphi_0))|_{t\to t+s,\tau\to 0} = \mathcal{U}(t+s+t_n,t_n) = \\ &= \mathcal{U}(t+s+t_n,t+t_n)\mathcal{U}(t+t_n,t_n) = \pi(\varphi_{t+t_n}(\varphi_0))|_{t\to s,\tau\to 0} \times \\ &\times \pi(\varphi_{t_n}(\varphi_0))|_{t\to t,\tau\to 0} = \Omega_0^s(\varphi_t(\varphi_{t_n}(\varphi_0)))\Omega_0^t(\varphi_{t_n}(\varphi_0)),\end{aligned}$$

it remains to note that the multiplication operator from $\mathcal{L}(X)$ is continuous in the strong topology.

Bibliographical notes

Section A.1. In this section we follow traditions of Kiev school of impulsive systems and reduce the initial value problem $x(0) = x_0$ for the system

$$\frac{dx}{dt} = f(t,x); \quad \Delta x|_{t_i} = h_i(x), \tag{A.57}$$

to the Volterra-Stieltjes integral equation

$$x(t) = x_0 + \int_0^t F(t,x)d\mu(t), \tag{A.58}$$

and its solution is considered to be a solution of (A.57). (The impulsive system is sometimes written in the form of a Caratheodory-Radon equation $dx/d\mu = F(t,x)$ or in the form of the measure equation $dx = F(t,x)d\mu$ [37].) The general theory of equation (A.58), which is also called a generalized differential equation, (and, in connection with it, the theory of the generalized Kurzweil-Perron integral) was developed in detail in works of Ya. Kurzweil and Š. Schwabik [71, 154, 155]. At the beginning of these works there was a study of certain effects in convergence of solutions of differential systems (for the solutions that converge to a solution of an impulsive system). Particular properties of the Perron integral itself were not actually used in [154, 155] for the assumptions made imply existence of the Lebesgue integral. As opposed to this, it was necessary in [73, 144] that the sequence t_i in A.57 would not have finite limit points.

Results given in this section are chiefly based on [162, 163] (monograph [55], fundamental to the theory of impulsive systems, also contains a discussion of these problems; see also [37, 158]). Systems similar to the first example were thoroughly studied in [94], the phenomenon described in this example became known as a "quasiplastic impact".

Section A.2. Theorem 60 is a particular case of Theorem 121 given in this section. There is large number of new works on periodic impulsive systems (included generalized impulsive systems [154]). Amongst them, the works of M.U. Akhmetov and N.A. Perestyuk should be especially noted. There for the first time, the authors have studied in detail the periodic problem (including the degenerate case) and the almost periodic problem for systems with impulses located on a surface (see [6, 7, 8, 9]). The scheme introduced there was used in a particular case ($x(t) \equiv 0$) in [123], and the works [5, 73], published almost simultaneously, for the first time contain explicitly complete variational equations for $\varphi(t,x) = t - t(x)$. Semilinear parabolic T-periodic systems

$$\frac{dx}{dt} = -Ax + f(t,x), \quad \Delta x|_{t_i(x)} = B_i x + g_i(x)$$

with a sectorial operator A and unbounded closed operators B_i were studied in [121, 122]. Finally, a significant number of works devoted to new modifications of the numerical-analytical method for studying periodic impulsive systems should be noted (see, for example, [161, 165]).

Section A.3. It should noted that in Chapter 4 of this book, as in almost all works on almost periodic impulsive systems, the requirement that the sequences $\{t_{n+j} - t_n\}$ be equipotentially almost periodic (often with an additional assumption that $t_n^1 \geq \theta > 0$ for all n) is considered as an indispensable condition for the system to be almost periodic. It is interesting that the function $F(t) = mt + f(t)$ can also be singled out as a function normal with respect to the operation of a symmetric translation (see [81]). This section is based on [168, 149].

Section A.4. For a study of an almost periodic subset of the line by using methods of topological dynamics, see also [4]. The most important in this section for the subsequent considerations is the theorem on closure of the almost periodic Wexler set with respect to taking the free union. The given proof is rather laborious and it would be interesting to find another more simple proof. The work [3] should be noticed. It contains a simple proof of a particular case of Theorem 124. This section is based on article [148].

Section A.5. Almost periodic impulsive systems were studied at the same time as a theory of piecewise continuous almost periodic functions has been constructed. As a foundation for this theory, the definition of D. Wexler [55] was taken. Fundamentals of this definition were singled out and taken as axioms for a definition of piecewise continuous almost periodic functions in the first edition of this book. However, now there are many definitions of piecewise continuous almost periodic functions. For example, two different definitions (it is possible to show that they are nonequivalent) are given in [2, 4] but they are equivalent to the definition given in Chapter 4 if $\inf_k\{t_i^1\} > 0$. To our opinion, studying piecewise continuous almost periodic functions is of a separate interest because to study systems with impacts, it is sufficient to use Stepanov almost periodic functions. It seems that an excessive interest to the notion of Wexler almost periodicity had a negative influence on development of the theory of almost periodic systems. Finally we remark that this section is based on [150, 151].

Sections A.6 – A.8. Similarly to the theory of ordinary differential equations, one of the most important problems in the theory of impulsive systems is the problem of studying almost periodic solutions. We do not touch here a similar problem of studying discontinuous toroidal invariant manifolds (see, for example [3, 160, 167]) as well as the problems concerning almost periodic motions in discontinuous dynamical systems (see [124, 125, 126, 127, 166]). We remark that by using methods of topological dynamics, the analysis of differential equations with almost periodic coefficients can be reduced to a study of a certain semigroup, which is an extension of a compact minimal isometric flow, and the difficulties that appear in the transition from periodic systems to proper almost periodic systems originate in the extension of the base space becoming more complex. As we saw, even for very simple almost periodic impulsive systems, it is possible that the corresponding continuous semigroup does not exists. The problem, when one can construct it and what to do if such a construction is impossible, is being solved in this section. It became clear that the obstacle for constructing a theory similar to that in [75, 94] is noncommutativity of the mappings $H_i(x) = x + h_i(x)$.

A natural reference point for us is the theory of almost periodic differential equations which was developed to a great extend by many mathematicians during the last decades. Reading the last two sections, one can easily see the influence of work of V.V. Zhikov [75], A.M. Samoilenko [135] (for the results from the theory of almost periodic systems used here, see the end of Bibliographical notes).

Proper almost periodic solutions of impulsive systems (A.57) were also studied by D. Wexler [55, 174, 175, 176, 177], A.M. Samoilenko, N.A. Perestyuk, M.U. Akhmetov [2, 3, 4, 109, 110, 143, 144], V.E. Slyusarchuk [157], A.A. Pankov [160]. All these works (with an exception of works by D. Wexler and V.E. Slesarchuk) were based on the notions of Wexler almost periodicity (for sets and functions) and this significantly restricted the class of considered systems for it would be more natural to require that only the distribution, generated by the right-hand side of (A.57), be almost periodic. It should be remarked that, for the first time, almost periodic distributions were considered by L. Schwartz in the monograph [156, §9, Ch.6, Vol.2] (see also [55], where the exposition basically follows [156]). It is curious to notice the following remark from the introduction to Chapter 6, " Nous ignorons si ces distributions peuvent avoir des usages pratiques". We follow [55, 164] to obtain auxiliary results on almost periodic distributions in Section A.6.

In Section A.7 we study those classes of almost periodic systems, for which one can introduce the notion of a hull of the system, standard in the theory of almost periodic solutions. In this case, it becomes possible to establish all (V.V. Zhikov) elementary principles of the theory of almost periodic impulsive systems. Here we give the most general principles due to V.V. Zhikov and B.M. Levitan. The example of a simple exponentially stable almost periodic linear system with a bounded but not S-bounded solution shows that bounded solutions of impulsive equations belong to a set larger than a space of S-almost periodic functions. This set turned out to be the space of Levitan N-almost periodic functions and it became possible to describe the class of systems (A.57), for which the principle of Levitan holds. These results were first given in [151].

In Section A.8 we extend Favard theory to abstract linear impulsive systems. For one particular case, a similar problem was considered by A.A. Pankov [99, 100]. Here we suggest another approach, which allows to consider the general case and, in particular, to answer the questions posed in [99, 100]. Here we follow [164].

Now we give the statements from the theory of almost periodic systems ([75]), which were used without a proof.

In the first three theorems, we assume that (X, S^t) is a compact semi-flow extending a compact minimal isometric flow (H, h^t) with respect to a mapping $j : X \to H$.

Theorem 136 (Zhikov) *If the flow (X, S^t) is distal and a certain fiber $j^{-1}(h)$ has dimension zero, then all its trajectories are almost periodic.*

Theorem 137 (Levitan) *If a fiber $j^{-1}(h)$ consists of a single element, then the corresponding trajectory is a Levitan N-almost periodic function.*

Theorem 138 (Zhikov) *Let the trajectory $S^t x_0$ be uniformly Lyapunov stable in the fiber $j^{-1}(\{h^t(jx_0), t \in \mathbf{R}\})$. If, moreover, it is asymptotically stable, the it will also be almost periodic.*

Let now B be a separable Banach space and there be defined semigroups of transformations $S^t x = (S^t(m)p, h^t(m))$, $L^t x = (L^t(m)p, h^t(m))$[22] in the product $X = B \times H = \{x = (p, m)\}$ and $S^t(m)p = L^t(m)p + N^t(m)0$,

[22] which clearly extend the flow (H, h^t) with respect to a projection onto H

$L^t(m)p \in \mathcal{L}(B)$ for all t, m. Suppose that $S^t x$ and $L^t x$ satisfy the following conditions of "weak continuity" (formulated only for $S^t x$): for any $p \in B$ and any $h \in H$, the function $S^t(h)p : \mathbf{R}^+ \to B$ is strongly continuous and if the space B is endowed with the weak topology on every bounded set, the mapping $S^t(h)p : X \to B$ is continuous for each $t \geq 0$.

Theorem 139 (Favard-Zhikov) *If all nonzero bounded trajectories $L^t(m)p$ of the semigroup (X, L^t) are semiseparated from zero, $\lim_{t \to -\infty} |L^t(m)p| > 0$ and there exists at least one weakly compact semitrajectory $S^t(h)p : \mathbf{R}^+ \to B$, then (X, S^t) has scalar almost periodic trajectories.*

Bibliography

[1] M.U. Akhmetov. On periodic solutions of certain systems of differential equations. *Vestnik Kievskogo Universiteta. Matematika i Mekhanika*, (24):3–7, 1982.

[2] M.U. Akhmetov. Recurrent and almost periodic solutions of nonautonomous impulsive systems. *Izvestia Acad. Sci. Kazakhskoi SSR, Seria phys.-math*, (3):8–10, 1988.

[3] M.U. Akhmetov. Quasiperiodic solutions of impulsive systems. In *Asymptotic methods in equations of mathematical physics*, pages 12–18. Institute of Mathematics of Ukrainian Academy of Sciences, Kiev, 1989.

[4] M.U. Akhmetov and N.A. Perestyuk. Almost periodic solutions of nonlinear impulsive systems. *Ukrainskii Matematicheskii Zhurnal*, 41(3):291–296, 1989.

[5] M.U. Akhmetov and N.A. Perestyuk. On differentiable dependence of impulsive systems on initial conditions. *Ukrainskii Matematicheskii Zhurnal*, 41(8):1028–1033, 1989.

[6] M.U. Akhmetov and N.A. Perestyuk. On comparison method for differential systems with pulse effect. *Differential Equations*, 26(9):1079–1081, 1990.

[7] M.U. Akhmetov and N.A. Perestyuk. Differential properties of solutions and integral surfaces of nonlinear impulsive systems. *Differential Equations*, 28(4):445–453, 1992.

[8] M.U. Akhmetov and N.A. Perestyuk. Periodic and almost periodic solutions of strongly nonlinear impulsive systems. *Prikladn. matem. mekhan.*, 56(6):926–934, 1992.

[9] M.U. Akhmetov and N.A. Perestyuk. Periodic solutions of strongly nonlinear systems with nonclassical right-hand sides in the case of a family of generating solutions. *Ukrainskii Matematicheskii Zhurnal*, 45(2):215–222, 1993.

[10] M.A. Amatov. On stability of motion of impulsive systems. *Differentsial'nye Uravneniya*, (9):21–28, 1977.

[11] A.A. Andronov, A.A. Vitt, and S.E. Khaikin. *Oscillation theory.* Nauka, Moscow, 1981. (in Russian).

[12] A.A. Andronov, A.A. Vitt, and S.E. Khaikin. *Oscillation theory.* Nauka, Moscow, 1981. (in Russian).

[13] V.I. Arnol'd. *Ordinary differential equations.* Nauka, Moscow, 1975. (in Russian).

[14] A.A. Aslanyan. Conditions for optimum in problems of control of impulsive systems. *Dokl. Acad. Nauk Ukrain. SSR, Ser A*, (11):3–6, 1982.

[15] A.A. Aslanyan. Maximum principle for discontinuous dynamical systems. *Teoriya Funktsii, Funktsional'nii Analiz i ikh Prilozhenia*, (37):132–137, 1982.

[16] A.A. Aslanyan. Necessary conditions for optimum in problems of control of systems differential equations with an impulsive effect occurring at variable times. *Dokl. Acad. Nauk Ukrain. SSR, Ser. A*, (9):58–61, 1982.

[17] R. Baire. *Theory of discontinuous functions.* GTTI, Moscow, Leningrad, 1932. (in Russian).

[18] E.A. Barbashin. *Introduction to stability theory.* Nauka, Moscow, 1967. (in Russian).

[19] Yu.M. Berezanskii, G.F. Us, and Z.G. Sheftel'. *Functional analysis.* Naukova Dumka, Kiev, 1990.

[20] A. Blaquiere. Differential games with piecewise continuous trajectories. *Lecture Notes in Control and Information Sciences*, (3):34–69, 1977.

[21] S. Bochner. A new approach to almost periodicity. *Proc. Nat. Acad. Sci. USA*, 48:195–205, 1962.

[22] Yu.S. Bogdanov. An estimate of impulsive reaction of continuous finite dimensional systems. In *Third scientific conference*, volume 4, pages 26–30, Tbilisi, 1971. Institute of applied mathematics of Tbilisi State University. Abstracts.

[23] N.N. Bogolyubov and Yu.A. Mitropol'skii. *Asymptotic methods in nonlinear mechanics.* Nauka, Moscow, 1974. (in Russian).

[24] N.N. Bogolyubov and Yu.A. Mitropol'sky. *Asymptotic methods in the theory of nonlinear oscillations.* Nauka, Moscow, 1974. (in Russian).

[25] N.N. Bogolyubov, Yu.A. Mitropol'sky, and A.M. Samoilenko. *An improved convergence method in nonlinear mechanics.* Naukova Dumka, Kiev, 1969. (in Russian).

[26] G. Bohr. *Almost periodic functions.* Gostekhizdat, Moscow-Leningrad, 1934. (in Russian).

[27] V.G. Boltyansky. *Mathematical methods of optimum control.* Nauka, Moscow, 1969. (in Russian).

[28] N. Bourbaki. *Integration. Vector integration, Haar measure, convolution and representations.* Nauka, Moscow, 1970. (in Russian).

[29] G.N. Boyarkin. On stability of solutions of a certain class of nonlinear differential equations with impulse type effects. *Differentsial'nye Uravneniya*, 14(6):1128–1130, 1978.

[30] N.G. Brujin. Functions whose difference belong to a given class. *Nieuw. Arch. Wisk.*, 23:194–218, 1951.

[31] A.E. Bryson. *Applied optimal control.* Blaisdell Publ. Company, Waltham, Mass., 1969.

[32] N.V. Butenin. *Elements of theory of nonlinear oscillations.* Sudprogiz, Leningrad, 1962. (in Russian).

[33] B.F. Bylov, R.E. Vinograd, D.M. Grobman, and V.V. Nemytskii. *Theory of Lyapunov exponents.* Nauka, Moscow, 1966. (in Russian).

[34] O.S. Chernikova. Reduction principle for impulsive differential systems. *Ukrainskii Matematicheskii Zhurnal*, 34(5):601–607, 1982.

[35] O.S. Chernikova. On dissipativity of impulsive differential systems. *Ukrainskii Matematicheskii Zhurnal*, 35(5):656–660, 1983.

[36] C. Corduneanu. *Finktii aproape-periodice.* Edit. Acad. RPR, Bucharest, 1961.

[37] P.C. Das and R.R. Sharma. Existence and stability of measure differential equations. *Chech. Math. J.*, 22(97):145–158, 1972.

[38] B.P. Demidovich. *Lecture on mathematical theory of stability.* Nauka, Moscow, 1967. (in Russian).

[39] I.Z. Shtokalo (editor). *A course in ordinary differential equations.* Vyshcha Shkola. Golovnoe Isdatel'stvo, Kiev, 1974. (in Russian).

[40] A.F. Filippov. Differential equations with discontinuous right-hand side. *Matematicheskii Sbornik*, 51(1):99–128, 1960.

[41] R. Gabasov and F.M. Kirillova. *Special optimum controls.* Nauka, Moscow, 1973. (in Russian).

[42] F.R. Gantmakher. *Theory of matrices.* Nauka, Moscow, 1967. (in Russian).

[43] H.P. Geering. Continuous-time optimal theory for cost functionals including discrete state penalty terms. *JEEE Transactions on Automatic Control*, AC-21(5):866–869, 1976.

[44] I. Gelfand. Abstrakte functionen und lineare operatoren. *Matematicheskii Sbornik*, 4(42)(2):235–286, 1938.

[45] W.M. Getz and D.H. Martin. Optimal control systems with state variable jump discontinuities. *Journal of Optimization Theory and Applications*, 31(2):195–205, 1980.

[46] V.K. Gorbunov. Differential-impulsive games. *Izv. Acad. Nauk SSSR, Tekhnicheskaya Kibernetika*, (4):80–84, 1973.

[47] V.K. Gorbunov and F.L. Chernous'ko. An optimum multi-impulse correction of perurbations. *Avtomatika i Telemekhanika*, (8):59–66, 1970.

[48] E.A. Grebenikov and Yu.A. Ryabov. *Constructive methods for nonlinear system analysis.* Nauka, Moscow, 1971. (in Russian).

[49] E.A. Grebenikov and Yu.A. Ryabov. *Constructive analysis methods for nonlinear systems.* Nauka, Moscow, 1979. (in Russian).

[50] S.I. Gurgula. A study of stability of solutions of impulsive systems by using the second Lyapunov method. *Ukrainskii Matematicheskii Zhurnal*, 34(1):100–103, 1982.

[51] S.I. Gurgula and N.A. Perestyuk. On stability of solutions of impulsive systems. *Vestnik Kievskogo Universiteta. Matematika i Mechanika*, (3):33–40, 1981.

[52] S.I. Gurgula and N.A. Perestyuk. On stability of equillibrium of impulsive systems. *Matematicheskaya Fizika*, (3):9–14, 1982.

[53] S.I. Gurgula and N.A. Perestyuk. On the second Lyapunov method in impulsive systems. *Dokl. Acad. Nauk Ukrain. SSR, Ser. A*, (10):11–14, 1982.

[54] B.I. Gurman. *Degenerate optimum control problems.* Nauka, Moscow, 1977. (in Russian).

[55] A. Halanay and D. Wexler. *Teoria calitativă a sistemelor cu impulsuri.* Academiei Republicii Socialiste România, Bucuresti, 1968.

[56] Rao V. Sree Hari. Asymptotically of measure differential equations. *Nonlinear Anal.: Theory Math and Appl.*, (2):483–489, 1978.

[57] Ph. Hartman. *Ordinary differential equations.* John Wiley & Sons, New York, 1964.

[58] C.S. Hcu. Impulsive parametric excitation. NWA/APM 19, ASME, 1971.

[59] C.S. Hcu and W.H. Cheng. Applications of the theory of impulsive parametric excitation and new treatements of general parametric excitations problems. *Trans. ASME*, 40(1):78–86, 1973.

[60] B.S. Kalitin. On pendulum oscillations with a shock impulse. *Differentsial'nye Uravneniya*, 6(12):2174–2181, 1970.

[61] B.S. Kalitin. On limit cycles of pendulum systems with impulsive perturbation. *Differentsial'nye Uravneniya*, 7(3):540–542, 1971.

[62] I.K. Karkinbaev and N.A. Perestyuk. On application of the averaging methods to study impulsive systems. In *Asymptotic methods in the theory of nonlinear oscillations*, pages 43–50. Naukova Dumka, Kiev, 1979. (in Russian).

[63] J.L. Kelley. *General topology.* Springer-Verlag, Ney York, 1955.

[64] V. Kh. Kharacakhal. *Almost periodic solutions of ordinary differential equations.* Nauka, Alma-Ata, 1979. (in Russian).

[65] A.Ya. Khinchin. *Continuous fractions.* Nauka, Moscow, 1967. (in Russian).

[66] I.T. Kiguradze. *Certain singular boundary value problems for ordinary differential equations.* Izdatel'stvo Tbiliskogo Universiteta, Tbilisi, 1975. (in Russian).

[67] A.E. Kobrinskii and A.A. Kobrinskii. *A theory of shock vibration systems.* Nauka, Moscow, 1973. (in Russian).

[68] N.N. Krasovskii. *Theory of motion control (linear systems).* Nauka, Moscow, 1968. (in Russian).

[69] N.N. Krasovskii and A.I. Subbotin. *Positional differential games.* Nauka, Moscow, 1974. (in Russian).

[70] N.M. Krylov and N.N. Bogolyubov. *Introduction to nonlinear mechanics.* Academiya Nauk Ukrin. SSR, Kiev, 1937. (in Russian).

[71] Ya. Kurtsveil'. On generalized ordinary differential equations which possess discontinuous solutions. *Applied Mathematics and Mechanics*, 22(1):27–45, 1958.

[72] A.B. Kurzhanskii. *Control and observation in uncertainty conditions.* Nauka, Moscow, 1977. (in Russian).

[73] V. Lakshmikantham, D. Bainov, and P.S. Simeonov. *Theory of impulsive differential equations.* World Scientific, Singapore, 1989.

[74] B.M. Levitan. *Almost periodic functions.* Gostekhizdat, Moscow, 1953. (in Russian).

[75] B.M. Levitan and V.V. Zhykov. *Almost periodic functions and differential equations.* Izdatel'stvo Moskovskogo Universiteta, Moscow, 1978. (in Russian).

[76] E.B. Lie and L. Markus. *Fundamentals of optimum control theory.* Nauka, Moscow, 1972. (in Russian).

[77] I.V. Livartovskii. On stability of discontinuous systems with almost reducible linear approximations. *Differentsial'nye Uravneniya*, 1(9):1131–1139, 1965.

[78] L.H. Loomis. Spectral characterization of almost periodic functions. *Ann. Math.*, 72:362–368, 1960.

[79] I.G. Malkin. *Motion stability theory.* Nauka, Moscow, 1966. (in Russian).

[80] E.Yu. Mamsa. Integral sets of a certain class of impulsive differential system. *Vestnik Kievskogo Universiteta. Matematika i Mehanika*, (24):61–68, 1982.

[81] V.A. Marchenko. On functions which are normal with respect to symmetric operation of translation. *Zapiski NII Matematiki i Mekhaniki Kharkovskogo Gos. Universiteta*, XX:33–42, 1950.

[82] A.A. Martynyuk and R. Gutovski. *Integral inequalities and motion stability.* Naukova Dumka, Kiev, 1979. (in Russian).

[83] B.M. Miller. Nonlinear impulsive control probles for processes described by ordinary differential equation. *Avtomatika i Telemehanika*, (1):75–86, 1978.

[84] V.D. Mil'man and A.D. Myshkis. On motion stbility with shocks. *Sibirskii Matematicheskii Zhurnal*, 1(2):233–237, 1960.

[85] E.F. Mishchenko and N.Kh. Rozov. *Differential equations with small parameter and relaxational oscillations.* Nauka, Moscow, 1975. (in Russian).

[86] Yu.A. Mitropol'skii. *Averaging method in nonlinear mechanics.* Naukova Dumka, Kiev, 1977. (in Russian).

[87] Yu.A. Mitropol'skii and O.B. Lykova. *Integral manifolds in nonnlinear mechanics.* Nauka, Moscow, 1973. (in Russian).

[88] Yu.A. Mitropol'skii and D.I. Martynyuk. *Periodic and quasi-periodic oscillations in systems with time delay.* Vyshcha Shkola. Golovnoe Izdatel'stvo, Kiev, 1979. (in Russian).

[89] Yu.A. Mitropol'skii, N.A. Perestyuk, and O.S. Chernikova. Convergence of systems of impulsive differential equations. *Doklady Academii Nauk Ukrain. SSR. Seria A*, (11):11–15, 1983.

[90] Yu.A. Mitropol'skii, A.M. Samoilenko, and N.A. Perestyuk. On substantiation of the averaging method for an impulsive second order equation. *Ukrainskii Matematicheskii Zhurnal*, 29(6):750–762, 1977.

[91] Yu.A. Mitropol'skii, A.M. Samoilenko, and N.A. Perestyuk. On substantiation of the averaging method for second order impulsive equations. *Ukrainskii Matematicheskii Zhurnal*, 29(6):750–762, 1977.

[92] Rao M. Rama Mohana and Rao V. Sree Hari. Stability of impulsively perturbed systems. *Bull. Austal. Math. Soc.*, 16(1):99–110, 1976.

[93] A.D. Myshkis and A.M. Samoilenko. Systems with shocks at given moments. *Matematicheskii Zbornik*, 74(2):202–208, 1967.

[94] R.F. Nagaev. *Mechanical processes with repeating decaying impacts.* Nauka, Moscow, 1985. (in Russian).

[95] Yu.I. Neimark. *Method of point mappings in nonlinear oscillation theory.* Nauka, Moscow, 1972. (in Russian).

[96] V.V. Nemytskii and V.V. Stepanov. *Qualitative theory of differential equations.* Gostehizdat, Moscow-Leningrad, 1949. (in Russian).

[97] S.G. Pandit. On the stability of impulsively perturbed differential systems. *Bull. Austral. Math. Soc.*, 17(3):423–432, 1977.

[98] S.G. Pandit. Differential systems with impulsive perturbations. *Pacif. J. Math*, 86(2):553–560, 1980.

[99] A.A. Pankov. On Favard theory for impulsive evolution equations. *Rev. Roum. math. pure et appl.*, XXV(3):385–401, 1982.

[100] A.A. Pankov. *Bounded and almost periodic solutions of nonlinear differential operator equations.* Kiev, 1985. (in Russian).

[101] T. Pavlidis. Optimal control of pulse frequency modulated systems. *IEEE Trans. Automat. Control*, AC-11(1):35–43, 1966.

[102] T. Pavlidis. Stability of a class of discontinuous dynamical systems. *Infor. Control*, 9(6):298–322, 1966.

[103] T. Pavlidis and E.I. Jury. Analysis of a new class of pulce-frequency modulated feedback systems. *IEEE Trans. Automat. Control*, AC-10(1):35–43, 1965.

[104] N.A. Perestyuk. On stability of solutions of impulsive differential equations. In *Conference on differential equations and their application*, page 35, Bulgary, 1975. (Absract).

[105] N.A. Perestyuk. On stability of equilibrium of impulsive systems. *God. na VUZ. Prilozhenia Matematiki*, 11:145–150, 1976.

[106] N.A. Perestyuk. Stability of solutions of linear impulsive systems. *Vestnik Kievskogo Universiteta. Matematika i Mehanika*, (19):71–76, 1977.

[107] N.A. Perestyuk. On integral sets of a certain class of differential equations subjected to an impulsive effect. In *III State symposium on differential and integral equations*, pages 33–34, Odessa, 1982. (Abstract).

[108] N.A. Perestyuk. Invariant sets of a certain class of discontinuous dynamical systems. *Ukrainskii Matematicheskii Zhurnal*, 36(1):63–68, 1984.

[109] N.A. Perestyuk and M.U. Akhmetov. On almost periodic motions of impulsive systems. *Ukrainskii Matematicheskii Zhurnal*, 39(1):74–80, 1987.

[110] N.A. Perestyuk and M.U. Akhmetov. On almost periodic motions of impulsive systems. *Ukrainskii Matematicheskii Zhurnal*, pages 94–98, 1989.

[111] N.A. Perestyuk and O.S. Chernikova. On stability of solutions of impulsive differential systems. *Ukrainskii Matematicheskii Zhurnal*, 36(2):190–195, 1984.

[112] N.A. Perestyuk and V.N. Shovkoplyas. Periodic solutions of nonlinear imppulsive differential equations. *Ukrainskii Matematicheskii Zhurnal*, 31(5):517–524, 1979.

[113] G.N. Petrovskii. On a certain system of linear differential equations with shocks at fixed moments. *Vestnik Beloruskogo Universiteta*, (2):7–10, 1978.

[114] I.G. Petrovskii. *Lectures on theory of ordinary differential equations.* Nauka, Moscow, 1964. (in Russian).

[115] V.A. Pliss. *Integral sets of periodic systems of differential equations.* Nauka, Moscow, 1977. (in Russian).

[116] L.S. Pontryagin. *Ordinary differential equations.* Nauka, Moscow, 1970. (in Russian).

[117] L.S. Pontryagin, V.G. Boltyanskii, R.V. Gamkrelidze, and E.F. Mishchenko. *Mahtematical theory of optimum processes.* Nauka, Moscow, 1969. (in Russian).

[118] L.A. Prokhorova. Linear systems with concentrated perturbations i, ii. *Vestnik Beloruskogo Universiteta. Seriya 1*, (2):3–5, 1975.

[119] L.A. Prokhorova. Linear systems with concentrated perturbations. ii. *Vestnik Beloruskogo Universiteta. Seriya 1*, (1):37–42, 1979.

[120] V. Raghavendra and Rao Rama Mohana. On the stability of differential systems with respect to impulsive perturbations. *J. Math. Anal. and Appl.*, 48(2):515–526, 1974.

[121] Yu.V. Rogovchenko and S.I. Trofimchuk. Periodic solutions of weakly nonnlinear impulsive partial differential equations of parabolic type and their stability. Preprint 86.65, Institute of Mathematics, Ukrainian Academy of Sciences, Kiev, 1986. 44 p.

[122] Yu.V. Rogovchenko and S.I. Trofimchuk. Bounded and periodic solutions of weakly nonlinear impulsive evolution systems. *Ukrainskii Matematicheskii Zhurnal*, 39(2):260–264, 1987.

[123] Yu.V. Rogovchenko and S.I. Trofimchuk. Periodic solutions of weakly nonlinear impulsive systems. *Ukrainskii Matematicheskii Zhurnal*, 41(5):622–626, 1989.

[124] V.F. Rozhko. On theory of discontinuous dynamical systems i, ii, iii. *Matematicheskie Issledovaniya, Akademiya Nauk Mold. SSR*, 4(3):63–73, 1969.

[125] V.F. Rozhko. On theory of discontinuous dynamical systems i, ii, iii. *Matematicheskie Issledovaniya, Akademiya Nauk Mold. SSR*, 5(1):102–117, 1970.

[126] V.F. Rozhko. On theory of discontinuous dynamical systems i, ii, iii. *Matematicheskie Issledovaniya, Akademiya Nauk Mold. SSR*, 5(2):157–167, 1970.

[127] V.F. Rozhko. On a class of almost periodic motions in systems with shocks. *Differentsial'nye Uravneniya*, 8(11):2012–2022, 1972.

[128] V.F. Rozhko. Lyapunov stability in discontinuous dynamical systems. *Differentsial'nye Uravneniya*, 11(6):1005–1012, 1975.

[129] N.Kh. Rozov and T.R. Gichev. A controled nonlinear impulsive system i. *Differentsial'nye Uravneniya*, 15(11):1933–1939, 1979.

[130] N.Kh. Rozov and T.R. Gichev. A controled nonlinear impulsive system i. *Differentsial'nye Uravneniya*, 16(2):208–213, 1980.

[131] A.M. Samoilenko. Numerical-analitical method for studying periodic systems of ordinary differential equations i, ii. *Ukrainskii Matematicheskii Zhurnal*, 17(4):82–93, 1965.

[132] A.M. Samoilenko. Numerical-analitical method for studying periodic systems of ordinary differential equations i, ii. *Ukrainskii Matematicheskii Zhurnal*, 18(2):50–59, 1966.

[133] A.M. Samoilenko. On preservation of invariant torus under a perturbation. *Izvestia Akademii Nauk SSSR, Seriya Matematicheskaya*, 34(5):1219–1240, 1970.

[134] A.M. Samoilenko. Averaging method for systems with shocks. *Matematicheskaya Fizika*, (9):101–117, 1971.

[135] A.M. Samoilenko. *Elements of the mathematical theory of multi-frequency oscillations.* Kluwer Academic Publishers, 1991.

[136] A.M. Samoilenko and N.A. Perestyuk. On averaging method for impulsive systems. *Ukrainskii Matematicheskii Zhurnal*, (3):411–418, 1974.

[137] A.M. Samoilenko and N.A. Perestyuk. The second bogolyubov theorem for impulsive differential systems. *Differentsial'nye Uravneniya*, 10(11):2001–2010, 1974.

[138] A.M. Samoilenko and N.A. Perestyuk. Second N.N. Bogolubov's theorem for impulsive differential systems. *Differentsial'nye Uravneniya*, 10(11):2001–2010, 1974.

[139] A.M. Samoilenko and N.A. Perestyuk. Stability of solutions of impulsive differential equations. *Differentsial'nye Uravneniya*, 13:1981–1992, 1977.

[140] A.M. Samoilenko and N.A. Perestyuk. Periodic solutions of weakly nonlinear impulsive systems. *Differentsial'nye Uravneniya*, 14(16):1034–1045, 1978.

[141] A.M. Samoilenko and N.A. Perestyuk. *Impulsive differential equations.* Kievskii Gosudarstvennyi Universitet, Kiev, 1980. (in Russian).

[142] A.M. Samoilenko and N.A. Perestyuk. On stability of solutions of impulsive systems. *Differentsial'nye Uravneniya*, 17(11):1995–2002, 1981.

[143] A.M. Samoilenko and N.A. Perestyuk. Periodic and almost periodic solutions of impulsive differential equations. *Ukrainskii Matematicheskii Zhurnal*, 34(1):66–73, 1982.

[144] A.M. Samoilenko and N.A. Perestyuk. *Impulsive differential equations.* Vishcha Shkola, Kiev, 1987. (in Russian).

[145] A.M. Samoilenko, N.A. Perestyuk, and M.U. Akhmetov. Almost periodic solutions of impulsive differential equations. Preprint 83-26, Institute of Mathematics of Acad. Sci. of Ukrain. SSR, Kiev, 1983.

[146] A.M. Samoilenko and N.I. Ronto. *Numerical-analytical methods for studying periodic solutions.* Vushcha Shkola. Izdatel'stvo Kievskogo Universiteta, Kiev, 1976. (in Russian).

[147] A.M. Samoilenko and T.G. Strizhak. On motion of an oscillator under an effect of an instantaneous force. In *Mathematical physics and nonlinear oscillations seminar works*, number 4, pages 213–218, Kiev, 1968.

[148] A.M. Samoilenko and S.I. Trofimchuk. On spaces of piecewise continuous almost periodic functions and on almost periodic sets on the line. i. *Ukrainskii Matematicheskii Zhurnal*, 43(12):1613–1619, 1991.

[149] A.M. Samoilenko and S.I. Trofimchuk. Unbounded functions fith almost periodic differences. *Ukrainskii Matematicheskii Zhurnal*, 43(10):1409–1413, 1991.

[150] A.M. Samoilenko and S.I. Trofimchuk. On spaces of piecewise continuous almost periodic functions and on almost periodic sets on the line. ii. *Ukrainskii Matematicheskii Zhurnal*, 44(3):412–423, 1992.

[151] A.M. Samoilenko and S.I. Trofimchuk. Almost periodic impulsive systems. *Differentsial'nye Uravneniya*, 29(5):799–808, 1993.

[152] Š. Schwabic. Bemercunden zu stabieitäts fragen für verallgemeinerte differentialgleichungen. *Časop. pěstov. mat.*, 96(2):57–66, 1971.

[153] Š. Schwabic. Verallgemeinerte lineare differentialgleichungsysteme. *Časop. pěstov. mat.*, 96(2):183–211, 1971.

[154] Š. Schwabic. *Generalized differential equations. Fundamental results.* Rozpravy CSAV (rada MPV), Praha, 1985.

[155] Š. Schwabic. *Generalized differential equations. Special results.* Rozpravy CSAV (rada MPV), Praha, 1989.

[156] L. Schwartz. *Theorie des distributions.* Hermann, Paris, 1966.

[157] V.E. Slusarchuk. Bounded solutions of impulsive systems. *Differentsial'nye Uravneniya*, 19(4):588–596, 1983.

[158] F.W. Stallard. Functions of bounded variations as solutions of differential systems. *Proc. Amer. Math. Soc.*, 13(3):366–373, 1962.

[159] V.M. Starzhinskii and V.A. Yakubovich. *Linear differential equations with periodic coefficients.* Nauka, Moscow, 1972. (in Russian).

[160] V.I. Tkachenko. Green's function and conditions for existence of invariant sets for impulsive systems. *Ukrainskii Matematicheskii Zhurnal*, 41(10):1379–1383, 1989.

[161] E.P. Trofimchuk and A.V. Kovalenko. A numerical-analytical method of a.m. samoilenko without a defining equation. To be published in Ukrainskii Matematicheskii Zhurnal.

[162] E.P Trofimchuk and S.I. Trofimchuk. Impulsive systems with fixed times of generally distributed shocks: existence and uniqueness of solution and correctness of the cauchy problem. *Ukrainskii Matematicheskii Zhurnal*, 42(2):230–237, 1990.

[163] E.P Trofimchuk and S.I. Trofimchuk. Impulsive systems with fixed times of generally distributed shocks: structure of the set of times of shocks. *Ukrainskii Matematicheskii Zhurnal*, 42(3):378–383, 1990.

[164] S.I. Trofimchuk. Almost periodic solutions of abstract linear impulsive systems. To be published in Differentsial'nye Uravneniya.

[165] S.I. Trofimchuk. A study of improved convergence of solutions of a boundary value problem for an impulsive system by using a numerical-analytical method. In *Asymptotic methods in problems of mathematical physics*, pages 152–158. Institute of Mathematics, Ukrainian Academy of Sciences, Kiev, 1988.

[166] S.I. Trofimchuk. Piecewise continuous almost periodic functions and discontinuos dynamical systems. In *Nonlinear problems in the theory of differential equations*, pages 16–20. Institute of Mathematics, Ukrainian Academy of Sciences, Kiev, 1990.

[167] S.I. Trofimchuk. Impulsive linear extensions. In *Asymptotic solutions of nonlinear equations with a small parameter*, pages 124–131. Institute of Mathematics, Ukrainian Academy of Sciences, Kiev, 1991.

[168] S.I. Trofimchuk. Unbounded functions with the differences that are bounded and uniformly continuous on **R**. *Doklady Akademii Nauk Ukrain. SSR. Seriya A*, (5):24–25, 1993.

[169] K.V. Tsidylo and S.S. Gul'ka. On periodic solutions of nonlinear impulsive systems. *Doklady Akademii Nauk Ukrain. SSR. Seriya A*, (10):21–23, 1981.

[170] N.I. Vasil'ev. On stable solutions of Riccati equation with an impulsive effect. *Latviiskii Matematicheskii Ezhegodnik*, (21):26–30, 1977.

[171] V.S. Vladimirov. *Equations of mathematical physics.* Nauka, Moscow, 1981. (in Russian).

[172] Th. Vogel. *Théorie des systéms evolutifs.* Goutier-Villons, 1965.

[173] Reitmann Volcer. Swache instabietät im ganzen von nichtlineare impulssystemen. *Wiss. z. Techn Univ. Dresden*, 26(6):1055–1057, 1977.

[174] D. Wexler. Solutions periodiques des systems d'equations differentieles lineaires aux impulsions. *C. R.*, 259(2):287–289, 1964.

[175] D. Wexler. Solutions periodiques et presque periodiques des systems d'equations differentieles lineaires et distributions. *J. Diff. Equations*, 259(2):287–289, 1964.

[176] D. Wexler. Solutions periodiques et presque periodiques des systems d'equations differentieles aux impulsions. *Rev. Roumaine Math. pures et appl.*, 10(8):1163–1199, 1965.

[177] D. Wexler. Solutions presque periodiques des systems d'equations differentieles à perturbations distribution. *C.R.*, 262(2):436–439, 1966.

[178] L.Ch. Yang. *Lectures on variational calculus and optimal control theory.* Mir, Moscow, 1974. (in Russian).

[179] K. Yosida. *Functional analysis.* Springer-Verlag, Berlin, New-York, 1974.

[180] S.T. Zavalishchin. Stability of generalized processes i, ii. *Differentsial'nye Uravneniya*, 2(7):872–881, 1966.

[181] S.T. Zavalishchin. Stability of generalized processes, ii. *Differentsial'nye Uravneniya*, 3(2):171–179, 1967.

[182] S.T. Zavalishchin. Implementation of a given motion by impulsive correction in the case of constantly acting perturbations. *Differentsial'nye Uravneniya*, 8(3):435–442, 1972.

[183] S.T. Zavalishchin. Impulsive calculus for operators in the space of distributions. *Differentsial'nye Uravneniya*, 8(6):1098–1100, 1972.

[184] S.T. Zavalishchin and A.N. Sesekin. Minimization of integral esimate of kinetic energy of harmonic oscillator by using impulsive control. *Prikladnaya Matematika i Mekhanika*, 38(3):441–450, 1974.

[185] S.T. Zavalishchin, A.N. Sesekin, and S.E. Drozdenko. *Dynamical systems with impulsive structure.* Sredne-Ural'skoe Knizhnoe Izdatel'stvo, Sverdlovsk, 1983. (in Russian).

[186] B.M. Zhivikhin. On boundedness of solutions of impulsive differential systems. In *A study in theory of differential and difference equations*, pages 33–40, Riga, 1976.

*

Subject Index